Illustrated English-Chinese Dictionary

of Water Resources and Hydropower Engineering

水利水电工程
英汉图文辞典

中国电建集团成都勘测设计研究院有限公司　编著

（地质卷）

·北京·

图书在版编目（CIP）数据

水利水电工程英汉图文辞典. 地质卷 / 中国电建集团成都勘测设计研究院有限公司编著. -- 北京 : 中国水利水电出版社, 2021.9
ISBN 978-7-5226-0053-6

Ⅰ. ①水… Ⅱ. ①中… Ⅲ. ①水利水电工程—词典—英、汉 Ⅳ. ①TV-61

中国版本图书馆CIP数据核字(2021)第201359号

责任编辑：吴娟　王海琴

书　名	**水利水电工程英汉图文辞典**（地质卷） SHUILI SHUIDIAN GONGCHENG YING-HAN TUWEN CIDIAN (DIZHI JUAN)
作　者	中国电建集团成都勘测设计研究院有限公司　编著
出版发行	中国水利水电出版社 （北京市海淀区玉渊潭南路1号D座　100038） 网址：www. waterpub. com. cn E-mail：sales@mwr. gov. cn 电话：（010）68545888（营销中心）
经　售	北京科水图书销售有限公司 电话：（010）68545874、63202643 全国各地新华书店和相关出版物销售网点
排　版	中国水利水电出版社微机排版中心
印　刷	涿州市星河印刷有限公司
规　格	140mm×203mm　32开本　12.75印张　826千字
版　次	2021年9月第1版　2021年9月第1次印刷
定　价	**150.00**元

《水利水电工程英汉图文辞典》（地质卷）

编　委　会

凡例（User's Guide）

1. 本辞典供广大水利水电从业者和院校学生学习地质工程、岩土工程、水利水电工程等专业英语使用。

2. 本辞典词条收录单词和短语共3000余条。词条和相应例句、常用搭配等主要通过以下渠道获得：

（1）国内外通用工程地质工具书：《Engineering Geology》《Engineering Geology and Construction》《Engineering Geology for Society and Territory》《汉英水电工程常用词汇》等。

（2）国内外现行技术标准：《ASTM International Standards》《Eurocode7 Geotechnical Design》《水力发电工程地质勘察规范》《可再生能源工程勘察术语标准》《水电水利工程地下建筑物工程地质勘察技术规程》《水电水利工程坝址工程地质勘察技术规程》等中、英文版本。

（3）招标文件：被世界银行列为东南亚贷款项目范本的《二滩水电站土建工程第一招标文件》《二滩水电站土建工程第二招标文件》等。

（4）国际地质、岩土、结构等专业权威期刊：《International Journal of Rock Mechanics and Mining Sciences》《Rock Mechanics and Rock Engineering》《ASTM Geotechnical Testing Journal》《Journal of Structural Geology》《Journal of Civil Engineering and Architecture》《Geophysical Journal International》等。

（5）设计文件：中国电建集团成都勘测设计研究院有限公司开展国际工程勘测设计与咨询工作编写的百余项英文技术和商务文件。

3. 词条配图主要来自中国电建集团成都勘测设计研究院有限公司勘测设计的工程项目的实景照片，工程图纸和GOCAD、CATIA等三维模型。配图所涉及的工程项目包括二滩、溪洛渡、锦屏一级、瀑布沟、两河口等水利水电工程。

4. 单词采用美式英标进行注音。

5. 词条为短语的，未配音标，若短语有较为固定的缩写形式，在短语后的圆括号内列出。

6. 释义用逗号“，”或分号“；”进行分开，词性相同的用逗号，词性不同的用分号。

7. 词条下除【释义】条目外，按需设【常用搭配】【例句】和【同义词】条目。部分常用工程术语设【中文定义】条目以供参考。

8. 词性缩写表：

adj. =adjective 形容词　　*n*. =noun 名词

adv. =adverb 副词　　*v*. =verb 动词

目　录

A

abnormal fault

【释义】 逆断层，逆向断层

【同义词】 reverse fault; thrust fault; upthrow fault

【例句】 The pipe is damaged more seriously when the dip angle of the abnormal fault is larger. 断层面倾角越大的逆断层对管线的破坏越严重。

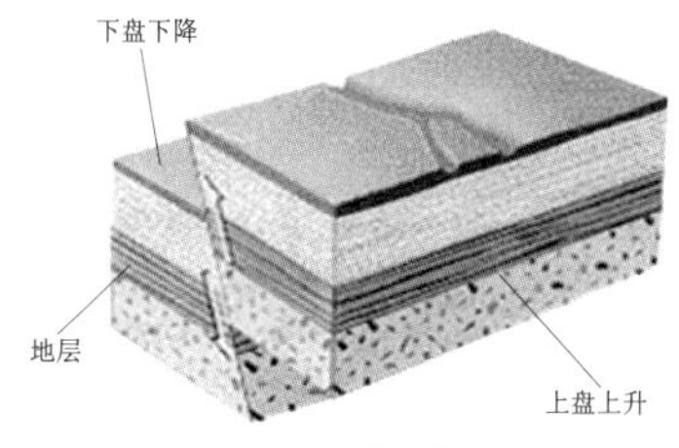

abnormal fault

【中文定义】 地质构造中断层的一种，为上盘上升、下盘相对下降的断层，主要由水平挤压与重力作用而形成。

absolute elevation

【释义】 绝对高程

【例句】 Traditional method to measure the elevation of underwater structures is to transmit the absolute elevation from the land-based control points to the underwater points to be measured by using leveling instrument or GPS together with leveling rod. 传统的水下结构高程测量方法是利用水准仪或GPS与水准标尺相结合，将地面高程控制点的绝对高程传递到水下点。

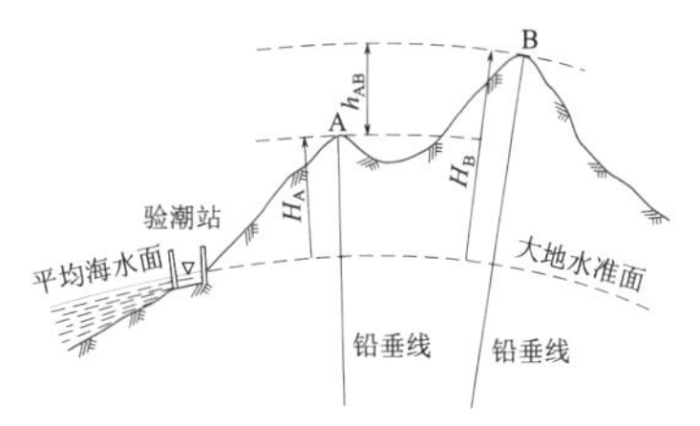

地面点A、B沿铅垂方向至大地水准面的距离H_A H_B为绝对高程。

absolute elevation

【中文定义】 或称海拔，是指地面点沿垂线方向至大地水准面的距离。我国在青岛设立验潮站，长期观测和记录黄海海平面的高低变化，取其平均值作为绝对高程的基准面。

abutment [ə'bʌtmənt]

【释义】 *n.* 坝肩

【常用搭配】 dam abutment 坝肩；left abutment 左岸坝肩；right abutment 右岸坝肩

【中文定义】 指水坝两端所依托的山体。水坝建设之初，对两侧山体进行开挖，以符合坝肩的设计要求，并最终与水坝连为一体，称"左岸坝肩"和"右岸坝肩"。

accelerated creep

【释义】 加速蠕变

【常用搭配】 accelerated creep damage 加速蠕变破坏

【例句】 When the stress reaches a certain level, the accelerated creep damage of rock occurred after the decay creep and steady creep. 当应力达到一定水平时，岩石的加速蠕变损伤发生在衰减蠕变和稳态蠕变之后。

【中文定义】 指应变随时间的延续而加速增加，直达材料的破裂点。

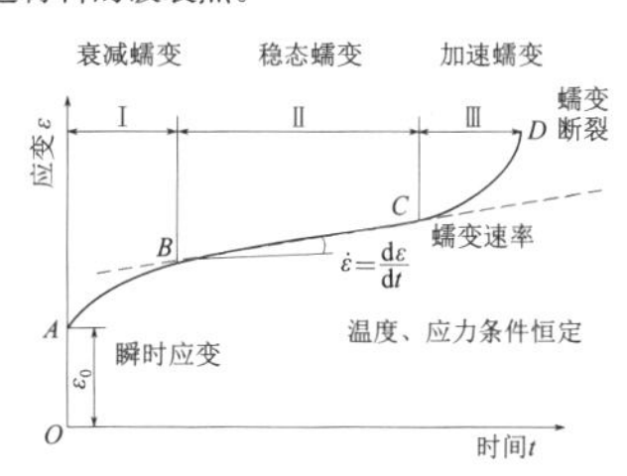

typical creep curve (accelerated creep)

accumulated region of debris flow

【释义】 泥石流堆积区

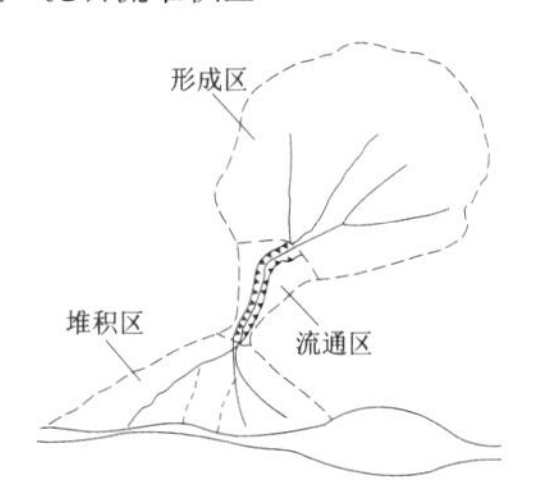

accumulated region of debris flow

【中文定义】 泥石流碎屑物质大量淤积的地区。泥石流堆积区位于泥石流下游或中下游。堆积活动有时发生在流通区内泥石流沟谷坡度急剧减小或转折处，有时发生在泥石流沟谷前端的开阔地带。

accumulation of debris flow

【释义】 泥石流堆积

【例句】 The model experiment forecast on the risk range of debris flow is to determine the possible maximum range, which is accumulation of debris flow. 泥石流危险范围的模型试验预测旨在确定一次泥石流堆积可能泛滥的最大范围。

【中文定义】 是指包含大量泥沙、石块的洪流堆积体。

accumulation of debris flow

accuracy [ˈækjərəsɪ]

【释义】 *n.* 精度

【常用搭配】 measuring accuracy 测量精度；survey accuracy 调查精度

【同义词】 precision [prɪˈsɪʒən]

【例句】 Finally, measuring accuracy in application is analyzed. 最后对应用中的测量精度进行了分析。

AC-electrical method

【释义】 交流电法

acid rock

【释义】 酸性岩

【常用搭配】 acid rock body 酸性岩体；acid intrusive rock 酸性侵入岩；intermediate-acid intrusive rock 中酸性侵入岩；acid rock reaction 酸岩反应

【例句】 The Late Jinninggian volcanic and intrusive rock in the area of Ciwu are consist of basic rock(Shangshu Formation lower-section basic volcanic rock and Ciwu diabase) and acid rock(Shangshu Formation upper-section acid volcanic rock and Daolinshan alkalic feldspar granite). 次坞地区晋宁纪晚期火山岩与侵入岩由基性岩（上墅组下段基性火山岩与次坞辉绿岩体）和酸性岩（上墅组上段酸性火山岩与道林山碱长花岗岩体）组成。

【中文定义】 二氧化硅含量大于65%的岩浆岩。

acidity [əˈsɪdəti]

【释义】 *n.* 酸度，酸性，酸味

【例句】 Both nitrogen and sulphur emissions are major contributors to soil acidity and environmental pollution. 氮和硫的排放物是造成土壤酸度和环境污染的两大主要来源。

【中文定义】 酸度表示中和1g化学物质所需的氢氧化钾（KOH）的毫克数。

acoustic emission

【释义】 声发射

【例句】 Acoustic emission is a nondestructive testing method. 声发射是一种无损检测方法。

acoustic speed test of rock mass

【释义】 岩体声波速度测试

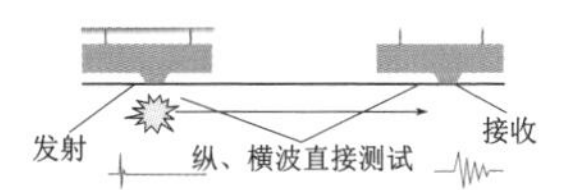

纵横波地面(表面)测试示意图

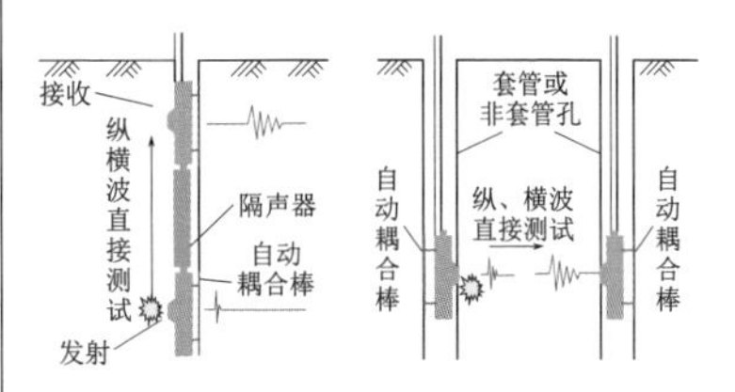

纵横波单孔测试示意图　纵横波跨孔测试示意图

acoustic speed test of rock mass

acoustic velocity logging

【释义】 声速测井

【例句】 In addition to calculating various elastic parameters, acoustic velocity logging can also be used to classify lithology and explain weak interlayers. It plays an important role in engineering investigation. 声速测井除了能够计算各种弹性参数外，还能够进行岩性划分、解释软弱夹层等，在工程勘察中发挥了重要作用。

actinolite [ækˈtɪnəˌlaɪt]

【释义】 *n.* 阳起石

【常用搭配】 actinolite asbestos 阳起石石棉；

actinolite schist 阳起片岩

【例句】 Actinolite is an end member of the tremolite-actinolite series. 阳起石是透闪石-阳起石系列的端员矿物。

actinolite

active fault

【释义】 活断层

【例句】 The surface rupture zone of Luobaoquan active fault is as long as 13km and the fault scarp is 3.6－14.8m in height, which was caused by the earthquake with M7.1 on Aug.5, 1914. 洛包泉活断层地表破裂带长 13km，断坎高 3.6～14.8m，是由 1914 年 8 月 5 日发生的 7.1 级地震造成的。

the site of active fault in Tancheng

【中文定义】 指现今在持续活动的断层，或在人类历史时期或近期地质时期曾经活动过，极有可能在不远的将来重新活动的断层。后一种也可称为潜在断层。

active fault detecting

【释义】 活动断层探测

【例句】 Therefore, the active fault detecting and the jeopardize evaluating are very important. 对活动断层进行探测并评价其危害性具有重要意义。

active fault detecting

【中文定义】 确定活动断层位置和产状，获取其晚第四纪活动性质、幅度、时代、速率及地震复发间隔等参数的技术过程。

active net

【释义】 主动网

【中文定义】 是以钢丝绳网为主的各类柔性网，覆盖包裹在所需防护的斜坡或岩石上，以限制坡面岩。

active net

active structure

【释义】 活动构造

【例句】 Location of line in a mountainous environment in which active structures are well developed is a major engineering-geological problem concerned in railway and highway construction. 在活动构造发育的山地环境进行选线是铁路、公路等工程建设关注的重大工程地质问题。

【中文定义】 活动构造与新构造在含义上有所不同。新构造通常指晚第三纪以来形成的地质构造。由于第三纪晚期的地壳运动比起第四纪来要强烈而广泛，因而新构造比活动构造要明显和多见。总的来说，活动构造也可理解为至今活动着的地质构造。

activity index

【释义】 活动性指数

【例句】 The experiments proved that relationship between the mortar strength of the cement based composite cementitious material and the activity index of several admixture is not a simple linear one. 试验证明水泥基复合胶凝材料的胶砂强度并不是简单地与几种掺合料活动性指数线性相关。

【中文定义】 是黏质土塑性指数与小于 2μm 颗粒含量百分率的比值。

additional stress

【释义】 附加应力

【例句】 The causes of such a direction difference may be associated with the tectonic setting and the additional stress field. 产生这种方向差异的原因可能是构造条件的差异以及附加应力场的不同。

【中文定义】 指荷载在地基内引起的应力增量。它是使地基失去稳定产生变形的主要原因，通常采用布辛尼斯克理论公式计算。

adit ['ædɪt]

【释义】 *n.* 平洞，入口，平坑

【常用搭配】 auxiliary adit 辅助平洞，辅助坑道；exploratory adit 勘探平洞，探洞

【例句】 These survey monuments are installed in the exploration adits to monitor deep seated deformation of the rock slopes at horizontal depths up to 200m into the slope. 在勘探平洞里安装测点以监测边坡水平深度达 200m 范围内的深部变形。

adit

adit exploration

【释义】 平洞勘探

【例句】 Methods of exploration for dam site area mainly involve drilling exploration, adit exploration, shaft exploration, pit and trench exploration. 坝址区勘探方法主要有钻探、洞探、井探、坑探和槽探。

adit exploration

admixture [æd'mikstʃə]

【释义】 *n.* 混合，添加物，掺合剂

【常用搭配】 antifreeze admixture 防冻添加剂，防冻剂；expanding admixture 膨胀剂，补偿收缩材料

【例句】 The kinds and quantity of additives and admixtures shall be determined through tests. 采用的外加剂和掺合料的种类及数量应通过试验确定。

【中文定义】 掺合料指为改善混凝土性能、减少水泥用量及降低水化热而掺入混凝土中的活性或惰性材料。

admixture

adsorb [əd'sɔrb]

【释义】 *v.* 吸附

【例句】 Charcoal can adsorb colored liquids. 木炭可以吸附有色的液体。

【中文定义】 指流体与多孔固体接触时，流体中某一组分或多个组分在固体表面处产生积蓄的现象，也指物质（主要是固体物质）表面吸住周围介质（液体或气体）中的分子或离子现象。

advance geological prediction

【释义】 超前地质预报

【例句】 Advance geological prediction, as an important guarantee means to avoid construction risk in the tunnel construction process, has been widely used. 超前地质预报作为隧道工程施工过程中规避施工风险的一项重要保证手段，已广泛运用。

【中文定义】 在地下洞室开挖时，运用钻探、物探等手段，探测掌子面前方相关地质信息，对掌子面前方地质情况做出的预报。

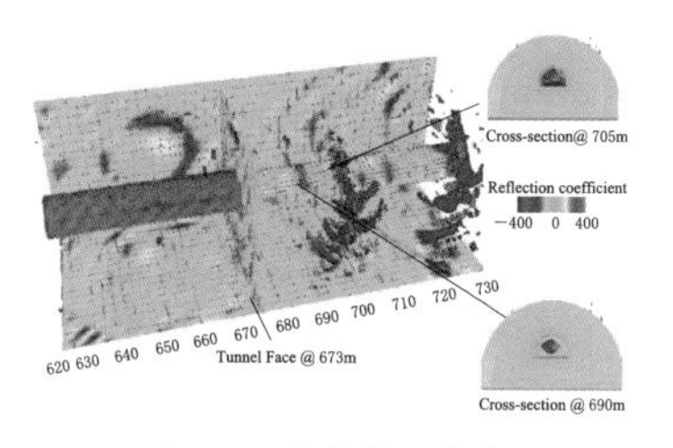

advance geological prediction

advanced anchor

【释义】 超前锚杆

【例句】 Working face with strong rock burst shall be pre-supported by advanced anchor so as to lock the front surrounding rock mass. 岩爆强烈的开挖面应采用超前锚杆预支护，以锁定前方的围岩。

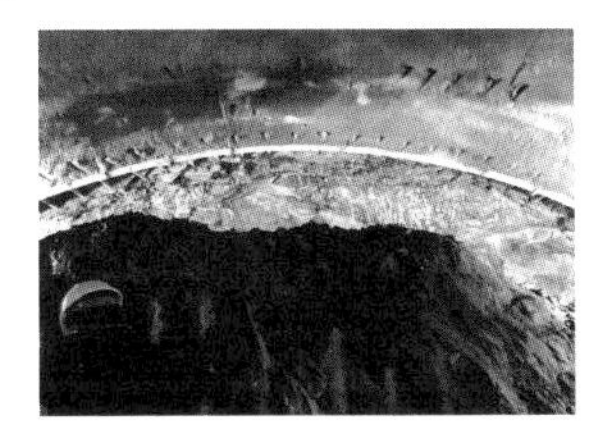

advanced anchor

advanced grouting

【释义】 超前灌浆

advanced grouting

aeolian [i'oʊliən]

【释义】 *adj*. 风成的，风的，风积的

【常用搭配】 aeolian erosion 风蚀作用；aeolian soil 风积土；aeolian landform 风成地形；aeolian desertification 风沙化

【例句】 Aeolian desertification is one of the major desertification types in Northern China. 风沙化是中国北方沙漠化的主要种类之一。

aeolian sandstone

aeolian monadnock

【释义】 风蚀残丘

【中文定义】 风蚀谷经长期风蚀，不断扩展，使风蚀谷之间的地面不断缩小而成为岛状高地或孤立小丘，称为风蚀残丘。

aeolian monadnock

aeolian soil

【释义】 风积土

【例句】 Aeolian soil is a kind of specific soil formed in arid or semiarid area. 风成土是形成于干旱区、半干旱区的一种特殊性质的土。

【中文定义】 风积土是指岩石风化碎屑物质经风力搬运作用至异地降落、堆积所形成的土。

aeolian soil

aerial photography

【释义】 航拍

【例句】 Patterns that are invisible on the ground can be the most striking part of an aerial photography. 地面上看不出来的图案可能会成为航拍照片中最引人注目的东西。

aerial photography

【中文定义】 从空中拍摄地球地貌，获得俯视图，此图即为航拍照片。

aerial remote sensing

【释义】 航空遥感

【例句】 The aerial remote sensing technique is one of important means in the survey of city land and resources. 航空遥感技术是城市国土资源调查的重要手段之一。

aerial remote sensing

aerial view

【释义】 鸟瞰图

【常用搭配】 aerial view of building 建筑鸟瞰图

【中文定义】 是根据透视原理，用高视点透视法从高处某一点俯视地面起伏，绘制成的立体图。

aerial view

African Plate

【释义】 非洲板块

【例句】 The African Plate's speed is estimated at around 2.15cm (0.85in) per year. 非洲板块每年的移动速度约为 2.15cm（约 0.85in）。

African Plate

【中文定义】 全球六大板块之一。其范围包括大西洋中脊南段以东、印度洋中脊以西，印度洋中脊西南支以北和阿尔卑斯山以南地区。

aftershock ['æftəʃɑːk]

【释义】 *n.* 余震

【例句】 Aftershocks can occur within a month or a few days after the main shock.

【中文定义】 余震是在主震之后接连发生的小地震。

age [edʒ]

【释义】 *n.* 期

【常用搭配】 Maastrichtian Age 马斯特里赫特期；Pragian Age 布拉格期

【中文定义】 期是最常用的基本地质年代单位，是一个统范围内生物演化阶段的更具体的划分，适用于同一生物地理区。

agglomerate [ə'glɑːməreɪt]

【释义】 *n.* 集块岩；*vt.* （使）聚集，（使）聚结，（使）凝聚，（使）结块；*adj.* 结块的，成团的

【常用搭配】 agglomerate lava 集块熔岩

【例句】 The Jinping area exposes Late Permian mafic lava, pillow breccia, agglomerate, and volcaniclastites. 锦屏地区出露晚二叠系基性熔岩、枕状角砾岩、集块岩及火山碎屑岩。

【中文定义】 由火山爆发作用产生粗大的带棱角的火山岩碎块和较小的碎块混合组成，常能在火山口附近找到。

agglomerate

aggregate ['ægrɪgət]

【释义】 *n.* 骨料；*vt.* 集合，聚集，合计

【常用搭配】 light weight aggregate 轻骨料；concrete aggregates 混凝土骨料

【例句】 Field investigations for concrete materials before construction are confined chiefly to existing aggregate sources and to locating, exploring, and sampling potential sources. 工程开工前

对混凝土材料的现场调查主要为对现存骨料源的调查以及对潜在骨料源的定位、探测以及试验取样。

【中文定义】 骨料是指在混凝土中起骨架或填充作用的粒状松散材料。骨料作为混凝土中的主要原料，在建筑物中起骨架和支撑作用。

aggregate

aggregate gradation

【释义】 骨料级配

【例句】 The characteristic curve of aggregate gradation may be described by the three parameters: maximum aggregate size, grading index and mineral filler content. 骨料级配曲线可用三个参数表征：骨料最大粒径、级配指数和填料用量。

【中文定义】 骨料级配就是组成骨料的不同粒径颗粒的比例关系。

aging ['edʒɪŋ]

【释义】 *n.* 老化，陈化

agricultural soil

【释义】 耕种土

airborne laser radar

【释义】 机载激光雷达

【例句】 Airborne laser radar can be used to rapidly measure the range of theater missile. 机载激光雷达可以用来快速测量导弹的射程。

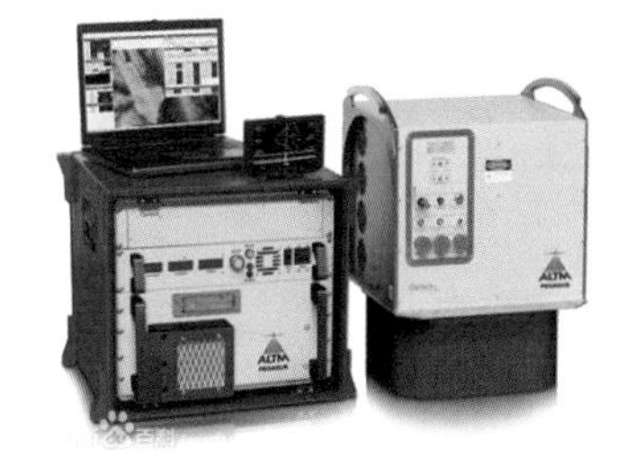

airborne laser radar

air-cushion mechanism

【释义】 气塑层机制

alcohol burning method

【释义】 酒精燃烧法

【例句】 Microwave drying method can used instead of alcohol burning method to determine the water content of soil samples. 可以用微波烘干法代替酒精燃烧法测定土样的含水量。

alkali activity

【释义】 碱活性

【例句】 The potential alkali activity of aggregates and the influence of alkali-silica reaction (ASR) on compactibility, impermeability and behavior of diffusivity of chloride of mortars were investigated by AC impedance spectroscopy. 用交流阻抗谱方法对几种集料的潜在碱活性和由此引起的碱硅酸反应对砂浆密实性、抗渗性和氯离子扩散性的影响进行了研究。

【中文定义】 碱活性指混凝土骨料与水泥中的碱起膨胀反应的特性。

alkali-aggregate

【释义】 碱性骨料

【例句】 The applications of X-ray microscopy in study of alkali-aggregate reaction and calcium distribution in cement were introduced. 在此主要介绍 X 射线显微术在水泥碱性骨料反应研究以及水泥中钙分布的应用。

【中文定义】 碱性骨料是指含碱性成分的骨料，会与水泥发生碱活性反应，从而引起混凝土内部疏松、强度降低开裂等。

alkali rock

【释义】 碱性岩

【常用搭配】 alkali rock series 碱性岩系；calc alkali rock 钙质碱性岩

【例句】 The discovery of ultrapotassic alkali rocks in this area is of far-reaching significance in the research on deep geological processes and tectonic evolution of the study area. 该超钾质碱性岩的发现对研究该区壳幔相互作用及岩石圈演化具有深远意义。

alkalinity [ˌælkə'lɪnəti]

【释义】 *n.* 碱度

【例句】 There is a possibility that alkalinity of ash is related to fiber fineness. 灰分的碱度可能与纤维的细度有关。

【中文定义】 表示水吸收质子的能力的参数，通常用水中所含能与强酸中和作用的物质总量来标定。

allowable bearing pressure

【释义】 容许承压力

【例句】 Allowable bearing pressure shall be defined and based on a factor of safety of 3 applied to the ultimate bearing capacity. 容许承载压力的定义是基于三向应力的极限承载力安全系数。

allowable deformation

【释义】 容许变形

【例句】 In the end the allowable deformation degree of some metal materials was enumerated. 文末列出了一些金属材料的许用变形程度。

【中文定义】 结构或构件在荷载作用下可容许的变形。

allowable factor of safety

【释义】 容许安全系数

【例句】 This paper discusses the allowable factor of safety obtained by the analysis using non-linear strength parameters. 本文论述了采用非线性强度参数分析得出的容许安全系数。

allowable settlement

【释义】 容许沉降

【常用搭配】 maximum allowable settlement 最大容许沉降；allowable settlement of building 建筑物容许沉降

【中文定义】 为保证建筑物安全而规定的地基临界沉降值。

allowable yield of groundwater

【释义】 地下水容许开采量，地下水可开采量

【例句】 The evaluation of the allowable yield of groundwater must be based on that the disastrous ground subsidence don't take place. 地面沉降条件下的地下水容许开采量评价必须以灾害性地面沉降不发生为前提。

alluvial fan

【释义】 冲积扇

alluvial fan

【常用搭配】 alluvial fan deposit 冲积扇层

【例句】 Alluvial fans have a steep slope, coarse-grained sediments which may include boulders, and are usually caused by large flash-floods or debris flows. 冲积扇一般坡度较陡，其含有的粗粒沉积物可能包括大圆石，通常是由大的山洪或泥石流引起的。

【中文定义】 山地河流在出口处由于散流速度降低，大量碎屑物质经分选、沉积而形成的扇形地带。

alluvial plain

【释义】 冲积平原

【常用搭配】 piedmont alluvial plain 山前冲积平原

【例句】 Commonly floods occur in the alluvial plains which are also rich agricultural lands. 洪水通常发生在冲积平原，那里也常常有肥沃的农田。

【中文定义】 由于河流泛滥在其下游地区堆积泥沙而形成的平原。

alluvial plain

alluvial soil

【释义】 冲积土

【例句】 Soft soil foundation is common in the eastern inshore plain areas in our country where alluvial soil strength is low and underground water level is high. Instability accidents happened frequently in soft soil foundation pits, and basal upheave is one of the most common form of destruction. 我国东部沿海平原地区多为软土地基，地下水水位高，冲积土强度低，经常发生基坑失稳事故，坑底隆起破坏是其中常见的一种基坑失稳形式。

【中文定义】 指河流冲积物上发育的土壤。广泛分布于世界各大河流泛滥地、冲积平原、三角洲，以及滨湖、滨海的低平地区。地面平坦。一般成土时间较短，发育层次不明显，土壤肥力较高。

alluvial soil

alluvial terrace

【释义】 冲积阶地，冲积台地

【中文定义】 受河流下切侵蚀和堆积交替作用，河床加深，使原来的河漫滩抬高到洪水位以上，从而使靠河一侧形成了陡坎的河流阶地。

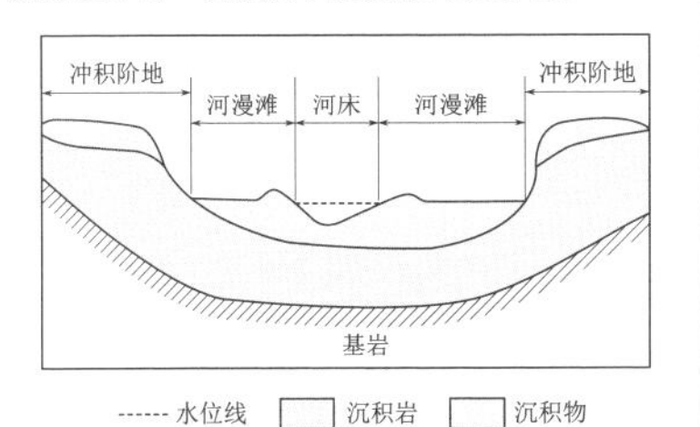

alluvial terrace

α-card survey

【释义】 α卡测量

α-track etch survey

【释义】 α径迹蚀刻测量，α径迹找矿法

alteration [ˌɔltəˈreʃən]

【释义】 *n*. 蚀变，蚀变作用

【常用搭配】 hydrothermal alteration 热液蚀变，水热蚀变

alteration

【例句】 Quartz, serpentine and chlorite are minerals commonly associated with hydrothermal alteration. 石英、蛇纹石及绿泥石均为常见的与热液蚀变作用伴生的矿物。

【中文定义】 岩石、矿物受热液作用产生新的物理化学条件，使原岩的结构、构造以及成分相应地发生改变，生成新的矿物组合的过程。

altered mineral

【释义】 蚀变矿物

【例句】 Altered mineral is a mineral that has undergone some changes in its chemical composition and mineralogical properties. 蚀变矿物是在化学成分和矿物性质上有些变化的一类矿物。

【中文定义】 在化学成分和矿物性质上有不同程度变化的矿物。

altered mineral

altered rock

【释义】 蚀变岩

【例句】 The discovery of Be bearing striation altered rock plays a vital role in ore prospecting and forecasting, and provides a new concept for ore searching. 含铍条纹状蚀变岩的发现在找矿预测中起到了至关重要的作用，为找矿工作开拓了新的思路。

altered rock

【中文定义】 蚀变岩是指受后期岩浆侵入、接触、热液及其他物理化学作用而使得原岩本身矿物成分发生一些改变的岩石，在结构构造上与变

质岩有一定差别。

altered zone

【释义】 蚀变带

【常用搭配】 structural altered zone 构造蚀变带

【例句】 The structural altered zone is formed by a long-term action of structure stress adding hydrothermal solution alteration owing to late structure rupture. 构造蚀变带是遭受长期构造应力作用，并叠加后期构造破裂，经热液蚀变形成的。

【中文定义】 地质体中发生蚀变作用及受其影响的区域或条带。

altered zone

alternately bedded structure

【释义】 互层状结构

【中文定义】 两种岩石交替沉积。

alternately bedded structure

alternating layers

【释义】 互层

alternating layers

【例句】 Mt. Hood is a stratovolcano: a steep-sloped, conical structure composed of alternating layers of hardened lava, solidified ash, and rocks thrown out by earlier eruptions. 胡德雪山是成层火山：陡峭的山坡，由变硬的岩浆交互形成的锥形结构，结晶的火山灰，还有由于早期喷发所喷射出的岩石。

【中文定义】 两种岩层反复出现，多次重复，表明沉积环境反复、重复变化。

American Plate

【释义】 美洲板块

【例句】 One recent study suggests that a mantle convective current is propelling the American Plate. 最近的一项研究表明，地幔对流正在推动美洲板块运动。

【中文定义】 是 1968 年勒皮雄首次提出的六大板块中美洲板块的一部分。范围覆盖了北美洲的大部，向东延伸至中大西洋海岭，向西延伸至东西伯利亚的切尔斯基山脉。后来，美洲板块又被细分为北部的北美洲板块、南部的南美洲板块和周围的一些小板块。

amorphous [ə'mɔrfəs]

【释义】 *n.* 非晶质

【常用搭配】 amorphous alloy 非晶质合金；amorphous silicon 非晶硅

【例句】 Experiment tested the nanometer amorphous diamond film can be used in prosthodontics field. 实验证明先进的纳米非晶金刚石薄膜技术，可应用于义齿（假牙）修复领域。

【中文定义】 又称玻璃质，组成物质的原子或离子呈不规则排列，因而不具备格子构造的固态物质，不能自发的成长为几何多面体，因而被称为无定形体。

amygdaloidal structure

【释义】 杏仁状构造

【中文定义】 岩浆岩中的气孔由次生矿物充填成像杏仁形状的构造。

amygdaloidal structure

anchor ['æŋkə]

【释义】 *n.* 锚，锚杆

【常用搭配】 anchor bolt 锚杆

【例句】 The displacement of excavation decreases with the increasing of anchor prestress. However, when the anchor prestress reach a big value, its influences is small. 基坑位移随着锚杆预应力的增加而减小，当预应力施加到一定数值之后，对基坑位移的影响幅度较小。

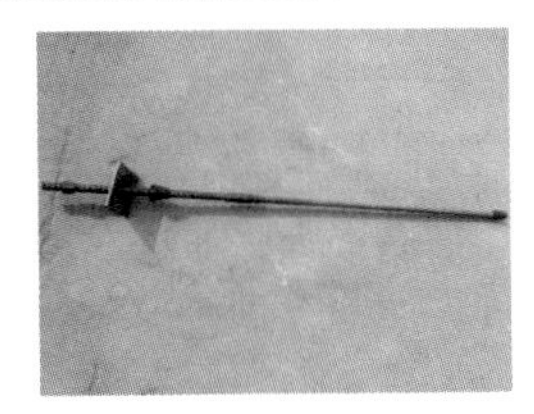

anchor

【中文定义】 由锚固体、锚杆体、外锚头组成的将拉力传递到岩土体的锚固体系。当采用钢绞线或高强钢丝束作杆体材料时，也可称为锚索。

anchor and shotcrete support

【释义】 锚喷支护

anchor and shotcrete support

anchor cable

【释义】 锚索

【常用搭配】 pre-stressed anchor cable 预应力锚索

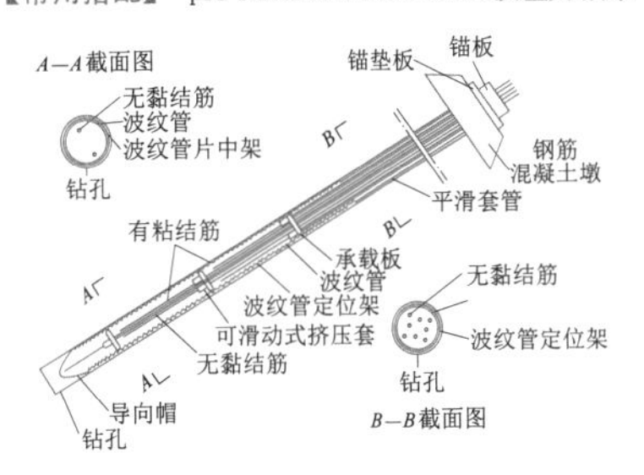

anchor cable

【例句】 Here is extensive application of antislide pile in slide treatment in resent years, especially those with prestressed anchor cables. 抗滑桩尤其是预应力锚索抗滑桩已广泛地应用于滑坡防治工程中。

anchor cavern

【释义】 锚固洞

【例句】 The ground coefficient method is commonly used for internal force calculation for anti-slide piles and anchor caverns. 工程中常用的抗滑桩和锚固洞的内力计算方法多为地基系数法。

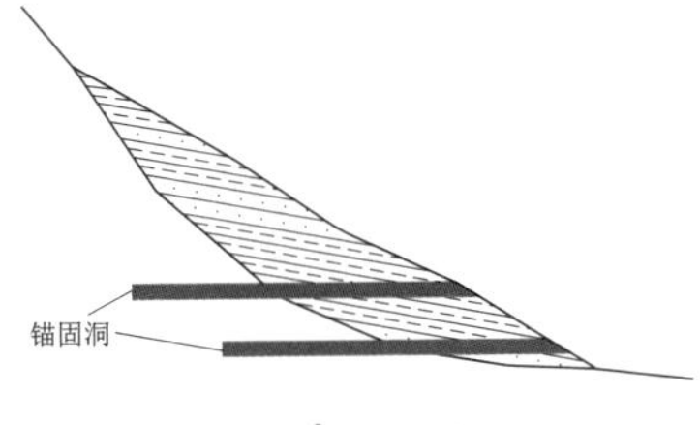

anchor cavern

anchoring section

【释义】 锚固段

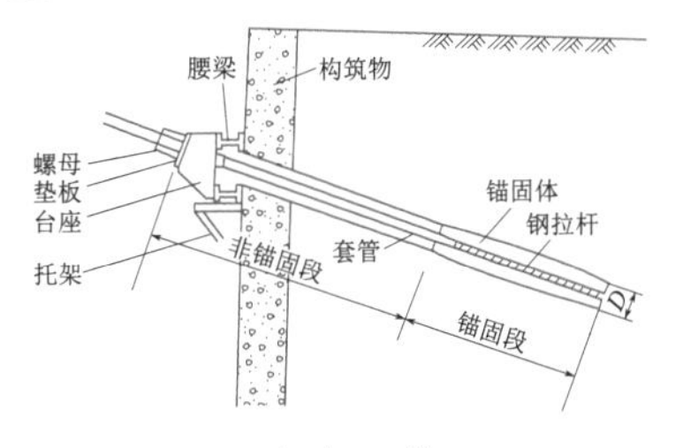

anchoring section

ancient landslide

【释义】 古滑坡

【常用搭配】 ancient debris landslide 碎石土古滑坡；ancient landslide revivification 古滑坡复活

ancient landslide

【例句】 Based on the analysis of landform, structure of land body and stability of slope, it can draw a conclusion that Youyiqiao landslide is a giant ancient landslide. 根据地貌形态、坡体结构以及坡体稳定性等因素综合分析，友谊桥滑坡为一处

巨型的古滑坡。

【中文定义】 全新世以前发生滑动、现今整体稳定的滑坡。

ancient river course

【释义】 古河道

【中文定义】 河流改道后废弃的河道。

ancient river course

andesite [ˈændɪˌzaɪt]

【释义】 *n.* 安山岩

【常用搭配】 basaltic andesite 玄武安山岩；pyroxene andesite 辉安山岩；trachy andesite 粗面安山岩

【例句】 Early Cretaceous volcanic rocks from Songliao basin, Northeast China, are characterized with basic rocks (BRS) which include dorgalite and basalt, intermediate rocks (IRS) which include basaltic andesite, andesite and trachyandesite, and acid rocks (ARS) include trachyte, trachydacite, dacite and rhyolite. 东北松辽盆地早白垩系火山岩具有基性岩类的橄榄玄武岩、玄武岩，中性岩类的玄武安山岩、安山岩、粗安岩，酸性岩类的粗面岩、粗面英安岩、英安岩和流纹岩的特征。

andesite

angle of repose

【释义】 休止角

【中文定义】 无黏性土松散或自然堆积时，其坡面与水平面形成的最大夹角。

angular [ˈæŋgjələr]

【释义】 *adj.* 棱角状

【中文定义】 碎屑颗粒具尖锐的棱角，原始形状基本未变或变化很小。

anhydrite [ænˈhaɪdraɪt]

【释义】 *n.* 硬石膏

【常用搭配】 anhydrite formation 硬石膏层；anhydrite cement 硬石膏灰泥，无水石膏胶凝材料

【例句】 Anhydrite will swell because when exposed to water it slowly converts to gypsum, which has a larger molar volume. 硬石膏遇水膨胀是因为硬石膏在水化作用下慢慢转变为具有较大摩尔体积的石膏。

【中文定义】 是一种硫酸盐矿物。它的成分为无水硫酸钙，与石膏的不同之处在于它不含结晶水。

anhydrite

anionic detergent

【释义】 阴离子洗涤剂

【例句】 Based on this sensitive displacement reaction, a method for the determination of anionic detergents (AD) in wastewater was established. 根据这一灵敏的置换反应，建立了污水中阴离子洗涤剂的测定方法。

【中文定义】 又称阴离子表面活性剂，是一种混合物，主要成分是烷基苯磺酸钠，还有一些增净剂、漂白剂、荧光增白剂、抗腐蚀剂、泡沫调节剂、酶等辅助成分。阴离子洗涤剂不是单一的化合物，可能包括具有不同链长和异构体的几个或全部有关的 26 个化合物。

anisotropy [ˌænaɪˈsɒtrəpɪ]

【释义】 *n.* 各向异性，异向性，非均质性

【例句】 It was found that the foundation stiffness and foundation anisotropy play an important role in the seismic response of the dam. 研究表明，地基的刚度和各向异性对坝体地震反应有重要影响。

【中文定义】 各向异性是指物质的全部或部分化学、物理等性质随着方向的改变而有所变化，在不同的方向上呈现出差异的性质。

annual average uplift rate

【释义】 年平均隆升速率

【例句】 Fission track ages from the O'Xi-Daban granite pluton, located between the south foothill of the Tian Shan and the northern margin of the Kuche Basin, show that the rapid uplift of the Tian Shan occurred with the average rate of 0.13mm/a during the early Cretaceous period and the slow uplift with the average rate of 0.03mm/a from the late Cretaceous to Present. 取自天山南侧库车盆地北缘欧西达坂花岗岩深层岩体裂变径迹年龄给出了初步结果：中生代天山板内造山带在早白垩纪发生了明显快速隆升，平均隆升速率0.13mm/a；而从晚白垩纪到现在，天山板内造山带的平均隆升速率为0.03mm/a。

annual incidence rate

【释义】 年平均发生率

【例句】 The relation among probability of exceedance, recurrence period and annual incidence rate is deduced. 推导出烈度的超越概率与重现期、年平均发生率三者间的关系。

【中文定义】 地震每年发生的频率。

anomaly [ə'nɑməli]

【释义】 *n.* 异常，反常，异常现象；*adj.* 反常性

【常用搭配】 geothermal anomaly 地热异常；electromagnetic anomaly 电磁异常

Antarctic Plate

【释义】 南极洲板块

【例句】 The Antarctic Plate has an area of about $1690 \times 10^4 km^2$. It is the Earth's fifth largest plate. 南极洲板块面积约1690万km^2。它是地球的第五大板块。

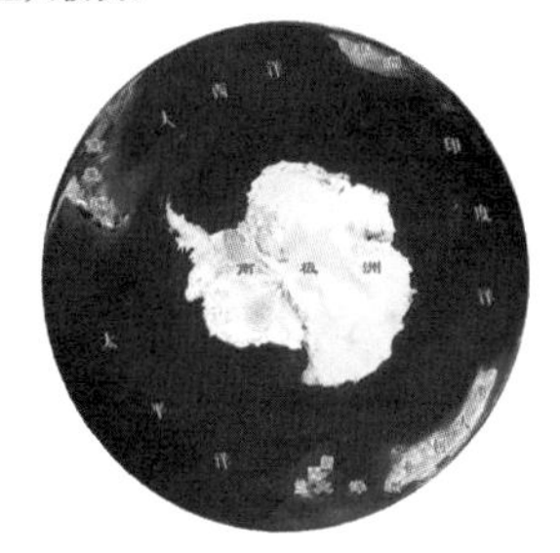

Antarctic Plate

【中文定义】 南极洲板块简称南极板块，是一块包括南极洲和周围洋面的板块，面积约1690万km^2。每年正以1cm的速度向大西洋移动。

anticline ['æntiklaɪn]

【释义】 *n.* 背斜，背斜层

【常用搭配】 residual anticline 残余背斜

【例句】 The low dipping strata and structural disposition of the anticline present a fine opportunity for detailed field analyses. 背斜构造的缓倾斜地层和构造配置，有利于进行详细的现场分析。

anticline

【中文定义】 指岩层发生折曲时，形状向上凸起的岩层形态者。在一般平地上，背斜地层上半部受到侵蚀变平，会形成中间古老、两侧较新的地层排列方式。

anticlinorium [ˌæntiklaɪ'nɔːrɪəm]

【释义】 *n.* 复背斜，复背斜层

扇形复背斜

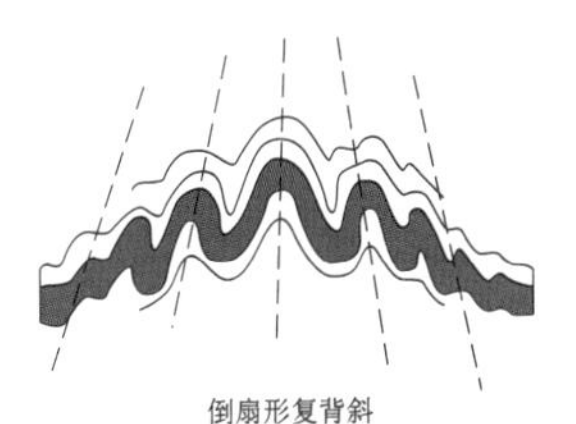

倒扇形复背斜

anticlinorium structure diagram

【例句】 The Gengma Basin located in the Gengma lantsang anticlinorium of Baoshan fold belt, Western Yunnan is 195km^2 in area, accepted widespread fluvial lacustrine deposit, and is typically Tertiary small continental basin. 滇西耿马盆地位于三江褶皱系保山褶皱带的耿马—澜沧复

背斜，面积 195km^2，接受广泛的河湖相沉积，为典型的第三纪小型陆相盆地。

【中文定义】 由若干次级褶皱组合而成的大型背斜构造，它规模大，需经过较大范围的地质制图才能了解其全貌。

anti-dip structural plane

【释义】 反倾结构面

【例句】 The slope acceleration amplification coefficient of bedding structural planes is integrally larger than the slope acceleration amplification coefficient of anti-dip structural planes. 顺倾结构面边坡加速度放大系数整体大于反倾向结构面边坡加速度放大系数。

schematic diagram of anti-dip structural plane

anti-scouring velocity

【释义】 抗冲流速

anti-slide pile

【释义】 抗滑桩

anti-slide pile

anti-sliding stability

【释义】 抗滑稳定

【例句】 Based on the friction theory and the vector geometry, the formula of anti-sliding stability safety coefficient is presented. 基于摩擦理论和矢量几何条件，给出了计算抗滑稳定安全系数的计算式。

apatite ['æpətaɪt]

【释义】 *n.* 磷灰石

【常用搭配】 apatite group 磷灰盐类，磷灰石族

【例句】 In the new study, researchers located grains of the mineral apatite in thin sections from the moon rocks and meteorite. 在新的一项研究中，研究人员在月岩和陨石中找到了磷灰石的矿物颗粒。

【中文定义】 磷灰石是一系列磷酸盐矿物的总称，它们有很多种，如黄绿磷灰石、氟磷灰石、氧硅磷灰石、氯磷灰石、锶磷灰石等。

apatite

aperture ['æpətʃə]

【释义】 *n.* 开度

【常用搭配】 average aperture 平均张开度

【例句】 The joint aperture is an important parameter in the study of fissured rock mass hydraulics. 节理张开度是裂隙岩体水力学研究中的一个重要参数。

【中文定义】 指相邻岩壁间的垂直距离。

aperture deformation method

【释义】 孔径变形法

【例句】 The aperture deformation method is used to test the stress field in deep slope. 采用孔径变形法对斜坡深部地应力场进行测试。

aplite ['æplaɪt]

【释义】 *n.* 细晶岩

aplite

【例句】 Alkaline rocks in the area are mainly composed of gray black trachyte, and some breccialava, syenite porphyry and syenite aplite. 在这区域碱性岩的岩性组合主要由灰黑色粗面岩、角砾熔岩、正长斑岩、正长细晶岩组成。

aplite vein

【释义】 细晶岩脉

aplite vein

apparent density

【释义】 表观密度

【例句】 The PVC paste blending resin is a kind of suspension polymerized resin with spheric particle morphology, high apparent density and low oil absorption. 聚氯乙烯掺混树脂是一种颗粒外观球形化、表观密度较高、吸油率低的悬浮聚合树脂。

【中文定义】 表观密度是骨料颗粒单位体积（包括颗粒内封闭孔隙）的质量。

apparent dip

【释义】 视倾角，假倾角

【例句】 Study indicates that high dip dissolution fracture is presented with lower apparent dip and high true dip in FMI data. 研究表明，高倾角溶蚀裂缝在水平井中 FMI 资料上的表现为视倾角较低，真倾角高。

【中文定义】 指视倾斜线与其在水平参考面上的投影线的夹角。

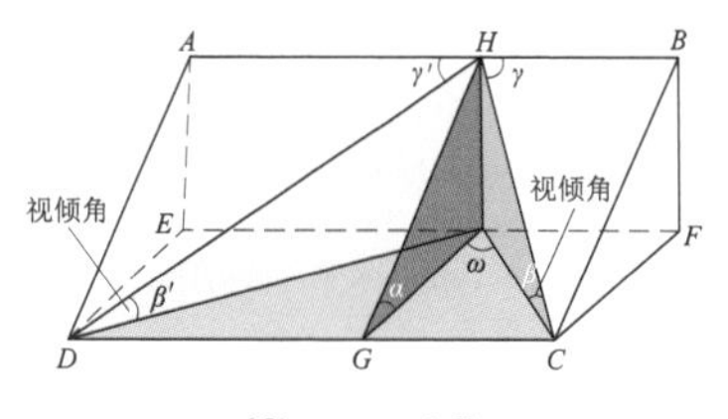

（β）apparent dip

apparent velocity

【释义】 表观流速，视流速

【例句】 The apparent velocity of the continuous phase(water) had a greater effect on the mass transfer coefficient than that of the dispersed phase. 连续相的表观速度对传质系数的影响大于分散相的表观速度。

【中文定义】 在多相流或多孔介质流动工程学上，假定一个虚拟的（人造的）流动或者说假定流体是单一一种流体流过所在区域时的速度称为表观流速。

aquiclude [ˈækwikluːd]

【释义】 *n.* 相对隔水层

【例句】 An aquiclude stores water but does not transmit significant amounts. 一个相对隔水层可以含水但不能大量导水。

【中文定义】 允许地下水以极小速度流动的弱导水岩层。在多层含水层叠置的含水系统中，弱透水层与隔水层的作用不同，后者起隔水作用，前者则构成其上、下含水层间的水交换通道，而且上、下含水层的水头压差越大，通过弱透水层的水量也越大。

aquifer [ˈækwəfə]

【释义】 *n.* 含水层

【常用搭配】 confined aquifer 承压含水层；perched aquifer 上层滞水含水层

【例句】 The Nubian Sandstone Aquifer System, the world's largest, is located under the eastern apron extension part of the Sahara Desert and spans the political boundaries of Libya, Chad, Sudan and Egypt. 世界上最大的努比亚砂岩蓄水系统，位于撒哈拉沙漠东部地区地下，跨越利比亚、乍得、苏丹和埃及。

【中文定义】 凡透水性能好、空隙大的岩石以及卵石、粗沙、疏松的沉积物、富有裂隙的岩石，岩溶发育的岩石均可为含水层。

aquifuge [ˈækwɪfjuːdʒ]

【释义】 *n.* 隔水层，不透水层

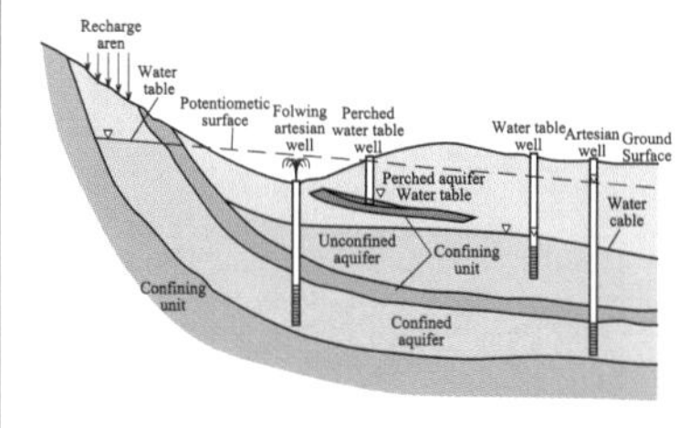

aquifuge

【例句】 There are 4 aquifers and 3 aquifuges in the area. 区内由 4 个含水层和 3 个隔水层组成。

【中文定义】 重力水流不能透过的土层或岩层。如黏土、重亚黏土以及致密完整的页岩、火成岩、变质岩等。

aquitard ['ækwtəd]

【释义】 *n.* 弱透水层

【中文定义】 允许地下水以极小速度流动的弱导水岩层。在多层含水层叠置的含水系统中，弱透水层与隔水层的作用不同，后者起隔水作用，前者则构成其上、下含水层间的水交换通道，而且上、下含水层的水头压差越大，通过弱透水层的水量也越大。

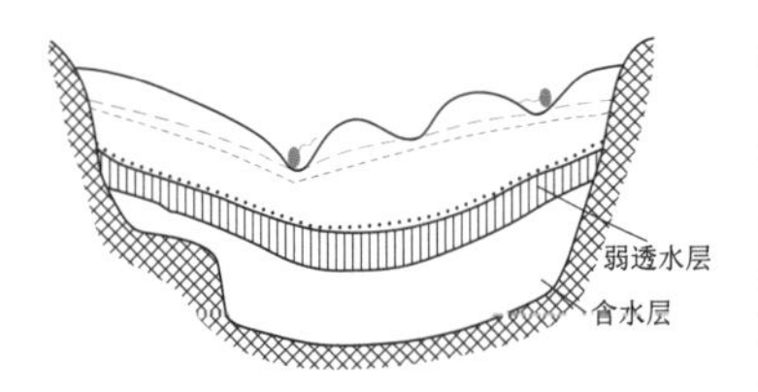

schematic diagram of aquitard

Archean (AR) [ɑr'kiən]

【释义】 *n.* 太古宙（宇）

【中文定义】 地质时代中的一个宙，元古宙之前。起始于 38 亿年前，地球岩石开始稳定存在并可以保留到现在，结束于 25 亿年前的大氧化事件。以甲烷为主的还原性的太古宙原始大气转变为氧气丰富的氧化性的元古宙大气，并导致了持续 3 亿年的休伦冰河时期。太古宙是原始生命出现及生物演化的初级阶段。

arenaceous-pelitic texture

【释义】 砂泥质结构

arete [æ'ret]

【释义】 *n.* 刃脊，陡峭的山脊

【常用搭配】 arete climbing 岩角

arete

【中文定义】 两个冰斗或冰谷间所夹的山峰，被侵蚀而成的尖锐陡峻的山脊。

argilization interlayer

【释义】 泥化夹层

argillaceous infilled

【释义】 泥质充填

【例句】 The friction coefficient of the weak structural plane is relatively small and extends longer, and it is generally clay filling or argillaceous infilled. 软弱结构面的摩擦系数相对较小，延伸较长，且普遍为黏土充填或泥质充填。

argillaceous infilled

【中文定义】 指结构面中的充填物主要由泥质矿物组成。

argillization [ɑːd'ʒɪlaɪzeɪʃn]

【释义】 *n.* 泥化，泥化作用，黏土化作用

【常用搭配】 secondary argillization 二次泥化；mudstone argillization 泥岩泥化

【例句】 The most frequent reaction causing the self-sealing of rocks with high secondary permeability is argillization. 这种发生频率很高的、使次生渗透性能良好的岩体产生自封闭现象的反应就是泥化。

arithmetic average thickness

【释义】 算术平均厚度

【例句】 The arithmetic average thickness is the average thickness of the ore body obtained by the arithmetic mean calculation method when the mineral reserves are calculated. 算术平均厚度是矿产储量计算时，用算术平均数的计算方法求得的矿体平均厚度。

arrangement of exploration works

【释义】 勘探工作布置

artesian head

【释义】 承压水头，自流水压头

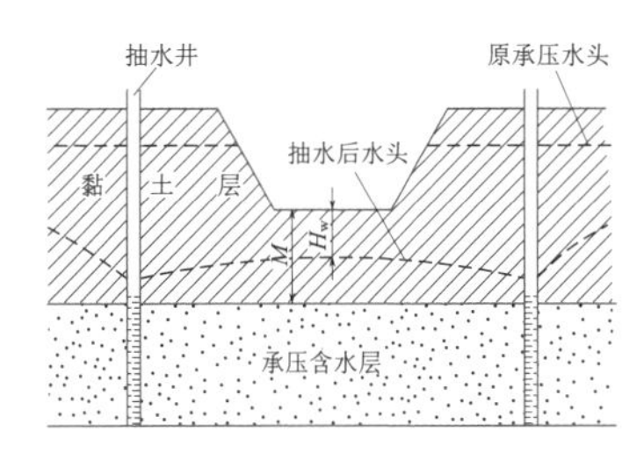

pumping water to reduce artesian head

artificial accumulation

【释义】 人工堆积

【例句】 The Heping Square slip mass is a loose accumulation body, which comprised artificial accumulation, collapse-slide accumulation, weathering residual deposits and alluvium by making use of comprehensively investigation methods. 通过综合勘察手段的运用，查明了和平广场滑坡体是包含人工堆积、崩滑堆积、残坡积与河流冲积等物质的松散堆积体。

【中文定义】 指由人工堆积的各种沙土、岩屑、矿渣等。如修渠打井时堆积在地面的沙土；修路、平整土地时填入洼地的沙土，开采煤矿时堆积的矸石；烧制石灰时堆积的灰渣等。有时把已毁坏废弃的土堤、土墙、土烽火台等构筑物也看成是一种人工堆积物。在野外工作时，注意不要把一些人工堆积物误认为天然的第四纪沉积物。

artificial accumulation

artificial aggregate

【释义】 人工骨料

【例句】 Artificial aggregates are used for Jiangya RCC dam. 江垭碾压混凝土大坝采用人工骨料。

【中文定义】 指采用爆破等方法开采岩石作为原料，经过破碎、碾磨、筛分而成的混凝土骨料。

artificial aggregate

artificial fill

【释义】 人工填土

【例句】 Highway subgrade is soil body structure including natural earth foundation and artificial fill embankment. 公路路基是土体构筑物，含天然土地基和人工填土路堤两大部分。

【中文定义】 人工填土是由于人类活动而形成的堆积土。物质成分较杂乱，均匀性差，根据物质组成和堆填方式，又可分为素填土（碎石、砂土、黏性土等），杂填土（含大量建筑垃圾及工业、生活废料），冲填土（水力充填）三类。

artificial fill

artificial recharge of groundwater

【释义】 地下水人工补给

【例句】 Artificial recharge of groundwater with urban rainwater runoff has become one of the effective measures for urban water resources man-

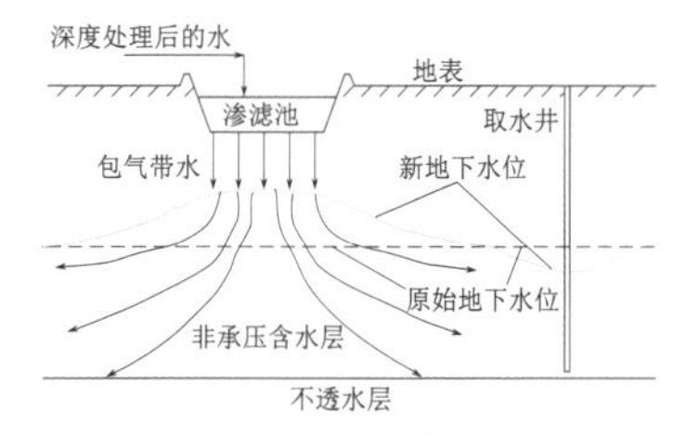

schematic diagram of artificial recharge of groundwater

agement. 利用雨水补给地下水已成为城市水资源管理的有效措施。

【中文定义】 地下水含水层自外界或相邻含水层获得水量的过程。地下水的补给方式有降雨入渗、灌溉入渗、河渠渗漏、人工回灌、山前和邻区侧向补给，以及相邻含水层的水量转移等。降雨入渗补给，降雨入渗的水量，首先补充地下水面以上土层。

artificial sand

【释义】 人工砂

【例句】 With gradually increasing of the civil construction and reducing of useable natural sand in our country, the artificial sand will be inevitably selected as fine aggregate used in concrete. 随着我国土木工程建设力度的逐渐加大和可用自然砂资源的日益匮乏，人工砂将是混凝土细骨料的必然选择。

【中文定义】 人工砂指经除土处理的机制砂、混合砂的统称，包括机制砂、混合砂。

artificial sand

as-built geological map

【释义】 竣工地质图

【中文定义】 依一定比例尺和图例绘制，表示竣工后地质情况的地质图件。

ascending spring

【释义】 上升泉

【例句】 A few of ascending spring are recharged by pressure water coming from deep basin and fault zone. 少部分上升泉通过盆地水或断裂带承压水补给。

【中文定义】 是由承压水形成的泉水。

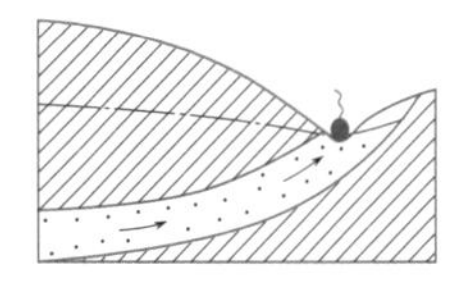

schematic diagram of ascending spring

assaying [ə'seɪŋ]

【释义】 *n.* 含量测定，试料分析，试金

【常用搭配】 assaying method 检测方法；wet assaying 湿分析

【例句】 The variation and influence factor of soilwater mineral degree is analyzed in this paper by using the assaying result of water sample. 在文中通过对水样含量测定结果分析了试验土壤水的矿化度变化规律及其影响因素。

assessment [ə'sesmənt]

【释义】 *n.* 评价，估价

【常用搭配】 preliminary assessment 初步评价；qualitative assessment 定性评价；quantitative assessment 定量评价

【例句】 Based on the comparative study and assessment, the research directions of design flood hydrograph are suggested. 在对比评价和分析的基础上，提出设计洪水过程线的研究方向的建议。

associated mineral

【释义】 伴生矿物

【例句】 The phase transformation sequence resulting from heating and, on the other hand from dry grinding of each of the associated minerals, depends on the structure and physicochemical properties of the starting materials. 对不同伴生矿物加热和干磨所产生的相变序列，取决于原始材料的结构和物理化学性质。

associated minerals in natural calcite mineral

【中文定义】 不同成因或不同矿化期（或矿化阶段）生成的矿物组合。

atlas ['ætləs]

【释义】 *n.* 地图集

【例句】 The design and formation of comprehensive atlas is a complex system engineering. 大型综合性

地图集的设计与编制是一项复杂而烦琐的系统工程。

【中文定义】 具有统一的设计原则和编制体例、协调的地图内容、规定的比例尺、分幅系统和装帧形式的多幅地图的汇集。

atmospheric influence depth

【释义】 大气影响深度

【中文定义】 是指在自然气候作用下，由降水、蒸发、地温等因素引起土的升降变形的有效深度。

atmospheric precipitation recharge

【释义】 大气降水补给

【例句】 The groundwater in Hebei Plain is recharged from the precipitations. 河北平原地下水是由大气降水补给的。

【中文定义】 大气降水一部分转化为地表径流，一部分被蒸发，仅有部分渗入地下，渗入地下的部分在到达潜水面以前，必须经过由土颗粒、空气和水三相组成的包气带，故入渗过程中水的运动极其复杂。

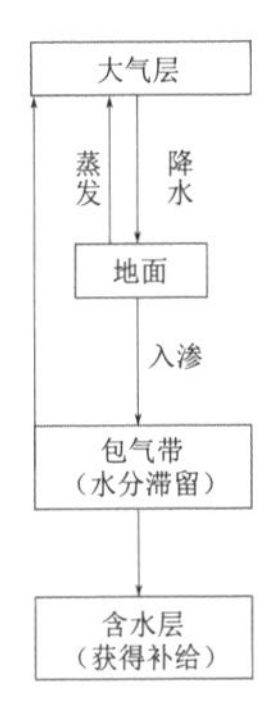

block diagram of atmospheric precipitation recharge

Atterberg limits

【释义】 阿太堡限度，界限含水量

【常用搭配】 Atterberg limits value 阿氏限度；Atterberg limits test 界限含水量试验

【例句】 The plasticity index of silt is in general less than 10, but whether Atterberg limits can be used as indices for such soils is doubtful. 粉土的塑性指数一般小于 10，但阿太堡界限是否适用于作为评价粉土的指标是值得怀疑的。

【中文定义】 黏性土的重要物理性质指标。黏性土从一个稠度状态过渡到另一个稠度状态时的分界含水量。

Atterberg test (boundary water content test of soil)

【释义】 阿太堡试验（土的界线含水量试验）

【例句】 According to the Atterberg test of soil, the relationships between organic matter ratio and liquid limits, plastic limits, and plasticity index have been studied. 通过阿太堡试验研究有机质含量与液限、塑限、土壤的塑性指数的关系。

Atterberg test apparatus

attitude [ˈætɪtuːd]

【释义】 *n.* 产状

【常用搭配】 attitude of bed 岩层产状；attitude of stratum 地层产状

【同义词】 occurrence

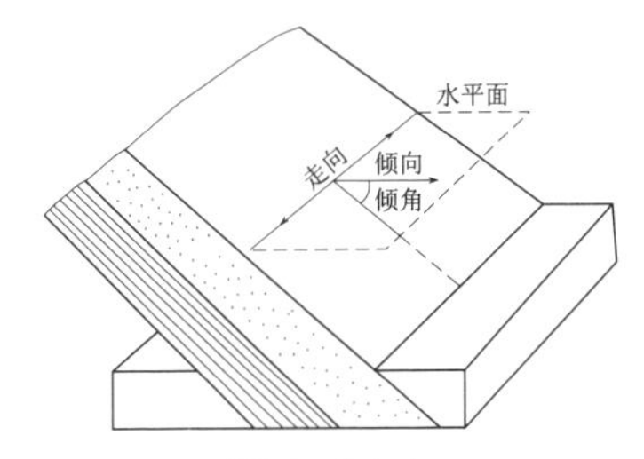

attitude of stratum

【例句】 In the typical geological researches, the attitude of bed is characterized by using its strike angle, dip direction angle and dip angle. 在经典的地质研究中，岩层的产状都采用走向角、倾向角和倾角这三个参数进行表征。

【中文定义】 由岩层面或节理面的三度空间的延伸方位及其倾斜程度来确定，即采用走向、倾向和倾角三个要素的数值来表示。

augen structure

【释义】 眼球状构造

【中文定义】 斑晶较大，呈眼球状或透镜体团状，断续分布于具良好片理的基质中，晶体大小不一，大的可达几厘米至十几厘米。

augen structure

augite [ˈɔdʒaɪt]

【释义】 *n*. 辉石，普通辉石

【常用搭配】 basaltic augite 玄武辉石；aegirine augite 霓辉石；titan augite 钛辉石

【例句】 The rock stack is made up of various sized black or dark grey olivine, augite basalt, amphibole and andesite. 此岩系由黑色或暗灰色橄榄石、普通辉石玄武岩、角闪石安山岩等大小石块构成。

augite

available layer

【释义】 有用层

available layer

【中文定义】 指质量技术指标能满足水利水电工程天然建筑材料要求的岩土层。

average grain size of aggregate

【释义】 骨料平均粒径

【中文定义】 骨料平均粒径是指骨料粒径的平均值。

average of the higher half values

【释义】 大值平均值

【例句】 Average of the higher half values is the average of all values that are greater than the mean. 大值平均值就是大于平均值的所有值的平均值。

average thickness method

【释义】 平均厚度法

【例句】 The average thickness method is the method that multiplying the total area within the calculated range by the average thickness of the available layer. 平均厚度法指用储量计算范围内的总面积乘以有用层的平均厚度的方法。

【中文定义】 指用储量计算范围内的总面积乘以有用层的平均厚度的方法。

axial plane cleavage

【释义】 轴面劈理

【例句】 The origin of shear fold and slip plane of axial plane cleavage have been a control-versial problem for a long time. 剪切褶皱的成因及轴面劈理的滑动面是长期以来颇有争议的问题。

【中文定义】 是指其产状平行于或者大致平行于褶皱轴面的劈理，这类劈理主要发育在强烈褶皱的地质体中。

axial plane cleavage

azimuth [ˈæzəməθ]

【释义】 *n*. 方位角

【常用搭配】 azimuth angle 方位角；magnetic azimuth 磁方位角；coordinate azimuth 坐标方位角

【例句】 The measurement of the inclination angle and the azimuth angle in the drilling process is important. 钻探过程中倾斜角和方位角的测量至关重要。

【中文定义】 是从某点的正北方向线起，依顺时针方向到目标方向线之间的水平夹角。

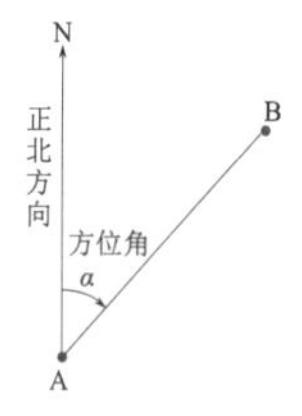

azimuth

B

background earthquake
【释义】 本底地震
【中文定义】 本底地震是指一定区域内没有明显构造标志的最大地震。该“一定区域”即为地震区或地震带。

backwater length
【释义】 壅水长度

backwater level
【释义】 壅高水位

ball structure
【释义】 球状构造
【同义词】 globular structure; orbicular structure
【中文定义】 斑晶较大，呈眼球状或透镜体团状，断续分布于具良好片理的基质中，晶体大小不一，大的可达几厘米至十几厘米。

bamboo-leaf texture
【释义】 竹叶状结构
【中文定义】 石灰岩的一种结构，其特点为截面有砾石呈竹叶状。

bamboo-leaf texture

banded structure
【释义】 带状构造
【中文定义】 在变质岩石中，各种矿物成分分布不均匀，以石英、长石、方解石等粒状矿物为主的浅色条带和以黑云母、角闪石、磁铁矿等为主的暗色条带，各以一定的宽度成互层状出现，形成颜色不同的条带状构造。

bank deposits
【释义】 岸边淤积

bank erosion
【释义】 岸边侵蚀

bank erosion

barbotage [ˈbɑːbətɑːʒ]
【释义】 *n*. 起泡作用
【例句】 Through the experiments on flotation with different samples, the barbotage of each component in the pine oil for flotation was revealed. 通过试样对铁矿的浮选实验研究，揭示了松油中各组分对浮选的起泡作用。

barbotage

bare karst
【释义】 裸露型喀斯特，裸露型岩溶
【例句】 The water resources evaluation of substreams in bare karst area is the hot topic of present karst water research. 裸露岩溶地区地下河系统的水资源评价一直是目前喀斯特水研究的热点。
【中文定义】 指可溶性岩石裸露地表，经地表水溶蚀、机械侵蚀后形成的各种岩溶地貌。

landform of bare karst

barite [ˈbɛraɪt]
【释义】 *n.* 重晶石
【常用搭配】 barite group 重晶石群
【例句】 Drilling mud is a mixture of water and sodium montmorillonite (bentonite), often with additives like barite (barium sulphate) to give the required properties of density and viscosity. 钻孔冲洗液是一种钠基蒙脱土水溶液，通常加入添加剂如重晶石以取得密度和黏性方面的特性。
【中文定义】 重晶石系硫酸盐矿物。成分为 $BaSO_4$。自然界分布最广的含钡矿物。

barrier lake
【释义】 堰塞湖
【例句】 Debris from the earthquake has blocked rivers and streams, creating 34 barrier lakes that could become unstable. 由于地震松动形成的山崩滑坡体阻塞河流，造成了 34 处很不稳定的堰塞湖。
【中文定义】 由火山熔岩流或由地震活动等原因引起山崩滑坡体等堵截河谷或河床后贮水而形成的湖泊。

basalt [bəˈsɔlt]
【释义】 *n.* 玄武岩
【例句】 Under the water, volcanoes continued to erupt sending up magma and forming mountains of basalt. 在水下，火山依旧持续喷发，向上喷射岩浆并形成了一座座玄武岩山峰。

basalt

basic earthquake intensity
【释义】 基本地震烈度
【例句】 Then, the paper evaluated the anti-seismic performance of the structure under basic and rare earthquake intensity. 然后文章对结构在基本烈度和罕遇烈度地震作用下的抗震性能进行评估。
【中文定义】 指地震引起的地面震动及其影响的强弱程度。当以地震烈度为指标，按照某一原则，对全国进行地震烈度区划，编制成地震烈度区划图，并作为建设工程抗震设防依据时，区划图可标志烈度便被称之为"地震基本烈度"。

basic rock
【释义】 基性岩
【常用搭配】 basic rock body 基性岩体；basic rock zone 基性岩带
【例句】 The basic rock intrusive activity implied continuation and development of the tension process. 基性岩的侵入活动代表张拉过程的持续和进一步发展。

basically stable
【释义】 基本稳定
【中文定义】 根据《水力发电工程地质勘察规范》，Ⅱ类围岩稳定性多为基本稳定，表现为围岩整体稳定，不会产生塑性变形，局部可能产生掉块。

basin [ˈbeɪsn]
【释义】 *n.* 盆地，流域
【常用搭配】 river basin 流域；intermontane basin 山间盆地；karst basin 喀斯特盆地
【例句】 The pore confined aquifers are mostly distributed in the deep overburden of the mountainous rivers, the quaternary faulted basin, and the thick sediments of piedmont alluvial-proluvial fan. 孔隙承压含水层多分布于山区河流深厚覆盖层、第四纪断陷盆地和山前冲洪积扇的巨厚沉积层内。
【中文定义】 陆地上中间低四周高的盆状地形。

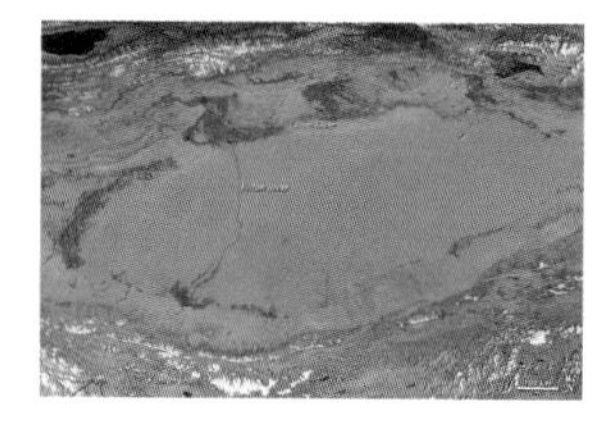

Tarim Basin

batholith [ˈbæθəlɪθ]
【释义】 *n.* 岩基
【常用搭配】 concordant batholith 整合岩基；discordant batholith 不整合岩基
【例句】 The Xiangride batholith consists of variable rock types, including quartz-diorite, monzogranite and granodiorite, with an intrusive period from 258Ma to 218Ma. 香日德复式花岗岩类岩基主要由石英闪长岩、二长花岗岩和花岗闪长

岩等多种岩性单元构成，侵入期为258～218Ma。
【中文定义】 一种大规模的岩浆岩侵入体。

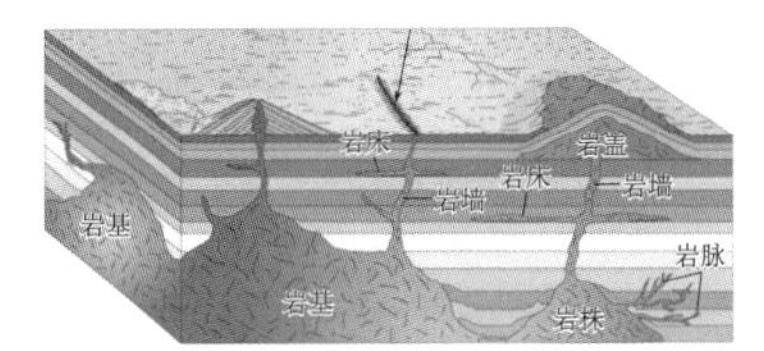

batholith

bathymetry [bə'θɪmɪtri]
【释义】 *n*. 水深测量
【例句】 In this paper a method is proposed for improving the measurement precision of bathymetry by utilizing ray tracing in the depth direction. 本文提出了一种在深度方向上采用射线跟踪技术提高水深测量精度的方法。

bathymetry

bearing plate test
【释义】 承压板试验
bed [bɛd]
【释义】 *n*. 层
【中文定义】 是最小的岩石地层单位，是段的再分。
bedding ['bediŋ]
【释义】 *n*. 底层，底部，基础；层理，层面
【常用搭配】 bedding plane 层面，层理面；graded bedding 粒级层，序粒层，粒级层理；cross-bedding 交错层理；herringbone cross-bedding 羽状交错层理
【例句】 Bedding is often the most obvious feature of a sedimentary rock and consists of lines called bedding planes, which mark the boundaries of different layers of sediment. 层理是沉积岩最显著的特征，由表现出线状特点的层面组成，层面是不同沉积层之间的边界。

bedding plane

【中文定义】 指岩层中物质的成分、颗粒大小、形状和颜色在垂直方向发生改变时产生的纹理。一般厚几厘米至几米，其横向延伸可以是几厘米至数千米。
bedding cleavage
【释义】 层面劈理，顺层劈理
【例句】 Note the bedding cleavage relationship on this outcrop, the steep dip of the cleavage is indicative of the upright folds in the region. 注意这个露头上层面劈理的产状，劈理的陡倾角说明这个地区的直立褶皱。
【中文定义】 指宏观上与岩性层界面近平行的区域性劈理。

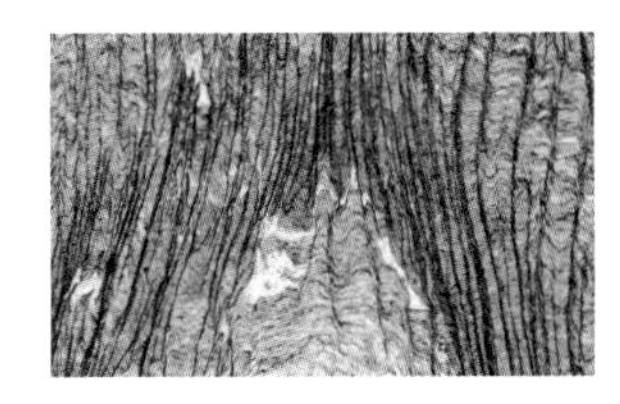

bedding cleavage

bedding joint
【释义】 顺层节理，层面裂隙

bedding joint

【例句】 The boulder, rock wall, bottom toe always appear in limestone mine mid deep bench blasting because of bedding joint, fault and cave. 石灰岩矿山中深孔台阶爆破常因层理节理、断层、溶洞等原因出现大块石、岩壁、底趾等。

【中文定义】 顺层节理是在沉积岩中平行于层面发育的节理。

bedding plane structure

【释义】 层面构造

【中文定义】 沉积岩的一种原生构造，未固结的沉积物由于机械原因或由于生物活动在其表面造成的痕迹，被后来的沉积物覆盖面保留在层面上的构造现象。

bedding structure

【释义】 层理构造

【例句】 Research on coal bedding structure is the theory foundation and precondition of coal seam gas exploitation. 煤体层理结构的研究是煤层气开发开采的理论基础和前提条件。

【中文定义】 是沉积物最重要的特征之一，是由沉积物的成分、颜色、粒度在垂直于沉积物表面的方向上（即层理面）显示出来的特征。

bedrock landslide

【释义】 基岩滑坡（岩质滑坡）

【同义词】 rock landslide

【例句】 The bedrock landslide is an important appearance of the seismicity. 基岩滑坡是该区地震活动的重要表现。

bedrock landslide

bedrock seated terrace

【释义】 基座阶地

【中文定义】 阶地表面有较厚的冲积层，但地壳上升、河流下切较深，以致切透了冲积层，切入了下部基岩以内一定深度，从阶地斜坡可明显地看出，阶地由上部冲积层和下部基岩两部分构成。

bedrock surface contour map

【释义】 基岩等值线图

【例句】 Bedrock surface contour map is one of the most important data for 3D geological model, and also an important task in 3D geological survey. 基岩等值线图是建立三维地质模型所需的一项重要数据，也是三维地质调查工作的一项重要任务。

【中文定义】 以基岩分布高程相等数值点的连线反映基岩分布情况的地质图件。

beds alternation

【释义】 岩层交互

【中文定义】 岩层交互相间组成的岩层。

Beijing coordinate system 1954

【释义】 1954 北京坐标系

【例句】 In order to find the better solution for the conversion between the old Beijing coordinate system 1954 and the new Xi'an 1980 coordinate system, the author chooses three parameters for conversion between these two coordinate systems by using coordinate system conversion function of the MAPGIS software in this paper. 为了更好地解决 1954 北京坐标系与 1980 西安坐标系之间的新旧坐标成果转换问题，作者通过采用 MAPGIS 软件下的坐标系转换功能，采用 3 种参数进行 1954 北京坐标及 1980 西安坐标系之间转换。

bench mark

【释义】 水准基点

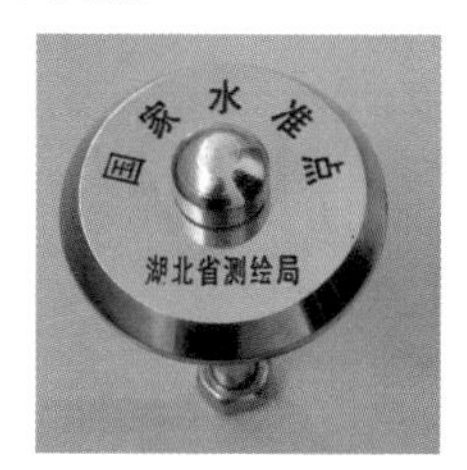

bench mark

bend fold

【释义】 弯曲褶皱

【例句】 The in-situ thrust system, with duplex or stacking structure in its south, a simpler deformation of propagation fold and fault bend fold in its northeast, and a pop-up in its front, is under the short distance thrust system and the Quaternary. 原地冲断系统隐伏在近距离冲断系统和第

四系之下，其南部的变形主要表现为双冲构造或堆垛构造，往北东方向主要表现为传播褶皱和断层弯曲褶皱的简单变形，前锋地带表现为三角带的突起构造。

【中文定义】 又称纵弯褶皱或弯褶皱，是在平行层理的侧向挤压力作用下形成的褶皱。

bend fold

bending strength

【释义】 抗弯强度，抗折强度

【例句】 The results show that the dry density, bending strength and compressive strength of foam concrete block decreased with the increasing of the amount of foam. 结果表明：泡沫混凝土砌块的干密度、抗弯强度、抗压强度随泡沫含量的增加而减小。

【中文定义】 指材料抵抗弯曲不断裂的能力，主要用于考察陶瓷等脆性材料的强度。

bending test

【释义】 抗弯试验

【例句】 The technique of acoustic emission has been employed in the SENB bending test for the thermal damaged materials. 在热震损伤材料的SENB抗弯试验中采用了声发射技术。

bid letter

【释义】 投标函

【中文定义】 指投标人按照招标文件的条件和要求，向招标人提交的有关报价、质量目标等承诺和说明的函件。

big data

【释义】 大数据

【例句】 Big data is exploding as more and more information is collected and stored daily. 每天随着越来越多的信息被收集和存储起来，大数据正呈爆炸式增长。

【中文定义】 指无法在一定时间范围内用常规软件工具进行捕捉、管理和处理的数据集合，是需要新处理模式才能具有更强的决策力、洞察发现力和流程优化能力的海量、高增长率和多样化的信息资产。

bioclastic texture

【释义】 生物碎屑结构

【中文定义】 大多见于石灰岩及铁、锰、硅质岩中，含大量生物，但不完整。

bioclastic texture

biological skeleton texture

【释义】 生物骨架结构

【中文定义】 由原地生长的造礁生物形成的礁灰岩所具有的结构。

biotite [ˈbaɪəˌtaɪt]

【释义】 *n.* 黑云母

【常用搭配】 biotite granite 黑云母花岗岩

【例句】 Rock-forming minerals include olivine, pyroxene, hornblende, orthoclase feldspar, plagioclase, mica, quartz, biotite, calcite and other common metal and nonmetal minerals. 造岩矿物包括橄榄石、辉石、角闪石、正长石、云母、石英、黑云母、方解石和其他常见金属及非金属矿物。

【中文定义】 云母类矿物中的一种，颜色从黑到褐、红色或绿色都有，具有玻璃光泽，为硅酸盐矿物。黑云母主要产于变质岩中，在花岗岩等其他一些岩石中也有。

biotite

bird's eye structure

【释义】 鸟眼构造

【中文定义】 碳酸盐岩中的一种微小的空洞构造，空洞的形状像鸟眼。

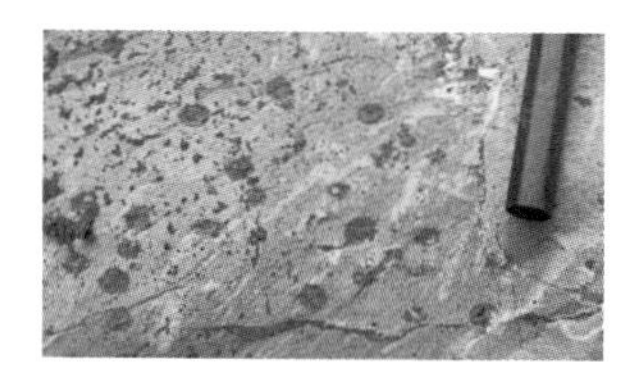

bird's eye structure

blasting [ˈblɑːstɪŋ]

【释义】 *n.* 爆炸，由爆破而产生的碎石块，枯萎，破坏；*v.* 爆炸（blast 的现在分词形式）

【常用搭配】 pre-splitting blasting 预裂爆破；smooth blasting 光面爆破

【例句】 Caution in blasting is most important within the immediate vicinity of buildings and installations. 在十分靠近建筑物和设备的地方爆破时，谨慎是最重要的。

barrier lake blasting

blasting induced fissure

【释义】 爆破裂隙

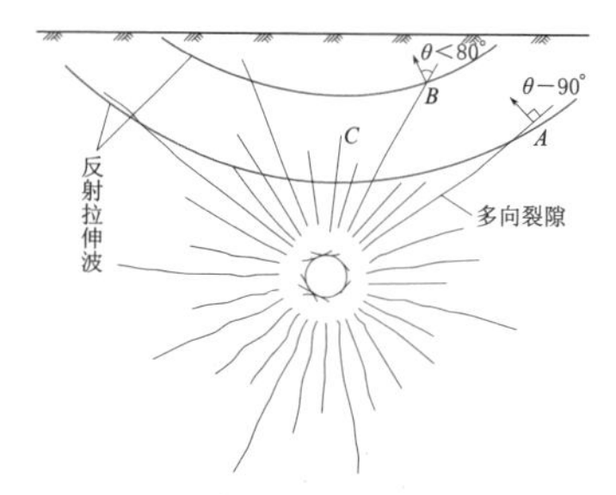

blasting induced fissure

【例句】 To control the rock mass damage under blasting and to minimize the range of blasting induced fissures in tunnel excavation are great significance for project safety. 隧洞开挖过程中控制爆炸对岩体的损伤，减小爆破裂隙范围，对保证工程安全具有重要意义。

【中文定义】 指在爆破过程中由爆破作用在岩土体中所形成的裂隙。

blasting parameters

【释义】 爆破参数

【例句】 The model of controlling the fragmentation is built according to blasting parameters that were tested in previous implementation. 根据施工前进行试爆取得的爆破参数，建立了岩石破碎块度控制模型。

blasting vibration monitoring

【释义】 爆破震动监测

【例句】 Blasting Vibration Monitoring and Control Technology Applied in Construction of Huaishuping Tunnel. 槐树坪隧道爆破震动监测与控制技术。

blind fault

【释义】 隐伏断层

【例句】 The blind faults and traps have been recognized by seismic profiles. 根据地震资料解释出隐伏断层及构造。

【中文定义】 是在地表无出露、潜伏在地表以下的断层。

blind fault

block limit equilibrium method

【释义】 分块极限平衡法

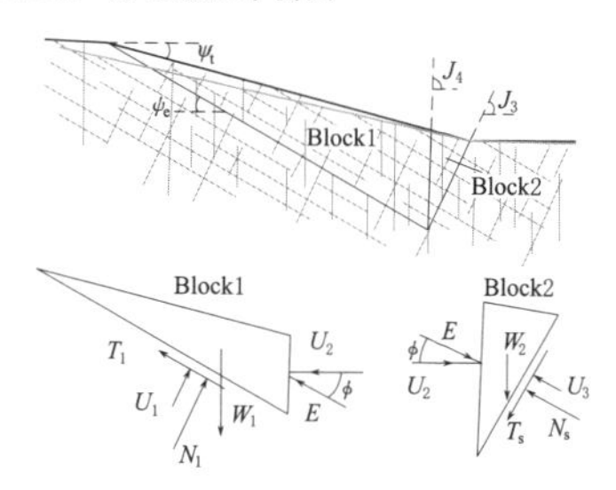

block limit equilibrium method

block toppling

【释义】 块状倾倒

【例句】 Before introducing the probabilistic analysis procedure, it is valuable to review the conventional deterministic approach for analyzing slopes to block toppling. 在引入概率分析方法之前，有必要回顾一下块状倾倒边坡的传统确定性分析方法。

【中文定义】 指陡倾的板状岩体被与其正交的一组缓倾横向节理切割成块状，坡体底部块状岩体被其后方倾倒岩体推动，发生向后、向上扩展的倾倒破坏，倾倒岩体的底部沿横向节理形成斜线排列的台阶状。

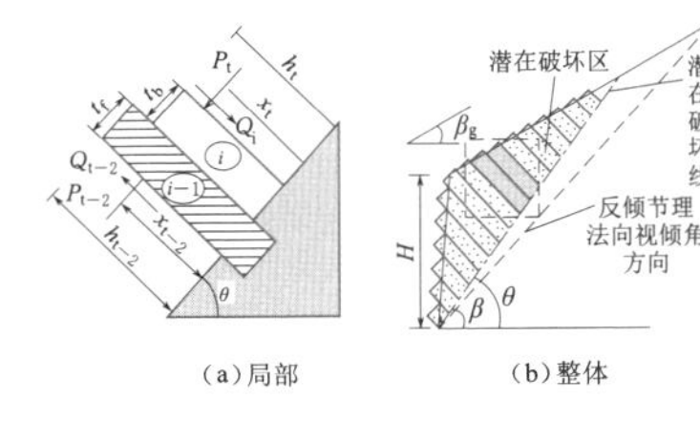

block toppling failure mechanism

block-up coefficient

【释义】 堵塞系数

【中文定义】 参与计算泥石流流量的一个系数，反映了泥石流阵流堵塞现象对流量的影响。

blocky structure

【释义】 块状结构

【同义词】 cloddy structure

【例句】 Fault zones in shales are brecciated and clayey, while those in limestones are of blocky structure. 页岩的断层带主要是由亚黏土和角砾岩组成，而石灰岩主要呈块状结构。

blocky-fractured structure

【释义】 块裂结构

blocky-fractured structure

【中文定义】 是由软弱结构面切割面成，其变形和破坏受软弱结构面控制。

body wave

【释义】 体波

【例句】 The major methods of measuring Q-value by the natural earthquakes are the ones of using the data of free vibration of earth, body wave, surface wave and coda. 利用天然地震波测量 Q 值的主要方法有：利用地球自由振荡资料、体波、面波和尾波来研究。

【中文定义】 是由震源振动直接产生的、在地球内部传播的地震波。体波分为纵波（P）和横波（S）。地震波在地下的反射和折射蕴含着丰富的信息。

boiling spring

【释义】 沸泉

【例句】 That boiling spring can ease one's fatigue. 那个沸泉能够解除人体的疲劳。

【中文定义】 泉口温度约等于当地沸点的地热水露头。

bolt (cable) stress monitoring

【释义】 锚杆（索）应力监测

【例句】 In this article, authors also state some construction technologies of the prestressed cable arch structure including the steel cable tension technology, and bolt stress monitoring and control technology and so forth. 在文中，作者还介绍了预应力索拱结构的施工技术，包括钢索张拉技术、锚杆应力监控技术等。

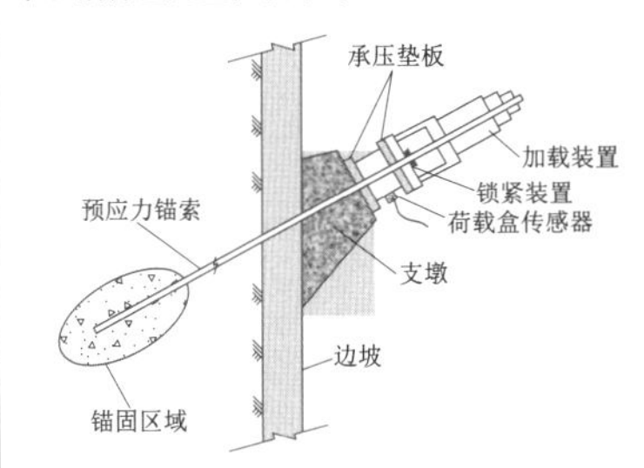

bolt (cable) stress monitoring

borehole camera

【释义】 钻孔照相

borehole camera

borehole deformation test

【释义】 钻孔变形试验

borehole log

【释义】 钻孔柱状图

【常用搭配】 borehole log through soil 土层钻孔柱状图；borehole log through rock 岩层钻孔柱状图

【例句】 Borehole log of the quaternary system is significant，that can be used to carry out the stratigraphic correlation and 3D modeling process. 第四系钻孔柱状图很重要，被用于进行地层对比和三维建模过程。

钻孔柱状图

工程编号	199433						
工程名称	商北					钻孔编号	ZK1
孔口标高	27.91m	坐标	X=303885.25m	开工日期		稳定水位深度	3.65m
钻孔直径	50mm		Y=319069.56m	竣工日期		测量水位日期	

地层编号	地质时代				柱状图 1:100	地质摘建	厚位测试深度 m		取排编号/深度 m
1		26.01	1.90	1.50					
2		25.51	2.40	0.50					
3		24.61	3.10	0.70					
4		24.11	3.80	0.70					
5		21.61	6.30	2.50					
6		21.31	6.60	0.30					
7	Q^{al}	20.71	7.20	0.60					
8		20.01	7.90	0.70					
9		19.31	8.40	0.50					
10		17.21	10.70	2.30					
11		16.51	11.40	0.70					
12		15.61	12.30	0.90					
13		14.91	13.00	0.70					
说明									

记录______ 制图______ 检查______ 年 月 日 图号

borehole log

【中文定义】 根据对勘探钻孔岩芯的观察鉴定、取样分析及在钻孔内进行的各种测试所获取资料编制而成的一种原始图件。

borehole televiewer

【释义】 电视测井

borehole (well) pumping test

【释义】 钻孔（井）抽水试验

【例句】 A single well pumping test can only be used to estimate the T-value and K-value. 单孔抽水试验只能被用于估计渗透系数和导水系数。

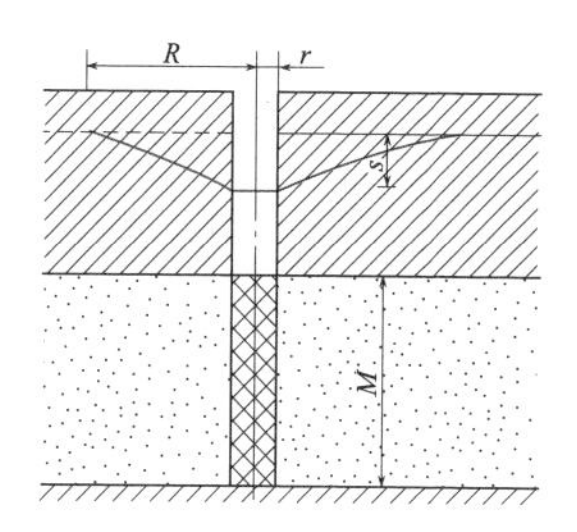

diagram of borehole pumping test in confined aquifer

bottom heave

【释义】 坑底隆起

【例句】 This paper discusses the mechanical behavior of ground bottom heave due to excavation. 文章研究了由于开挖引起的坑底隆起力学机制。

【中文定义】 基坑开挖后，由于卸荷回弹而发生的基坑底面向上鼓起的现象。

boulder [ˈboʊldər]

【释义】 *n.* 孤石，大圆石，巨砾

【例句】 The construction technology of well point unwatering method，manual pore-forming，manual cut deep concrete and boulder，static blast are applied in the foundation project，which solves the problem of the new tower built in the old site. 在基础工程中采用井点降水、人工成孔、深层混凝土和大漂石人工切割与静力爆破施工工艺，成功解决了原塔旧址新建问题。

【中文定义】 指坡面上零星分布、具有一定体积的（一般大于1m^3）、孤立的岩石块体。若以单体形式呈现则称为孤石；若以群体形式呈现，块石之间相互叠置、集中连片分布则称为孤石群。

erratic boulder

bound water

【释义】 结合水，化合水

【例句】 Relations between the shrinkage and chemical bound water of medium strength self-compacting concrete. 中低强度自密实混凝土干缩与化学结合水的关系研究。

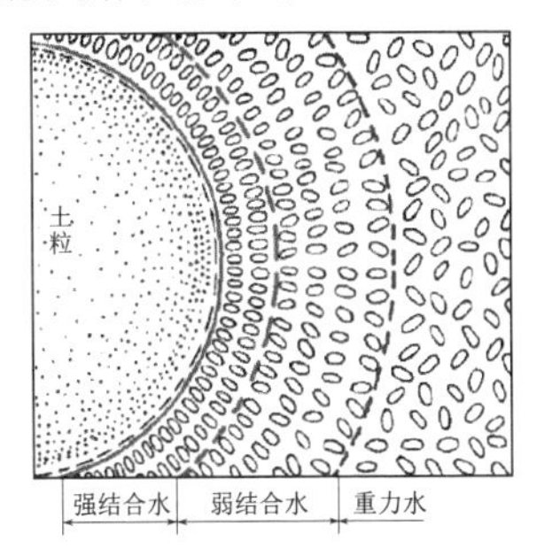

schematic diagram of bound water

【中文定义】 系指受电分子吸引力吸附于土粒表面的土中水，这种电分子吸引力高达几千到几万个大气压，使水分子和土粒表面牢固的黏结在一起。处于土颗粒表面水膜中的水，受到表面引力的控制而不服从静水力学规律，其冰点低于零度。

boundary condition

【释义】 边界条件

【例句】 In the seepage theory, the current expression of boundary condition on phreatic surface is aimed at a kind of seepage. 渗流理论中现有的潜水面边界条件表达式是针对特定的一类潜水渗流问题而给出的。

boundary of weathered zone

【释义】 风化界线

【常用搭配】 boundary of completely weathered zone and highly weathered zone 全风化与强风化界线; boundary of highly weathered zone and weakly weathered zone 强风化与弱风化界线; boundary of weakly weathered zone and slightly weathered zone 弱风化与微风化界线

【例句】 Rock boundary of weathered zone includes lower boundary of completely weathered zone, highly weathered zone, weakly weathered zone and slightly weathered zone. 岩体风化界线包括全风化带下限、强风化带下限、弱风化带下限及微风化下限。

【中文定义】 指地质剖面中表示岩体不同风化程度的地质分界线。

box foundation

【释义】 箱型基础

【例句】 The large scale model test proves that the rigidity of frame structure with large thick raft foundation is approximately equal to the rigidity of box foundation. 大型模型试验证明，高层框架及厚筏的刚度近似于箱型基础。

box foundation

brachy fold

【释义】 短轴褶皱

【例句】 Dacom uplift is a wide and gentle anticlinorium trending NNE and superimposed by brachy folds, which was cut by NWW, NE and NNE fault systems. 大康隆起是北北东向展布的宽缓复式背斜，叠加短轴褶皱，被北西西向、北东向和北北东向断裂切割。

brachy fold

【中文定义】 短轴褶皱是指褶皱枢纽向两端倾伏，在平面上呈长圆形的褶皱，其长宽比在10∶1到3∶1之间。

braided channel

【释义】 *n.* 多汊河道，游荡型河道

【例句】 The braided channels may flow within an area defined by relatively stable banks or may occupy

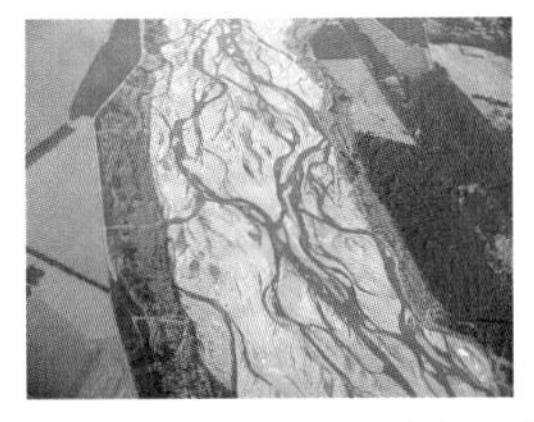

Waimakariri River (braided channel)

an entire valley floor. 多汉河道可以在相对稳定的岸坡区域内流动或者可以在整个谷底流动。

【中文定义】 河道宽浅，江心多浅滩和沙洲，水流散乱，支流和废河床密布的地形平缓区的河道。

breccia ['bretʃə]

【释义】 *n.* 角砾岩

【常用搭配】 ablation breccia 剥蚀角砾岩；alloclastic breccia 火山碎屑角砾岩；avalanche breccia 岩崩角砾岩；clastic breccia 碎屑角砾岩；crush breccia 压碎角砾岩；dislocation breccia 断层角砾岩；eruptive breccia 火成角砾岩；explosion breccia 爆发角砾岩

【例句】 That mylonite belts, breccia belts and schistosity belts occur along the fault. 糜棱岩带、角砾岩带和片理化带沿断层分布。

breccia

【中文定义】 是粒径大于 2mm、具棱角的岩石或矿物碎块。主要由暴露在地表的岩石经机械风化作用形成的粗碎屑，未经搬运或只有短距离搬运堆积。

brecciform texture

【释义】 角砾状结构

Brinell (hardness) test

【释义】 布氏（硬度）试验

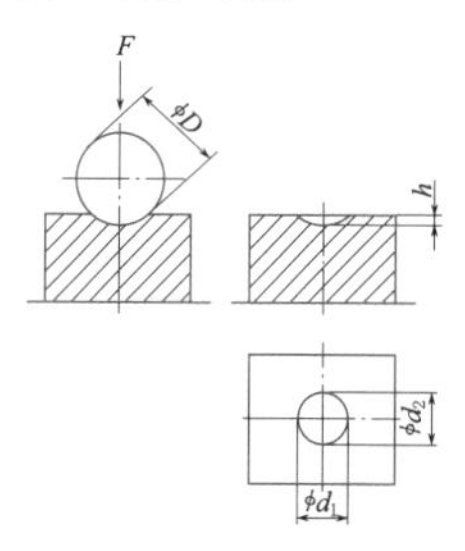

Brinell (hardness) test

brittle failure

【释义】 脆性破坏

【例句】 The phenomenon is harmful to full elaborate of material latent capacity and makes the section occur brittle failure code. 这种现象的存在不利于充分发挥材料潜能，并且使截面最终呈现脆性破坏特征。

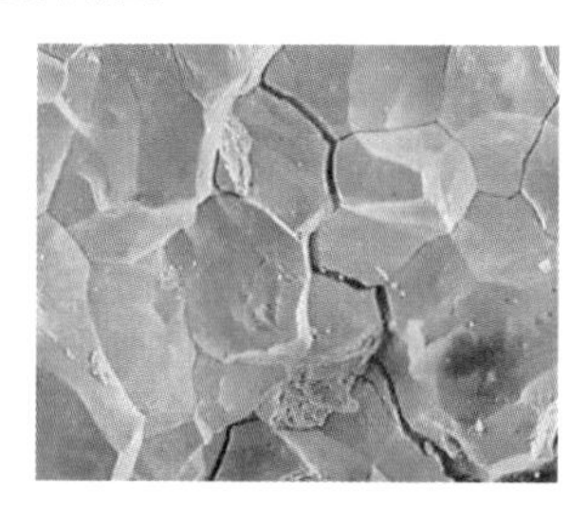

brittle failure

brittle rupture

【释义】 脆性断裂

【例句】 The results show that the broken type is brittle rupture, there are slip bands, and the reason is the expanding of graphite. 结果发现因为石墨膨胀造成的破坏为脆性断裂，且有滑移带。

【中文定义】 构件未经明显的变形而发生的断裂。断裂时材料几乎没有发生过塑性变形。如杆件脆断时没有明显的伸长或弯曲，更无缩颈，容器破裂时没有直径的增大及壁厚的减薄。脆断的构件常形成碎片。材料的脆性是引起构件脆断的重要原因。

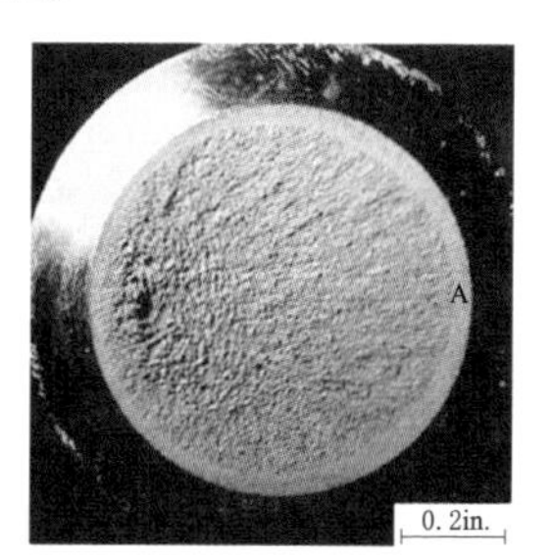

the surface of brittle rupture

buchite ['bʊtʃaɪt]

【释义】 *n.* 玻化岩；玻辉岩

【例句】 Buchite is a very uncommon hornfels containing glass. 玻化岩是一种非常罕见的含玻

璃质角岩。

【中文定义】 是一种由高热变质作用形成的玻璃质岩石。它们是火山岩和次火山岩中的捕虏体或悬挂体，在高温条件下发生部分熔融后快速冷却所形成。在泥质玻化岩中可含有堇青石、尖晶石、多铝红柱石等矿物。在砂质玻化岩中，石英碎屑因熔蚀而圆化，钾长石碎屑常沿解理发生玻璃化，基质为褐色玻璃和长柱状多铝红柱石等。

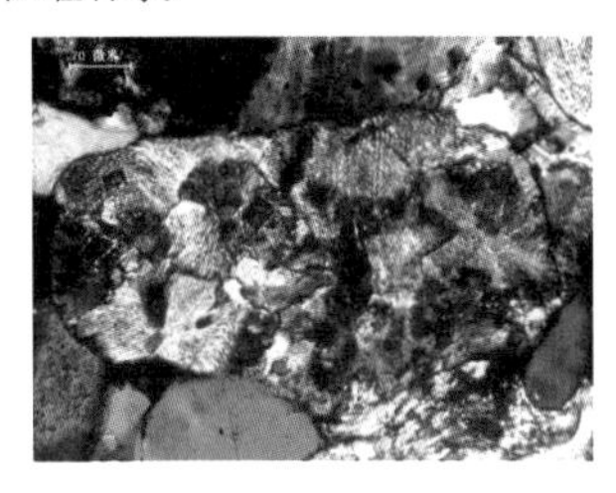

buchite

buckling [ˈbʌkliŋ]

【释义】 *n.* 溃曲

【常用搭配】 buckling failure 溃曲破坏

【例句】 This paper analyzes failure modes of buckling of consequent rock slopes, establishes mechanical model and analyzes failure mechanism of sliding and bending deformation of slopes by means of beam and plate theory. 本文通过分析顺层岩质斜坡的溃曲破坏模式，建立力学模型，运用梁板理论分析斜坡滑移弯曲变形破坏的机理。

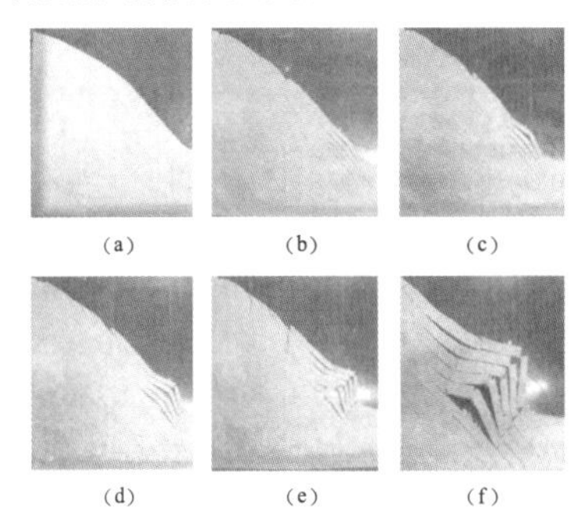

physical simulation process of buckling failure of bedding slope

bulk density

【释义】 堆积密度

buried conditions of groundwater

【释义】 地下水埋藏条件

【中文定义】 指含水岩层在地质剖面中所处的部位及受隔水层限制的情况，可将地下水分为包气带水、潜水及承压水，按含水介质类型，可将地下水区分为孔隙水、裂隙水及岩溶水。

buried karst

【释义】 埋藏型岩溶

【例句】 Ground collapse is a common geological hazard in shallow buried karst area. 地面塌陷是浅埋岩溶区常见的一种地质灾害。

buried karst

【中文定义】 是指可溶性岩石表面被红土等松散土层覆盖，渗透水流对可溶岩进行溶蚀，形成埋藏于松散土层下的岩溶，如溶沟、石芽、溶斗、溶蚀洼地等。

buried terrace

【释义】 埋藏阶地

【中文定义】 是堆积阶地的一种特殊类型，原有阶地被新的沉积物所掩埋，在地形上不再有阶梯状形态。

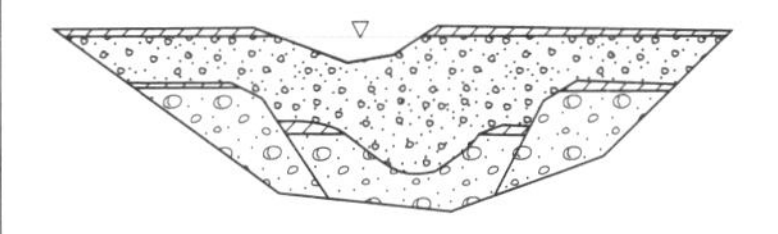

buried terrace

bury of groundwater

【释义】 地下水埋深

【同义词】 groundwater depth (GWD)

【例句】 Revealing the spatiotemporal variation characteristics of shallow bury of groundwater can provide scientific basis for evaluating and managing groundwater resources. 揭示区域地下水埋深的时空变异规律，可为地下水资源评价和管理提供科学依据。

【中文定义】 指地下水的水面到地表的距离。

Cadastral Information System (CIS)

【释义】 地籍信息系统

【例句】 Cadastral Information System (CIS) is cadastral data management, information service, and business operational system, which is widely used by the government, social or economic management department, and the public personnel. 地籍信息系统是各国政府用来为政府机关、社会经济管理部门以及个人提供地籍信息管理、信息服务、业务运营的系统。

【中文定义】 在计算机软硬件支持下，把各种地籍信息按照空间分布及属性以一定的格式输入、处理、管理、空间分析、输出的技术系统。

cadastral map

【释义】 地籍图

【中文定义】 描述土地及其附着物的位置、权属、数量和质量的地图。

caisson foundation

【释义】 沉井基础

【例句】 The caisson foundation is a kind of foundation that sink to the designed elevation by self-weight and surmount the side-wall friction, then filled the caisson tank as pier and foundation of bridge or other structures. 沉井基础是依靠自身重力克服井壁阻力后下沉到设计高程，然后经过混凝土封底并填塞井孔，使其成为桥梁墩台和其他结构物的基础。

【中文定义】 沉井基础：以沉井作为基础结构，将上部荷载传至地基的一种深基础。

calcite ['kælsaɪt]

【释义】 *n.* 方解石

【常用搭配】 calcite twinning 方解石双晶

calcite

【例句】 The joint surfaces are slightly iron stained and calcite filled. 接触面有轻微锈染及方解石充填。

【中文定义】 一种最常见的天然碳酸钙矿物，分布范围很广。

calcite infilled

【释义】 方解石充填

【例句】 In the limestone area, due to the action of groundwater, calcite infilled easily formed in the structure planes. 在石灰岩地区，由于地下水作用，容易在结构面中形成方解石充填。

【中文定义】 由于地下水活动，各种岩石的裂隙中经常充填有方解石脉。

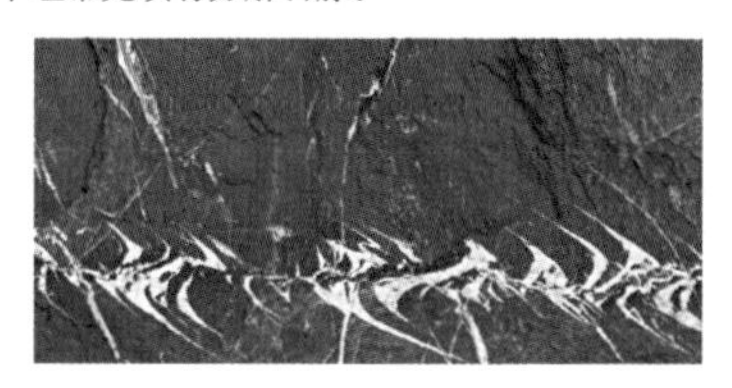

calcite infilled

calcite vein

【释义】 方解石脉

calcite vein

calcium coated

【释义】 钙质薄膜

【例句】 In general, calcite fine veins, calcium coated, or calcareous cements as filling materials appear in the structure surface as the infiltration diameter. 一般而言，方解石细脉、钙质薄膜或作为充填物的钙质胶结物出现于作为渗径的结构面中。

【中文定义】 结构面上附着的薄层钙质组成物。

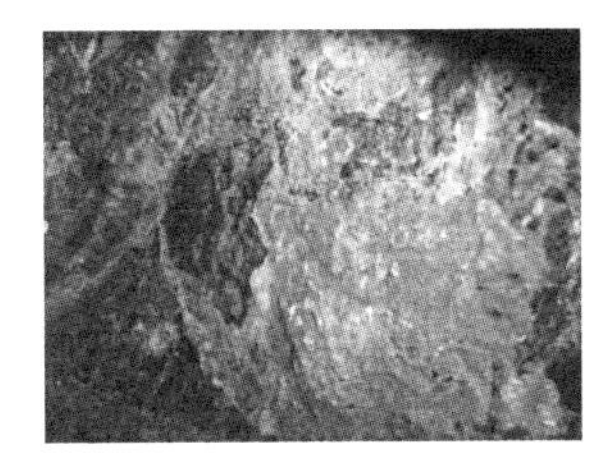

calcium coated

california bearing ratio (CBR)
【释义】 加州承载比
【例句】 California bearing ratio is a technical judgment indicator of the intension of subgrade and pavement strength. 加州承载比试验是评价路基土和路面材料强度及稳定性的重要技术指标。

caliper logging
【释义】 井径测量
【例句】 The maximum well diameter, minimum well diameter and average well diameter are calculated by the caliper logging curve. 利用井径测量曲线能计算出最大井径、最小井径及平均井径。

Cambrian (∈) ['kæmbriən]
【释义】 *n.* 寒武纪（系）
【常用搭配】 Cambrian Explosion 寒武纪爆发
【例句】 The Cambrian is the first geological period of the Paleozoic Era. 寒武纪是古生代的第一个地质时期。
【中文定义】 寒武纪是显生宙最早的地质时代，距今约5.42亿～4.88亿年。寒武纪的开始，标志着地球进入了生物大繁荣的新阶段。

Cambrian trilobite fossile

cap rock
【释义】 盖层，冠岩
【例句】 The combination evaluation of cap rock and reservoir is often neglected in the conventional reservoir evaluation. 在常规油藏评价中常常忽略对盖层和储层组合的评价。

cap rock

capillary rising height
【释义】 毛细上升高度
【同义词】 rising height of capillary water
【例句】 The results show that the relationship between capillary rising height and water content similar to the soil water characteristic curve. 结果表明，毛细上升高度与含水率的关系类似于土-水特征曲线。
【中文定义】 毛细水从地下水的水面沿土层或岩层空隙上升的最大高度。

capillary water
【释义】 毛细管水
【例句】 Porous media wettable by water can form capillary water and capillary pressure accelerates evaporation of water from the interface. 易被水润湿的多孔介质可形成毛细管水，并且毛细压力具有促进界面蒸发的作用。
【中文定义】 又称毛细水，指由于毛细作用保持在土层或岩层毛细空隙中的地下水。

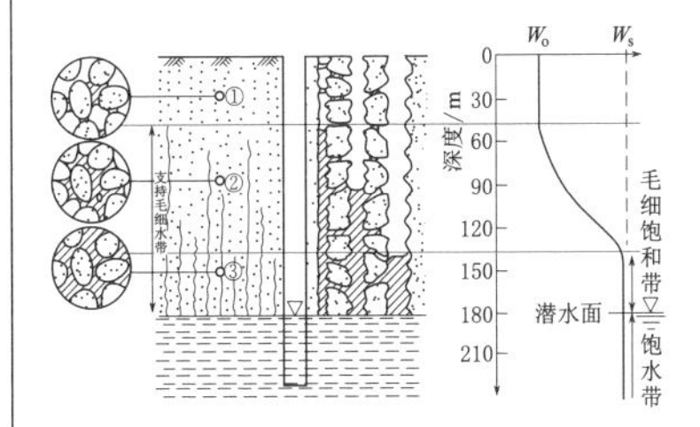

schematic diagram of capillary water

capillary water height
【释义】 毛细水上升高度
【例句】 The relationship between the rising capillary water height of coarse grained soil and time could be drawn by the double-unit regression

equation in the logarithmic coordinate. 粗粒土毛细水上升高度与时间过程可用对数坐标下二次多项式回归方程进行模拟和预测。

【中文定义】 毛细管水从地下水的水面沿土层或岩层空隙上升的最大高度。

Carbon-14

【释义】 *n.* 碳-14

【例句】 Carbon-14 decays into nitrogen-14 through beta decay. 碳-14 通过 β 衰变为氮-14。

【中文定义】 碳-14 是碳元素的一种具放射性的同位素，它是透过宇宙射线撞击空气中的氮原子产生的。碳-14 原子核由 6 个质子和 8 个中子组成。其半衰期约为（5730±40）年，衰变方式为 β 衰变，碳-14 原子转变为氮-14 原子。

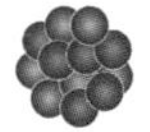  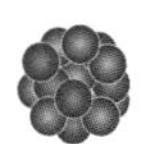

Carbon-14

Carboniferous (C) [ˌkɑːbəˈnifərəs]

【释义】 *n.* 石炭纪（系）

【中文定义】 古生代的第五个纪，距今 3.55 亿～2.95 亿年，延续约 6500 万年。石炭纪是地壳运动非常活跃的时期，古地理的面貌有着极大的变化，也是地壳发展史上重要的造山时期。

cartographic database

【释义】 地图数据库

【例句】 A major issue in the design of a cartographic database is the question of how linear features should be stored so that efficiency is maintained in terms of both generalisation and spatial access. 地图数据库设计的一个主要问题是线性特征应如何存放，从而使得在大众化和受限准入两个方面均保持效率。

【中文定义】 利用计算机存贮的各种地图要素的数据及数据管理软件的文件集合。

cartography [kɑrˈtɑgrəfi]

【释义】 *n.* 地图制图，制图，制图学

【常用搭配】 engineering cartography 工程制图；automated cartography 自动制图学

【例句】 Map model is a foundation of map use, mathematical models of map and computer-assisted cartography. 地图模型是地图应用、地图数学模型和计算机地图制图的基础。

【中文定义】 地图的设计、编制、复制以及建立地图数据库的技术、工艺和方法。

casing [ˈkesɪŋ]

【释义】 *n.* 套，套管

【常用搭配】 casing log 套管测井；casing wall 套管壁；casing tube 套管，井壁管

【例句】 The company makes casing, tubing and drill pipes for the natural gas exploration industry. 公司主要生产天然气勘探行业所用的套管、油管和钻杆。

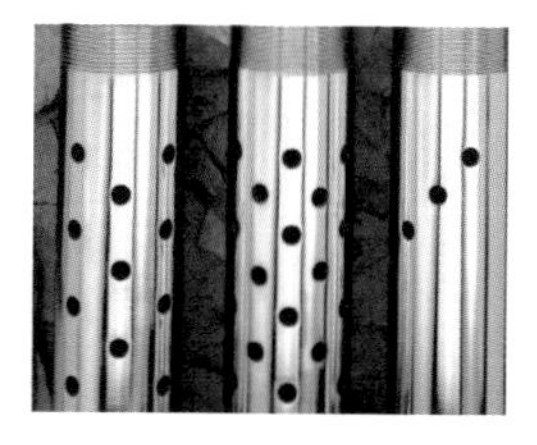

casing

cast [kæst]

【释义】 *n.* 印模

【中文定义】 由于水流的作用，或水流携带某种物质的刻划作用，在柔软的泥质沉积物表面形成的一些具有特殊形状的坑凹痕迹，并被上覆砂质岩层底面复印下来。

cast

cast-in-place pile

【释义】 灌注桩

cast-in-place pile

cataclasite [ˈkætəklæsait]

【释义】 *n.* 碎裂岩

【例句】 The fault rock is characterized as a low

temperature deformation of about 200℃ and it is mainly a type of cataclasite. 断层岩的变形特征为低温变形，断层岩主要为碎裂岩类，变形温度在200℃左右。

【中文定义】 具有碎裂结构或碎斑结构的岩石称为碎裂岩。

cataclasite

cataclastic structure

【释义】 碎裂结构

【中文定义】 层状岩体被层理、片理、节理、断层、层间错动面等切割呈碎块状、片状结构体组成的岩体结构类型。

cementation [ˌsimɛn'teʃən]

【释义】 *n.* 黏结，胶结，水泥结合，渗碳处理

【常用搭配】 degree of cementation 胶结度；cementation factor 胶结系数；cementation method 水泥灌浆法

【例句】 The main diagenesis in the studied area included the mechanical compaction, cementation, infilling, dissolution, and authigenic mineral precipitation. 区内主要的成岩作用有机械压实、胶结、充填、溶解及自生矿物沉淀等。

【中文定义】 在将沉积物压在一起的过程中，受压力的作用，岩石的一些矿物慢慢溶解在水里形成含有矿物的水溶液，随着水溶液渗入沉积物颗粒间的空隙，当水溶液中的矿物结晶时，沉积物颗粒被结晶体黏在一起，这一过程就叫胶结。

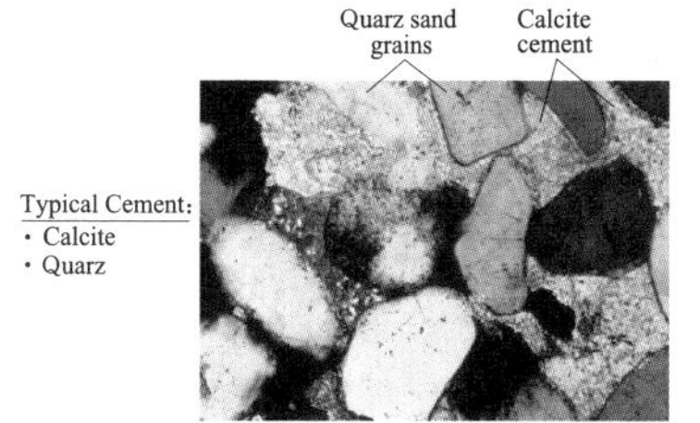

cementation

Cenozoic (Kz) [siːnəu'zəuik]

【释义】 *n.* 新生代（界）

【中文定义】 地球历史上最新的一个地质时代。随着恐龙的灭绝，中生代结束，新生代开始。新生代包含三个纪：古近纪、新近纪和第四纪。新生代以哺乳动物和被子植物的高度繁盛为特征。

central meridian

【释义】 中央子午线，中央经线

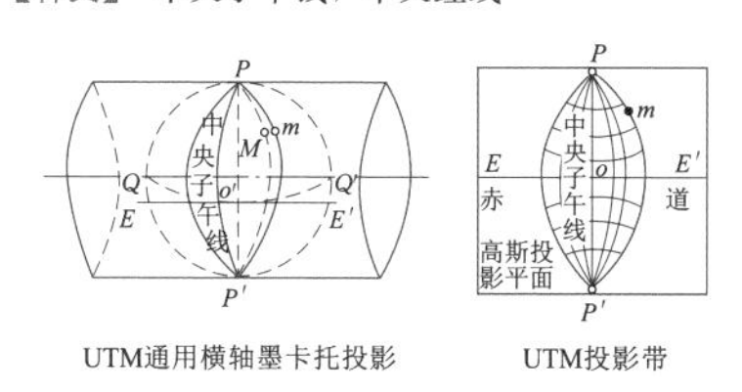

central meridian

【中文定义】 地图投影中投影带中央的子午线。

chalcopyrite [ˌkælkə'paɪraɪt]

【释义】 *n.* 黄铜矿

【常用搭配】 chalcopyrite ore 黄铜矿，原生硫化铜矿，黄铜矿矿石；chalcopyrite compound 黄铜矿类化合物

【例句】 The principal ore minerals are nickeline, gersdorffite, pyrrhotine, pentlandite and chalcopyrite with minor amounts of molybdenite, tellurobismuthite, gold, sphalerite and argento pentlandite. 主要矿石矿物为红砷镍矿、辉砷镍矿、磁黄铁矿、镍黄铁矿、黄铜矿，以及少量辉钼矿、碲铋矿、金、闪锌矿和银镍黄铁矿。

【中文定义】 一种铜铁硫化物矿物，常含微量的金、银等，铜黄色，常有暗黄或斑状锖色，条痕为微带绿的黑色。

chalcopyrite

channel roughness

【释义】 沟床糙度

【中文定义】 沟床底部的平整程度或堆积物质的粗糙程度。按沟床是否平整及有无石块、砂石、

植物阻塞等情况定出的沟床的粗糙系数，可用于计算泥石流的流速。

channelling [ˈtʃænlɪŋ]

【释义】 溶沟

【例句】 The key points of the construction are leakage stoppage during the pit excavation and foundation treatment when the ship dock is constructed on the complicated geology with growing fissures, karst caves, channellings, fluid bowls and many beaded karst caves directly linking up with external sea water. 在有较大裂隙，溶洞、溶沟、溶槽发育，有串珠状溶洞并多层处与外部海水直接贯通的复杂地质上建造船坞，基坑开挖堵漏和基础处理是施工的关键。

【中文定义】 指石灰岩表面上的一些沟槽状凹地。它是由地表水流，主要是片流和暂时性沟状水流顺着坡地，沿节理溶蚀和冲蚀的结果。沟槽深度不大，一般数厘米至数米。

channelling

characteristic curve

【释义】 特征曲线

【例句】 The soil-water characteristic curve is very important for studying the physical and mechanical characteristics of unsaturated soils. 土-水特征曲线对于研究非饱和土的物理力学特性至关重要。

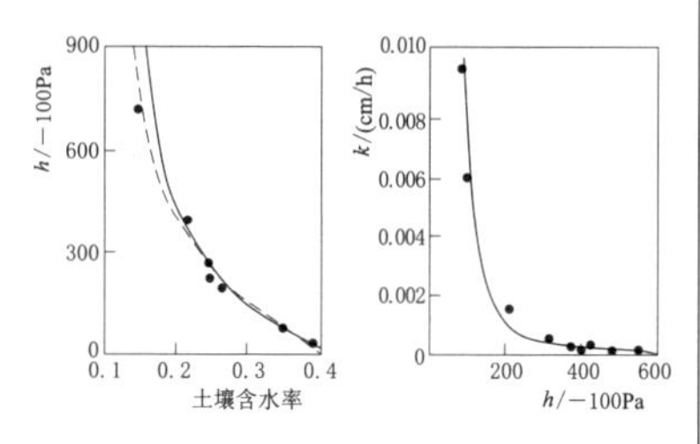

characteristic curve

characteristic parameter

【释义】 特性参数，特征参数

【例句】 The proposed main characteristic parameters of NT1HPP are: the full supply level is 292m with a corresponding storage capacity of 2921MCM; minimum operation level is 250m with a corresponding storage capacity of 961MCM. The reservoir has the ability to regulate according to seasons. The installed capacity of power station is 600MW, with three 200MW mixed flow turbine generator sets installed. 推荐NT1HPP主要特征参数：正常蓄水位292m，相应库容29.21亿m^3，死水位250m，相应库容9.61亿m^3，水库具有季调节能力；电站装机容量600MW，安装3台单机容量200MW混流式水轮发电机组。

characteristic value

【释义】 特征值，标准值

【同义词】 eigenvalue, standard value

【例句】 In fact, there are a large number of relations in all kinds of characteristic value of quality. 实际上，各种质量特性值之间存在着大量的关联关系。

characteristics of rock mass deformation

【释义】 岩体变形特性

【例句】 On the basis of analyzing the characteristics of rock mass deformation and general rheological models, a new nonlinear viscous substance which is related to stress state and time is presented. 在分析岩体变形特点和常用流变模型变形特性的基础上，提出与应力状态和时间相关的非线性黏性体。

checking datum point

【释义】 校核基准点

【例句】 In the system application, it places the target calibration board in the position which should be monitored (for the tunnel project, it usually choose tunnel crown as the position), and place the reference calibration board in the checking datum point. 在系统应用中，将目标标定板安置在需要监测的位置处（对于隧道工程来说，通常为拱顶位置），并在校核基准点处安置参考标定板。

chemical alteration

【释义】 化学蚀变

【例句】 In general, physical alteration and chemical alteration are associated with each other. 一般而言，

物理蚀变和化学蚀变是相互伴生的。
【中文定义】 由于化学变化而导致的蚀变过程。

chemical deposition
【释义】 化学沉积作用
【例句】 Karst cave minerals are products of hypergene chemical deposition under the action of groundwater. 喀斯特洞穴矿物是在洞穴环境下由地下水活动所产生的表生化学沉积作用产物。
【中文定义】 在地壳表层，在化学和物理化学规律支配下，物质以离子或胶体状态迁移、再结合成固态物质的过程。

chemical grouting
【释义】 化学灌浆
【常用搭配】 chemical compound grouting 化学复合灌浆
【例句】 This paper discusses the application and effect of chemical grouting in seepage proof curtain of interception-pollutant dam of Weihai waste treatment plant. 文章讨论了化学灌浆在威海市垃圾处理厂截污坝防渗帷幕中的应用及效果。

chemical grouting

chemical piping effect
【释义】 化学管涌
【中文定义】 化学管涌是地下水把土层中可溶盐溶解带走的现象。

chemical weathering
【释义】 化学风化，化学风化作用
【例句】 The ratios of Al_2O_3/SiO_2 and kaolinite + illite vs quartz for whole rock indicated the intensities of both chemical weathering and erosion. Al_2O_3 与 SiO_2 的比值和高岭石加伊利石的总量与石英的比值在全岩研究中代表化学风化与侵蚀作用的强度。
【中文定义】 岩石在水、二氧化碳、氧气等多种因素作用下，改变化学成分和形成新物质的过程。

chemical weathering

chervon fold
【释义】 尖棱褶皱
【中文定义】 又称为尖棱褶皱形褶皱、锯齿状褶皱（zigzag fold），指两翼较平直、转折端急剧转折甚至成尖顶的褶皱。

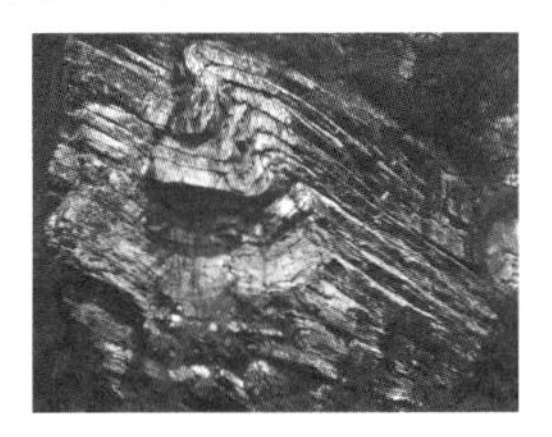

chervon fold

chlorite [ˈklɔraɪt]
【释义】 *n.* 绿泥石
【常用搭配】 chlorite schist 绿泥石片岩
【例句】 Members of the chlorite group of clay minerals also have a capacity for swelling. 黏土矿物中的绿泥石也具有膨胀性。

chlorite

circular slide
【释义】 圆弧滑动
【例句】 Circular slide method and bearing capability method are two frequently used methods in stability

analysis. 稳定分析常用的方法是圆弧滑动法和地基承载力法。

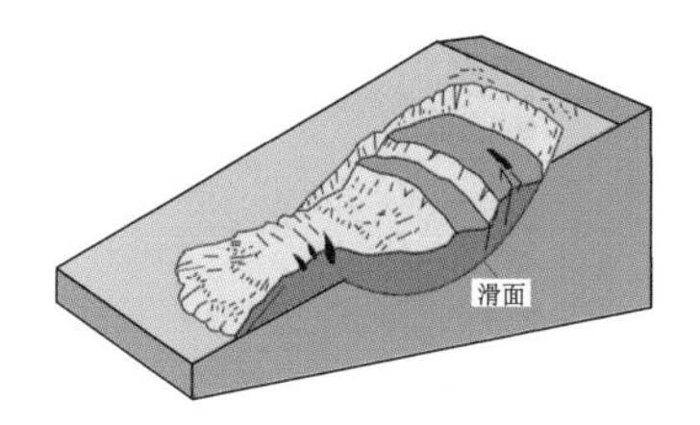

circular slide

cirque [sɜːk]

【释义】 *n.* 冰斗

【常用搭配】 cirque glacier 冰斗冰川；active cirque 活动冰斗

【中文定义】 冰斗是指三面为陡崖包围的簸箕状盛雪洼地、由冰斗底、冰斗肩、冰斗壁和冰斗坎儿部分组成，多发育在雪线附近。

glacial cirque

classification of geological disaster

【释义】 地质灾害分类

【例句】 Based on the previous studies of the classification of geological disasters,the concept of coastal geological hazards is re-examined. 文章在总结前人关于地质灾害分类的基础上，重新探讨了海岸带地质灾害的含义。

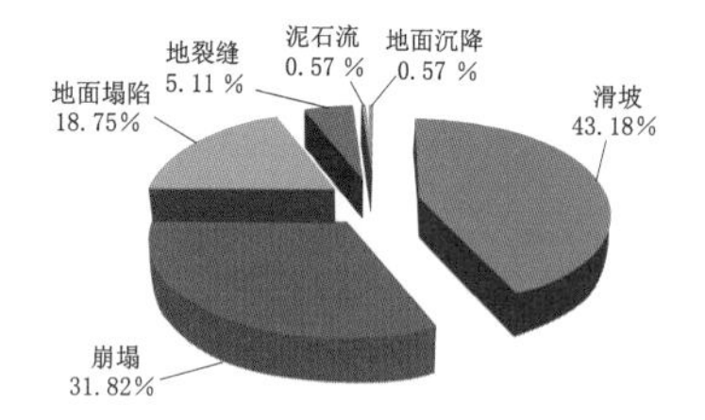

classification of geological disaster

classification of rock mass basic quality

【释义】 岩体基本质量分级

【中文定义】 将岩体基本质量的定性特征和岩体基本质量指标两者相结合的岩体分级方法。

岩体基本质量级别	岩体基本质量的定性特征	岩体基本质量指标（BQ）
Ⅰ	坚硬岩，岩体完整	＞550
Ⅱ	坚硬岩，岩体较完整； 较坚硬岩，岩体完整	550～451
Ⅲ	坚硬岩，岩体较破碎； 较坚硬岩，岩体较完整； 较软岩，岩体完整	450～351
Ⅳ	坚硬岩，岩体破碎； 较坚硬岩，岩体较破碎～破碎； 较软岩，岩体较完整～较破碎； 软岩，岩体完整～较完整	350～251
Ⅴ	坚硬岩，岩体破碎； 软岩，岩体较破碎～破碎； 全部极软岩及全部极破碎岩	≤250

classification of rock mass basic quality

clastic flow mechanism

【释义】 碎屑流机制

clastic rock

【释义】 碎屑岩

【常用搭配】 clastic reservoir rock 碎屑储集岩，碎屑岩储层

【例句】 There is a close relation between the terrigenous clastic rock and the plate tectonics. 陆源碎屑岩特征与板块构造性质有着密切关系。

【中文定义】 是母岩机械破碎的产物经搬运、沉积、压实、胶结而成的岩石。在沉积区外的陆地上搬运来的碎屑称为陆源碎屑或外碎屑，是碎屑的主要来源。在沉积区内形成的碎屑称为内碎屑（十分少见）。

clastic rock

clastic sediment

【释义】 碎屑沉积

【例句】 Clastic sediments bears much information about the nature of source rocks and their tectonic evolution,which is significant for sedimentary basin analysis. 碎屑沉积物记录着有关源岩性质和构造演化等诸多重要的信息，对于沉积盆地分析和理解

区域构造演化都有重要的指示意义。

clastic sediment

【中文定义】 在机械力（风力、水力）的破坏作用下，原来岩石破坏后的碎屑经过搬运和沉积的作用。

clastic structure

【释义】 碎块状结构

【例句】 Their main mineral is kaolinite, usually in pisolitic, oolitic or clastic structure. 主要矿物是高岭石，常见豆状、鲕状和碎块状结构。

【中文定义】 岩体较破碎，形成碎裂的块状。

clay [klei]

【释义】 *n.* 黏土，泥土；*vt.* 用黏土处理

【常用搭配】 soft clay 软黏土，软质黏土；clay mineral 黏土矿物；red clay 红黏土；clay content 黏粒含量，黏土含量

【例句】 The results indicate that less dosage of water, clay, bentonite and larger dosage of cement, and mixing coal ash and admixture can increase compressive strength of plastic concrete. 结果表明，减少水及黏土和膨润土的用量，增加水泥用量，掺加粉煤灰和外加剂，均可提高塑性混凝土的抗压强度。

clay

clay infill

【释义】 黏土充填

【例句】 The friction coefficient of the weak structural plane is relatively small and extends longer, and it is generally clay infill or argillaceous infill. 软弱结构面的摩擦系数相对较小，延伸较长，且普遍为黏土充填或泥质物充填。

【中文定义】 结构面中的充填物主要由黏土矿物组成。

clay mineral

【释义】 黏土矿物

clay mineral

【例句】 Clay minerals result mainly from the weathering of other rock forming minerals. 黏土矿物主要是其他成岩矿物通过风化作用形成的。

【中文定义】 是构成黏土岩、土壤的主要矿物，是一些含铝、镁等为主的含水硅酸盐矿物。

clay stone

【释义】 黏土岩

clay stone

【例句】 Application of Controlled Blasting in the Tunnel Excavation in the Area of Sandstone and Siltstone Intercalated with Clay Stone. 砂岩和粉砂岩夹黏土岩地区洞挖光面爆破技术应用。

【中文定义】 主要由黏土矿物组成的沉积岩。

cleavage ['kliːvɪdʒ]

【释义】 *n.* 劈理

【常用搭配】 fracture cleavage 破劈理；flow cleavage 流劈理

【例句】 The stress field analysis showed that the cleavages formed under horizontal, S-N trending compressive stress. 应力场的分析表明，劈理构造由近南北向的地壳水平挤压应力作用形成。

cleavage

【中文定义】 是一种由潜在分裂面将岩石按一定方向分割成平行密集的薄片或薄板的次生面状构造。

cliff [klɪf]

【释义】 *n*. 悬崖

【中文定义】 水电工程实践中，习惯上将地形坡度近垂直的岸坡称为悬崖。

cliff

clint monadnock

【释义】 石芽残丘

【中文定义】 石灰岩溶沟之间凸起的石脊称为石芽，石芽分布在裸露的地面上，成为石芽残丘。

clint monadnock

close fold

【释义】 闭合褶皱

【中文定义】 指翼间角小于70°、大于30°的褶皱。

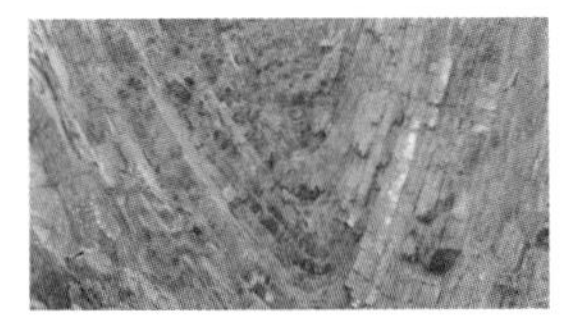

close fold

cloud computing

【释义】 云计算

【例句】 Cloud computing and big data are changing the enterprise. 云计算和大数据正在改变企业现状。

【中文定义】 云计算是分布式计算的一种，指的是通过网络中央的一组服务器，将巨大的数据计算处理程序分解成无数个小程序，然后，通过多部服务器组成的系统进行处理和分析这些小程序得到结果并返回给用户。

coarse aggregate

【释义】 粗骨料

【例句】 Coarse aggregate shall be rewashed on rinsing screens with spray bars immediately prior to elevating into batching plant bins. 粗骨料在储料仓以前需用带喷水管的淋水筛再次清洗。

【中文定义】 在混凝土中，砂、石起骨架作用，称为骨料，其中粒径大于5mm的骨料称为粗骨料。

coarse aggregate

coarse grained texture

【释义】 粗粒结构

【中文定义】 粒径为粗粒（颗粒直径大于5mm）的结构。

coarse gravel

【释义】 粗砾

coarse gravel

【例句】 The thermo-mechanism was studied to protect permafrost by the effects of Balch and natural convection in the coarse gravel layer. 通过对粗粒层 Balch 效应和自然对流效应的热学工作机制研究，从而保护多年冻土。

【中文定义】 指风化岩石经水流长期搬运而成的粒径为 2～60mm 的无棱角的天然粒料，10～100mm 的，称粗砾；砾石经胶结成岩后，称砾岩或角砾岩。

coarse sand

【释义】 粗砂

【例句】 The natural sand is the non-renewable resources, the coarse sand were already getting fewer and fewer. 天然砂是不可再生资源，粗砂已经越来越少。

【中文定义】 粒径大于 0.5mm 的颗粒含量超过全重 50%的土，是砂土的一种分类。

coarse sand

coarse-grained

【释义】 *adj*. 粗粒的

【常用搭配】 coarse-grained granodiorite 粗粒花岗闪长岩；coarse-grained granite 粗粒花岗岩；coarse-grained soil 粗粒土

【例句】 Coarse-grained soils exhibit evident dilatancy, and they do not obey the Hooke's law. 粗粒土具有明显的剪胀剪缩性，为非虎克定律。

【中文定义】 粗粒土是指大于 0.1mm 颗粒含量较多的土，大致相当于砂类。砾石土、砂卵石、残坡积碎石土和风化岩石渣等统称为粗颗粒土，简称粗粒土。它由大小不等的粗细颗粒组成，最大颗粒可达 1000mm 以上，最细可小于 0.1mm，粒径变化范围很大，粗细颗粒特性相差悬殊。

粗粒coarse-grained

细粒fine-grained

coarse-grained vs. fine-grained

coast [kəʊst]

【释义】 *n*. 海岸

【常用搭配】 coast erosion 海岸侵蚀；rocky coast 岩石海岸；Ivory Coast 象牙海岸；coast line 海岸线

【例句】 Tuffaceous and sandy shales of Miocene age dipping towards the coast are interbedded with several bentonite beds. 这组倾向海岸线的中新世凝灰质砂质页岩与几种膨润土层互层。

【中文定义】 由海水的侵蚀和堆积作用，在海洋与大陆的交界地带形成的地貌，称海岸地貌。

gold coast

coast erosion

【释义】 海岸侵蚀，海岸蚀退

【例句】 There is a close relationship between coast erosion and hydrodynamics. 海岸侵蚀与水动力之间有着密切的关系。

【中文定义】 是指在自然力（包括风、浪、流、潮）的作用下，海岸供沙少于海岸失沙而引起的沉积物净损失的过程，即海水动力的冲击造成海岸线的后退和海滩的下蚀。

coast erosion

coastal plain

【释义】 海岸平原，海滨平原

【中文定义】 海岸平原是地势低平，向海缓缓倾斜的沿海地带，主要由海蚀平台或水下浅滩相对上升露出海面形成。

coastal plain on the northern coast of the Mediterranean

coastal terrace

【释义】 海岸阶地，海岸台地

【例句】 Radiocarbon dates from coastal terraces of the island belt were used to compute uplift rates. 来自岛带海岸阶地的放射性碳数据被用于隆升率的计算。

coastal terrace prairie south of Goat Rock

【中文定义】 海岸阶地是指由海蚀作用形成的海蚀平台或由海积作用形成的海滩，以及因海平面的相对升降而被抬升或下沉后的海蚀平台和海滩。

cobbly soil

【释义】 粗砾质土

cobbly soil

cobble [ˈkɑːbl]

【释义】 *n.* 卵石，鹅卵石，圆石

【例句】 The large-diameter slurry shield is the first time to be used in sandy cobble stratum in Beijing Underground Zhijing Line. 北京地铁至京线是北京地区首次在砂卵石地层中采用大直径泥水盾构。

【中文定义】 卵石是自然形成的无棱角岩石颗粒，可形成砾岩。分为河卵石、海卵石和山卵石。卵石的形状多为圆形，表面光滑，与水泥的黏结较差，拌制的混凝土拌和物流动性较好，但混凝土硬化后强度较低。

cobble

code of geological age

【释义】 地质年代代号

【中文定义】 用来表示地质年代的符号。

新生代花岗岩 γ_6	晚第三纪 γ_6^3	喜山期	晚期
	早第三纪 γ_6^2		中期
	早第三纪 γ_6^1		早期
中生代花岗岩 γ_5	白垩纪 γ_6^3	燕山期	晚期
	侏罗纪 γ_5^2		早期
	三叠纪 γ_5^1	印支期	

example for code of geological age

code of rock

【释义】 岩石代号

【中文定义】 用来表示岩石名称的符号。

岩石名称	代号	岩石名称	代号
砾岩	Cg	卵石	Cb
砂砾岩	Scg	砾	G
砂岩	Ss	砂	S
粉砂岩	St	砂砾石	Sgr
黏土岩	Cr	粉砂	Sis
页岩	Sh	粉土	M
泥灰岩	Ml	黏土	C
石灰岩	Ls	黄土	Y
白云岩	Dm	淤泥	Sil

code of rock

coefficient of collapsibility

【释义】 湿陷系数

【例句】 This paper analyzes the collapsibility of loess roadbed in Taijiu Expressway, and probes into the variation regularities of vertical and horizontal coefficient of collapsibility. 对太旧高速公路黄土路基的湿陷性进行了分析，探讨了湿陷系数的纵向和横向变化规律。The plate-like saline soil is collapsible soil, and the coefficient of collapsibility increases with the increase of the salinity. 板块状盐渍土属湿陷性土，随着含盐量的增大，湿陷系数增大。

coefficient of consolidation

【释义】 固结系数

coefficient of correction

【释义】 修正系数

【例句】 For the three different types of containment sump strainers, coefficient of correction is between 0.5 and 0.6. 对于三种不同型号的地坑过滤器，修正系数在 0.5～0.6 之间。

coefficient of curvature

【释义】 曲率系数

【中文定义】 反映颗粒级配优劣程度的一个参数，以颗粒级配曲线上粒径累积质量占总质量30%的粒径平方值除以限制粒径与有效粒径的乘积所得的比值。

【例句】 It is related to uniformity coefficient and coefficient of curvature of the soil. 该土体的不均匀系数和曲率系数有关。

coefficient of friction

【释义】 摩擦系数

【例句】 Distinctive characteristics of this material include high impact and abrasion resistance, and a low coefficient of friction. 这种材料的特性包括高抗冲击性、耐磨损性，以及低摩擦系数。

【中文定义】 摩擦系数是指两表面间的摩擦力和作用在其一表面上的垂直力之比值。

coefficient of frost resistivity

【释义】 抗冻系数

【例句】 Finally, a new approach (BP neural Network) to predicting the coefficient of frost resistivity of concrete has been proposed here. 最后，本文提出用 BP 神经网络预测混凝土抗冻系数的方法。

coefficient of mean value

【释义】 均值系数

【例句】 Coefficient of mean value represents the relative concentration of random variable values. 均值系数表示随机变量取值的相对集中位置。

coefficient of rock mass weathering

【释义】 岩体风化系数

【中文定义】 新鲜岩石与风化岩石的超声波速度值之差与新鲜岩石的超声波速度值之比值。

coefficient of secondary consolidation

【释义】 次固结系数

【例句】 The test results show that with the change of consolidation pressure, the coefficient of secondary consolidation of compacted loess shows a change law related to the current stress state. 试验结果表明，压实黄土的次固结系数随固结压力的变化呈现出与当前应力状态相关的变化规律。

coefficient of self weight collapsibility

【释义】 自重湿陷系数

coefficient of uniformity

【释义】 不均匀系数

【例句】 In the situation of that coefficient of uniformity is bigger than 5, the degree of compaction increases and then decreases with the increase of the coefficient of uniformity. 不均匀系数在大于 5 的情况下，压实度随不均匀系数的增大，先增大后减小。

coefficient of variation

【释义】 变差系数/变异系数

【例句】 The smaller the coefficient of variation is, the more efficient is the intensity measure for super high-rise building. 变异系数越小，超高层建筑的强度测量越有效。

coefficient of volume compressibility

【释义】 体积压缩系数

【例句】 The re-consolidation deformation law of soil is analysed and a method of determining coefficient of volume compressibility is suggested. 讨论土体的再固结变形规律，提出再固结体积压缩系数的确定方法。

cohesion [kou'hiːʒn]

【释义】 *n.* 凝聚，黏聚力

【例句】 The cohesion and internal friction angle increase with the increase of compaction under the same water content. 在同一含水率下，黏聚力和内摩擦角随试样压实度的增加而增大。

【中文定义】 指同种物质内部分子间的相互吸引力。

cohesionless soil

【释义】 无黏性土

【例句】 The axial force of the circular supporting system in the cohesionless soil is greater than that in the clayey soil. 无黏性土中圆形支护结构环向

钢管的轴力大于黏性土中环向钢管的轴力。

【中文定义】 无黏性土一般指碎石（类）土和砂（类）土。这两大类土中一般黏粒含量甚少，呈单粒结构，不具有可塑性。

cohesionless soil

cohesive soil

【释义】 黏性土

【中文定义】 黏性指数 I_p 大于 10 的土。

cohesive soil

collaborative modeling

【释义】 协同建模

【例句】 Collaborative modeling and simulation takes strong advantages of resource sharing across the internet for the virtual enterprise. 协同建模与仿真技术为虚拟企业实现资源共享和管理提供了强大的优势。

【中文定义】 为了提高建模效率，将工作团队通过网络和软硬件环境组成协作体，通过分工和协作一起完成某个建模工作的过程。

collapse [kə'læps]

【释义】 *n*. 崩塌，倒塌，瓦解

collapse

【同义词】 avalanche

【例句】 The earthquake caused the collapse of several homes. 地震造成几处房屋倒塌。

【中文定义】 地质体在重力作用下，从高陡坡突然加速崩落或滚落（跳跃）。具有明显的拉断和倾覆现象。

collapse deformation

【释义】 湿陷变形

【例句】 The collapse deformation of collapsible loess is an important factor to the ground stability. 湿陷性黄土的湿陷变形是影响地基稳定性的一个重要因素。

collapse deformation

collapse doline

【释义】 塌陷漏斗

【中文定义】 溶洞顶板塌陷而成的漏斗。

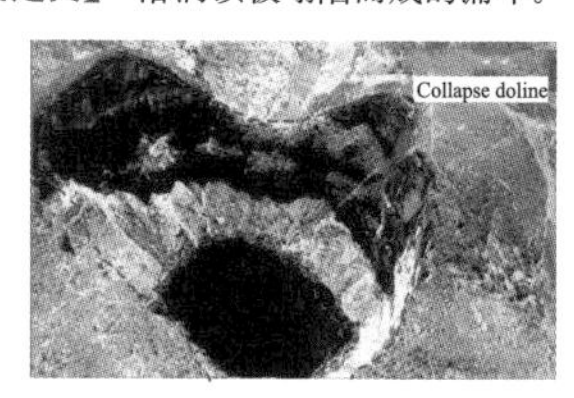

collapse doline

collapse hazard

【释义】 崩塌灾害

collapse hazard

【例句】 Collapse hazard will take place inevitably under the condition of steeper hillside after rocks collapse or cave to ground surface caused by exploitation with caving method. 在较陡的山体上利用崩落开采法进行围岩开采时，围岩崩落或塌陷到地面后将不可避免地产生崩塌灾害。

collapsibility [kəlæpsə'bɪlɪtɪ]

【释义】 *n*. 湿陷性，崩散性，易坍塌性

【常用搭配】 collapsibility evaluation 湿陷性评价

【例句】 There are many factors that affect the loess collapsibility, and loess collapsibility commonly have biggish otherness in a small area. 影响黄土湿陷性的因素很多，而同一场地黄土的湿陷性往往具有较大的差异性。

collapsibility test of loess

【释义】 黄土湿陷试验

【例句】 The judgment of the collapsibility test of loess and its application in engineering are discussed, and the author's own proposals and views are proposed, which have some reference value. 本文就黄土湿陷试验在湿陷类别的判定及其工程中的应用进行深入的探讨，提出了自己的建议和看法，具有一定的参考价值。

collapsible loess

【释义】 湿陷性黄土

【例句】 The application of dynamic compaction in loess area shows that it has good effect on compaction of collapsible loess. 强夯技术在黄土地区的应用表明，强夯对湿陷性黄土的压实具有良好的效果。

【中文定义】 指在上覆土层自重应力作用下，或者在自重应力和附加应力共同作用下，因浸水后土的结构破坏而发生显著附加变形的土，属于特殊土。

collapsible loess

collimation line method

【释义】 视准线法

【例句】 A basic formula for deciding weight by the maximum quasi probable method and the collimation line method for great dam observation is derived in this article. 本文用极大似然法和大坝观测的视准线法推导了基本的定权公式。

colluvial soil

【释义】 坡积土

【例句】 Determination of design parameters of composite ground including colluvial soil layer is far difficult because the maximum particle size of such a soil is remarkably large and particle distribution may vary from area to area. 含坡积土层的复合地基设计参数确定困难是因为不同部位土的最大粒径以及粒径分布存在差异。

【中文定义】 位于山坡上方的碎屑物质，在流水或重力作用下运移到斜坡下方或坡麓处堆积形成的土。

colluvial soil

colluvium [kə'lʊvɪəm]

【释义】 *n*. 崩积物，崩积层

【常用搭配】 colluvium soil 崩积土

【中文定义】 堆积于坡麓地带的崩落物。

color of geological map

【释义】 地质图色标

岩石地层用色	色标编号	
	常规	微机
新近纪	1～40	601～640
古近纪	41～66	641～666
白垩纪	67～94	667～694
侏罗纪	99～138	699～738
三叠纪	143～184	743～784
二叠纪	189～220	789～820
石炭纪	225～255	825～855
泥盆纪	260～307	860～907
志留纪	312～351	912～951
奥陶纪	356～388	956～988
寒武纪	393～426	993～1026
震旦纪	431～466	1031～1066
晚元古代*	467～480	1067～1080
中元古代	481～502	1081～1102
早元古代	503～510	1103～1110
太古宙	511～530	1111～1130

* 不含震旦纪

color of geological map

【中文定义】 编制地质图件的色相标准。地质图色标是地质体年代符号的补充，地质体年代和地层单位的划分，可以从图上的不同色相得到体现。

columnar joint

【释义】 柱状节理

【常用搭配】 radiating columnar joint 辐射状柱状节理；columnar joint structure 柱状节理构造

【例句】 Columnar joint, microfissure and gently dipping structural surface are the three fundamental factors which lead to the small deformation modulus of basaltic mass. 柱状节理玄武岩中发育的柱状节理、微裂隙及缓倾角结构面是导致岩体变形模量较低的主要因素。

【中文定义】 几组不同方向的节理将岩石切割成多边形柱状体，柱体垂直于火山岩的基底面。如熔岩均匀冷却，形成六方柱状，上细下粗，二者由顶柱盘面隔开。

columnar joint

columnar structure

【释义】 柱状节理构造

【同义词】 prismatic structure

【中文定义】 熔岩由规则的多边形柱体组成，是熔浆均匀而缓慢地冷却收缩形成的。

columnar structure

combustible gas

【释义】 可燃气体

【常用搭配】 combustible gas；combustible gas detector 可燃气体探测仪

【例句】 The combustible gas detector is widely used in gas, chemical and oil industries. 可燃气体探测仪在天然气、化工和石油工业中广泛应用。

【中文定义】 指甲类可燃气体或甲、乙$_A$类可燃液体汽化后形成的可燃气体。

compactability [kəmpæktəˈbɪlɪtɪ]

【释义】 *n.* 压实性，紧实性

compacted fill

【释义】 压实填土

【例句】 The evaluation on quality of compacted fill ground is directly influenced by the reliability of compaction factors. 压实系数指标的可靠性直接影响压实填土地基的质量评价。

compacted fill

compaction density

【释义】 紧密密度，压实密度

【例句】 Slag has many good features such as high strength and compaction density, small settlement, strong water permeability and scouring-resistance, and it is also easily available. 矿渣的优点是强度高、压实密度大、沉降变形小、透水性强、抗冲刷性能高、可就近取材等。

【中文定义】 材料体内固体物质充实的程度。

compaction test

【释义】 击实试验

compaction test

【例句】 The accuracy of the compaction test results directly affects the engineering quality. 击实试验结果准确与否直接影响到工程质量。

comparison method of precipitation infiltration coefficient

【释义】 降水渗入系数比拟法

complete specimen

【释义】 完整试样

completely penetrating borehole (well)

【释义】 完整孔（井）

【例句】 While the gravel layer is relatively thin, the occurrence and development of the boiling of sand correspond to completely penetrating well model. 当砂砾石层相对较薄时，涌砂的发生、发展过程符合完整井模型。

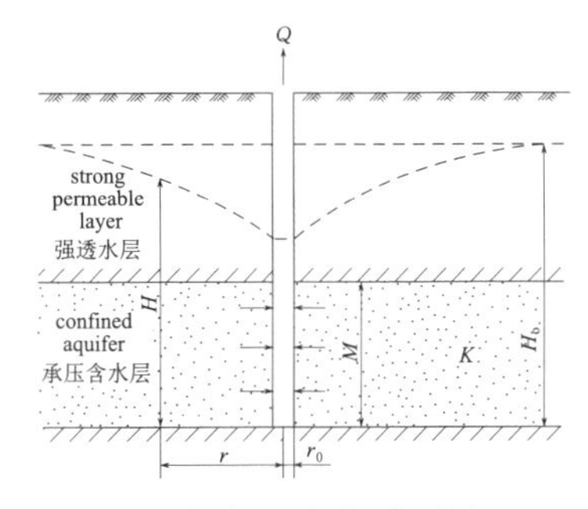

completely penetrating borehole

completely weathered

【释义】 全风化的

【常用搭配】 completely weathered granite 全风化花岗岩；completely weathered zone 全风化带；completely weathered rock 全风化岩石

completely weathered

【例句】 This tunnel's entrance is a heading slope which covered soil and completely weathered rock, and the slope is easy to lose the stability when perturbed. 隧道洞口段为仰坡，主要为覆盖层和全风化岩石，开挖扰动很可能引起仰坡的失稳。

【中文定义】 全风化是岩石受风化作用影响最剧烈的表现，岩石的组织结构完全破坏，已崩解和分解成松散的土状或砂状，有很大的体积变化。

completion geological report

【释义】 竣工地质报告

compliance evaluation

【释义】 合规性评价

composite columnar section

【释义】 综合柱状图

【例句】 The paper has mainly introduced the composite columnar section of exposed paleozoic and cenozoic rocks in the Pahranagat Range, Lincoln County, Nevada. 论文主要介绍了内华达州林肯县 Pahranagat 山脉出露的古生代和新生代岩石综合柱状图。

【中文定义】 综合测区的露头和钻孔资料编制而成的反映测区地层特征的柱状图件。

composite geological model

【释义】 复合地质模型

【中文定义】 表示地质体被勘探孔洞、地下洞室、基础边坡开挖形成的地质表面或实体模型。复合地质模型通常用于表达地层岩性、风化卸荷、岩体分类等地质实体模型中被勘探孔洞、地下洞室、基础边坡等工程对象开挖分割出来的部分，它即保留了地质实体模型的属性，又继承了开挖表面或实体的属性，有利于工程地质条件分析和展示。

composite ground

【释义】 复合地基

【常用搭配】 rigid pile composite ground 刚性桩复合地基

composite ground

【例句】 In the design of composite ground under the flexible foundation, the composite ground theory for rigid foundation is employed, so that great difference between measured value and design value occurs. 目前在柔性基础下复合地基的设计中

仍沿用刚性基础下复合地基的设计理论，因而造成实测值与设计值相差甚远。

composite subgrade

【释义】 复合地基

【中文定义】 部分土体被增强或被置换而形成的由地基土和增强体共同承担荷载的人工地基。

composition [ˌkaːmpəˈzɪʃn]

【释义】 *n.* 成分，构成，混合物

【常用搭配】 chemical composition 化学成分，化学组成；composition analysis 成分分析，组成分析

composition analysis

【释义】 成分分析，组分分析

【例句】 The composition of NiAl alloys have great effects on the structure and phase transformation temperatures, but the composition analysis is often disturbed by the difficulties in the separation of Ni and Al. NiAl 合金的组分对其结构和相变温度有着重要的影响，然而其组分分析受Ni、Al 难以分离的困扰。

comprehensive evaluation on seismic settings

【释义】 地震环境综合评价

comprehensive geological map of reservoir area

【释义】 水库区综合地质图

【中文定义】 反映水库区基本地质条件（地形地质、地层岩性、地质构造、物理地质现象等）及工程地质问题（如水库渗漏、固体径流、坍岸、浸没、水库触发地震等）等内容的综合性地质图件。

comprehensive hydrogeological map

【释义】 综合水文地质图

【例句】 For the geological information of Longquan mine in Linshui county, such as geological structure, hydrological information, pi tin flow, ect, the comprehensive hydrogeological map of the Longquan mine in Linshui is made by using MapGIS tool. 针对邻水县龙泉煤矿的地层构造、水文信息、矿坑涌水量等多种地质信息，采用了 MapGIS 软件编制了该煤矿综合水文地质图。

【中文定义】 把水文地质调查工作中所获得的各种水文地质现象和资料，用各种代表符号按一定比例尺绘制在图纸上所编制的一种综合性图件。

comprehensive logging

【释义】 综合测井

comprehensive result chart of construction material

【释义】 料场综合成果图

【例句】 In hydropower engineering, comprehensive result chart of construction material should include plan, survey profile, testing and storage capacity calculation result sheet. 水力发电工程中的料场综合成果图应包含平面图、勘探剖面图、试验和储量计算成果表。

【中文定义】 反映料场基本地质条件及料源相关特性的综合性地质图件，通常由地质平面图和地质剖面图组成。

compressibility [kəmˌpresəˈbɪləti]

【释义】 *n.* 压缩性，压缩系数，压缩率

【例句】 Regarding the compressibility of the mixed fluid in pores, mixed fluid continuity equations are established. 考虑孔隙中混合流体的压缩性，建立了混合流体的连续方程。

compression coefficient

【释义】 压缩系数

【例句】 The sensitivity of one-dimension consolidation degree to the uncertainty of seepage coefficient, void ratio and compression coefficient is analyzed. 分析了固结度对渗透系数、孔隙比和压缩系数三个参数不确定性的敏感性。

compression index

【释义】 压缩指数

【例句】 The study shows that besides boundary conditions, the factors influencing the nonlinear consolidation behavior of soils are the compression index, the permeability index, the level and the rate of loading, the thickness of soil, etc. 研究表明，除边界条件外，影响软土地基一维非线性固结性状的主要因素是压缩指数、渗透指数、荷载大小与加荷速率、土层厚度等。

compression modulus

【释义】 压缩模量

【例句】 Elastic resistant coefficient, deformation and compression modulus of surrounding rock mass in large-section loess tunnel are very important for design of tunnel. 围岩的弹性抗力系数、变形模量和压缩模量对大断面黄土隧道的设计来讲十分重要。

【中文定义】 物体在受三轴压缩时应力与应变的比值。

compressional fault

【释义】 压性断层

【例句】 The tunnel body passes through several compressional faults. 隧道洞身穿越数条压性断裂。

【中文定义】 由断层两盘相对运动引起的派生分

支构造，压性分支构造与主干断层所夹锐角指向对盘相对运动方向。

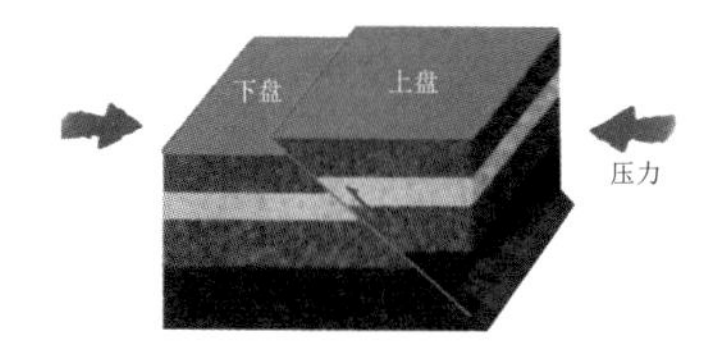

compressional fault

compressive deformation

【释义】 压缩变形

【例句】 Moreover, the uniaxial compressive deformation and damage of thenardite are different from other rocks during four deformation phases. 此外，芒硝在单轴压缩变形破坏过程中，具有与普通岩石试件不同的四阶段性特征。

compressive strength

【释义】 抗压强度

【例句】 The compressive strength and other mechanical performance can be affected by temperature stress induced under variable temperature. 混凝土在温变疲劳作用下产生的温度应力对混凝土的抗压强度等力学性能均会产生一定的影响。

【中文定义】 抗压强度是指在无侧束状态下所能承受的最大压力。

compressive structural plane

【释义】 压性结构面

【例句】 Single or double fold axial plane, thrust fault or thrust plane, and a part of regional foliation cleavage planes are compressive structural plane. 单式或复式褶皱轴面、冲断层或逆掩断层面、区域片理面和一部分劈理面等都是压性结构面。

compressive structural plane

【中文定义】 简称挤压面，是走向垂直主压应力方向、具有明显挤压特征的结构面。

compresso-shear fault

【释义】 压扭性断层

【例句】 Shear faults are also very developed assorted with compresso-shear faults. 扭性断层亦很发育，与压扭性断层配套出现

compresso-torsion structural plane

【释义】 压扭性结构面

【例句】 Compresso-torsion structural planes are characterized by both compressive and torsional properties. 压扭性结构面是既有压性特征又具扭性特征的结构面。

【中文定义】 简称压扭面，指既具有压性又具有扭性的结构面。

compresso-torsion structural plane

computed tomography of acoustic velocity

【释义】 声波速度层析成像

computed tomography of electromagnetic

【释义】 电磁波层析成像

computed tomography of seismic refraction velocity

【释义】 折射层析成像

【例句】 With computed tomography of seismic refraction velocity, we can gain velocity distributing, identify location and dimension of abnormal concrete regions, conclude the type of defect and the intensity of defect. 利用折射层析成像技术可以获得测区混凝土的波速分布图，确定异常区的位置、尺寸，推断缺陷的类型、强度。

computed tomography of seismic velocity

【释义】 地震波速度层析成像

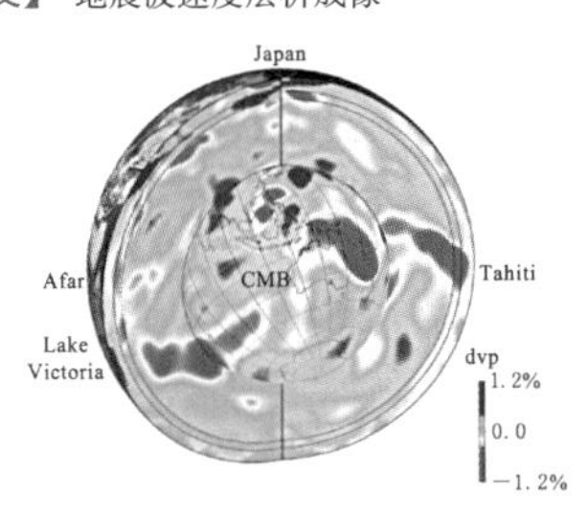

computed tomography of seismic velocity

computer-aided cartography

【释义】 机助地图制图

【中文定义】 利用电子计算机及外围设备和相应软件，进行地图信息的采集、存储、处理、管理、显示、绘图和制版的技术与方法。

concave bank

【释义】 凹岸

【中文定义】 河流弯曲河段岸线内凹的一岸。凹岸通常受主流冲刷，水深、流速较大。

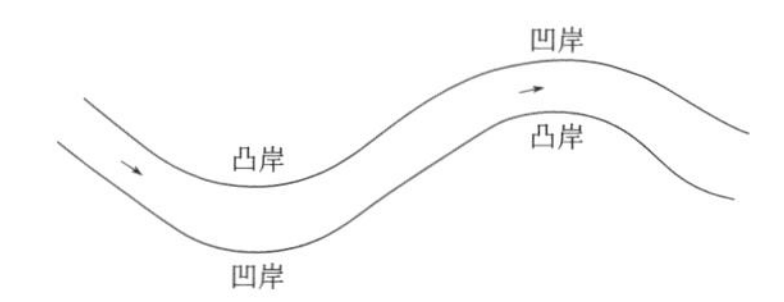

convex and concave bank

concave slope

【释义】 凹坡，凹形坡

【例句】 On compound slope field, the soil erosion displayed strong, week and depositional on steep, gentle and concave slopes, respectively. 在复合坡面，随坡面的陡、缓、凹，土壤侵蚀表现为强、弱、沉积。

【中文定义】 在岸坡形态上，凹坡是下部坡度缓，上部坡度陡；在等高线地形图上，凹坡等高线下部稀疏，上部密集。

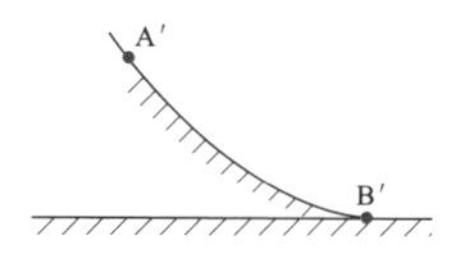

concave slope

concealed structure fracture zone detection

【释义】 隐伏构造破碎带探测

conchoidal fracture

【释义】 贝壳状断口

conchoidal fracture

【例句】 Onyx has a fat or wax-like luster, translucent, conchoidal fracture. 玛瑙具有脂肪或蜡状光泽，半透明，贝壳状断口。

【中文定义】 断裂面呈具有同心圆纹的规则曲面，似贝壳的壳面。

concrete diaphragm wall

【释义】 混凝土防渗墙

【例句】 In Earth-Rock Dam, there is situation that two kinds of deformation properties materials contact, for instance, concrete diaphragm wall and the body of dam on both sides. 土石坝中常存在变形性能相差很大的两种材料相接触的情况，如混凝土防渗墙与两侧坝体的接触等。

concrete diaphragm wall

concretion [kən'kriʃən]

【释义】 *n.* 结核

【中文定义】 指沉积岩中与围岩成分有明显区别的某种矿物质团块，其形态有球状、卵状及其他不规则状。

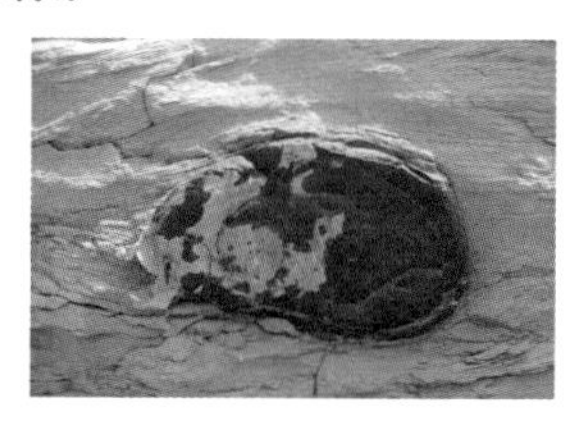

concretion

conductivity [ˌkɑːndʌk'tɪvəti]

【释义】 *n.* 电导率，电导性，传导性

【例句】 The conductivity of an isotropic medium is scalar, and the conductivity of anisotropic media is tensor. 各向同性介质的电导率是标量，而各向异性介质的电导率是张量。

【中文定义】 也可以称为导电率。在介质中该量与电场强度之积等于传导电流密度。对于各向同性介质，电导率是标量；对于各向异性介质，电导率是张量。生态学中，电导率表示溶液传导电流的能力。单位为西门子每米（S/m）。

cone balance method

【释义】 漏斗均衡法

cone penetration test (CPT)

【释义】 静力触探试验（CPT）

【例句】 Cone penetration test is one of the common in-situ test in geotechnical investigation. 静力触探试验是岩土工程勘察中最常用的原位测试手段之一。

cone penetration test (CPT)

confined water

【释义】 承压水，自流水

【例句】 With the increase of foundation pits depth, confined water problem has become more and more prominent in deep foundation pit projects in Shanghai. 随着基坑开挖深度的增加，上海地区深基坑工程的承压水问题日益突出。

【中文定义】 承压水是充满两个隔水层之间的含水层中的地下水。

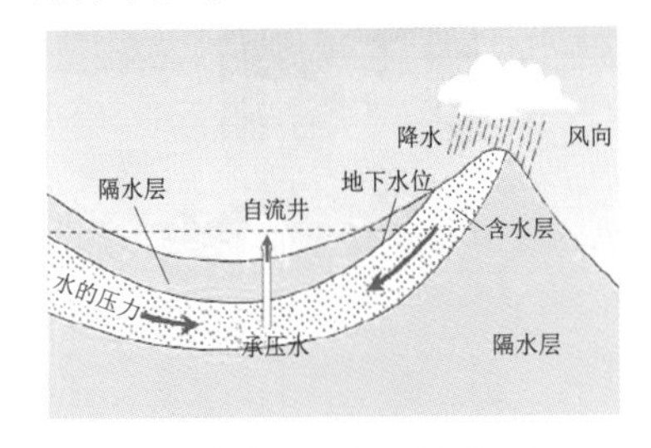

schematic diagram of confined water

conformity [kən'fɔrməti]

【释义】 *n.* 整合；整合接触

【中文定义】 地层接触关系的一种，指岩层连续沉积，新老地层产状一致，层序无间断，时代连续。

conglomerate [kən'glɑːmərət]

【释义】 *n.* 砾岩

【常用搭配】 volcanic conglomerate 火山砾岩；basal conglomerate 底砾岩；top conglomerate 顶盘砾岩；auriferous conglomerate 含金砾岩；epiclastic conglomerate 外力碎屑砾岩；intraformational conglomerate 建造内砾岩；monogenetic conglomerate 单成砾岩

【例句】 The results show that Hanjiagou conglomerate is mainly composed of fine gravel. 结果表明韩家沟砾岩以细砾为主。

【中文定义】 粒径大于 2mm 的圆状和次圆状的砾石占岩石总量 30%以上的碎屑岩。砾岩中碎屑组分主要是岩屑，只有少量矿物碎屑，填隙物为砂、粉砂、黏土物质和化学沉淀物质。

conglomerate

conjectural boundary

【释义】 推测界线

【常用搭配】 conjectural boundary of stratum 推测地层界线

【例句】 Conjectural boundary of stratum should be drawn with dotted line on geological section. 地质剖面图上推测的地层界线应用虚线表示。

【中文定义】 推测的各种地质界线，如推测地层界线、推测地层不整合线等。

conjugate fault

【释义】 共轭断层

【例句】 Moreover, in the paper further discussions of the characteristics of earthquake conjugate ruptures, its relations with the geologic structures and preparatory action of the conjugate faults, etc. were made. 文中进而讨论了地震共轭破裂特征及其与地质构造的关系，以及共轭断层的孕震作用等问题。

【中文定义】 又称共轭剪切带（conjugated shear zone），是在统一构造应力场作用下形成的两组方向不同、剪切方向相反、大体同时发育的交叉剪切带。

conjugate joint

【释义】 共轭裂隙，共轭节理

【常用搭配】 conjugate shear joints 共轭剪节理

【例句】 Inversion of the conjugate joints and striation is used to study orientation and timing of the tectonic stress field and define the stages of the field. 利用共轭节理和断层擦痕反演，对区域构造应力场进行了定向、定时研究，确定了各期构造应力场。

【中文定义】 两组剪切节理交叉，呈“X”形，称共轭节理。

consequent fault

【释义】 顺向断层

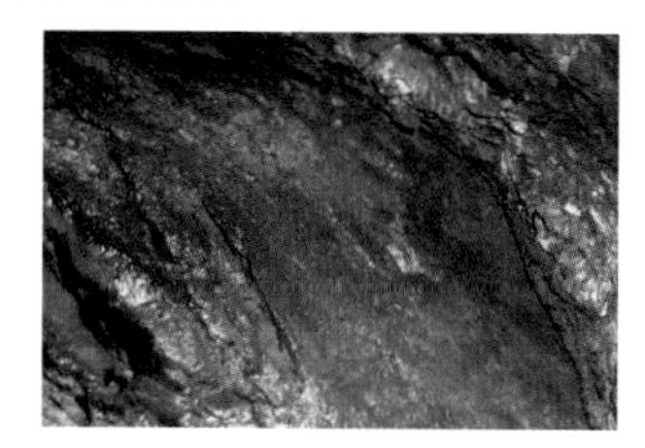

consequent fault

consequent landslide

【释义】 顺层滑坡

【例句】 Consequent landslide is a kind of common engineering geologic hazard, which often occurs suddenly with strong destructive power. 顺层滑坡是一种常见的工程地质灾害，往往具有突发性，且破坏力强。

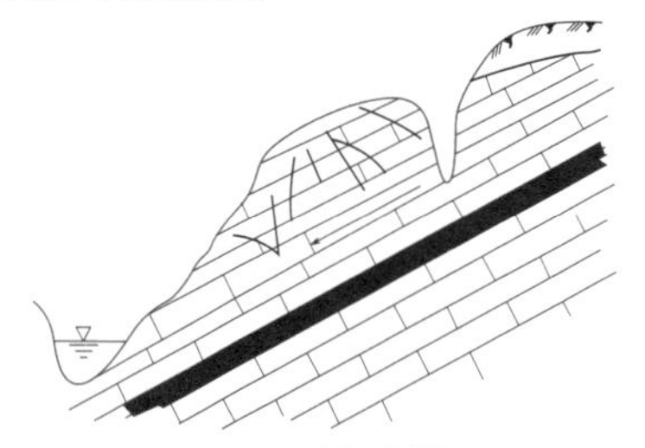

consequent landslide

consistency limit

【释义】 稠度界限

【例句】 It was found that the earth consistency limit was increased, along with the salt content increase. 发现随着黏粒含量的增加，土的稠度界限有增加的趋势。

consistency state

【释义】 稠度状态

consolidated drained triaxial test

【释义】 三轴固结排水试验

【例句】 The large-scale consolidated drained triaxial tests have been carried out for geobelt rein forced crushed gravel soil and non-rein forced crushed gravel soil. 大型三轴固结排水试验表明：土工带加筋后土体的破坏强度和破坏应变均得到提高。

consolidated quick shear test

【释义】 固结快剪试验

【例句】 Besides, correlation between the strength indexes of direct shear test and consolidated quick direct shear tests were studied. 此外，还对典型土层直剪试验和固结快剪强度指标间的相关性进行了研究。

consolidated undrained shear test (CU)

【释义】 固结不排水剪试验（CU）

consolidation [kənˌsɑlə'deʃən]

【释义】 *n.* 排水固结法

【中文定义】 排水固结法：通过预压使软黏土地基中孔隙水排出，土体发生固结，孔隙体积逐渐减小，抗剪强度提高，达到解决建筑物地基稳定和变形问题的地基处理方法。又称预压法。

consolidation drained shear test (CD)

【释义】 固结排水剪试验（CD）

consolidation drained shear test

consolidation grouting

【释义】 固结灌浆

【常用搭配】 high pressure consolidation grouting 高压固结灌浆

【例句】 The supersonic wave value and the dynamic modulus of elasticity are increased by high pressure consolidation grouting. 高压固结灌浆结果达到了提高坝基岩石超声波值和动弹模量的目的。

consolidation grouting

consolidation pressure

【释义】 固结压力

【例句】 The strength resisting liquefaction increases along with the increase of consolidation pressure, sample density initial shear stress, but the increase of initial shear stress is limited. 抗液化强度随固结压力、试样密度、初始剪应力的增加而增加，但初始剪应力的增加量是有一定限度的。

consolidation settlement

【释义】 固结沉降

consolidation test

【释义】 固结试验

【常用搭配】 triaxial consolidation tests 三轴固结试验

consolidation test apparatus

【例句】 First, Based on the consolidation test of Wenzhou soft clay, the relation between structure character and mechanical properties of structured soft soil is examined. 首先，通过温州软土的固结试验，探求土体结构性与土体力学性质之间的关系。

consolidation under K0 condition

【释义】 K0 固结

constant head permeability test

【释义】 常水头渗透试验

【例句】 To study soil permeability of soil element within the high earth - rock dam under complex stress state, constant head permeability tests are carried out for three kinds of soil with different gradation from the same material site. 为研究高土石坝坝体内土体单元在复杂应力状态下的渗透性能，对同一土料场 3 种不同级配土样进行了常水头渗透试验。

constrained grain size

【释义】 限制粒径

construction detailed design stage

【释义】 施工详图设计阶段

【例句】 The project is currently under construction detailed design stage. 该项目目前处于施工图设计阶段。

construction geological mapping

【释义】 施工地质编录图

【例句】 In construction detailed design stage of hydropower engineering, construction geological mapping shall be submitted as appropriate in report for engineering geological investigation. 水力发电工程施工详图设计阶段工程地质勘察报告中可根据需要施工地质编录图。

【中文定义】 将施工地质编录过程中直接观察和综合整理的地质信息系统地用文字和图表的方式编制而成的图件。

construction geology

【释义】 施工地质

constructional terrace

【释义】 堆积阶地

【中文定义】 由河流冲积物组成的阶地。根据河流下切程度不同，形成阶地的切割叠置关系不同又可分为：内叠阶地、嵌入阶地、埋藏阶地。

types of terrace

contact grouting

【释义】 接触灌浆

【常用搭配】 pipe guiding contact grouting 埋管

法接触灌浆；steel lining contact grouting 钢衬接触灌浆

【中文定义】 指在岩石上或钢板结构物四周浇筑混凝土时，混凝土干缩后，对混凝土与岩石或钢板之间形成的缝隙灌浆。

contact grouting

contact loss

【释义】 接触流失

【例句】 Contact loss is a form of seepage deformation. 接触流失是渗透变形一种形式。

【中文定义】 渗流垂直于渗透系数相差较大的两相邻土层流动时，将渗透系数较小的土层中的细颗粒带入渗透系数较大的土层中的现象。

contact stress monitoring

【释义】 接触应力监测

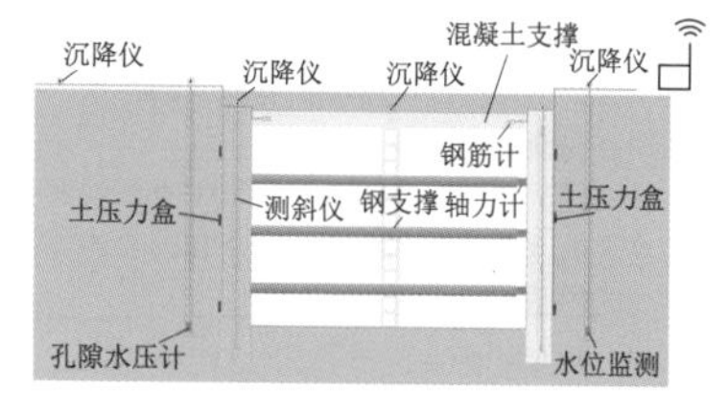

contact stress monitoring

contact thermal metamorphic rock

【释义】 热接触变质岩

contact thermal metamorphic rock

【例句】 Based on the degree of crystallization of andalusite, it is believed that the anthracite is a low-grade contact thermal metamorphic rock. 从红柱石的结晶程度分析，这种无烟煤是一种低级热接触变质岩。

【中文定义】 由热接触变质作用（也称热变质作用）形成，是在岩浆体散发的热量和挥发分作用下，使原岩发生重结晶变质结晶形成的。

contaminated soil

【释义】 污染土

【例句】 Nowadays contaminated soil treatment technologies can be sorted as six categories, including, bioremediation, chemical treatment, physical separation, solidification/(stabilization,) high-temperature technology and phyto-remediation. 污染土处理技术目前可归纳为 6 类，即微生物修复技术、化学处理技术、物理分离技术、固化/安定化技术、高温处理技术、植物修复技术等。

contaminated soil

contorted fold

【释义】 扭曲褶皱

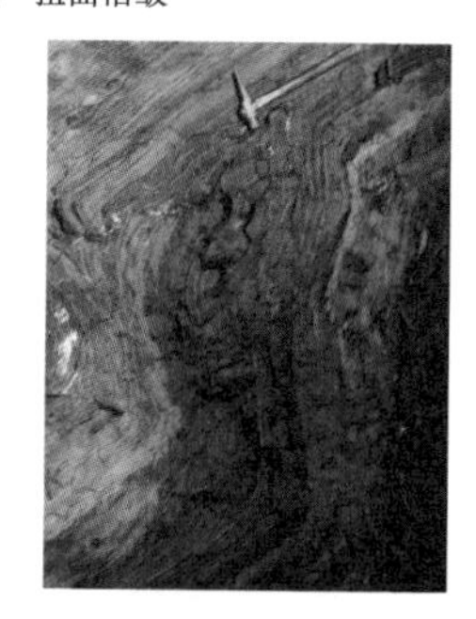

contorted fold

continental drift theory

【释义】 大陆漂移学说

【例句】 Plate tectonics integrates two older explanations for the distribution of the earth's geophysical features the theories of continental drift and sea-floor spreading. 板块构造学说整合了两个旧有的地球物理分布特征阐述——大陆漂移学

说和海底扩张学说。

【中文定义】 大陆漂移假说是解释地壳运动和海陆分布、演变的学说。大陆彼此之间以及大陆相对于大洋盆地间的大规模水平运动，称大陆漂移。大陆漂移说认为，地球上所有大陆在中生代以前曾经是统一的巨大陆块，称之为泛大陆或联合古陆，中生代开始分裂并漂移，逐渐达到现在的位置。

continental facies

【释义】 陆相

【常用搭配】 red bed of continental facies 陆相红层

【例句】 The depositional environment of the study area is mainly marine facies in the south, continental facies in the north. 研究区南部的沉积环境以海相为主，北部以陆相为主。

【中文定义】 陆相是大陆环境中形成的沉积物。在中国陆相沉积主要在中生代以后，分布面积也较广，厚度也大，对找水找油都有重要意义。但其相变大，岩性不均一，常见的岩石有碎屑岩及黏土岩。

continuous gradation

【释义】 连续级配

【中文定义】 无中间粒级缺失的级配称为连续级配。

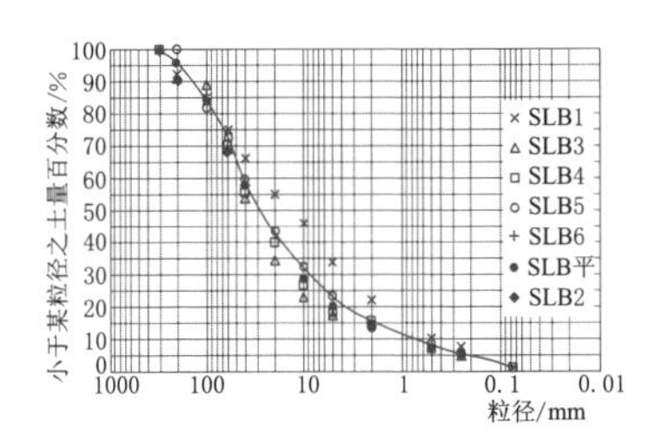

continuous gradation

continuous profiling

【释义】 连续剖面法

【常用搭配】 continuous seismic profiling 连续地震剖面法；continuous electromagnetic profiling (CEMP) 连续电磁剖面法

【例句】 The theory and the practice show that the inversion method has better results than synchronized array MT sounding(SAMT) and continuous electromagnetic profiling(CEMP). 理论和实践都表明，该反演方法比同步阵列大地电磁测深（SAMT）和连续电磁剖面法（CEMP）等MT方法的效果更佳。

continuous structural plane

【释义】 贯通结构面

【例句】 The spatial combination relation of the continuous structural plane in the slope has a direct influence on the stability of the slope. 边坡中贯通结构面的空间组合关系直接影响边坡的稳定性。

【中文定义】 在考察范围内，贯穿岩体或块体，构成岩体、块体边界的结构面称为贯通结构面。

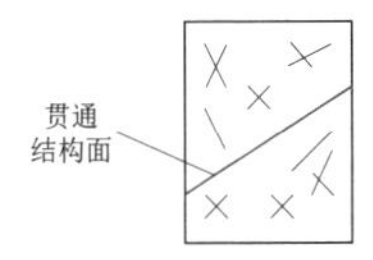

continuous structural plane

contour diagram

【释义】 等密图

【常用搭配】 joint contour diagram 节理等密图

【例句】 Joint contour diagram is a commonly used drawing in geological work, and it can intuitively reflect the advantages of joint orientation and development degree. 节理极点等密图是地质工作中的一种常用图件，它能直观形象地反映节理的优势方位和发育程度。

【中文定义】 等密图是在极点图的基础上，用等值线来表示图内极点密度分布的特征和规律的岩石组构图。极点密度集中的区域称极密或最密区，反映所研究构造发育的优选方位。

contour interval

【释义】 等高距

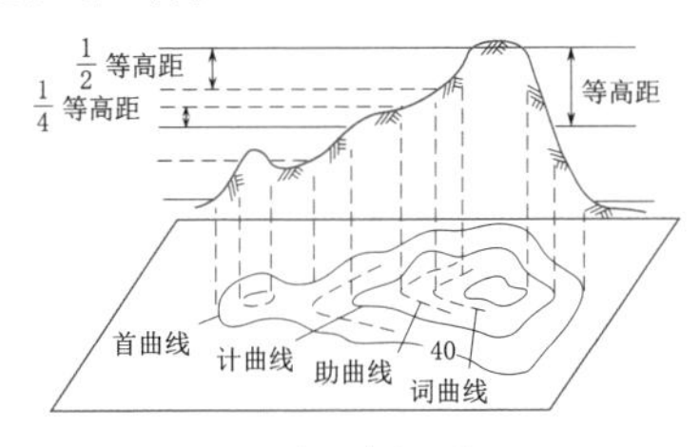

contour interval

【例句】 The accuracy of flood area is mostly dependent on the contour interval of map, accuracy of digitizing map, accuracy of DEM, and interval of

GRID, detecting resolution of pixels. 计算淹没区范围的准确性受地形图等高距大小、数字化采集精度、数字地面模型（DEM）高程精度以及格网间隔大小、像素探测分辨率等因素的影响。

【中文定义】 地图上相邻等高线的高差。

contour line

【释义】 等高线

【同义词】 isohypse

【中文定义】 地图上地面高程相等的各相邻点所连成的曲线。

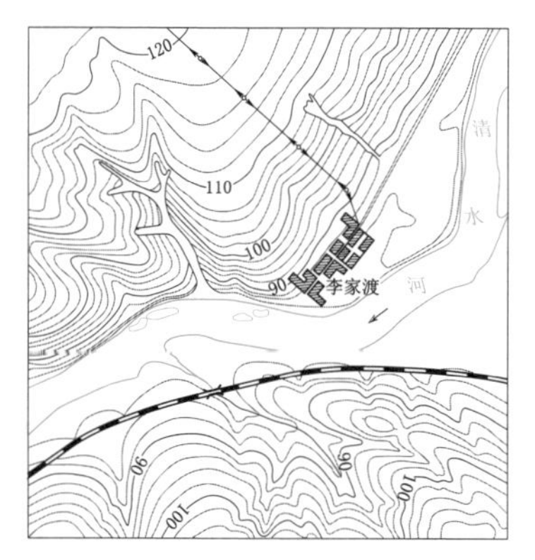

contour line

contour plot of the poles

【释义】 极点等值线图

control blasting

【释义】 控制爆破

control blasting

control point survey

【释义】 控制点测量

【例句】 Photograph control point survey is an essential part of aerophotogrammetry. 像控点测量是航空摄影测量工作中必不可少的环节。

control survey

【释义】 控制测量

【常用搭配】 topographic control survey 地形控制测量；horizontal control survey 平面控制测量；vertical control survey 高程控制测量

【例句】 This paper introduces the outline of inside tunnel survey, and expounds the survey control in tunneling works from aspects of the construction traverse survey, midline layout survey, midline detection and tolerance, elevation control survey, vertical section survey and cross section survey, and breakthrough survey, etc. 本文介绍了洞内测量纲要，并从施工导线测量、中线施工放样、中线检测与限差、高程控制测量、纵断面测量和横断面测量、贯通测量等方面阐述了隧洞工程测量控制。

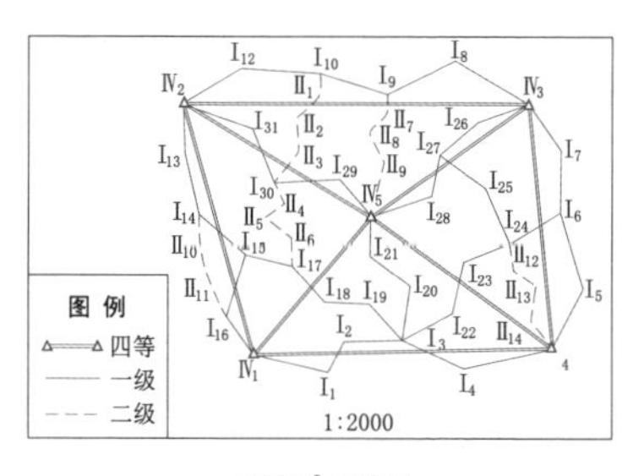

control survey

controlled source magnetotellurics (CSMT)

【释义】 可控源大地电磁测深法

convergence test

【释义】 收敛检验，收敛试验

【常用搭配】 integral convergence test 积分收敛判别法

【例句】 A mesh convergence test of the 3D finite difference method is presented for computing the electromagnetic response of a buried conductivity anomaly. 本文用三维有限差分法，在计算电导率异常的电磁响应时做了一次网格收敛试验。

convex bank

【释义】 凸岸

【中文定义】 河流弯曲河段岸线外凸的一岸。凸岸水流速度小，泥沙易沉积，河岸一般较缓。

cooling joint

【释义】 冷缩节理

【中文定义】 由于熔岩流冷缩而形成的次生结构面。

coordinate azimuth

【释义】 坐标方位角，格网方位角

【中文定义】 从过某点平行于纵坐标轴的方向线（正值方向）起，按顺时针方向旋转至目标方向线的水平夹角。

coordinate grid

【释义】 坐标格网

【例句】 The analysis technique based on coordinate grid, as a kind of physical method for simulating forming process, has been widely applied in the production of sheet metal parts. 基于坐标网格的分析技术作为一种模拟成形过程的物理方法，在钣金件的生产中得到了广泛的应用。

【中文定义】 按一定纵横坐标间距，在地图上划分的格网。

core acquisition rate

【释义】 岩芯获得率

core drilling

【释义】 岩芯钻进

core recovery percent

【释义】 岩芯采取率

【例句】 The effect of water to rock mechanics characteristic has something to do with anhydrite's content and core recovery percent. 水对岩石力学特性的影响主要与岩石的硬石膏含量和岩芯采取率有关。

correlation analysis

【释义】 相关分析

【例句】 The chloride as correlation analysis example results showed that groundwater chloride had the maximum correlation with observation well distance and correlation with time and monitoring value of surface water was very small. 氯作为相关分析的算例结果表明，地下水氯化物与观测井的距离相关性最大，与时间和地表水的监测值之间的相关性很小。

corrosion analysis of soil

【释义】 土的腐蚀性分析

corrosive action

【释义】 腐蚀作用

corrosive action

【例句】 The distribution of sulfate reduction bacteria is higher in the soil surface than in 1.5m deep soil, that indicates faintly corrosive action of microbe in the depth which cable are buried. 土壤表层硫酸盐还原菌的含量超过 1.5m 深土壤还原菌含量，表明在光缆的埋设深度上微生物腐蚀作用轻微。

corrosive carbon dioxide

【释义】 侵蚀性 CO_2

【例句】 The paper states the present research on the concrete carbonation in the context of atmospheric environment, the durability of the metro concrete and the neutralization of the concrete in the context of corrosive carbon dioxide. 文章阐述了大气环境下混凝土碳化研究现状、地铁混凝土耐久性研究现状及侵蚀性二氧化碳作用下混凝土中性化研究现状。

【中文定义】 侵蚀性 CO_2 是指超过平衡量并能与碳酸钙起反应的游离 CO_2。

corundum [kə'rʌndəm]

【释义】 *n.* 刚玉

corundum

【例句】 The corundum in the Kangjinla chromitite might be regarded as a new mineral index of a high-pressure environment. 康金拉铬铁矿石中的刚玉可以认为是一种显示高压环境下的新指示矿物。

Coulomb-Navier strength theory

【释义】 库仑-纳维强度理论

covered karst

【释义】 覆盖型岩溶

【例句】 Both exposed or covered karst water systems in fault basins are major hydrogeological type in the area of East Yunnan. 岩溶断陷盆地裸露-覆盖型岩溶水系统是滇东地区的主要水文地质类型。

【中文定义】 指被松散堆积物覆盖的岩溶。

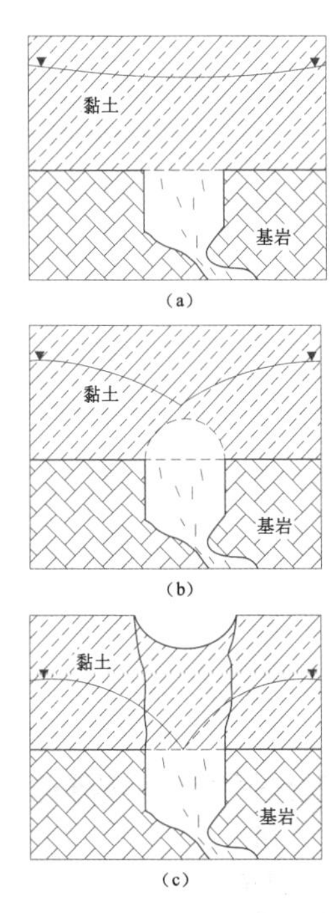

collapse of covered karst formation

craton [ˈkreɪˌtən]

【释义】 *n.* 克拉通；稳定地块

【常用搭配】 craton block 克拉通地块；craton basin 克拉通盆地；craton crust 克拉通地壳

【例句】 A continental interior that has been structurally inactive for hundreds of millions of years is called a craton. 亿万年内部不具活动特性的大陆称为克拉通。

【中文定义】 大陆地壳上长期稳定的构造单元，即大陆地壳中长期不受造山运动影响，只受造陆运动发生过变形的相对稳定部分，常与造山带对应。

creep [krip]

【释义】 *n.* 徐动，蠕动

【常用搭配】 creep strength 蠕变强度；creep limit 蠕变极限；creep test 蠕变试验

【例句】 Unlike in other creep mechanisms, the dislocation density here is constant and independent of the applied stress. 不同于其他蠕变机制，位错密度不受施加应力的影响，是恒定的和独立的。

【中文定义】 蠕变：固体材料在保持应力不变的条件下，应变随时间延长而增加的现象。

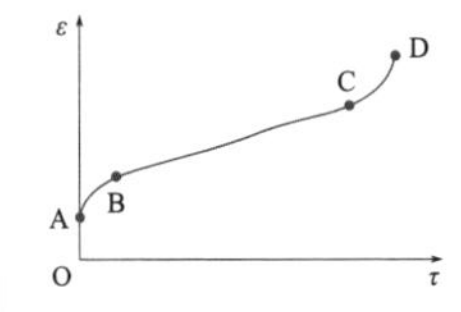

creep curve

creep curve

【释义】 蠕变曲线

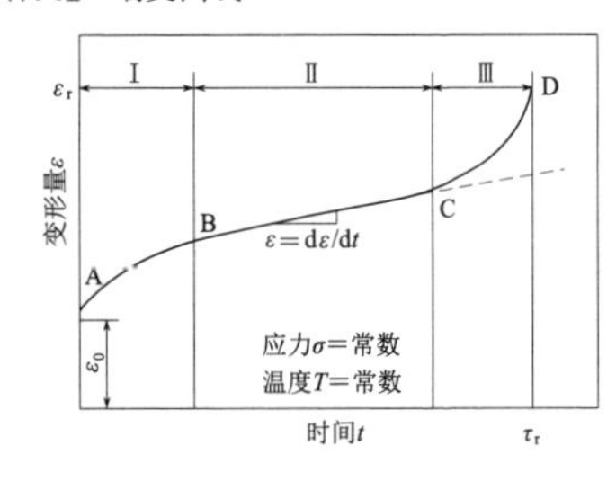

creep curve

【例句】 The characteristic of displacement creep curve in deep soil can response the initial creep and stabilized creep. 深层土体位移蠕变曲线特征反映了初始蠕变阶段和稳定蠕变阶段的变形特征。

【中文定义】 材料的蠕变过程常用变形与时间之间的关系曲线来描述，这样的曲线称为蠕变曲线。

creep deformation

【释义】 蠕变变形

【例句】 Numerical analysis can only simulate the former two phases of creep deformation. 数值分析只能模拟蠕变变形的前两个阶段。

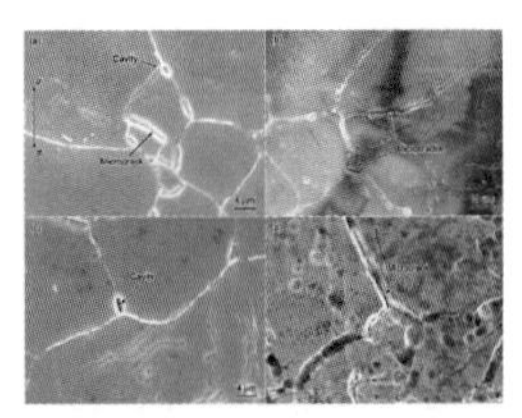

creep deformation

【中文定义】 在恒定温度和低于材料屈服极限的恒定应力下，随着时间的延长材料发生不可恢复的塑性变形。

creep landslide

【释义】 蠕动型滑坡

creep limit

【释义】 蠕变极限

【常用搭配】 conventional creep limit 公称蠕变极限；true limiting creep stress 真极限蠕变应力；creep rate limit 蠕变率极限

【例句】 Initial creep limit is a power function of steady state creep strain rate. 初始蠕变极限可以表示成稳定蠕变率的函数。

【中文定义】 蠕变极限指固体材料在一定温度和规定时间内的蠕变变形量或蠕变速度不超过某一规定值时所能承受的最大应力，单位为帕斯卡。

creep rate

【释义】 蠕变速度，蠕变应变速率，蠕变速率

【例句】 The accelerating stage in the radial creep course appears earlier than that in the axial creep course, and the primary creep rate, steady-state creep rate and accelerating creep rate in the radial creep course are larger than that in the axial creep course respectively for the specimen under the fracture stress level. 在破裂应力水平下，岩石径向蠕变比轴向蠕变先进入加速蠕变阶段，径向的初始蠕变速率、稳态蠕变速率以及加速蠕变速率均高于轴向相应的蠕变速率。

【中文定义】 蠕变速率指蠕变试验中单位时间的蠕变变形，即给定时间内蠕变曲线的斜率。

creep rupture

【释义】 蠕变断裂，蠕变破坏

【例句】 The P92 steel is a steel sort being used for main steam pipelines of ultra-supercritical units at present, its creep rupture life has widely drawn close attention. P92 钢是目前超超临界机组主蒸汽管道的应用钢种，其蠕变断裂寿命被广泛关注。

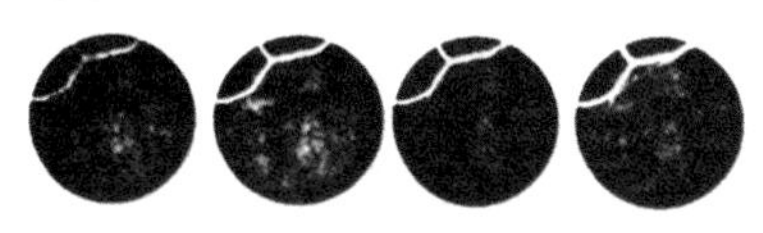

(a)蠕变开始 ($\Delta t=0$)　(b)蠕变过程 ($\Delta t=58.1$min)　(c)蠕变过程 ($\Delta t=50$min)　(d)蠕变过程 ($\Delta t=75.7$min)

creep rupture

【中文定义】 零件由于蠕变变形而引起的断裂称为蠕变断裂。

creep rupture strength

【释义】 持久强度，蠕变断裂强度

【常用搭配】 notch rupture strength 切口试样持久强度

【中文定义】 持久强度是指金属材料、机械零件和构件抗高温断裂的能力，常以持久极限表示。

creep test

【释义】 蠕变试验

【例句】 A laboratory triaxial creep test to on-site specimens can be used to study the creep properties of the carbonaceous phyllite. 现场岩样的室内三轴蠕变试验可以用来研究炭质千枚岩的蠕变特性。

【中文定义】 测定金属材料在长时间的恒温和恒应力作用下发生缓慢的塑性变形现象的一种材料机械性能试验。

creep test

creeping rock mass

【释义】 蠕变岩体

creeping rock mass

【例句】 This paper briefs the deformation mechanism and inversion collapse mode of the non-excavated slope with creeping rock mass. 这篇文章简要介绍了蠕变岩体非开挖区边坡变形机制和倾倒破坏模式。

【中文定义】 蠕变岩体指岩体在恒定载荷持续作用下，其变形随时间逐渐缓慢增长的现象。

crenulation cleavage

【释义】 折劈理（褶劈理）

【常用搭配】 extensional crenulation cleavage 伸展褶劈理

【例句】 In the second phase of deformation, crenulation cleavage (S _ 2) mainly occurred in metapelites, and the degree of deformation is various within the entire area. 第二期构造变形形成的折劈理（S _ 2）主要发育在变泥质岩石中，折劈理的强度在整个地区中有差异。

【中文定义】 是切过先存连续劈理的一种不连续劈理。由先存连续劈理形成紧密间隔、平行排列的微褶皱发展而成。

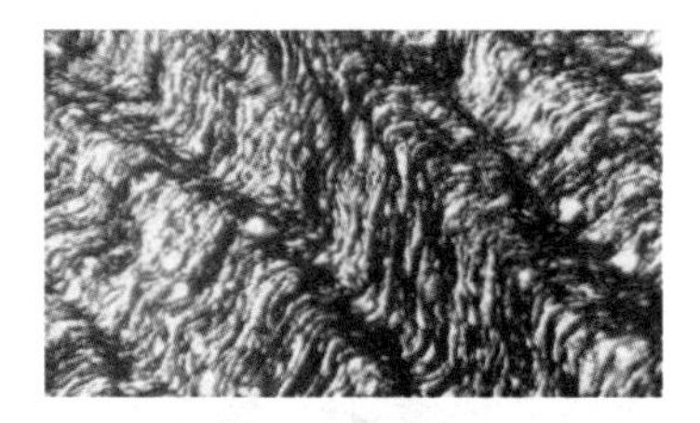

crenulation cleavage

crest of fold

【释义】 褶皱脊

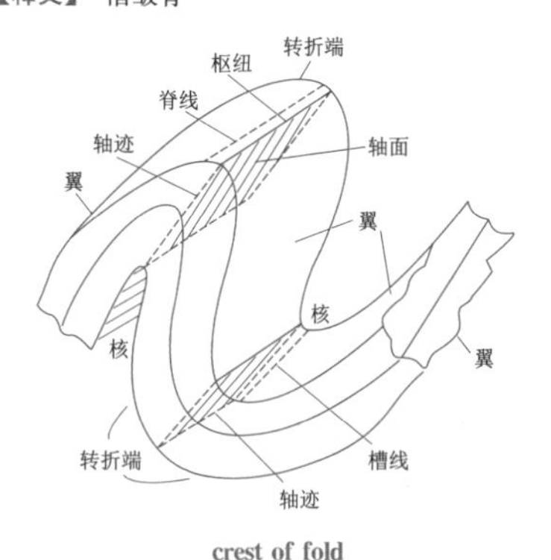

crest of fold

【中文定义】 褶皱脊是在背形横剖面中褶皱面的最高点。除轴面直立的背形外，其他背形中的脊和枢纽均不重合。

Cretaceous (K) [krɪ'teʃəs]

【释义】 *n.* 白垩纪（系）

【中文定义】 白垩纪是地质年代中中生代的最后一个纪，因欧洲西部该年代的地层主要为白垩沉积而得名，始于1.45亿年前，结束于6500万年前，历经8000万年，是显生宙的最长一个阶段。白垩纪时期，大陆被海洋分开，地球变得温暖、干旱，开花植物首次出现。

critical hydraulic gradient

【释义】 临界水力坡降，临界水力梯度

【例句】 The critical hydraulic gradient of tailing silt is greater than that of conventional sand soil, which is determined by the tailing silt's structure characteristics. 尾粉砂的结构构造特点决定了其临界水力梯度大于常规的砂性土。

critical void ratio

【释义】 临界孔隙比

【例句】 The value of critical void ratio is affected by soil kinds, water content and test conditions. 临界孔隙比的大小与土的类别、含水率和试验条件有关。

cross bedding

【释义】 交错层理

【例句】 The sedimentary structure mainly includes unidirectional and bidirectional cross bedding, interference and curved ripple mark, and wavy, flaser and lenticular bedding. 主要发育双向和单向交错层理、曲线型和干涉波痕及波状、脉状、透镜状层理。

【中文定义】 沉积岩细层与层系界面成角度相交的一种层理。

cross bedding

cross borehole acoustic detection

【释义】 跨孔声波检测

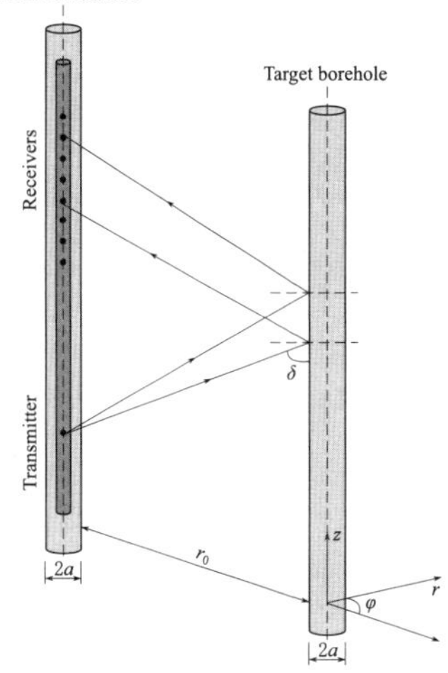

cross borehole acoustic detection

cross borehole ground penetrating radar

【释义】 跨孔对穿法

cross joint

【释义】 横节理

【例句】 In general, the rock mass structure planes are divided into three major types, including the primary structural plane, tectonic structural plane and secondary structural plane. longitudinal joint, cross joint, bedded joint, diagonal joint and columnar joint belong to the primary structural plane. 一般而言，岩体结构面分为原生结构面、构造结构面和次生结构面三大类，纵节理、横节理、层节理、斜节理和柱状节理都属于原生结构面。

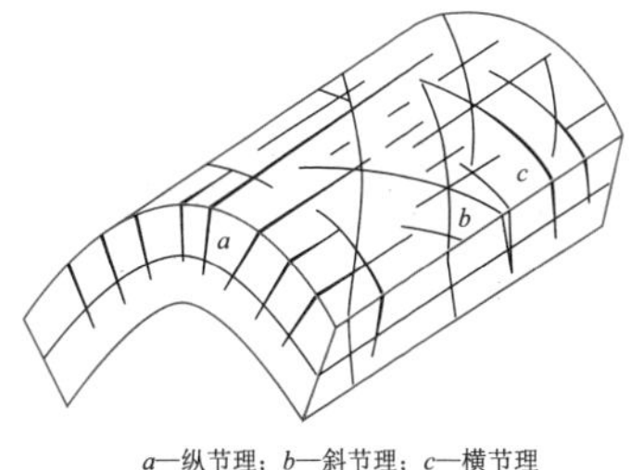

a—纵节理；b—斜节理；c—横节理

cross joint

【中文定义】 横节理是指走向与褶皱轴或区域构造线走向大致垂直的节理。

cross section

【释义】 横剖面图，横剖面

【常用搭配】 symmetrical cross section 对称横剖面；middle cross section 中央横剖面

【同义词】 transverse section

【例句】 This may enable the geologist to draw cross sections through areas to show the underground geology along that line of section. 这能够使地质学家绘制区域的横剖面，从而了解某个剖面地下的地质情况。

cross-fault deformation monitoring

【释义】 跨断层变形监测

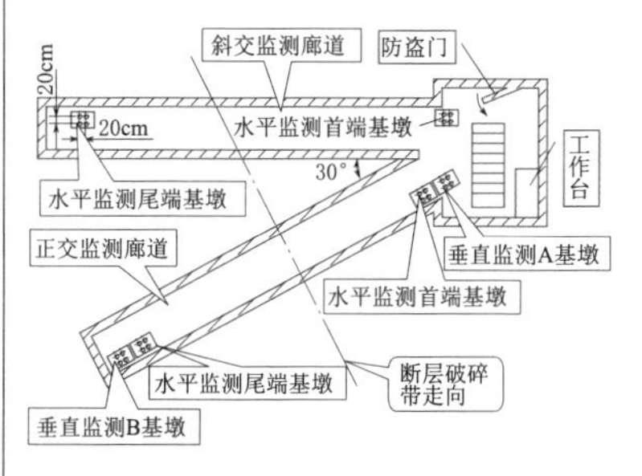

cross-fault deformation monitoring

cross-fault leveling

【释义】 跨断层水准测量

【例句】 As one of the main approaches to attain geomorphologic information, the anomaly of cross-fault leveling recording is always regarded as one important indicator for the possible occurrence of imminent earthquakes. 跨断层水准测量作为获取断层形变信息的最主要手段之一，其异常现象常常被看作地震前兆信息的一个非常重要指标。

crumb texture

【释义】 团块状结构

【中文定义】 由若干土壤单粒黏结在一起形成团聚体的一种土壤结构。

crumby ['krʌmɪ]

【释义】 *adj*. 团块状，柔软的

【常用搭配】 crumby soil 团块状土壤

【例句】 Through blank experiment of terraces. it shows that the soil fertility of newly built terraces is distributed in a crumby shape which is remarkably different than that of the ordinary farm soil. 通过梯田土壤空白试验，证明新修梯田土壤肥力呈现团块状分布规律，这与一般农地土壤肥力分布具有明显的区别。

crumpled texture

【释义】 揉皱结构

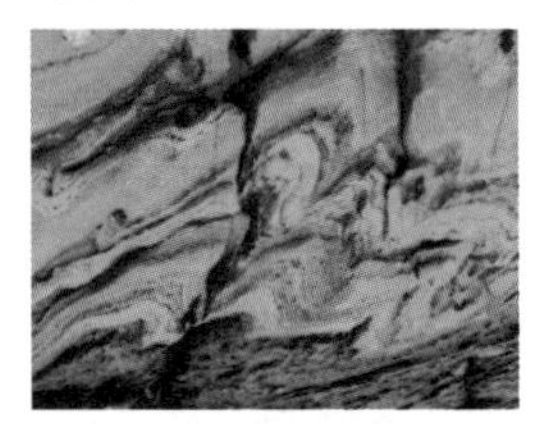

crumpled texture

【中文定义】 矿石中的矿物受力后发生塑性变形形成弯曲皱纹的一种结构。如方铅矿、辉锑矿等塑性矿物受构造变动后，往往扭成揉皱结构。

crushed [krʌʃt]

【释义】 *adj.* 破碎的

【中文定义】 根据《水力发电工程地质勘察规范》的规定，当岩体完整性系数 $K_V \leqslant 0.15$ 时，则该岩体的完整程度为破碎。或根据岩体中结构面发育程度判断，破碎岩体中结构面发育密集、杂乱无序。

crushed

crushed stone

【释义】 碎石，砾石

【例句】 Based on the cracking mechanism and FEM, the crackings on the grade crushed stone pavement and the semirigid base asphalt pavement are compared and analyzed. 基于断裂力学，采用有限元程序，对两种级配碎石夹层沥青路面结构和常规半刚性基层沥青路面结构的断裂进行了对比分析。

【中文定义】 碎石是由天然岩石、卵石或矿石经机械破碎、筛分制成的，粒径大于 4.75mm 的岩石颗粒。碎石多棱角，表面粗糙，与水泥黏结较好，拌制的混凝土拌合物流动性差，但混凝土硬化后强度较高。

crushed stone

crustal fault-block

【释义】 地壳断块

【例句】 Blocks confined by various crustal fractures are crustal fault-blocks. 由不同的地壳断裂所限制的断块称为地壳断块。

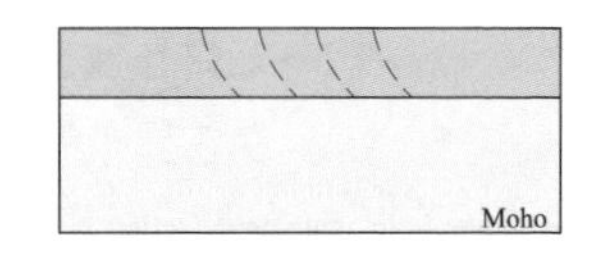

Developing normal faults with extensional basins

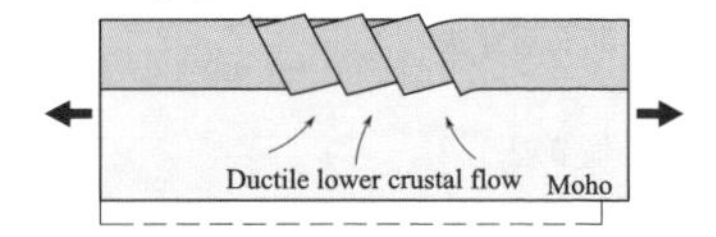

Upper crust fault block rotation

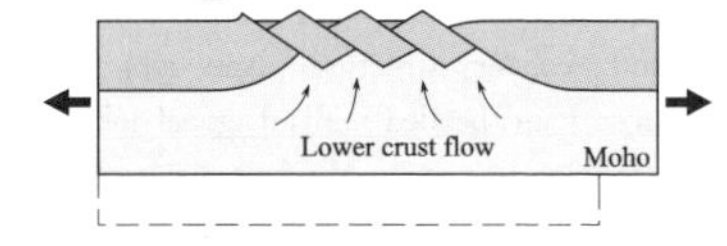

crustal fault-block

【中文定义】 地壳断块是地球表层的第二级断块，也是岩石圈断块内部的次一级断块。

crystalline [ˈkrɪstəlɪn]

【释义】 *n.* 结晶质

crystalline

【常用搭配】 crystalline polymers 结晶质聚合物；crystalline aggregate 结晶质集料；crystalline mineral 结晶质矿物

【中文定义】 组成物质的原子或离子都有规律地在三维空间呈周期性重复排列的，即具有格子构造的固态物质。

cryptocrystalline texture

【释义】 隐晶质结构

【中文定义】 晶粒小于0.1mm，岩石呈致密状，矿物颗粒用显微镜才能辨别的一种岩石结构。

crystalline granular texture

【释义】 结晶粒状结构

【中文定义】 是化学岩和生物化学岩结构类型之一。岩石全部由结晶颗粒组成，按晶粒大小分为粗晶结构、细晶结构和隐晶结构。

crystalloblastic texture

【释义】 变晶结构

【中文定义】 指岩石在固体状态下，通过重结晶和变质结晶而形成的结构。变晶结构要素包括：自形程度、颗粒大小（相对大小，绝对大小）、变质矿物的形状、颗粒间的相互关系。

cuesta [ˈkwɛstə]

【释义】 *n*. 单斜脊，单面山

【常用搭配】 cuesta slope 单面山斜坡

【中文定义】 在单斜构造地区，岩层倾角较缓，软硬相间，受侵蚀切割后，软岩层被蚀成谷地，硬岩层突露成山岭，单面山山体延伸方向与构造线一致，山脊往往成锯齿形，两坡明显不对称。又称单斜山。

Cuesta in Crimea

cumulative percent passing

【释义】 过筛累积百分率

cumulosol [kˈjuːmjʊləsɒl]

【释义】 *n*. 泥炭质土

【例句】 The paper introduces the engineering characteristics of cumulosol, the analysis of the stabilization and the settlement of the embankment of peat soil as well as the cement agitation peat soil and so on. 本文介绍了泥炭质土的工程性质、含泥炭层软土路堤稳定和沉降分析以及水泥搅拌泥炭质土的特性等。

【中文定义】 是在某些河湖沉积低平原及山间谷地中，由于长期积水，水生植被茂密，在缺氧情况下，大量分解不充分的植物残体积累并形成泥炭层的土壤。泥炭地可分为水藓泥炭地和沼泽泥炭地，这两类泥炭地的主要区别在于泥炭地形成的条件不同。

curtain grouting

【释义】 帷幕灌浆

【例句】 In the design of the impermeable curtain grouting engineering, high-pressure cement grouting with orifice-closed method was adopted. 工程防渗帷幕灌浆采用了孔口封闭法高压水泥灌浆。

curtain grouting

cutting plane

【释义】 切割面

【例句】 According to the relationship with the sliding direction of rock and soil, it can be divided into horizontal cutting plane and vertical cutting surface. 根据与岩土体滑动方向的关系，又可分为横向切割面和纵向切割面。

【中文定义】 切割面是指将松动岩土体与四周岩土体分离开的各种结构面。

D

dacite [ˈdeɪˌsaɪt]

【释义】 *n*. 英安岩

【例句】 Early Cretaceous volcanic rocks from Songliao basin, Northeast China, are characterized with basic rocks (BRS) which include dorgalite and basalt, intermediate rocks (IRS) which include basaltic andesite, andesite and trachyandesite, and acid rocks (ARS) include trachyte, trachydacite, dacite and rhyolite. 华北松辽盆地早白垩纪地层发育基性岩类的橄榄玄武岩、玄武岩，中性岩类的玄武安山岩、安山岩、粗安岩，酸性岩类的粗面岩、粗面英安岩、英安岩和流纹岩。

dacite

dam deformation survey

【释义】 大坝变形测量

【例句】 This paper introduces the measurement principle of Laser Alignment Automation Observation Instrument and its structure Characteristic, and gives a case study in dam deformation survey. 本文阐述了激光自动观测仪测量原理及结构特点，并且介绍了它在大坝变形测量中的应用。

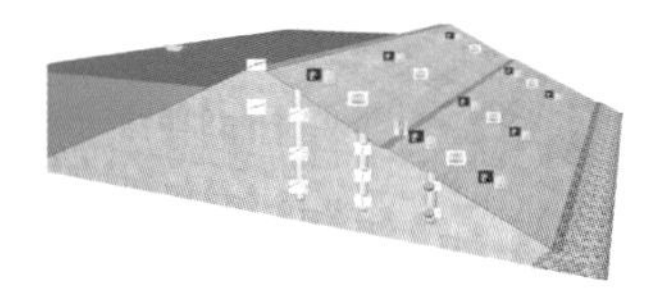

dam deformation survey

dam foundation

【释义】 坝基

【中文定义】 指堤坝的根基。包括河床和两岸放置坝体的部位，及其邻近承受坝体及水体等作用的部位。

dam leakage detection

【释义】 坝堤渗漏检测

【例句】 Artificial tracing methods have been widely used in dam leakage detection. 人工示踪方法已在堤坝渗漏检测中得到很多应用。

dangerous rock (group)

【释义】 危石（群），危岩

【例句】 Fragmentation and destruction of dangerous rock by blasting in the high and steep slope will bring about the flying-rock and rolling-rock. 高边坡上的危石爆破常常容易产生飞石和滚石。

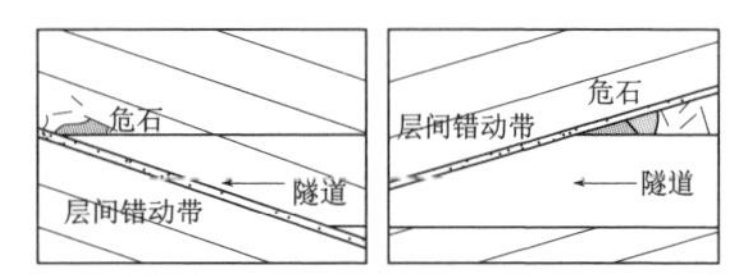

forming mechanism of dangerous rock around the interlayer shear belt

dangerous rock mass

【释义】 危岩体

【常用搭配】 dangerous rock mass deformation analysis 危岩体变形分析

【同义词】 potential unstable rock mass, dangerous rockmass

【例句】 Usually, there are densely distributed small dangerous rock mass in a low-angled stratofabric rock slope. 缓倾角层状岩质边坡是小危岩体出露的主要坡型之一。

dangerous rock mass

【中文定义】 危岩体指被多组不连续结构面切割分离，稳定性差，可能发生倾倒、坠落或滑塌等形式崩塌的岩体。

Danxia Landform

【释义】 丹霞地貌

【例句】 Danxia Landform is the special landform type characterized by red cliffs, which is highly valued in tourism development. 丹霞地貌是一种以赤壁丹崖群为特征的特殊地貌类型，具有很高的旅游开发价值。

【中文定义】 丹霞地貌是指以陆相为主（可能包含非陆相夹层）的红层（不限制红层年代）发育的具有陡崖坡的地貌，也可表述为“以陡崖坡为特征的红层地貌”。

Danxia Landform

Darcy Law

【释义】 达西定律（线性渗流定律）

【例句】 Darcy Law is used to analyze the stability of the existing slope after vainfall. 采用达西定律对现状边坡进行降雨后的稳定分析。

【中文定义】 描述饱和土中水的渗流速度与水力坡降之间的线性关系的规律，又称线性渗流定律。

dark mineral

【释义】 暗色矿物

dark mineral hornblende

【例句】 Very widespread, biotite gneisses are normally fine-grained with pronounced foliation caused by the parallel arrangement of alternating dark melanocratic ferromagnesian minerals and light leucocratic quartzo-feldspathic minerals. 黑云母片麻岩普遍明显具有的叶理结构（片麻理结构），一般是由暗色铁镁质矿物和亮色长英质矿物交替平行排列形成的。

【中文定义】 岩浆岩矿物中的黑云母、角闪石、辉石、橄榄石颜色较深，称为暗色矿物。

data acquisition

【释义】 数据采集，参数采集，信号采集

【常用搭配】 data acquisition system 数据采集系统

datum point

【释义】 基准点

【例句】 The cause and rule of deformation of datum point are analysized by data of repetition, and some useful conclusions and suggestions are also given. 根据复测数据分析了基准点的变形原因和规律，并得出了一些有益的结论和建议。

datum point

DC-electrical method

【释义】 直流电法

【例句】 In the South-North Water Diversion Project, weathered layer and buried structure grow complicatedly, so it's necessary to use seismic prospecting and DC-electrical method to improve the precision in explanation and the result reliability. 在南水北调工程中岩层风化带及隐伏构造的发育情况复杂，必须综合运用地震波法和直流电法来提高资料的解释精度和成果的可靠性。

debris flow

【释义】 泥石流

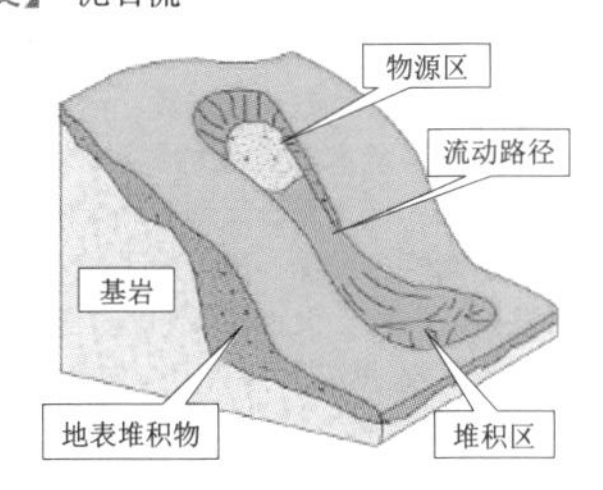

debris flow element

【中文定义】 由于降水在沟谷或山坡上产生的一种挟带大量泥沙、石块和巨砾等固体物质的特殊洪流。

debris flow fan

【释义】 泥石流堆积扇

【例句】 Debris flow fan is an important indication to discriminate tile characteristics, developmentand risk extent of debris flow. 泥石流堆积扇是判别泥石流性质、发育阶段和危险度的重要依据。

【中文定义】 泥石流冲出沟口后，固体物质堆积形成的扇形地貌。

debris flow fan

debris flow hazard

【释义】 泥石流灾害

【例句】 In the United States, landslide and debris flow hazards result in 25 to 50 deaths each year. 在美国，每年有25～50人死于滑坡与泥石流灾害。

debris flow hazard

debris flow investigation

【释义】 泥石流勘查

【中文定义】 在收集已有资料的基础上，对泥石流活动区域进行的有关泥石流形成、活动、堆积特征、发展趋势与危害等方面的各种实地调查、综合分析与评判，结合泥石流调查确定的防治工程方案，采用测绘、勘探、试验等手段，查明对应的可行性论证阶段、设计阶段和施工阶段防治工程所需要的工程地质条件的工作过程。

debris flow preventing

【释义】 泥石流防治

【例句】 The discharge of debris flow is a basic parameter in debris flow preventing and controlling engineering. It is very important for the hazardous evaluation of debris flow. 泥石流流量是泥石流防治工程设计的基本参数，对泥石流危害性的评价研究具有重要意义。

【中文定义】 指泥石流预防（避让、监测、预警预报）和工程治理。

debris flow terrace

【释义】 泥石流阶地

debris soil

【释义】 碎石土

【例句】 In the case of being difficult of drilling with common pneumatic drill, about $100m^3$ of debris soil is blasted out successfully by using sprung method through man-made setting up bulled holes. 在普通风动凿岩机钻孔困难的情况下，通过人工开设药孔，用药壶法成功爆破了$100m^3$的碎石质土壤。

【中文定义】 指粒径大于2mm的颗粒质量超过全重的50%的土。

debris soil

debris-gravelly soil material

【释义】 碎（砾）石土料

【中文定义】 碎（砾）石土料指粒径大于5mm颗粒的质量占总质量的20%～60%的宽级配砾类土。

debris-gravelly soil material

decomposition weathering

【释义】 风化分解

【中文定义】 由于风化作用引起岩石矿物成分的化学分解。

deceptive conformity

【释义】 假整合

【中文定义】 又称平行不整合，指同一地区新老两套地层间有沉积间断面相隔但产状基本一致的接触关系。

declivity survey

【释义】 倾斜测量

【例句】 The paper proposes a method of construction declivity survey based on the intelligent non-reverberation total station. 本文提出一种基于无反射棱镜全站仪的建筑物倾斜测量方法。

deep buried tunnel

【释义】 深埋隧道

【例句】 The geology condition of karst water bursting disaster in the course of deep buried tunnel is the technique foundation for the work of tunnel geology prediction. 岩溶山区深埋隧道施工突水灾害形成的地质条件是隧道施工地质超前预报工作的基础。

deep buried tunnel

deep seated landslide

【释义】 深层滑坡

【例句】 Under the condition of longer rainfall duration, the high stability response tends to occur at greater depth, which may lead to deep seated landslide. 在持续降雨的作用下，深层土的高度稳定性可能被影响，并引发深层滑坡。

deep-circulating hot water

【释义】 深循环热水

【例句】 The thermal sedimentation represents the abnormal sedimentation caused by exhalation of deep-circulating hot water on the background of normal sedimentation. 热水沉积事件是在正常沉积背景下深循环热水喷流插入的非正常沉积事件。

deep-focus earthquake

【释义】 深源地震

【例句】 A deep-focus earthquake in seismology is an earthquake with a hypocenter depth exceeding 300km. They occur almost exclusively at oceanic-continental convergent boundaries in association with subducted oceanic lithosphere. 在地震学中，深源地震是指震源深度超过 300km 的地震，几乎只发生在海洋大陆交汇的边界及下沉的岩石层。

【中文定义】 震源深度超过 300km 的地震，称为深源地震。到 2014 年为止，已知的最深的地震震源是 720km。深源地震约占地震总数的 4%，所释放的能量约占地震总释放能量的 3%。深源地震大多分布于太平洋一带的深海沟附近。深源地震一般不会造成灾害。

defects notification period

【释义】 缺陷通知期

deflection observation

【释义】 挠度观测

deflection survey

【释义】 挠度测量

【例句】 It demands no circumstance requirements in common optical axis and is especially effective inflexibility deflection survey of bridge and cloverleaf junction. 该方法无须共轴光路的测量环境要求，特别适用于桥梁、高速公路立交桥的静载挠度测量等工程应用。

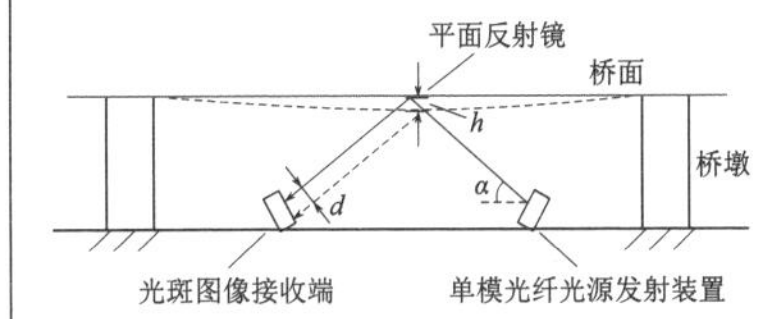

deflection survey

deformation [ˌdifɔr'meʃən]

【释义】 *n.* 变形，金属等（在压力作用下）变形

【常用搭配】 elastic deformation 弹性变形

【例句】 Depending on the intensity of the deformation the folds may be gentle, sharp, overturned, or overthrust. 根据变形的强度，褶皱可以是平缓的、陡峭的、倒转的或逆掩的。

【中文定义】 物体受外力作用而产生体积或形状的改变。

deformation analysis

【释义】 变形分析法

【中文定义】 变形分析法：采用数值分析方法，研究边坡的应力场和变形规律，评价边坡的稳定性的方法。

deformation bedding

【释义】 变形层理

【例句】 In the middle stage, with the deepening of the sea water, horizontal bedding, deformation bedding and convolution bedding developed, suggesting a platform-marginal slope facies. 中期海水加深，发育了水平层理、变形层理及包卷层理等，近于台地边缘斜坡相。

【中文定义】 岩石受力后发生塑性变形，形成弯曲皱纹的层面。

deformation bedding

deformation measurement

【释义】 变形测量

【常用搭配】 dam deformation measurement 大坝变形测量

【例句】 During the deformation measurement of welding contraction for stator in a large hydropower station, the method is used well. 在对某大型水电站焊接收缩变形测量中，该方法得到了很好的应用。

deformation modulus

【释义】 变形模量

deformation modulus

【例句】 The deformation modulus of rock mass is an important input parameter in analyzing deformation, which should be determined by field tests. 在分析岩体的变形特性时，岩体变形模量是一个非常重要的参数，一般要通过现场试验来确定。

【中文定义】 土的变形模量是通过现场载荷试验求得的压缩性指标，即在部分侧限条件下，其应力增量与相应的应变增量的比值。

deformation monitoring

【释义】 变形监测

【常用搭配】 dam deformation monitoring 大坝变形监测

【例句】 This paper presents the process of landslide monitoring and the design of deformation monitoring in Kala reservoir area. 本文介绍了卡拉电站库区滑坡监测数据处理过程和变形监测的设计。

deforming slope

【释义】 变形边坡

【中文定义】 已经变形或正在发生变形的边坡。

degree of compaction

【释义】 *n.* 紧密，密实度，紧密度

【例句】 Degree of compaction and strength of clay subgrade can be determined by ball and drop test. 土基的压实度和强度可用落球仪快速测定。

degree of consolidation

【释义】 固结度

【例句】 The degree of consolidation of soils is the function of water discharging distance, duration and coefficient of consolidation. 土的固结度是排水距离、固结时间和固结系数的函数。

degree of dryness

【释义】 干度，干燥度

【例句】 During the process, the pulp sheet department adopts interval cleaning and online cleaning method to increase the pulp sheet degree of dryness, decrease the steam consumption and save the produce cost. 在生产过程中，浆板机压榨部采用停机清洗和在线清洗相结合的方式，以提高湿浆板干度，降低蒸汽用量，节约生产成本。

【中文定义】 是指每千克湿蒸汽中含有干蒸汽的质量百分数。

degree of safety

【释义】 安全度

【例句】 Reliability theory considered the randomness and variability, and measured the degree of safety of slope by strict probability. 可靠性理论通过精确的概率来考虑其随机性和变异性，并计算出边坡的安全度。

degree of saturation

【释义】 饱和度

【例句】 However, this experimental study shows

that, for fine-grained cohesionless soils, degree of saturation can not be overlooked. 然而本文试验研究表明，对于细粒无黏性土，饱和度不能忽视。
【中文定义】 土中孔隙水体积与孔隙体积的比值。

degree of weathering
【释义】 风化度
【例句】 A good correlation exists between amounts of joints, block size in rockmass and degree of weathering. 岩体中裂隙数量、岩体块度的变化与岩体的风化程度具有较好的对应关系。
【中文定义】 是风化作用对岩体的破坏程度，它包括岩体的解体和变化程度及风化深度。

deloading [diːˈloudɪŋ]
【释义】 *n.* 减载
【常用搭配】 cutting and deloading 削坡减载
【例句】 Wall cutting and deloading is a task of top priority of comprehensive harness in the north slope of Jinduicheng open pit stope. The vibration by blasting influenced the stability of slope. 削坡减载是露天采场北部边坡综合治理的当务之急，在边坡上进行爆破作业产生震动，对边坡稳定有一定影响。

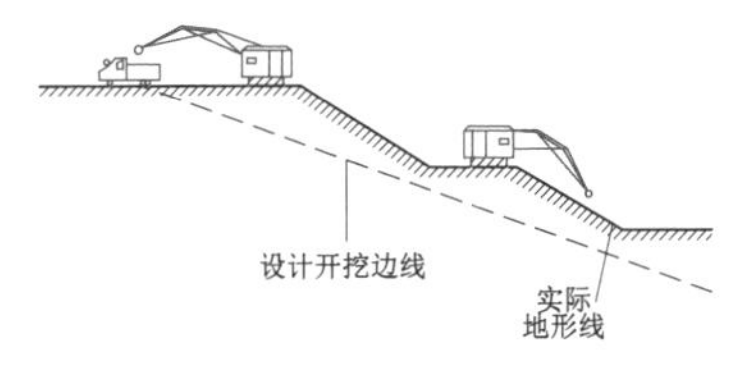

deloading

dense degree
【释义】 密实度

densimeter method
【释义】 密度计法

densimeter method

【例句】 The testing result of densimeter method is credible. 密度计法的试验结果是可靠的。

density [ˈdɛnsəti]
【释义】 *n.* 密度
【例句】 Mercury has a much greater density than water. 水银的密度比水大得多。
【中文定义】 单位体积所具有的质量。

density test
【释义】 密度试验
【例句】 Maximum dry density test of oversized earth-rock mixture is studied in this paper. 本文对超粒径土石混合料最大干密度进行了试验研究。

density test

denudation [ˌdiːnjuːˈdeiʃən]
【释义】 *n.* 剥蚀，剥蚀作用
【常用搭配】 circum denudation 环蚀；plain of denudation 剥蚀平原

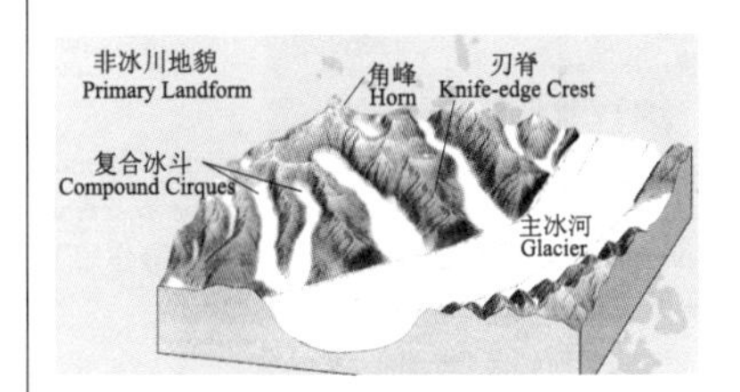

denudation

【例句】 Relaxed zone of rock mass is the epigenetic broken or loose range formed by the stress release caused by erosion, denudation and excavation. 围岩松弛区是指由于侵蚀、剥蚀和开挖造成应力释放而形成的后生破坏或松散范围。
【中文定义】 剥蚀作用就是指各种运动的介质

（如流水、风等）在其运动过程中，使地表岩石产生破坏并将其产物剥离原地的作用。

denudation monadnock

【释义】 剥蚀残丘

【中文定义】 低山在长期的剥蚀过程中，大部分的山地都被夷平为准平原，但在个别地段形成了比较坚硬的残丘，称为剥蚀残丘。

denudation monadnock

denudation peneplain

【释义】 剥蚀准平原

【中文定义】 剥蚀准平原是低山经过长期的剥蚀和夷平，外貌显得更为低缓平坦，具有微弱起伏的地形，其分布面积一般不大。

denudation peneplain

deposit body

【释义】 堆积体

【例句】 It is indicated that deformation mainly occurred in the deposit body while the reservoir level fluctuating. 结果表明，在库水升降作用下变形主要发生在堆积体内部。

deposit body

【中文定义】 是在陡坡下方或洞穴底部由于重力作用而堆积的物质。

depression [dɪ'prɛʃən]

【释义】 *n*. 洼地，凹陷，低压区

【常用搭配】 karst depression 溶蚀洼地；depression spring 低地泉；seismic depression 震陷

【例句】 In engineering geological investigation, factors of soft soil, such as bearing capacity, deformation, strength, seismic depression, and its suitability as natural foundation shall be specifically studied. 工程地质勘察中对软土的承载、变形、强度和震陷特征及其作为天然地基的适宜性，应予以专门研究。

【中文定义】 近似封闭的比周围地面低洼的地形。

depression landform

depression cone

【释义】 降落漏斗

【例句】 In reality, the depression cone will continue to expand until the recharge of the aquifer, if any, equals the discharge. 实际上，降落漏斗会持续扩大到补给水量等于排放水量的含水层位置。

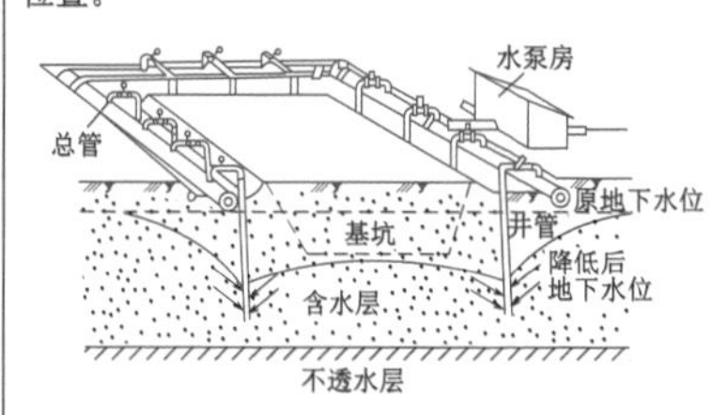

depression cone

depression cone method

【释义】 降落漏斗法

depression curve

【释义】 降落曲线

【中文定义】 渗透水流沿流向因摩擦损失不断产生水位下降，所以潜水面和承压水的测压水面总是带有一定坡度的曲面。曲面与沿流向方向的剖面相交的曲线，称降落曲线。

depth of foundation

【释义】 基础埋置深度

【中文定义】 基础埋于土层的深度，一般指从室外地坪至基础底面的垂直距离。

depth unloading

【释义】 深卸荷

【常用搭配】 deep unloading zone 深卸荷带

【中文定义】 深卸荷是谷坡山体中储存应变能的岩体在河谷下切演化过程中卸荷回弹而形成的破裂结构。

descending spring

【释义】 下降泉

【中文定义】 下降泉是由潜水或上层滞水形成的泉水。

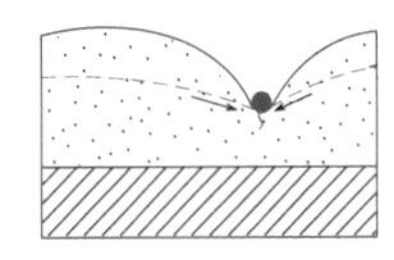

Schematic diagram of descending spring

desert ['dɛzət]

【释义】 *n*. 沙漠，荒漠

【常用搭配】 desert island 荒岛；gobi desert 戈壁滩；desert lake 沙漠湖

【例句】 The zonation of vegetations in Xizang changes from the southeast to the northwest as follows: forest-meadow-steppe-desert. 西藏植被成带现象自东南向西北的变化如下：森林—草甸—草原—荒漠。

【中文定义】 主要是指地面完全被沙所覆盖、植物非常稀少、雨水稀少、空气干燥的荒芜地区。

desert

design [dɪ'zaɪn]

【释义】 *v*. 绘图，制图，设计

design earthquake

【释义】 设计地震

【例句】 The modification was due to the concern with adequacy of the dam when subjected to the design earthquake or overtopping by large floods. 这次改造设计目的是使它能经受住达到设计震级的地震和巨大洪水漫顶。

【中文定义】 设计地震是实际结构设计中工程师使用的术语，相当于地质学家和地震学家建议的“最大地震”。

design factor of safety

【释义】 设计安全系数

design flood level

【释义】 设计洪水位

【例句】 Design flood level is one of characteristic water levels of reservoir. 设计洪水位是水库特征水位之一。

【中文定义】 当遇到大坝设计标准洪水时，水库经调洪后（坝前）达到的最高水位，称为设计洪水位。

design stage for landslide prevention project

【释义】 滑坡防治工程设计阶段

destructive earthquake

【释义】 破坏性地震

【例句】 The region is seismically active and has been subject to destructive earthquakes in the past, most recently in 1963 when Skopje was heavily damaged by a major earthquake. 斯科普里地震频发且历史上已经历多次破坏性地震，最近的一次是在 1963 年，遭到地震的严重破坏。

【中文定义】 震级大于 5 级，造成一定的人员伤亡和建筑物破坏或造成重大的人员伤亡和建筑物破坏的地震灾害称为破坏性地震。

detailed foundation design

【释义】 详细基础设计

detailed geological recording

【释义】 地质编录

【同义词】 geological logging

【例句】 Detailed geological mapping and recording is one of the main tasks in the construction geology of water conservancy projects. 地质测绘与编录是水利水电工程施工地质工作的一项主要内容。

detailed investigation

【释义】 详查

【例句】 Based on detailed investigation information of subgrade in Qingshuihe experimental section, the occurrence reason of cracking and classification of the cracking are studied. 通过清水河试验段路基裂缝

的详查资料，分析了裂缝产生的原因，并对裂缝进行了分类。

detection of rock mass relaxation

【释义】 岩体爆破松弛检测

detection of rock mass unloading zones

【释义】 岩体卸荷带探测

deterministic seismic hazard analysis (DSHA)

【释义】 地震危险性确定性分析法

【中文定义】 根据历史地震重演和地质构造外推的原则，利用区域历史地震活动特征、地震地质构造背景、地震烈度衰减关系等资料，估计某一区域未来遭遇的地震烈度水平，并以确定的数值来表达。

developed [dɪ'vɛləpt]

【释义】 *adj.* 发育的，发达的

【中文定义】 根据《水力发电工程地质勘察规范》的规定，当岩体内结构面发育组数2～3组，间距小于10cm时，其岩体结构面发育程度为发育。

developed

developed plan of exploratory adit

【释义】 探洞展示图

【例句】 Developed plan of exploratory adit is a kind of drawing which shows the geological conditions in a adit, and it is also the basic data for analysing and evaluating engineering geological problems. 探洞展示图是反应平洞中的地质情况的图件，也是分析评价工程地质问题的基本资料。

【中文定义】 依一定比例尺和图例，按平面连续展开的方式，根据勘探平洞中揭露的工程地质、水文地质现象和岩体原位试验成果编制而成的原始图件。

development degree of discontinuities

【释义】 结构面发育程度

【中文定义】 岩体中各种结构面发育的密集程度，多通过结构面组数、结构面间距或岩石质量指标（RQD）等指标进行划分。

deviation coefficient

【释义】 离差系数

Devonian (D) [dɪ'voniən]

【释义】 *n.* 泥盆纪（系）

【中文定义】 古生代的第四个纪，约开始于4.05亿年前，结束于3.5亿年前，持续约5000万年。从泥盆纪开始，地球发生了海西运动，许多地区升起，露出海面成为陆地，古地理面貌与早古生代相比有很大的变化。

dextral rotation fault

【释义】 右旋断层

【同义词】 right handed fault

【例句】 It is preliminarily believed that the seismogenic structure of the Jinggu earthquake is a newly generated nearly vertical dextral rotation fault. 初步认为景谷地震的发震构造为一条新生的近直立右旋断层。

【中文定义】 一般走滑断层都会伴生一系列雁列式断层，有的直接与主断层相交，有的不相交，察看伴生断层或者伴生断层延长线与主断层相交锐角所指方向，即为断层的走滑性质，逆时针方向为左旋，顺时针方向为右旋。

diabase ['daɪəˌbes]

【释义】 *n.* 辉绿岩

【同义词】 dolerite ['dɒləraɪt]

【例句】 The gabbro and diabase porphyry belong to the sub-alkaline series, and they both locate in the mafic accumulate region in the ACM diagrams. 辉长岩和辉绿玢岩均属于亚碱性系列，在ACM图解中均位于镁铁质堆积岩区域。

diabase

diagenesis [ˌdaɪə'dʒenɪsɪs]

【释义】 *n.* 成岩作用，岩化作用

【例句】 The main diagenesis types of sandstones are the mechanical compression, cementation and dissolution. 砂岩主要的成岩作用包括机械压实作用、胶结作用、溶蚀溶解作用。

【中文定义】 成岩作用是形成岩石的各种地质作用的统称。如岩浆成岩作用、变质成岩作用、沉积成岩作用、花岗岩化作用、混合岩化作用等。

diagram ['daɪəgræm]

【释义】 *n.* 图表，图解；*vt.* 用图解法表示

【常用搭配】 general arrangement diagram 总体布置图；mimic diagram 模拟图；scatter diagram 散布图、散点图、散形图

【同义词】 chart，graph

【例句】 Stability design diagrams for replace with out - soil are obtained after discussion on parameters of the equation. 通过讨论算式中的参数，给出客土的稳定设计图。

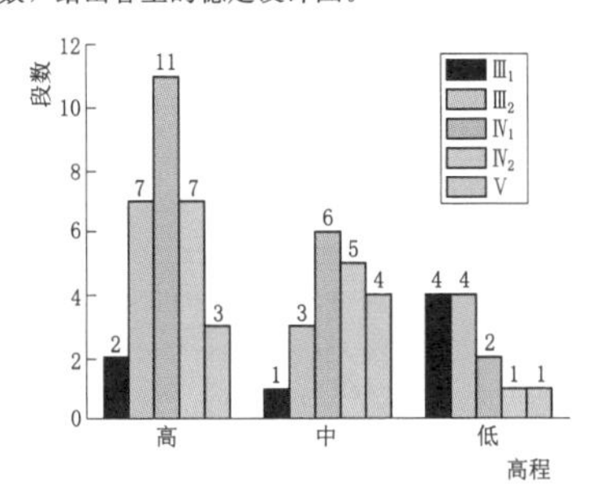

scatter diagram

diamond ['daɪəmənd]

【释义】 *n.* 金刚石

【例句】 As ninny unusual minerals including the typical press index minerals such as diamond and coesite have been discovered from Luobusa chromitite in the Yarlung Zangbo ophiolite belt, Tibet. 前人在雅鲁藏布江蛇绿岩带的罗布莎铬铁矿石中发现许多异常矿物，包括金刚石和柯石英等典型压力指示矿物。

diamond

diamond drilling

【释义】 金刚石钻进

【例句】 Diamond drilling is a kind of advanced way in drilling engineering. 金刚石钻进是钻探工程中一种比较先进的钻进方法。

dielectric constant

【释义】 介电常数

【例句】 The dielectric constant of the material is closely related to the frequency of the test. 材料的介电常数值与测试的频率密切相关。

differential pressure method

【释义】 差压法

【例句】 Differential pressure method is a rapid and accurate testing method. 压差法是一种快速、准确的测试方法。

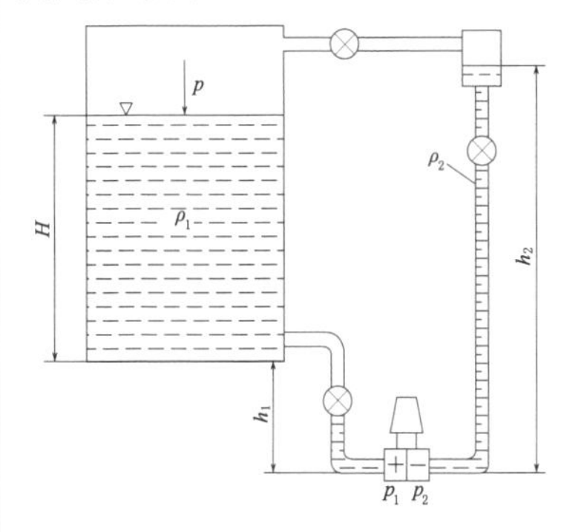

differential pressure method

differential settlement

【释义】 差异沉降

【常用搭配】 allowable differential settlement 容许不均匀沉降；differential settlement of foundation 地基不均匀沉降差；differential settlement of subgrade 路基不均匀沉降差

differential settlement

【例句】 Because of the stress diffusivity, reinforced sand blankets decreased the differential settlement on the bottom of the embankment. 由于砂垫层的应力扩散作用，加筋砂垫层减小了堤底的不均匀沉降。

【中文定义】 同一结构物基底的不同部位在同一时刻的沉降量的差值。

differential unloading rebound

【释义】 差异卸荷回弹

【中文定义】 卸荷作用引起卸荷面附近岩体内部应力重分布，造成局部应力集中效应，并且在卸荷回弹变形过程中还会因差异回弹而在岩体中形成一个被约束的残余应力体系。应力状态这两方面的变化所引起岩体在卸荷过程中的变形与破坏，即差异卸荷回弹。

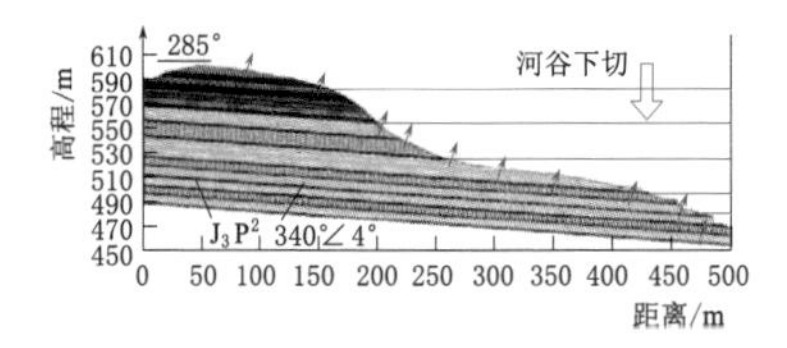

differential unloading rebound stage

differential uplift

【释义】 差异性隆升

【例句】 Emplacement depths estimated from aluminum-in-hornblende geobarometry indicate that the differential uplift and exhumation between the IMPU and the Yanshan fold-and-thrust belt was distinct during the Late Paleozoic to Early Mesozoic, but not distinct since the Early Jurassic. 运用斜长石-角闪石温压计测定的花岗质侵入岩的侵入深度表明，晚古生代-早中生代期间，在内蒙古隆起及燕山褶断带之间，存在有强烈的差异性隆升及剥露过程，但在早侏罗世以来表现得不明显。

differential weathering

【释义】 差异风化，分异化风化

【常用搭配】 differential weathering pothole 差异风化壶穴

differential weathering

【例句】 Rock cells formed by the effects of differential weathering of lithology of cliffs or steep slopes are the cause for collapse of perilous rock. 陡崖或陡坡下岩性差异风化形成的岩腔是危岩崩落的起因。

【中文定义】 由于组成岩石的矿物成分、结构构造的差异，不同岩石的风化速度和风化程度不同。在相同的风化条件下，常常在地表形成凹凸不平的地貌。

digital aerial meter

【释义】 数字航拍仪

digital cartography

【释义】 数字地图制图

【中文定义】 实现数字地图的设计、生产、管理与应用的技术与方法。

digital elevation model (DEM)

【释义】 数字高程模型（DEM）

【例句】 In TIN of digital elevation model constructed from contours, flat areas which are made up of flat triangles can not reflect the real shape of the surface, so that they need be corrected properly. 在基于等高线建立的数字高程模型 TIN 中，由平坦三角形连接成的平坦区域不能真实地反映地表的真实形状，需要进行适当地修正。

【中文定义】 数字高程模型是描述地表起伏形态特征的空间数据模型，由地面规则格网点的高程值构成的矩阵，形成栅格结构数据集。

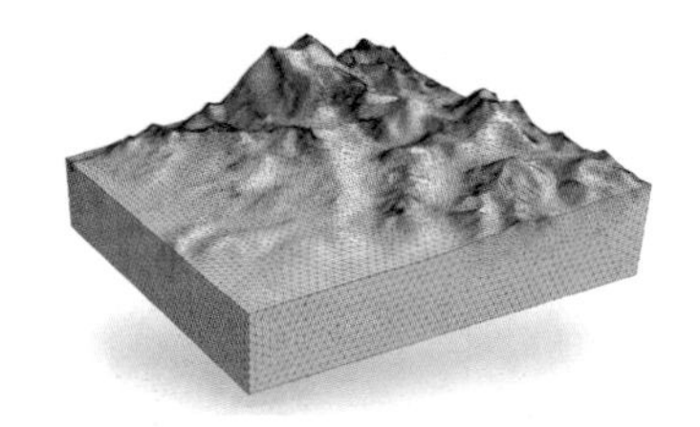

DEM

digital line graphic

【释义】 数字线划地图

【中文定义】 是现有地形图要素的矢量数据集，保存各要素间的空间关系和相关的属性信息，全面地描述地表目标。

digital map

【释义】 数字地图

【中文定义】 以数字形式存储在磁带、磁盘、光盘等介质上的地图。

digital orthophoto map

【释义】 数字正射影像图

【例句】 Combined with the practical engineering of Guilin, the manufacturing process of digital orthophoto map (DOM) and the quality control method are introduced in this paper. 结合桂林市数字正射影像图生产实践，介绍了数字正射影像图的生产工艺流程及质量控制方法。

【中文定义】 是利用数字高程模型（DEM）对经扫描处理的数字化航空相片，经逐像元进行投影差改正、镶嵌，按国家基本比例尺地形图图幅范围剪裁生成的数字正射影像数据集。它是同时具有地图几何精度和影像特征的图像，具有精度高、信息丰富、直观真实等优点。

digital raster graphic

【释义】 数字栅格地图

【例句】 Digital raster graphic(DRG) and digital line graphic(DLG) are two digital maps that can be taken cheaply, and DLG is a commonest data format now, which the main data source of GIS. 数字栅格图和数字线划图是可以低成本获取的两种数字地图，而DLG是目前最普遍的矢量数据格式，是GIS的主要数据来源。

【中文定义】 是现有纸质地形图经计算机处理后得到的栅格数据文件。每一幅地形图在扫描数字化后，经几何纠正，并进行内容更新和数据压缩处理，彩色地形图经色彩校正，使每幅图像的色彩基本一致。数字栅格地图在内容上、几何精度和色彩上与国家基本比例尺地形图保持一致。

digitized mapping

【释义】 数字化测图

【例句】 The main method for presenting high quality digitized mapping products is comprehensive, reasonable and quantitative evaluation. 全面合理、定量评价数字化测图的质量是确保提供高质量数字地图产品的主要方法。

digitized mapping

dike [daɪk]

【释义】 *n.* 岩墙

【常用搭配】 clastic dike 碎屑岩墙；ring dike 环状岩墙

【例句】 Dikes are discordant masses that are produced when magma is injected into fractures. 岩墙是岩浆沿裂隙贯入形成的板状不整合侵入体。

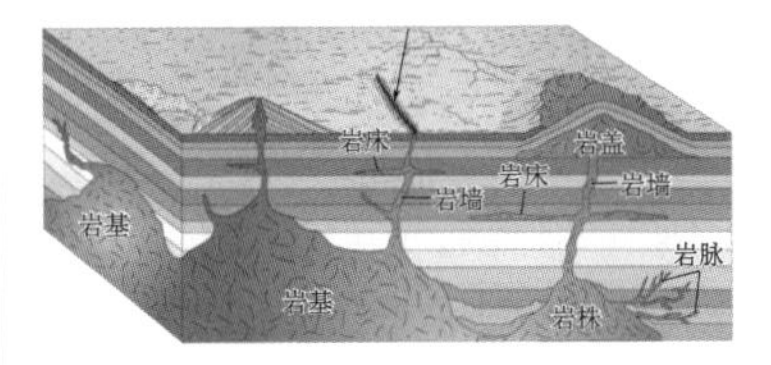

dike

dilatancy [daɪ'letnsi]

【释义】 *n.* 剪胀性

【例句】 Moreover, the paper discussed the influence of dilatancy on shear strength and poisson's ratio. 本文讨论了剪胀性对强度与泊松比的影响。

diluted debris flow

【释义】 稀性泥石流（紊流型泥石流）

【中文定义】 浆体由不含或少含黏性物质组成的泥石流。

diluvial soil

【释义】 洪积土

【例句】 At home and abroad, the native soil, the eolian soil, the moraine, the diluvial soil, the age-old river sediments and the recent river shoal sediments were researched as material core for ECRD. 国内外实践中曾经用过残积土、风成土、冰渍土、洪积土、古河川沉积及近代河滩沉积土作为土石坝防渗体的心墙土料。

diluvial soil

【中文定义】 由暂时性洪流将山区高地的碎屑物质携带至沟口或平缓地带堆积形成的土。

diorite [ˈdaɪəˌraɪt]

【释义】 *n.* 闪长岩

diorite

【例句】 Being one of the largest gold deposits in Ailaoshan gold belt, the Daping gold deposit is hosted in a ductile deformed and altered diorite batholith. 云南大坪金矿床是哀牢山金矿带中最大的金矿之一，主要赋存在受到强烈剪切和水-岩反应的闪长岩中。

dip [dɪp]

【释义】 *n.* 倾角

【常用搭配】 dip angle 倾角；true dip 真倾角；apparent dip 视倾角

【例句】 Generally speaking, the smaller the dip angle is, the more stable the slope is. 一般而言，岩层倾角越小，边坡越稳定。

【中文定义】 面状构造的产状要素之一。即在垂直地质界面走向的横剖面上所测定的此界面与水平参考面之间的两面角。也就是倾斜线与其水平投影线之间的夹角。这个倾角又称“真倾角”。

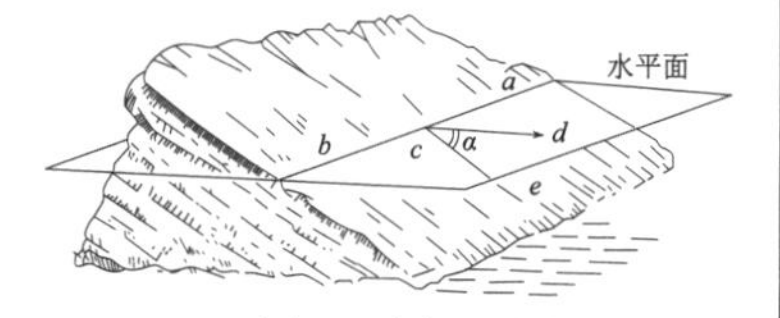

ab—走向；*cd*—倾向；*α*—倾角

dip (see **α**)

dip angle of drilling hole

【释义】 钻孔倾角

【例句】 Inclinometer is an instrument testing dip angle and azimuth of drilling hole. 测斜仪是一种测定钻孔倾角和方位角的仪器。

dip direction

【释义】 倾向

【常用搭配】 dip direction angle 倾向角；dip direction structure 倾向构造

【例句】 The paleocurrent direction can be deduced by dip direction of cross laminae of glutenite interpreted on images. 根据图像解释的砂砾岩交错层理倾向，可推断古水流方向。

【中文定义】 地质构造面由高处指向低处的方向。用构造面上与走向线垂直并沿斜面向下的倾斜线在水平面上的投影线（称作倾向线）表示。

dip fault

【释义】 倾向断层

【例句】 Dip fault of Longyong coal-fied is studied. 对龙永煤田之倾向断层进行研究。

【中文定义】 倾向断层是指断层走向与被断岩层走向基本直交的断层。

direct shear test

【释义】 直剪试验

【常用搭配】 field direct shear test 现场直剪试验

【例句】 The direct shear test on rigid presser was carried out to examine the load-deformation curve. 在刚性压力机上进行了直剪试验，以求得荷载变形全过程曲线。

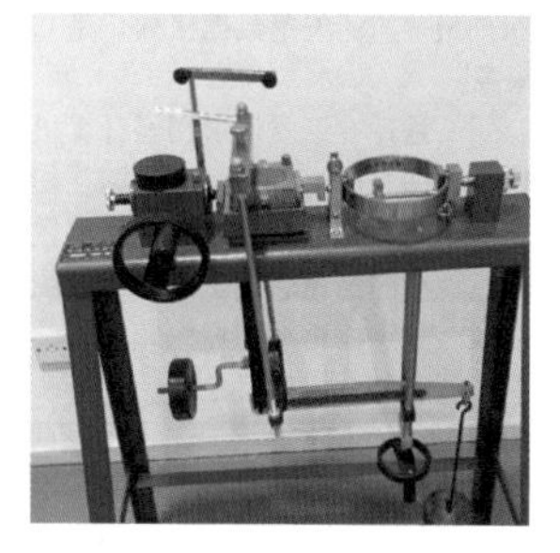

direct shear test

direct shear test of discontinuity plane

【释义】 结构面直剪试验

【例句】 On the base of engineering application, the select principle of method of direct shear test of discontinuity plane is discussed. 从工程实际出发，讨论了岩体结构面直剪试验方法的选取原则。

direction angle

【释义】 方向角

【常用搭配】 flying direction angle 飞行方向角；initial direction angle 起始方向角；breaking direction angle 断裂方向角

【例句】 The locating precision of direction of arrival is increased by improving the MUSIC and ESPRIT algorithm. 通过改进 MUSIC 算法和 ESPRIT 算法，提高了目标方向角的定位精度。

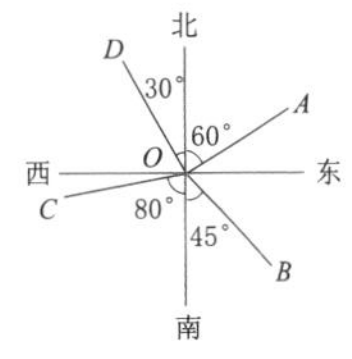

direction angle

【中文定义】 指采用某坐标轴方向作为标准方向所确定的方位角。方向角是从正北或正南方向到目标方向所形成的小于 90°的角。

disaster guarding preparatory

【释义】 防灾预案

【例句】 In consideration of deficiencise existing in China's disaster guarding preparatory, some suggestions are presented. 针对中国防灾预案中存在的不足提出建议。

disaster of surface collapse

【释义】 地面塌陷灾害

【例句】 Especially in recent years as the mining develops, human activities have been the main factor leading to geological hazards, for example, the disaster of surface collapse, the earth crevice, etc. 特别是近年来，随着开矿业的发展，人类活动已成为诱发地质灾害的主要因素，比如地面塌陷、地裂缝等。

disaster of surface collapse

discharge rate

【释义】 排泄量

【例句】 Groundwater discharge rates varied spatially and temporally. 地下水排泄量随时空变化而不同。

discontinuity [ˌdɪsˌkɑːntəˈnuːəti]

【释义】 *n.* 结构面

【常用搭配】 secondary discontinuity 次生结构面

【例句】 The geological formation movement produced many discontinuities with different characteristic. 岩体伴随地质构造运动产生了特征不同的结构面。

【中文定义】 指具有极低的或没有抗拉强度的不连续面。包括一切地质分离面。不同结构面的，力学性质不同，规模大小不一。

discontinuity strength

【释义】 结构面强度

【例句】 The discontinuity strength has an important influence on the stability of rock mass. 结构面强度对于岩体的稳定性具有重要的影响。

【中文定义】 结构面强度是指岩体中结构面在外力作用下抵抗破坏的能力。

discrete element method (DEM)

【释义】 离散元分析

【例句】 Combined with the discrete element method for deformable bodies, the technique has been well applied to the water intake high slope in Baihetan Hydropower Station. 该技术与三维可变形离散元分析法相结合，在白鹤滩水电站进水口高边坡三维可视化和应力变形分析中得到很好的应用。

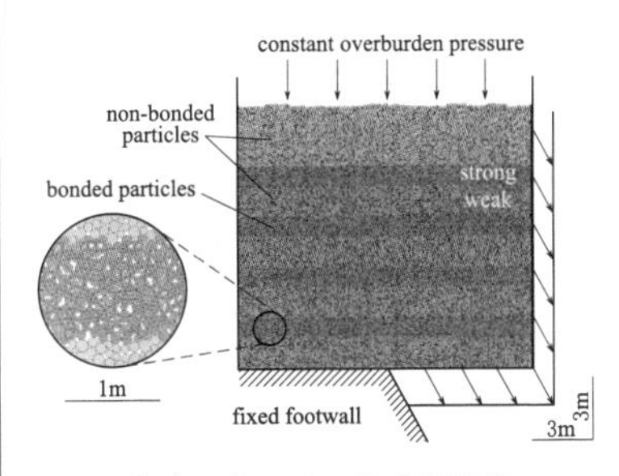

discrete element method (DEM)

discrete smoothing interpolation (DSI)

【释义】 离散平滑插值 (DSI)

【例句】 For the regional geology with complex stratigraphic structure, Discrete Smoothing Interpolation Based(DSI-Based) algorithm for three-dimensional geological modeling simplifies and effective the interpolationprocess, also can be used for non-continuous state of the interpolation. 针对地层数相对多的地质区域，基于离散平滑 (DSI) 插值改进算法的三维地质建模方法，使

得插值过程更为简单有效，并可用于非连续状态的插值。

【中文定义】 由法国南锡大学 J. L. Mallet 教授提出的，针对离散化的地质体模型，建立相互之间具有完备空间拓扑关系的网络，并且网络上的节点满足一定的约束条件后，通过解线性方程插值计算未知结点上值的一种插值算法。

disintegration [dɪsˌɪntɪ'greʃən]

【释义】 *n.* 崩解，瓦解，分解

【例句】 The testing results indicate that this mud does not behave in remarkable disintegration. Then the disintegration does not play an important role for the slope failure. 研究结果表明，该套泥岩崩解性并不大，对其边坡崩落的影响有限。

disintegration

disintegration weathering

【释义】 风化破碎

【中文定义】 由于风化作用引起岩石结构的解体破碎。

disintegration weathering

disintegration-resistance test of rock

【释义】 岩石耐崩解性试验

【例句】 The paper explores the suitability of using the disintegration-resistance test of rock, originally developed for argillaceous rocks, to evaluate the durability of low-cost walling materia. 浸泡耐崩解性试验原本是为泥质岩石开发的一种试验方法，本文旨在探索这种试验方法对预测低成本墙体材料耐久性的适用性。

disk-like rock core

【释义】 饼状岩芯

【例句】 One of the formation condition of disk-like rock core is a maximum principal stress of more than 30MPa. 饼状岩芯的形成条件之一是具有 30MPa 以上的最大主应力。

【中文定义】 岩芯饼化是一种岩体力学现象，高地应力区所特有的钻进过程中岩芯裂成饼状的现象。

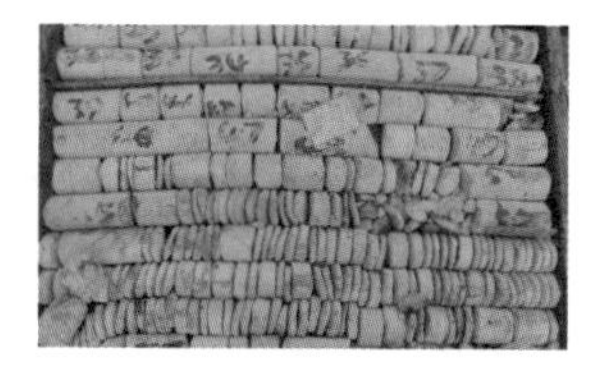

disk-like rock core

dispersive clay

【释义】 分散性黏土

【例句】 Dispersive clay has a very low erosion resistance capacity so as to cause serious threatening to the engineering safety such as hydraulic engineering and road engineering. 分散性黏土抗水蚀能力很低，对水利工程、道路工程等建筑物的安全造成严重威胁。

【中文定义】 指土中所含黏性土颗粒在水中散凝呈悬浮状，易被雨水或渗流冲蚀带走引起破坏的土。

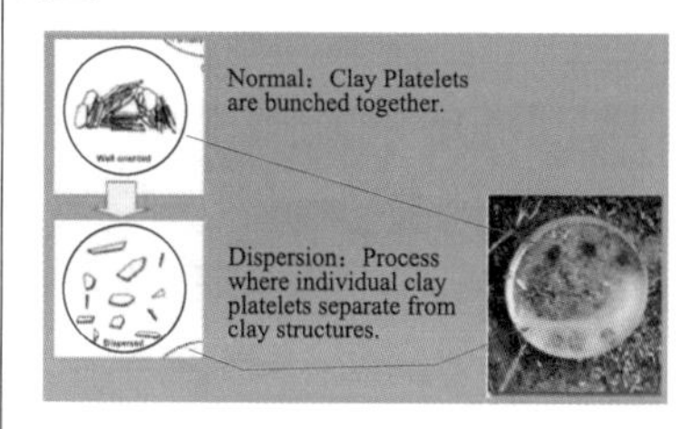

identification of dispersive clay

dispersive soil

【释义】 分散性土，散粒土

【例句】 This paper analyzes the origin of solution cavern of the dispersive soil dyke and points out the treatment measures utilizing complex geo membrane and lime soil to deal with dispersive soil dyke. 本篇论文分析了分散性土堤坝溶洞成因，提出了复合土工膜和白灰土处理分散性土堤

坝的措施。
【中文定义】 分散性土是土在低含盐量水中（或纯净水中）细颗粒之间的黏聚力大部分甚至全部消失，呈团聚体存在的颗粒体自行分散成原级的黏土颗粒的土。

dispersivity [dɪspɜːˈsɪvɪtɪ]
【释义】 *n.* 分散性
【例句】 The system disposition is characterized by good physical dispersivity, high reliability, light load for main machine, high precision, less construction cost and so on. 该配置具有物理分散性好、可靠性高、主机负荷轻、精度高、施工费用少的特点。
【中文定义】 分散性固体粒子的絮凝团或液滴，在水或其他均匀液体介质中，能分散为细小粒子悬浮于分散介质中而不沉淀的性能。

dissolution [ˌdɪsəˈluːʃn]
【释义】 溶解作用
【例句】 Surfactants can work in three different ways: roll-up, emulsification and dissolution. 表面活性剂可通过三种方式起作用：卷缩作用，乳化作用和溶解作用。
【中文定义】 指水溶液溶解岩石的某些易容成分，使其松软、破碎、崩解的过程。

dissolved oxygen
【释义】 溶解氧
【例句】 A fluorescence method is used to measure the partial pressure of dissolved or gaseous oxygen. 使用荧光方法测量溶解氧和气态氧局部压力。
【中文定义】 溶解在水中的空气中的分子态氧称为溶解氧，水中的溶解氧的含量与空气中氧的分压、水的温度都有密切关系。

schematic diagram of dissolved oxygen

distribution area
【释义】 分布区
【例句】 The line-structure node and color anomaly of the Archoeozoic beds distribution area are the key signal of remote information for gold exploration. 太古宙地层分布区的线性构造节点、色异常是找矿的重要遥感信息标志。

distribution map of natural construction material sources
【释义】 天然建筑材料分布图
【例句】 In planning stage, prefeasibility study stage and feasibility study stage of hydropower engineering, distribution maps of natural construction material sources shall be submitted in report for engineering geological investigation. 水力发电工程规划阶段、预可行性研究阶段和可行性研究阶段工程地质勘察报告中应附天然建筑材料分布图。
【中文定义】 反映天然建筑材料产地分布地理位置及料源基本概况的图件。

disturbed soil sample
【释义】 扰动土样，非原状土样

doline [dəˈliːnə]
【释义】 溶斗
【中文定义】 曾称喀斯特漏斗，指岩溶区呈碟状、漏斗状、井状洼地。在可溶岩裂隙发育的地区由于溶蚀作用或地下溶洞洞顶塌陷而成。

doline

dolomite [ˈdoʊləˌmaɪt]
【释义】 *n.* 白云石，白云质大理岩；白云（灰）岩
【常用搭配】 dolomite brick 白云石砖；dolomite limestone 白云石质石灰石，白云石质灰岩，白云灰岩；baroque dolomite 变态白云岩

dolomite

【例句】 Geological units containing materials such as limestone, dolomite, and gypsum may have natural solution cavities. 石灰岩、白云岩、石膏等分布地区可能发育有天然溶洞。

【中文定义】 白云石是碳酸盐矿物，分别有铁白云石和锰白云石。它的晶体结构像方解石，常呈菱面体。遇冷稀盐酸时会慢慢出泡。有的白云石在阴极射线照射下发橘红色光。白云石是组成白云岩和白云质灰岩的主要矿物成分。

dome [dəum]

【释义】 *n.* 穹隆

【常用搭配】 dome structure 穹隆构造

【例句】 The ore deposit displays a dome structure, bounded by the Pijiang fracture eastward. 矿床呈一穹隆构造，东界为批江断层。

【中文定义】 是一种由对称的背斜组成的变形地质构造，在地质图上一般以圆形或卵形表示。

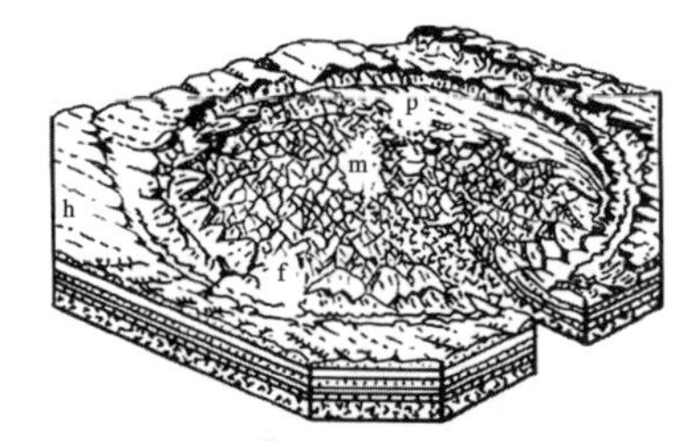

dome structure

down-the-hole hammer

【释义】 潜孔锤

【例句】 Down-the-hole hammer drilling is one of the high-efficieng drilling methods in hard rocks. 潜孔锤钻进是硬岩钻进效率较高的方法之一。

down-the-hole hammer

downstream view

【释义】 下游立视图

【常用搭配】 downstream view of dam 大坝下游立视图

【中文定义】 逆水流方向的视图称为下游立视图。

drag fold

【释义】 拖拽褶皱，牵引褶皱

【例句】 The structural style found in the basin includes imbricated thrust zone, trapdoor structure, drag fold, listric fault, rollover anticline, horst-graben structure and the like. 盆地中发育多种构造样式，主要有：叠瓦状冲断层、滚板构造、牵引褶皱、铲形正断层、滚动背斜、"垒—堑"构造等。

【中文定义】 断层两盘紧邻断层的岩层，常发生明显弧形弯曲，这种弯曲称为牵引褶皱。一般认为这是两盘相对错动对岩层拖曳的结果，并且以褶皱的弧形弯曲的突出方向指示本盘的运动方向。

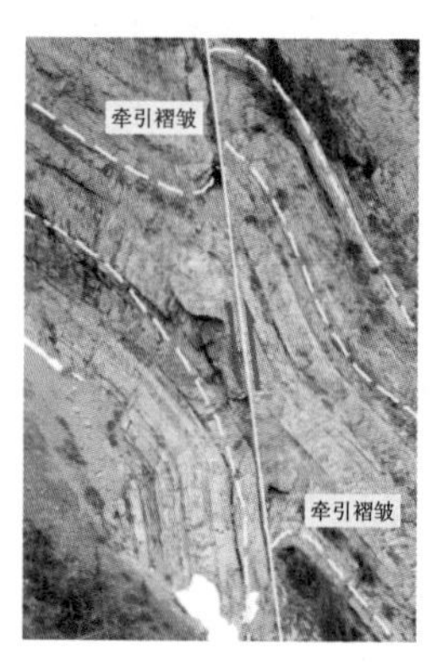

drag fold

drainage area

【释义】 流域面积，汇水面积，集水面积

【例句】 The drainage area of Xiaojiang is $3043km^2$. 小江的流域面积达 $3043km^2$。

【中文定义】 流域分水线所包围的面积。

drainage consolidation method

【释义】 排水固结法

【例句】 The article introduces the types of disease in expressway soft foundation, and proposes strengthening treatment measures for soft foundation such as earth replacing method, drainage consolidation method, compaction method, soil cement mixing method, and chemical strengthening method according to action mechanism of the diseases, in order to improve the bearing capacity of expressway soft foundation. 本文介绍了高速公路软土地基的病害类型，并根据

其作用机理提出了换填法、排水固结法、挤密法、水泥土搅拌法、化学加固法等软土地基的加固处理措施，以提高高速公路软土地基的承载力。

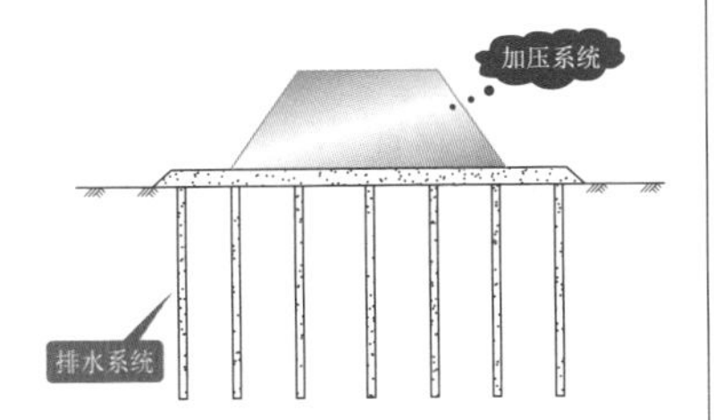

drainage consolidation method

drainage offset

【释义】 水系错位

【例句】 On the basis of analysis of the tectonic landform and geologic features and TL dating, as well as drainage offset, we can draw some conclusions as follows. 通过基于对构造地貌、地质特征、TL测年和水系错位的分析，我们可以得到如下结论。

【中文定义】 河流主、支流因所穿越的断层发生水平错动而随之错位的现象。

drainage system

【释义】 排水系统

drainage system

drawdown ['drɔːˌdaʊn]

【释义】 *n.* （抽水后）水位降低，水位降低量

【常用搭配】 drawdown curve 压降曲线；designed drawdown 设计水位降深；allowable drawdown 允许水位降深

【例句】 This assumption is necessary because after steady-state conditions have developed, the drawdowns at an observation well begin to fall along a straight line on a semi-log graph. 这个假设是十分必要的，因为当稳定状态持续一段时间之后，观测井中的水位降深将在半对数坐标图中呈直线下降的趋势。

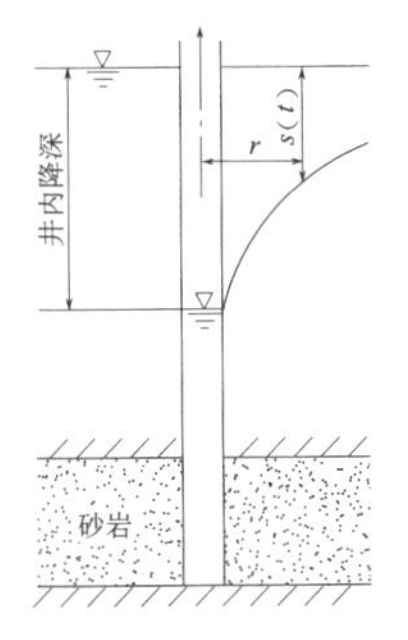

drawdown of bare wells in confined aquifers

drill bit

【释义】 钻头，钎头

【同义词】 bore bit

【例句】 As the drill bit breaks the rock into small cuttings, the gas is released into the mud stream. 当钻头破碎岩石时，气体进入泥浆。

drill rod burying

【释义】 埋钻

drill rod sticking

【释义】 卡钻

【例句】 When constituting releasing drill rod sticking, we must analyze the sticking reasons and confirm the depth of the free point at first. 处理卡钻问题时，首先要分析卡钻原因，确定卡点位置。

drilling ['drɪlɪŋ]

【释义】 *n.* 钻孔，训练，演练；*v.* 钻孔，训练

【常用搭配】 bucket drilling 带钻粉筒螺旋钻机

【同义词】 rock drill

【例句】 Bucket drillings are used to drill large diameter borings for disturbed sampling of overburden soil and gravel material. 带钻粉筒螺旋钻机用于钻进大孔径钻孔，获得扰动的覆盖层砾石土样本。

drilling

drilling and blasting method

【释义】 钻爆法

drilling exploration

【释义】 钻探

【例句】 Similar to the Seattle project, this one required new drilling exploration techniques, including the use of probe holes to assess ground conditions ahead of the huge tunnel-boring machines (TBM). 与西雅图的工程相似的是，这里的项目也需要新的钻探技术，包括在采用巨型隧道掘进机（TBM）掘进前使用探孔来评估地质条件。

drilling exploration

drilling fluid

【释义】 冲洗液

drilling machine

【释义】 钻孔机，凿岩机

【同义词】 jackdrill

【例句】 This new model of drilling machine can bore through solid rock ten metres deep. 这台新型钻孔机能钻透 10m 厚的坚固岩石。

drip [drɪp]

【释义】 *v.* 滴水，漏下

【中文定义】 地下水活动状态之一，根据《水力发电工程地质勘察规范》（GB 50287—2016）的规定，当地下洞室内地下水每 10m 洞长水量 $q \leqslant 25$L/min 或压力水头 $H \leqslant 10$m 时，地下水状态为渗水、滴水。

dripstone ['drɪpstəʊn]

【释义】 *n.* 滴石

【例句】 Stalagmite is a typical dripstone shape of $CaCO_3$ in caves and the most comprehensive and systematical paleoclimate environment information carrier too. With advances of dating techniques and analytic methods with high resolution, it is of great significance in the paleoclimate environment reconstruction. 石笋是洞穴碳酸钙沉积的典型滴石类型，也是岩溶记录中最全面、最系统的古气候环境信息载体，随着高分辨率测年技术和测试方法的提高，它在重建古气候环境方面具有重要的意义。

【中文定义】 由洞中滴水形成的碳酸钙沉积。滴石可形成各种形态，其具有代表性的有钟乳石、石笋、石柱等。

dripstone

dry [draɪ]

【释义】 *adj.* 干燥，干的

【中文定义】 地下水活动状态之一，地下洞室内地下水每 10m 洞长水量 $q=0$L/min 或压力水头 $H=0$m 时，地下水状态为干燥。

dry compressive strength

【释义】 干抗压强度

【中文定义】 指材料在干燥状态下的抗压强度。

ductile fault

【释义】 韧性断层

【例句】 By measuring the finite strain of deformed quartz under the microscope, it has been recognized that ductile fault was generated by combination of compression and shear stress. 根据镜下有限应变测量成果分析，该韧性断层是区域压扁加剪切应力背景下的产物。

【中文定义】 韧性断层是在剪切带中沿着微细滑动面做微小滑动，导致韧性断层两侧岩块有剪切错位，但却断而未破。

ductile fault

ductile shear zone

【释义】 韧性剪切带

【例句】 The Qiugamintashi-Huangshan large-scale ductile shear zone in Eastern Tianshan is 600 km long in east-west and 5－20km wide. 位于东天山的秋格明塔什-黄山韧性剪切带，东西长

600km，宽 5～20km。
【中文定义】 韧性剪切带是地壳中深层次呈带状展布的高应变的韧性断层。

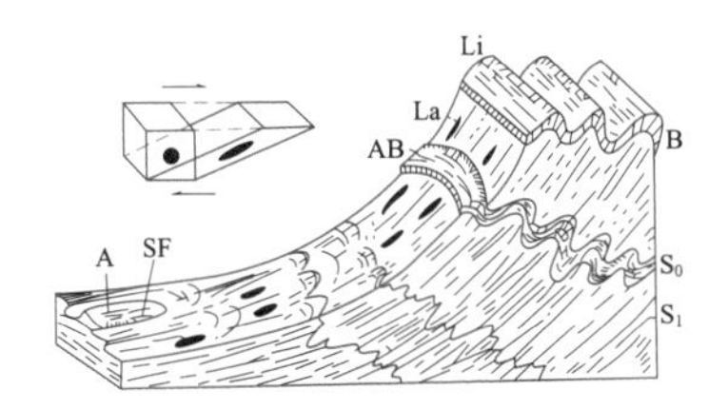

ductile shear zone

dune [dʊn]
【释义】 *n.* 沙丘
【常用搭配】 sand dune 沙丘；wandering dune 移动沙丘；transverse dune 横向沙丘
【例句】 Between the shore and the inland edge of the dune field, sand alternates with vegetated ground. 在海岸与内陆沙丘地带的边缘之间，沙地与植被交替出现。
【中文定义】 小山、沙堆、沙埂或由风的作用形成的其他松散物质叫沙丘。沙丘的存在是风吹移未固结的物质所致。沙丘主要有以下三大类：新月形沙丘、纵向沙垄、金字塔沙丘。

Indiana Sand Dunes

dust detector
【释义】 粉尘检测仪，灰尘探测器

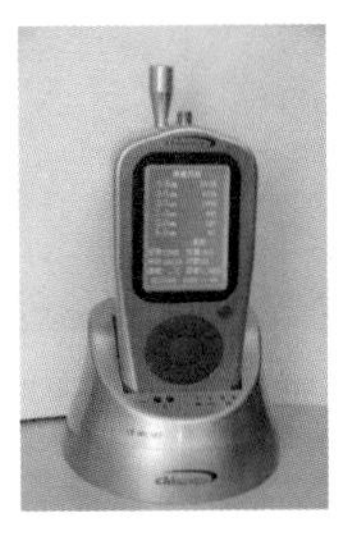

laser dust detector

【例句】 The dust detector on Cassini for the first time measured the dust charge directly. 卡西尼号上的粉尘检测仪首次测量到了尘埃电荷。

dynamic consolidation
【释义】 强夯法
【常用搭配】 dynamic replacement 强夯置换法
【例句】 Dynamic consolidation is used to treat wet-falling loess foundation of airport runway. 采用强夯法处理机场跑道湿陷性黄土地基。

dynamic design method
【释义】 动态设计法

dynamic penetration test (DPT)
【释义】 动力触探试验（DPT）
【常用搭配】 heavy cone dynamic penetration test 重型圆锥动力触探试验
【例句】 Dynamic cone penetration test show that the cement and sodium silicate double grout to improve the bearing capacity of foundation soil is obvious. 动力触探试验表明，水泥-水玻璃双浆液改善地基承载力的效果是明显的。

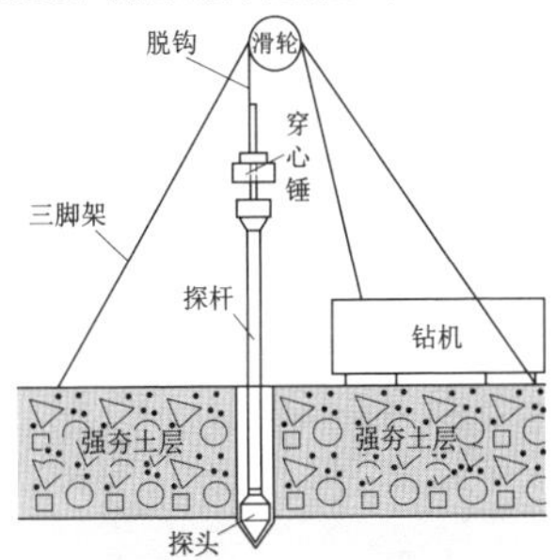

dynamic penetration test (DPT)

dynamic photographic resolution
【释义】 动态摄影分辨率

dynamic pressure
【释义】 动水压力
【常用搭配】 dynamic pressure field 动压力场
【例句】 The dynamic pressure on the original riverbed and the scour hole are measured. 文章对原始河床和冲坑的动水压力进行了测量。
【中文定义】 流体在流动过程中受阻时，由于动能转变为压力能而引起的超过流体静压力的压力。

dynamic reaction monitoring
【释义】 动反应监测

【例句】 Dynamic reaction monitoring is an effective way to monitor. 动反应监测是监测的一个有效办法。

dynamic replacement

【释义】 强夯置换法

【例句】 In Shenzhen area the subgrade of reclamation strengthened by dynamic replacement method has found widespread application, and has obtained a good processing effect in practice. 在深圳地区强夯置换法加固填海路基应用非常广泛，并在实践中取得了较好的处理效果。

【中文定义】 强夯置换法：采用边填碎石边强夯的方法在地基中形成碎石墩体，由碎石墩、墩间土以及碎石垫层形成复合地基，以提高承载力减小沉降。

dynamic replacement

dynamic simple shear test

【释义】 动单剪试验

dynamic test

【释义】 动力试验，动态测试，动载试验

【常用搭配】 dynamical test 动力特性测试；dynamic testing 动态试验

【例句】 Dynamic load testing is a high strain dynamic test which is applied after pile installation. 动态载荷试验是一种在桩安装后应用的高应变动态试验。

【中文定义】 ①施加负荷速率或变形速率随时间变化的破坏性试验，如疲劳试验、冲击试验、制品动态模拟试验等。②用周期应力或变形研究材料性能的非破坏性试验。

dynamic triaxial test

【释义】 动三轴试验

【例句】 The dynamic triaxial test is one of the most important methods all over the world, which is used to study ballast character. 动三轴试验是当今世界研究道床特性的最重要方法之一。

dynamic triaxial test

dynamic water level

【释义】 动水位

【例句】 Based on the theory of fluid dynamics, we can study the relationship between water level and flow rate in dynamic water level observation wells. 以流体动力学理论为基础，研究动水位观测井中水位与流量的关系。

dynamometamorphic rock

【释义】 动力变质岩

【例句】 Dynamometamorphic rocks include mylonitization rock, protomylonite, mylonite, ultramylonite and tectonic schist in ductile shear zone. 韧性剪切带内出露的动力变质岩主要包括糜棱岩化岩石、初糜棱岩、糜棱岩、超糜棱岩和构造片岩等。

【中文定义】 发生在强烈地壳错动带内，由机械作用占主导地位形成的变质岩。可分为构造或压碎角砾岩、碎裂岩类、糜棱岩类、千糜岩等。

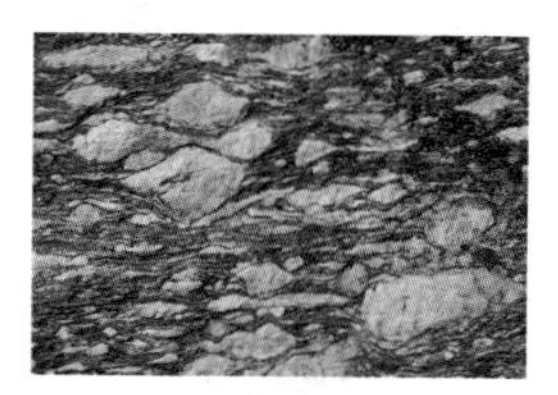

dynamometamorphic rock mylonite

E

earth ellipsoid

【释义】 地球椭球

【例句】 The elements of earth ellipsoid such as radius of curvature in meridian are the function of latitude. 如子午线曲率半径等的地球椭球要素是纬度的函数。

【中文定义】 又称地球椭圆体，代表地球大小和形状的数学曲面。一般采用旋转椭球。

earth material

【释义】 土料

【例句】 The method successfully solved the problem of high level of groundwater, soft foundation and complex earth materials. 该方法成功解决了高地下水水位、软土地基和复杂土料等问题。

earth material

earthquake caused debris flow

【释义】 地震泥石流

【中文定义】 指地震震动诱发的水、泥、石块混合物顺坡急速向下流动的混杂体。

earthquake caused debris flow

earthquake collapse

【释义】 地震崩塌

【中文定义】 是地震震动引起岩体或土体脱离母体，在重力作用下极其快速地下滑、堆积的过程。

earthquake collapse

earthquake crack

【释义】 地震裂缝

【中文定义】 指地震在地面上所造成的没有明显位移的裂隙。

earthquake hydrodynamic pressure

【释义】 地震动水压力

【例句】 Stress intensity factors KⅠ, KⅡ and K0 change in reverse proportion to the ratio of elastic module while in direct proportion to the crack length. Earthquake hydrodynamic pressure has a fairly strong influence on the cracks. 应力强度因子KⅠ、KⅡ、K0随弹模比的增加而减少，随裂缝长度的增加而增加，地震动水压力对重力坝裂缝影响较大。

【中文定义】 指地震使大坝产生振动时，在水库水体和大坝坝体之间会产生相互作用，此时水库水体对大坝产生的动水压力。

earthquake landslide

【释义】 地震滑坡

【常用搭配】 Mechanism of Earthquake Landslide 地震滑坡机理

【例句】 Vibration table simulation test is one of

earthquake landslide

methods to make research on earthquake landslide. 实验室振动台模拟试验是研究地震滑坡的手段之一。

【中文定义】 指地震作用诱发的滑坡。

earthquake magnitude

【释义】 地震震级

【例句】 Earthquake magnitude and timing are controlled by the size of a fault segment, the stiffness of the rocks, and the amount of accumulated stress. 地震的震级和持时受断层大小、岩石硬度以及累积起的压力多少的控制。

【中文定义】 是指地震大小，通常用字母 M 表示。地震愈大，震级数字也愈大，世界上最大的震级为 9.5 级。震级是根据地震波记录测定的一个没有量纲的数值，用来在一定范围内表示各个地震的相对大小（强度）。

earthquake precursor

【释义】 地震前兆

【例句】 Despite years of searching for earthquake precursors, there is currently no method to reliably predict the time of a future earthquake. 虽然对地震前兆的研究已有多年，但至今仍没有办法能够可靠地预计到未来地震发生的时间。

【中文定义】 地震前兆指地震发生前出现的异常现象，岩体在地应力作用下，在应力应变逐渐积累、加强的过程中，会引起震源及附近物质发生如地震活动、地表的明显变化以及地磁、地电、重力等地球物理异常，地下水水位、水化学、动物的异常行为等，并概括性地称这些与地震孕育、发生有关联的异常变化现象为地震前兆（也称地震异常）。它包括地震微观异常和地震宏观异常两大类。

earthquake risk analysis report

【释义】 地震危险性分析报告

earthquake subsidence

【释义】 震陷

【同义词】 seismic subsidence

【例句】 Based on the above analysis, the formation mechanism of the landslide is summarized as earthquake subsidence-liquefaction complex mechanism under earthquake. 根据以上分析及计算结果归纳出该滑坡形成机理为地震作用下的震陷-液化复合机理。

【中文定义】 在强烈地震作用下，由于土层加密、塑性区扩大或强度降低而导致工程结构或地面产生的下沉。

earthquake subsidence

earthquake-induced geological hazard

【释义】 地震地质灾害

【例句】 The strong effect of ground damage and the serious earthquake-induced geological hazard are the notable features of MS8.0 Wenchuan Earthquake. 地面破坏效应强烈和严重的地震地质灾害是汶川 MS8.0 地震震害效应的显著特点。

【中文定义】 在地震作用下，地质体变形或破坏所引起的灾害。地震地质灾害类型主要有：地基土液化、软土震陷、崩塌、滑坡、地裂缝和泥石流等。

earthquake-induced geological hazard (landslide)

easily occurring zone of geological hazard

【释义】 地质灾害易发区

【例句】 The exploitation of mineral resources should be restricted in easily occurring zone of geological hazard and prohibited in areas with real danger from geological hazard. 限制在地质灾害易发区开采矿产资源，禁止在地质灾害危险区开采矿产资源。

effective drainage porosity

【释义】 有效孔隙率

【例句】 There are phyteral cell remaining pore, matrix pore and new-forming pore in coal. The pores of remaining phyteral cell mainly are complicated dentritic structure, and the effective drainage porosity is high. 煤中发育植物细胞残留孔隙、基质孔隙和次生孔隙、植物细胞残留孔隙主要为复杂孔隙树结构，有效孔隙率高。

effective grain size

【释义】 有效粒径

【例句】 Smaller fine grain content and bigger effective grain size will lead to larger permeability coefficient. 细颗粒含量越少，有效粒径越大，相应的渗透系数越大。

effective porosity

【释义】 有效孔隙率，有效孔隙度，实际孔隙率

【例句】 First, an empirical relationship formula between the effective porosity and total porosity of cement paste is given based on the experimental results of concrete. 首先，根据混凝土的试验结果给出了水泥石有效孔隙率与总孔隙率关系的经验公式。

effective stress method

【释义】 有效应力法

【例句】 Consolidation pressure method and effective stress method were used to analyze the shear strength of the soil for different periods, respectively. And the results were compared with the value measured by field vane tests to verify the validity of the method. 可分别采用固结压力法和有效应力法对不同时期地基土体抗剪强度进行分析，并与现场十字板试验强度进行对比，以核实方法的有效性。

effective stress path

【释义】 有效应力路径

effective stress strength

【释义】 有效应力强度

【例句】 The pore water pressures of the refuses will increase with the increase of axial stress during the shear of the consolidation drained triaxial compression tests. The total stress strength parameters are not the effective stress strength parameters. 在固结排水三轴压缩试验的剪切过程中，孔隙水压力会随轴向应力的增加而增加，试验所得到的总应力强度参数并不就是有效应力强度参数。

【中文定义】 有效应力强度是指材料颗粒间接触应力抵抗外力作用的能力。

eigenperiod of seismic response spectrum

【释义】 反应谱特征周期

【同义词】 characteristic period of seismic response spectrum

【例句】 Since it is considered just the effect of engineering site condition and epicenter distance or only site condition to determine the characteristic period of acceleration response spectrum (T_g) in the antiseismic design criterion of our country at present, the eigenperiod of seismic response spectrum determined from those criterion is different from that got from special seismic hazard analysis. 由于我国现行的各种行业建筑物抗震设计规范中对标准加速度反应谱特征周期 T_g 作出的规定中一般只考虑场地和震中距的影响或只考虑场地影响，因此在某些情况下，由规范所得到的反应谱特征周期与由专门的地震危险性分析所得到的结果相差较大。

【中文定义】 地震动反应谱特征周期是指地震动加速度反应谱开始下降点的周期，是建筑场地自身的周期。

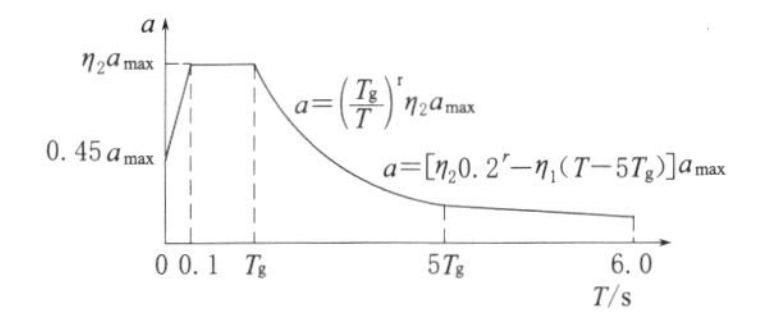

eigenperiod of seismic response spectrum

elastic aftereffect

【释义】 弹性后效

【例句】 Decelerating creep stage is caused by elastic aftereffect. 减速蠕变阶段是弹性后效的结果。

elastic deformation

【释义】 弹性变形

【例句】 Intact rocks around a roadway could carry more elastic-deformation than jointed rock mass. 与节理岩体相比，完整岩体在破裂前可以承受较大的弹性变形。

【中文定义】 材料在外力作用下产生变形，当外力取消后，材料变形即可消失并能完全恢复原来形状的性质称为弹性。这种可恢复的变形称为弹性变形。

elastic limit

【释义】 弹性极限，弹性权限，弹性限度

【常用搭配】 elastic limit for compression 压缩弹性极限；elastic limit under compression 抗压弹性极限；proportional elastic limit 比例弹性极限；

apparent elastic limit 表观弹性极限

【例句】 For elastomers, such as rubber, the elastic limit is much larger than the proportionality limit. 对于弹性体，比如橡胶，弹性极限相对于比例极限大得多。

【中文定义】 材料受外力（拉力）到某一限度时，若除去外力，其变形（伸长）即消失而恢复原状，这种不产生永久残余变形所能承受的最大应力称为材料的弹性极限。

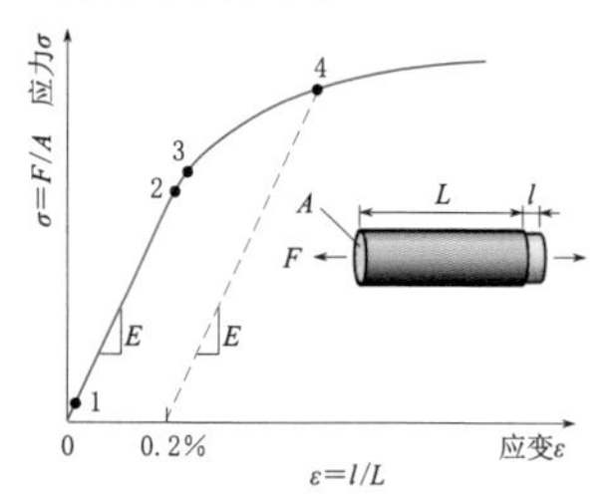

1—压密极限；2—弹性极限；3—屈服点；4—屈服极限

stress strain curve

elastic storage

【释义】 弹性储存量

【例句】 Part of the water in the aquifer belongs to elastic storage. 含水层内的水有一部分属于弹性储存量。

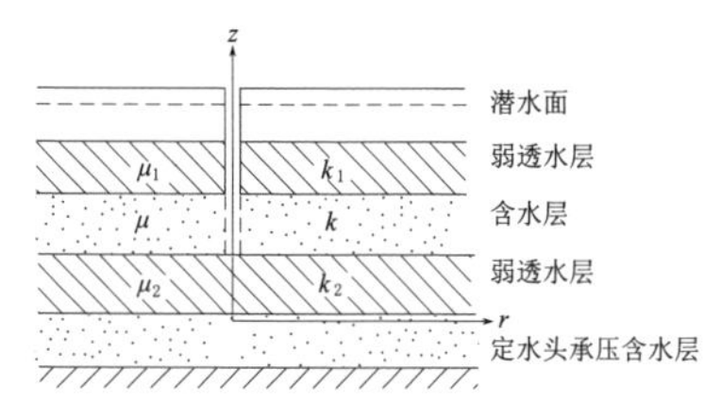

elastic storage

elastic storage coefficient

【释义】 弹性释水系数

【常用搭配】 elastic skeleton storage coefficient 弹性骨架释水系数

【同义词】 elastic storativity

【例句】 Elastic storage coefficient is the amount of water released from a confined aquifer in a unit area when the pressure drops by one unit. 弹性释水系数是单位面积承压含水层在压力下降一个单位时所释放出的水量。

【中文定义】 单位面积承压含水层在压力下降一个单位时所释放出的水量，又称弹性给水度。

elastic wave CT

【释义】 弹性波层析成像

【例句】 Elastic wave CT method is known as seismic CT. 弹性波 CT 方法，又称地震波层析成像技术。

elasticity modulus

【释义】 弹性模量

【例句】 Early concrete cracks because concrete's shrinkage deformation and concrete elasticity modulus and tensile strength does not adapt produces. 混凝土早期裂缝产生是因为混凝土的收缩变形与混凝土弹性模量和抗拉强度不相适应产生的。

【中文定义】 材料在弹性变形阶段，其应力和应变成正比，其比例系数称为弹性模量。

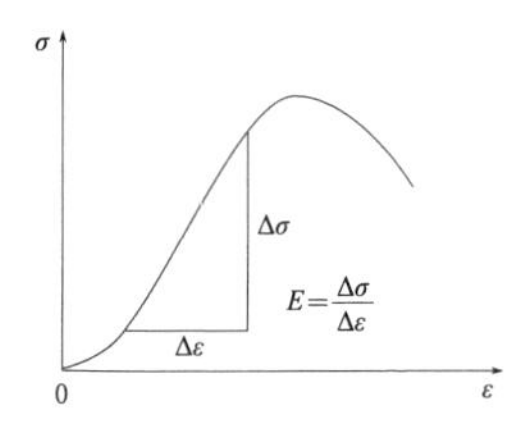

elasticity modulus

elasticity wave testing

【释义】 弹性波测试

elastic-plastic deformation

【释义】 弹塑性变形

【例句】 This method can account for the effect of elastic-plastic deformation. 这种模型可以考虑弹塑性变形影响。

【中文定义】 弹塑性体在外力施加的同时立即产生全部变形，而在外力解除的同时，只有部分变形立即消失，其余部分变形在外力解除后却永不消失。消失的那部分是弹性变形，永不消失的部

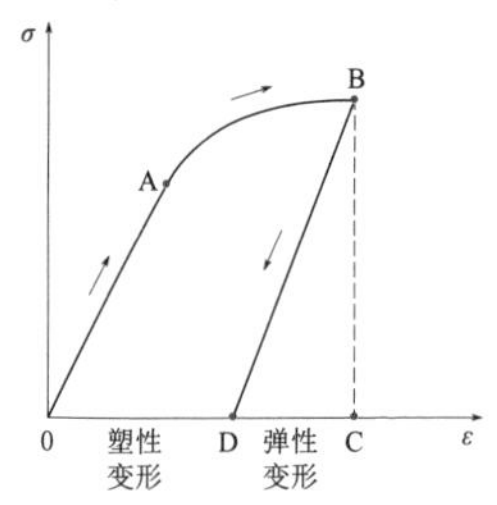

elastic-plastic deformation

分为塑性变形。

electrical exploration

【释义】 电法勘探

【例句】 Electrical exploration is widely used in fault surveys. 电法勘探被普遍应用于断层调查中。

electrical profiling method

【释义】 电剖面法

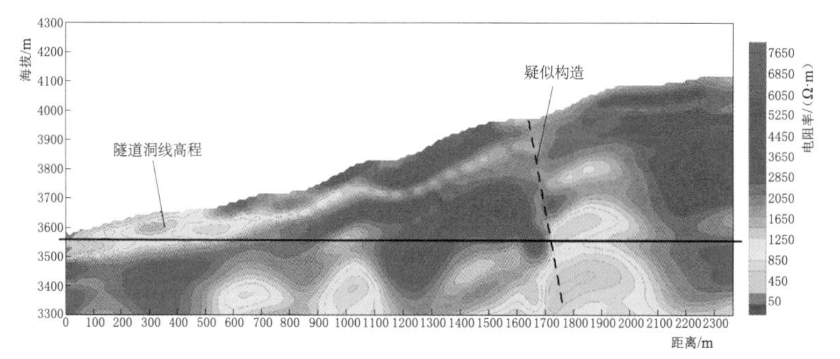

electrical profiling method

electrical sounding

【释义】 电测深法

electromagnetic exploration

【释义】 电磁法勘探

【例句】 Electromagnetic exploration has a unique advantage for the exploration of fine tectonics and stratigraphic division. 电磁法勘探对于勘探精细构造和地层划分具有独特的优势。

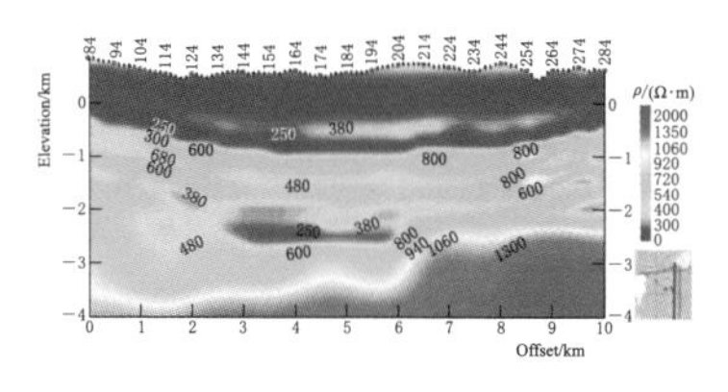

electromagnetic exploration

electron spin resonance (ESR)

【释义】 电子自旋共振法

【例句】 The opinions of further study on ESR dating of fault movement are also discussed in this paper. 本文还进一步讨论了运用电子自旋共振法测定断层活动年代的研究方向。

【中文定义】 指当外加具有与电子自旋能相等的频率电磁波时，便会引起能级间的跃迁。ESR测年法的基本原理就是利用电子自旋共振的方法直接测定样品自形成以来由于辐射损伤所产生的顺磁中心的数目（即所接受的放射性射线辐照和本身的累积效应）。

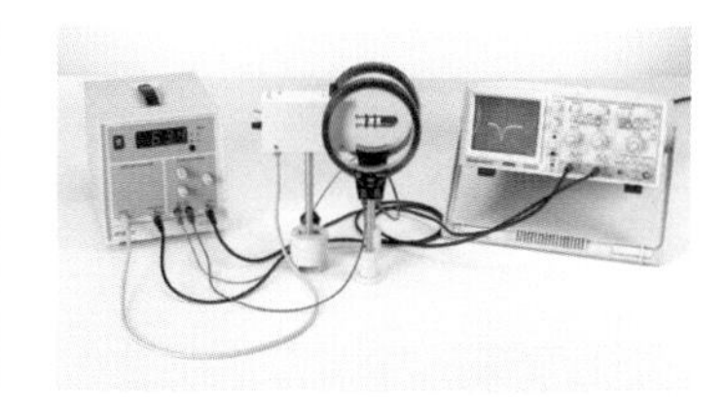

electron spin resonance

elevation [ˌɛlə'veʃən]

【释义】 *n.* 高程，海拔，标高

【常用搭配】 ground elevation 地面高程；absolute elevation 绝对高程；datum elevation 基准面标高

【中文定义】 地面点至高程基准面的垂直距离。

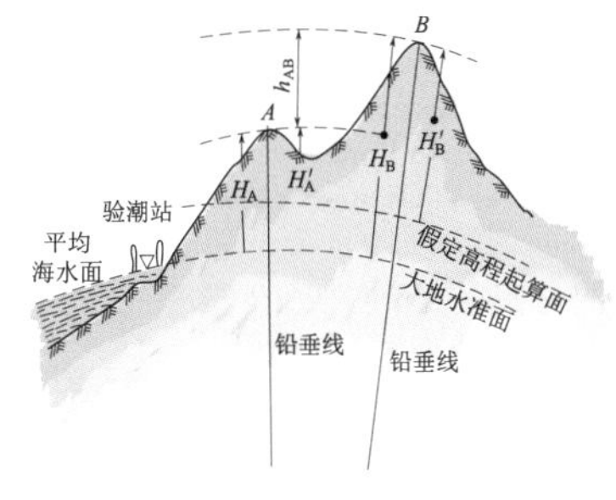

elevation

elevation error

【释义】 高程中误差

【例句】 The test results show that, on a tidal mudflat topographic map of 1 : 50000 with contour interval of one meter, the elevation error is 0.38 meter. 试验结果表明，在等高距 1m，比例尺 1 : 5 万的淤泥质海滩地形图上，其等高线高程中误差为 0.38m。

elevation view

【释义】 立视图

【常用搭配】 elevation view of dam 大坝立视图

【例句】 The elevation view of dam associated with flow direction includes an upstream view and a downstream view of the dam. 大坝与水流有关的立视图包括大坝上游视图和大坝下游视图。

【中文定义】 立视图是由建筑物或构建向垂直面投影所得的图形，一般只表示可见轮廓线，只在必要时以虚线绘出不可见轮廓。

elongated [ɪ'lɔːŋgeɪtɪd]

【释义】 *adj*. 长条状

embedment depth of foundation

【释义】 基础埋置深度

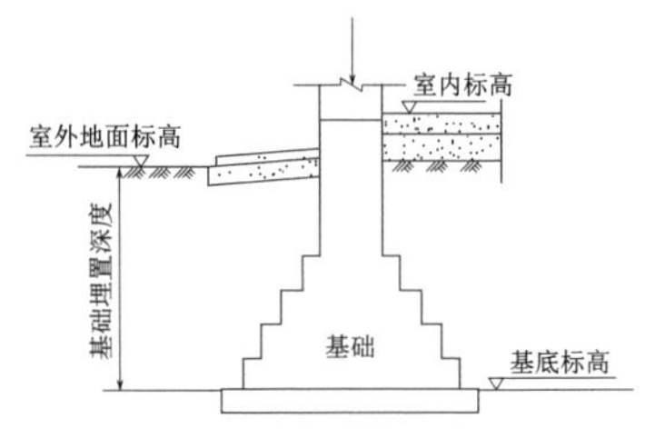

embedment depth of foundation

【例句】 As to the natural base, when the foundation soil meets the need of its bearing capacity and is under the influence of earthquake at 7 or 8 degree, a high-rise building which satisfies the height-width ratio provided in "Design & Construction Code for Reinforced Concrete High_Rise Building", won't overturn or slide or lose its base stability, and its inclination won't exceed the requirement in the code either, even when the embedded depth of the foundation is zero. 对于天然地基，当地基土满足承载力要求时，符合《钢筋混凝土高层建筑结构设计与施工规程》规定高宽比的高层建筑，在 7 度或 8 度地震作用下，即使基础埋置深度为零，也不会产生倾覆、滑移或地基失稳，且基础的倾斜度不会超过现行规范的规定。

emergency [ɪ'mɜːdʒənsi]

【释义】 *n*. 突发事件

emergency response

【释义】 应急处置

empirical coefficient

【释义】 经验系数

【例句】 Comparing the numerical results between the method of finite element analysis and the conventional empirical coefficient. 对有限元数值分析结果和常规经验系数法结果进行了数值比较。

empirical data

【释义】 经验数据

【例句】 Poisson's ratios should be determined from uniaxial compression tests pulse velocity tests, seismic field tests, or empirical data. 泊松比应该由单轴压缩试验，脉冲速度测试，地震现场测试，或经验数据确定。

empirical formula

【释义】 经验公式

【同义词】 empirical equation

【例句】 By use of the result, the calculating precision of numeric differential method and empirical formula is analyzed. 利用所得到的结果，分析了数值微分法和经验公式的计算精度。

empirical parameter

【释义】 经验参数

【例句】 Finally, it was shown by the comparison between the calculated results and the actual project that it was feasible to use the average value of the optimal parameter as empirical parameter. 最终，与工程实例进行了对比验证，结果表明，把最佳参数的平均值作为经验参数是可行的。

en echelon fault

【释义】 雁列断层

【例句】 Strike slip structural pattern mainly are flower-type structure and en echelon fault, additionally, there also developed structural pattern relevant to gravity and other factors, such as compactional structure and diapiric structure. 走滑构造样式主要有花状构造、雁列断层等；另外，还有与重力等其他因素有关的构造样式发育，如压实构造、底辟构造等。

【中文定义】 雁列构造是地壳中一种常见的构造形式，它是由一系列走向大致平行的断层或褶皱斜向排列而成的构造组合。

en echelon fault

endogenic force

【释义】 内营力，内力

【中文定义】 地壳运动产生强大水平挤压力，可以造成地壳运动、岩浆活动、变质作用和地震等。这种强大的力来自地球内部，称为内营力。它使地面变得高低不平，改变着地球表面的形态，也改变着部分岩性。

energy conservation and emission reduction

【释义】 节能减排

engineering geologic attribute

【释义】 工程地质属性

【例句】 Great achievement had been alive in the research about engineering geologic attribute, such as slope. But the research about society attribute is not enough. 以往对滑坡工程地质属性的研究取得了重大成绩，但对滑坡的社会属性研究仍然比较薄弱。

【中文定义】 地质体、勘探孔洞、物探、试验、边坡、地下洞室等工程地质对象所具有的本质特性和相互间关系。

engineering geological classification of dam-foundation rock mass

【释义】 坝基岩体工程地质分类

【中文定义】 中国国家标准《水力发电工程地质勘察规范》(GB 50287) 推荐采用的坝基岩体质量分类方法，该方法适用于坝高大于 70m 的混凝土坝。

类别	A：坚硬岩（R_b>60MPa）		
	岩体特征	岩体工程性质评价	岩体主要特征值
Ⅰ	$A_{Ⅰ}$：岩体呈整体状或块状、巨厚层状、厚层状结构，结构面不发育～轻度发育，延展性差，多闭合，岩体力学特性各方向的差异性不显著	岩体完整，强度高，抗滑、抗变形性能强，不需作专门性地基处理，属优良高混凝土坝地基	R_b>90MPa，V_p>5000m/s，RQD>85%，K_v>0.85
Ⅱ	$A_{Ⅱ}$：岩体呈块状或次块状、厚层结构，结构面中等发育，软弱结构面局部分布，不成为控制性结构面，不存在影响坝基或坝肩稳定的大型楔体或棱体	岩体较完整，强度高，软弱结构面不控制岩体稳定，抗滑、抗变形性能较高，专门性地基处理工程量不大，属良好高混凝土坝地基	R_b>60MPa，V_p>4500m/s，RQD>70%，K_v>0.75
Ⅲ	$A_{Ⅲ1}$：岩体呈次块状、中厚层状结构或焊合牢固的薄层结构。结构面中等发育，岩体中分布有缓倾角或陡倾角（坝肩）的软弱结构面，存在影响局部坝基或坝肩稳定的楔体或棱体	岩体较完整，局部完整性差，强度较高，抗滑、抗变形性能在一定程度上受结构面控制。对影响岩体变形和稳定的结构面应做局部专门处理	R_b>60MPa，V_p>4000～4500m/s，RQD>40%～70%，K_v>0.55～0.75
	$A_{Ⅲ2}$：岩体呈互层状、镶嵌状结构，层面为硅质或钙质胶结薄层状结构。结构面发育，但延展差，多闭合，岩块间嵌合力较好	岩体强度较高，但完整性差，抗滑、抗变形性能受结构面发育程度、岩块间嵌合能力，以及岩体整体强度特性控制，基础处理以提高岩体的整体性为重点	R_b>60MPa，V_p>3000～4500m/s，RQD>20%～40%，K_v>0.35～0.55
Ⅳ	$A_{Ⅳ1}$：岩体呈互层状或薄层状结构，层间结合较差。结构面较发育～发育，明显存在不利于坝基及坝肩稳定的软弱结构面、较大的楔体或棱体	岩体完整性差，抗滑、抗变形性能明显受结构面控制。能否作为高混凝土坝地基，视处理难度和效果而定	R_b>60MPa，V_p>2500～3500m/s，RQD>20%～40%，K_v>0.35～0.55
	$A_{Ⅳ2}$：岩体呈镶嵌或碎裂结构，结构面很发育，且多张开或夹碎屑和泥，岩块间嵌合力弱	岩体较破碎，抗滑、抗变形性能差，一般不宜作高混凝土坝地基。当坝基局部存在该类岩体时，需做专门处理	R_b>60MPa，V_p>2500m/s，RQD>20%，K_v>0.35

rock mass classification（1）

类别	A：坚硬岩（R_b>60MPa）		
	岩体特征	岩体工程性质评价	岩体主要特征值
Ⅴ	A_{V}：岩体呈散体结构，由岩块夹泥或泥包岩块组成，具有散体连续介质特征	岩体破碎，不能作为高混凝土坝地基。当坝基局部地段分布该类岩体时，需做专门处理	—
B：中硬岩（R_b=30～60MPa）			
Ⅰ	—	—	—
Ⅱ	B_{II}：岩体结构特征与 A_{I} 相似	岩体完整，强度较高，抗滑、抗变形性能较强，专门性地基处理工程量不大，属良好高混凝土坝地基	R_b=40～60MPa，V_p=4000～4500m/s，RQD>70%，K_v>0.75
Ⅲ	B_{III1}：岩体结构特征与 A_{II} 相似	岩体较完整，有一定强度，抗滑、抗变形性能一定程度受结构面和岩石强度控制，影响岩体变形和稳定的结构面应做局部专门处理	R_b=40～60MPa，V_p=3500～4000m/s，RQD=40%～70%，K_v=0.55～0.75
	B_{III2}：岩体呈次块或中厚层状结构，或硅质、钙质胶结构的薄层结构，结构面中等发育，多闭合，岩块间嵌合力较好，贯穿性结构面不多见	岩体较完整，局部完整性差，抗滑、抗变形性能受结构面和岩石强度控制	R_b=40～60MPa，V_p=3000～3500m/s，RQD=20%～40%，K_v=0.35～0.55
Ⅳ	B_{IV1}：岩体呈互层状或薄层状，层间结合较差，存在不利于坝基（肩）稳定的较弱结构面、较大楔体或棱体	同 A_{IV1}	R_b=30～60MPa，V_p=2000～3000m/s，RQD=20%～40%，K_v=0.35
	B_{IV2}：岩体呈薄层状或碎裂状，结构面发育～很发育，多张开，岩块间嵌合力差	同 A_{IV2}	R_b=30～60MPa，V_p=2000m/s，RQD=20%，K_v=0.35
Ⅴ	同 A_{V}	同 A_{V}	—
C：软质岩（R_b<30MPa）			
Ⅰ	—	—	—
Ⅱ	—	—	—
Ⅲ	C_{III}：岩体强度 15～30MPa，岩体呈整体状或巨厚层状结构，结构面不发育～中等发育，岩体力学特性各方面的差异性不显著	岩体完整，抗滑、抗变形性能受岩石强度控制	R_b<30MPa，V_p=2500～3500m/s，RQD>50%，K_v>0.55
	—	—	—

rock mass classification（2）

engineering geological classification of surrounding rock mass for underground chambers

【释义】 地下洞室围岩工程地质分类

【中文定义】 中国国家标准《水力发电工程地质勘察规范》（GB 50287）推荐采用的地下洞室岩体质量分类方法。

围岩类别	围岩稳定性评价	支 护 类 型
Ⅰ	1. 稳定； 2. 围岩可长期稳定，一般无不稳定块体	不支护或局部锚杆或喷薄层混凝土； 大跨度时，喷混凝土，系统锚杆加钢筋网
Ⅱ	1. 基本稳定； 2. 围岩整体稳定，不会产生塑性变形，局部可能产生组合块体失稳	
Ⅲ	1. 局部稳定性差； 2. 围岩强度不足局部会产生塑性变形，不支护可能产生塌方或变形破坏。完整的较软岩，可能短时稳定	喷混凝土，系统锚杆加钢筋网。大跨度时，并加强柔性或刚性支护
Ⅳ	1. 不稳定； 2. 围岩自稳时间很短，规模较大的各种变形和破坏都可能发生	喷混凝土，系统锚杆加钢筋网，并加强柔性或刚性支护，或浇筑混凝土衬砌
Ⅴ	1. 极不稳定； 2. 围岩不能自稳，变形破坏严重	

engineering geological classification of surrounding rock mass for underground chambers

engineering geological horizontal section

【释义】 工程地质平切图

【常用搭配】 engineering geological horizontal section of arch dam 拱坝的工程地质平切图；engineering geological horizontal section of underground powerhouse 地下厂房的工程地质平切图

【中文定义】 按一定比例尺表示地质水平剖面上的地形地貌、地质现象及其相互关系、地下水水位、工程地质分区界线与代号等内容的图件。

engineering geological investigation report

【释义】 工程地质勘察报告

【中文定义】 工程地质勘察报告是工程地质勘察工作的总结。根据勘察设计书的要求，考虑工程特点及勘察阶段，综合反映和论证勘察地区的工程地质条件和工程地质问题，做出工程地质评价。

engineering geological map

【释义】 工程地质图

【常用搭配】 engineering geological map of project area 枢纽区工程地质图；engineering geological map of dam site 坝址工程地质图

【例句】 Examples of small-scale engineering geological maps have been reviewed to provide a background for the content of small scale engineering geological maps. 为了描述小比例尺工程地质图的内容背景，本文综述了多个小比例尺工程地质图的实例。

【中文定义】 按一定比例尺表示工程地质条件在一定区域或建筑区内的空间分布及其相互关系的图件。

engineering geological map for specialized engineering geological problems

【释义】 专门性问题工程地质图

【例句】 In bidding design stage and construction details design stage of hydropower engineering, engineering geological map for specialized engineering geological problems shall be submitted in report for engineering geological investigation. 在水力发电工程招标设计阶段和施工详图设计段，工程地质勘察报告中应附专门性问题工程地质图。

engineering geological profile

【释义】 工程地质剖面图

【常用搭配】 engineering geological profile of alternative dam site 坝址工程地质剖面图

【例句】 Engineering geological profiles are basic plot in the engineering geology report. They express stratums' lithology and lithology characters, and intuitionisticly reflect stratums' lithology combination and variational direction with multifarious symbols, curves and characters. 工程地质剖面图是工程地质报告中广泛使用的一种基础性图件，它用各种符号、曲线、文字表示地层岩性、岩组特征，直观反映地层的岩性组合、变化趋势。

【中文定义】 按一定比例尺表示地质剖面（通常指垂直剖切）上的地形地貌、地质现象及其相互关系、地下水水位、工程地质分区界线与代号等内容的图件。

engineering geophysical exploration

【释义】 工程地球物理勘探

【例句】 Engineering geophysical exploration is a physical exploration method for solving engineering geology and hydrogeological problems in civil engineering survey. 工程地球物理勘探是解决土木工程勘察中工程地质、水文地质问题的一种物理勘探方法。

engineering measure

【释义】 工程处理措施

engineering remote sensing

【释义】 工程遥感

engineering remote sensing

engineering slope

【释义】 工程边坡

【例句】 Preventing rainfall infiltration from slope and draining of slope could improve the stability of engineering slope effectively. 防止坡面降雨入渗和坡体排水对提高工程边坡稳定性具有重要作用。

engineering slope

engineering surveying

【释义】 工程测量

【例句】 At present, the technology of GPS was applied to engineering surveying with broad prospects. 全球定位技术（GPS）目前已深入到工程测量的各个领域，应用前景十分广阔。

【中文定义】 工程建设和自然资源开发各阶段进行的控制测量、地形测绘、施工放样、变形监测等。

engineering-triggered landslide

【释义】 工程滑坡

engineering-triggered landslide

environmental hydrogeology

【释义】 环境水文地质

【例句】 On the basis of analysis of the environmental hydrogeological characteristics in Changqing mine field, the water environmental pollution comprehensive index method has been adopted for the evaluation of shallow ground water. 在分析长清井田矿区环境水文地质特征的基础上，采用水环境污染综合评价指数方法对浅层地下水环境质量的现状进行了评价。

environmental impact assessment of groundwater

【释义】 地下水环境影响评价

【例句】 The environmental impact assessment of groundwater on nuclear waste disposal site is a focal point in the course of disposing radioactive waste, and it also is one of the key problems about security evaluation of waste disposal field. 核废物处置场的地下水环境影响评价是放射性废物处置过程中的一个重要研究内容，也是处置场安全评价的关键问题之一。

【中文定义】 是环境影响评价的主要组成部分，主要工作内容包括：评价范围内水文地质条件的详细调查；评价范围内地下水开采利用价值、现状及规划、井位分布及水源地保护的调查；地下水质量目标的确定；评价范围现有地下水污染源、在建与拟建项目地下水污染源的调查；地下水环境质量现状检测；地下水污染途径的分析；地下水污染预测模式及参数的研究与确定；建设项目对地下水环境影响的预测评价：保护与改善地下水环境质量措施的分析等。

environmental isotope

【释义】 环境同位素

【例句】 The meteoric origin of geothermal water in study area is evidenced by its environmental isotope composition. 环境同位素研究结果表明，研究区地热水为大气降水成因。

【中文定义】 环境同位素是存在于自然环境中的同位素。主要是天然形成的，但也有一部分来源于核试验产生的人工放射性同位素（如氚）。它不包括作为示踪剂的人工投放的放射性同位素。

environmental management

【释义】 环境管理

environmental performance

【释义】 环境绩效

environmental policy

【释义】 环境方针

environmental radioactivity detection

【释义】 环境放射性检测

environmental radioactivity test

【释义】 环境放射性测试

environmental resources information system

【释义】 环境资源信息系统

【例句】 Remote sensing was applicated in Missouri environmental resources information system. 遥感技术在密苏里环境资源信息系统中的应用。

【中文定义】 在计算机软硬件支持下，把资源环境信息按照空间分布及属性，以一定的格式输入、处理、管理、空间分析、输出的技术系统。

environmental risk

【释义】 环境风险

Eocene [ˈiəˌsin]

【释义】 *n.* 始新世（统）

【中文定义】 第三纪的第二个世。

Eon [ˈiən]

【释义】 *n.* 宙

【常用搭配】 Cryptozoic Eon 隐生宙；Phanerozoic Eon 显生宙

【中文定义】 最大的地质年代单位，是根据生物出现显著与否进行划分的。

Eonothem [ənɔˈθem]

【释义】 *n.* 宇

【常用搭配】 Archean Eonothem 太古宇；Proterozoic Eonothem 元古宇；Phanerozoic Eonothem 显生宇

【中文定义】 最大的地层年代单位，是指在“宙”的时间内形成的地层，根据生物演化最大的阶段性，即生命的存在与否及存在方式划分。

epicenter [ˈεpɪˌsεntə]

【释义】 *n.* 震中，中心

【常用搭配】 distance of epicenter 震中距；epicenter map 震中图

【中文定义】 指震源在地表的投影点。震中也称震中位置，是震源在地表水平面上的垂直投影用经、纬度表示。实际上震中并非一个点，而是一个区域。震中也有一定范围，称为震中区，震中区是地震破坏最强的地区。从震中到任一地震台（站）的地面距离，称震中距。

epicenter distribution map of historical earthquakes

【释义】 历史地震震中分布图

【例句】 In prefeasibility study stage of hydropower engineering, epicenter distribution map of historical earthquakes shall be submitted in report for engineering geological investigation. 水力发电工程预可行性研究阶段工程地质勘察报告中应附历史地震震中分布图。

【中文定义】 表示地震震中位置、震级、发震时间、震源深度、烈度及地震分区等内容的图件。

epidote [ˈεpɪˌdoʊt]

【释义】 *n.* 绿帘石

【常用搭配】 epidote amphibolite facies 绿帘石闪石岩相；epidotization 绿帘石化

epidote

【例句】 Part eclogites in Qinglongshan are characterized by epidote, kyanite and talc porphyroblast. 青龙山部分榴辉岩以含绿帘石、蓝晶石和滑石变斑晶为主。

epidote vein

【释义】 绿帘石脉

【常用搭配】 quartz-epidote vein 石英-绿帘石脉

【例句】 There are 15 kinds veins in the Emeishan basalt such as: feldspar vein, hematite vein, quartz vein, epidote vein, chlorite vein, sphene vein, calcite vein, chalcedony-allophane vein, calcite-hematite vein, calcite-chlorite vein, quartz-calcite vein, quartz-epidote vein, quartz-bituminous vein, hematite-chalcedony-chlorite vein and epidosite-chlorite-chalcedony vein. 峨眉山玄武岩中的岩脉有：长石脉、赤铁矿脉、石英脉、绿帘石脉、绿泥石、榍石脉、方解石脉、玉髓-水铝英石脉、方解石-赤铁矿脉、方解石-绿泥石脉、石英-方解石脉、石英-绿帘石脉、石英-沥青脉、赤铁矿-玉髓-绿泥石脉、绿帘石-绿泥石-玉髓脉等 15 种。

epidote vein

epimetamorphic rock

【释义】 浅变质岩

【例句】 Paragentic gold deposits are controlled by shear fracture belt of epimetamorphic rock series of Precambrian Period. 共生金矿主要赋存于前寒武纪浅变质岩系中，受剪切断裂带控制。

Epoch ['ɛpək]

【释义】 *n.* 世

【常用搭配】 Holocene Epoch 全新世；Pleistocene Epoch 更新世

【中文定义】 世是常用的第四级地质年代单位，代表比纪次一级的生物界演化阶段，以古生物的科和目的更新作为依据。

equal-angle projection

【释义】 等角投影

【例句】 Equal-angle projection is one kind of map projection classified by transmutation. 等角投影是地图投影按变形方式划分中的一种。

【中文定义】 等角投影，又称正形投影，指投影面上任意两方向的夹角与地面上对应的角度相等。在微小的范围内，可以保持图上的图形与实地相似；不能保持其对应的面积成恒定的比例；图上任意点的各个方向上的局部比例尺都应该相等；不同地点的局部比例尺，是随着经度、纬度的变动而改变的。

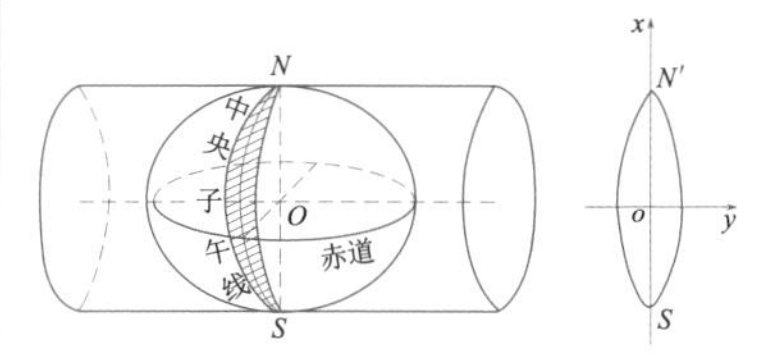

equal-angle projection

equal-area projection

【释义】 等面积投影

【例句】 Equal-area projection is one kind of map projection classified by transmutation. 等面积投影是地图投影按变形方式划分中的一种。

【中文定义】 地图上任何图形面积经主比例尺放大以后与实地上相应图形面积保持大小不变的一种投影方法。

equation of water balance

【释义】 水均衡方程

【例句】 Systematic management model is set up by time series analytic method based on equation of water balance to control, predict and manage groundwater resources system. 以水均衡方程为基础，应用时间序列分析法，建立系统管理模型，对地下水资源系统进行系统运行控制、预测和管理。

equidisplacement chart

【释义】 等位移量曲线图

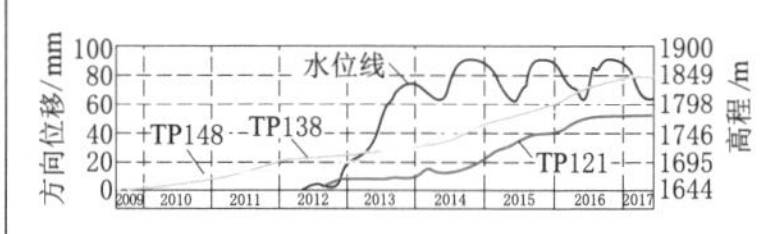

equidisplacement chart

equigranular texture

【释义】 等粒结构

【中文定义】 火成岩结构之一，岩石中同类矿物的颗粒大小相近的全晶质结构。

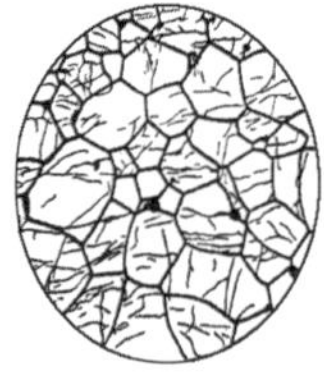

equigranular texture

Era [ˈɪrə]

【释义】 *n.* 代

【常用搭配】 Paleozoic Era 古生代；Mesozoic Era 中生代

【中文定义】 第二级地质年代单位，是根据全球生物界演化的重大变化进行划分的。

Erathem

【释义】 *n.* 界

【常用搭配】 Paleozoic Erathem 古生界；Mesozoic Erathem 中生界；Cenozoic Erathem 新生界

【中文定义】 第二级年代地层单位，是指在“代”的时间内形成的地层，根据生物界发展的总体面貌和地壳演化的阶段性划分。

erosion base level

【释义】 侵蚀基准面

【例句】 Erosion base level, deposition base level and stratigraphic datum are all potential energy levels, kinetic surfaces and are all variable, just differing from the criterion of analyses and the mechanism of formation. 侵蚀基准面、沉积基准面、地层基准面都是势能面，都是动力学界面，都是变化的，只是分析的基准不同、形成的机制不同。

【中文定义】 是河流垂直下切侵蚀的界限。其高低决定河流纵剖面的状态，其升降会引起长河段的冲淤和平面上的变化。在这个面上侵蚀停止或侵蚀与堆积达到平衡。

erosional terrace

【释义】 侵蚀阶地

【中文定义】 由基岩构成，阶面上往往没有或只有很少的残余冲积物分布。阶面是河流侧蚀造成的谷底侵蚀面，阶坡是河流下切造成的侵蚀阶梯。

esker [ˈɛskə]

【释义】 *n.* 蛇堤，蛇形丘，冰河沙堆

【常用搭配】 fault esker 断蛇丘；beaded esker 串珠蛇形丘；esker fan 蛇丘扇形地

【中文定义】 蛇形丘是一种狭长曲折的地形，呈

esker

蛇形弯曲，两壁陡直，丘顶狭窄，其延伸的方向大致与冰川的流向一致，主要分布在大陆冰川区。

estuarine delta

【释义】 河口三角洲

【中文定义】 河流在入海或入湖的地方堆积了大量的碎屑物，构成了一个三角形的地段，称为河口三角洲。

estuarine delta

Eurasian Plate

【释义】 欧亚板块

【例句】 The geodynamics of central Asia is dominated by the interaction between the Eurasian and Indian Plates. 中亚地区的地球动力学由印度板块与欧亚板块的相互作用主导。

【中文定义】 欧亚大陆板块为包括大部分欧亚地区的大陆板块，但不包括南亚的印度半岛（印度次大陆）、西南亚的阿拉伯半岛（阿拉伯次大陆）以及东西伯利亚的上扬斯克山脉以东的地区。

Eurasian Plate

evaporation rate of groundwater

【释义】 地下水蒸发量

【例句】 Drainage water volume connected negatively to evaporation rate of groundwater. 地下水蒸发量和退水量之间具有负相关性。

evaporation-concentration process

【释义】 蒸发浓缩作用

【例句】 In such vertical convection system, potassium ions were continuously enriched due to the evaporation-concentration process. The current K-rich brines during the glauberite intergranular were formed eventually. 在这种垂向对流系统中，蒸发浓缩作用导致钾离子不断富集，最后形成目前的钙芒硝晶间富钾卤水。

【中文定义】 地下水受蒸发而引起水中成分的浓缩过程。

excavated material

【释义】 开挖料

【例句】 The excavated material could be used as aggregate in concrete, filling and paving material in road construction. 公路建设中的开挖料可以作为混凝土、填料与铺路料的骨料。

excavated material

excavation [ˈɛkskəˈveʃən]

【释义】 *n*. 挖掘，发掘

【常用搭配】 foundation excavation 基础开挖；excavation depth 开挖深度

excavation

【例句】 The excavation of the penstock section has a horse shoe shape with width of 4.5m. 压力管道采用马蹄形开挖，宽 4.5m。

excavation gradient

【释义】 开挖坡比

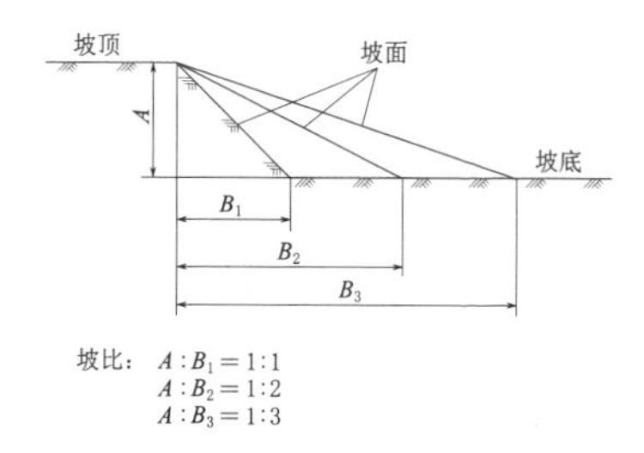

excavation gradient

【例句】 The excavation gradient of the side slopes at both sides of the connecting channel is 1∶1, so the supporting measures will not be considered. 鉴于 1∶1 的开挖坡比，可以不考虑支护措施。

excavation line

【释义】 开挖线

【常用搭配】 slope excavation line 边坡开挖线；subgrade excavation line 地基开挖线

【例句】 The slope line with a part allowed to indicate the excavation line in hydropower and water conservancy project. 水电水利工程中，沿开挖坡面顶部开挖线用示坡线表示坡面倾斜方向，并允许只绘制一部分示坡线。

【中文定义】 指开挖坡边线顶边线。

excavation plan

【释义】 开挖平面图

【中文定义】 依一定比例尺和图例绘制，表示地基、基坑、边坡开挖平面布置的图件。

excavation unloading

【释义】 开挖卸荷

【例句】 Based on the unloading region and the mechanical parameters of a typical cross section of shiplock slope of TGP given by in-situ test and geological survey, the rheological analysis on the stability of shiplock slope is made considering the effects of excavation unloading. 对三峡工程船闸高边坡的典型剖面，在地质调查和原位试验确定的边坡岩体开挖卸荷带及其参数的基础上，对边坡进行了施工开挖卸荷效应的流变稳定性分析。

excellent surrounding rock mass

【释义】 极好围岩

【中文定义】 地下洞室在采用 RMR（Rock Mass Rating system）分类时，当 RMR 评分值

介于 81～100 时，为Ⅰ类围岩，定性评价为极好围岩。

exfoliation [eksˌfəuliˈeiʃən]
【释义】 *n.* 剥落，表皮脱落
【常用搭配】 exfoliation corrosion 层蚀，剥离腐蚀；exfoliation mountain 页片剥离山；exfoliation joint 剥落节理
【中文定义】 小块岩石的崩塌。

exogenic force
【释义】 外营力
【例句】 By the action of exogenic force, the thin-skinned structure was subjected to denudation to varying degree, appearing as different folded forms. 在外营力的剥蚀与破坏影响下，薄皮构造受到程度不同的剥蚀破坏而呈现出不同的褶曲形态。
【中文定义】 外营力又称外动力。由地球外部所产生的改变地表形态、地壳结构构造和地壳岩矿成分的动力。

expansion [ɪkˈspænʃən]
【释义】 *n.* 膨胀；阐述；扩张物

expansion coefficient
【释义】 膨胀系数
【例句】 The calculation method is derived from the temperature dependence of the thermal expansion coefficient. 据热膨胀系数随温度变化的规律推导了热膨胀系数的计算方法。

expansion joint
【释义】 伸缩缝
【常用搭配】 expansion joint construction techniques 伸缩缝施工工艺
【例句】 The acceleration of the bridge when vehicle reaches the expansion joint is greater than that when vehicle passes the deck. 车辆通过伸缩缝时引起的冲击加速度响应大于通过桥跨时的结构振动。

expansion joint

expanded polystyrene (EPS)
【释义】 聚苯乙烯发泡材料，发泡聚苯乙烯
【例句】 Expanded polystyrene looks like a rigid white foam and is used as packing or insulation. 聚苯乙烯发泡材料像白色硬泡沫，用于包装或绝缘。

expanded polystyrene

expected value
【释义】 期望值
【例句】 The variance of a random variable measures how far its values are typically from the expected value. 随机变量的方差衡量随机变量值与期望值偏离的程度。

exploitable reserves
【释义】 可开采量
【例句】 Exploiting the thin oil-layer is an important measure to increase exploitable reserves and preserve steady production of Daqing oilfield. 开发薄油层是大庆油田增加可采储量、保持油田稳产的重要措施。

exploration items
【释义】 勘探项目

exploration layout
【释义】 勘探平面布置图

exploration management
【释义】 勘探管理

exploration method
【释义】 勘探方法
【例句】 With the development and profound use of remote sensing technique and surface geophysical prospecting, a modern comprehensive exploration method has come into being. 随着遥感技术和地面物探技术的发展和深入应用，产生了一种现代的综合勘探方法。

exploratory pit
【释义】 探坑，试坑
【同义词】 prospect pit

exploratory trench

【释义】 探槽

exploratory trench

external water pressure

【释义】 外水压力

【例句】 The external water pressure of lining under limited discharge in deep buried tunnel was tested. 对深埋岩溶隧道限量排放条件下的衬砌外水压力进行了测试。

【中文定义】 指地下洞室外地下水作用在洞室衬砌上的静水压力，单位为帕斯卡。

external water pressure observation

【释义】 外水压力观测

【例句】 The external water pressure observation refers to the observation of the external water pressure of the tunnel. 外水压力观测是指对隧洞外水压力进行的观测。

extrapolation method

【释义】 外延法

【例句】 The extrapolation method is used to obtain the SIF and stress triaxiality around the tip cracks. 应用外延法得到应力强度因子及裂尖周围的应力三维度。

extruded schistosity

【释义】 挤压片理

【中文定义】 又称片状构造，指岩石形成薄片状的构造。板状、千枚状、片状、片麻状构造可通称为片理，在变质岩中极为常见，是变质岩的重要特征之一。

extruded schistosity

extrusive rock

【释义】 喷出岩

【例句】 The basic extrusive rocks, diabase veins accompanied with above mentioned ophiolites could be classified as interplate rift-type alkaline basalts and other transitional series. 与之共（伴）生的基性喷出岩、辉绿岩脉属板内洋岛型、裂谷型碱性玄武岩及其过渡类型系列。

extrusive rock and intrusive rock

F

factor of safety (FOS)

【释义】 安全系数

【例句】 In the designing of the geotechnical engineering, the safe degree of the pile foundation is different by using the same factor of safety because of the different geology conditions, different soil strata, different pile types and different loads characters. 在岩土工程设计中，对于不同地质条件、不同土层、不同桩型和承受不同性质荷载的桩基，在取相同安全系数的条件下，其实际的安全度是不同的。

factor of safety against sliding

【释义】 抗滑安全系数

【例句】 When the seismic process finishes, the time-history curve of factor of safety against sliding of rock block can be obtained. 当地震过程结束时，块体的抗滑安全系数的时程曲线就可以得出了。

factor of stability

【释义】 稳定系数，稳定性系数

【例句】 The factor of stability is only 0.99 - 1.03 in the condition of self-weight and torrential rain and earthquake that is the most disadvantageous condition, the slope is in unstable condition and the situation is very dangerous. 在最不利组合工况（自重＋暴雨＋地震）下，滑坡体稳定系数仅为0.99～1.03，处于不稳定状态。

failure envelope

【释义】 破坏包线

【例句】 Only two yield Mohr circles are needed for determining the failure envelope just as for Mohr-Coulomb criterion. 与莫尔-库仑准则一样，确定此破坏包线只需要两个破坏莫尔圆。

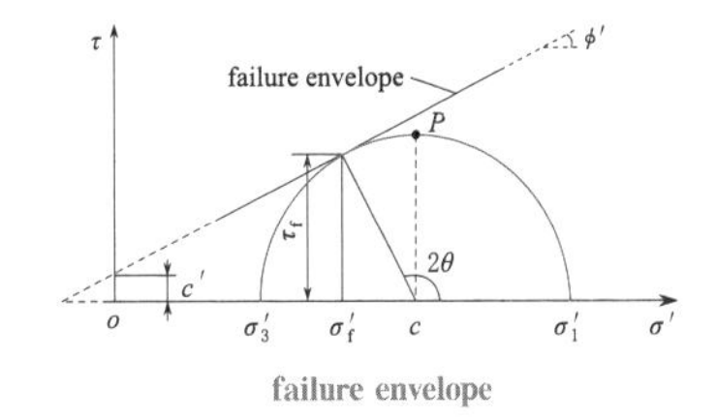

failure envelope

【中文定义】 破坏包线是指剪应力与法向有效应力所构成的一种表征材料简单破坏特性的曲线。破坏包线代表破坏时破坏面上的剪应力和有效法向应力之间的关系。

failure strength

【释义】 破坏强度

【例句】 Furthermore the relation of static preload and failure strength of red sandrock under combination loading has been founded. 实验还获得了红砂岩在组合加载情况下预静载应力与最终破坏强度的对应关系。

fair surrounding rock mass

【释义】 一般围岩

【中文定义】 地下洞室围岩采用 RMR（Rock Mass Rating system）法进行分类时，当 RMR 评分值介于 41～60 之间，为Ⅲ类围岩，定性评价为一般围岩。

falling [ˈfɔːliŋ]

【释义】 *n.* 坠落，崩落

【中文定义】 小型块石的崩落。

falling

falling head permeability test

【释义】 变水头渗透试验

【例句】 A multifunctional permeameter is designed and developed for performing constant head and falling head permeability tests on geotextiles or soil-covered geotextiles with and without vertical load. 设计开发了一种多功能渗透仪，可以开展纯土工织物以及覆土土工织物在垂向无压和有压情况下的常水头和变水头渗透试验。

fallout [ˈfɔːlaut]

【释义】 *n.* 掉块

【中文定义】 洞室局部岩块塌落，是洞室局部失稳类型之一。

fan fold

【释义】 扇形褶皱

【中文定义】 扇状褶皱是指褶皱面呈扇形弯曲、两翼都倒转的褶皱。

fan fold

far-field region

【释义】 远场区

【例句】 Theoretical analysis shows that the method was applicable whether the objects are located in the far-field region or near-field region. 理论分析表明，无论目标位于远场区还是近场区，本方法都是适用的。

fatigue strength

【释义】 疲劳强度

【常用搭配】 fatigue ultimate strength 疲劳极限强度；fatigue impact strength 疲劳冲击强度；fatigue dynamic strength 疲劳动载强度

【例句】 This paper describes the performance test, contact fatigue strength test and bending fatigue strength test on the gear reducer of involute point contact ferrographical analysis is made. 本文对渐开线点啮合齿轮减速器进行了性能试验、接触疲劳强度和弯曲疲劳强度试验，同时还进行了铁谱分析。

fault [fɔlt]

【释义】 *n.* 故障，断层；*vi.* 弄错，产生断层

【常用搭配】 Normal fault 正断层；Active fault 活动断层

【例句】 The scientists positioned the seismic activity as being along the San Andreas fault. 科学

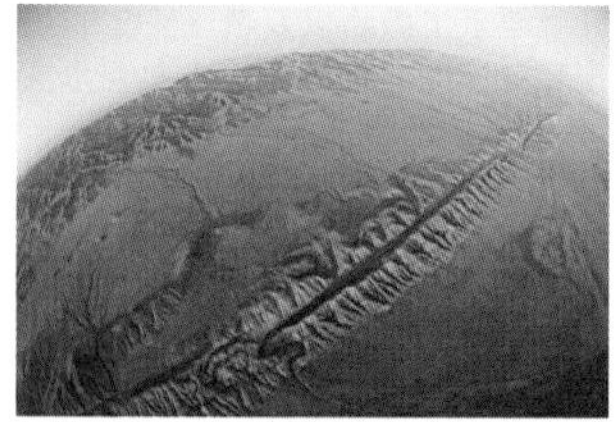

San Andre fault

家们找到了沿着圣安德烈亚斯断层的地震活动位置。

【中文定义】 指地壳岩层因受力达到一定强度而发生破裂，并沿破裂面有明显相对移动的构造。

fault breccia

【释义】 断层角砾岩

【例句】 The differences among cryptoexplosion breccia, fault breccia and volcanic explosion breccia, and the differences of the cryptoex plosion breccias in the upper, middle and lower of cryptoexplosion breccia pipe were discussed. The features of ore bearing cryptoex plosion breccia pipe and subvolcanic rocks (ultra-hypabyssal rocks) relate to cryptoexplosion and mineralization were also studied. 研究了隐爆角砾岩与断层角砾岩、火山爆破角砾岩的区别，隐爆岩筒上、中、下部隐爆角砾岩的区别，含矿隐爆角砾岩筒特征，以及与隐爆、成矿作用有关的次火山（超浅成）岩特征。

fault breccia

【中文定义】 断层角砾岩指在应力作用（断层作用）下，断层的两个断盘移动时，上、下两盘之间的岩石不断地被糅合，原岩破碎成角砾状，其砾石空隙间被水冲刷干净后被破碎细屑充填胶结或有部分外来物质胶结的岩石。

fault creep monitoring

【释义】 断层蠕变监测

fault fracture zone

【释义】 断层破碎带

【中文定义】 断层两盘相对运动，相互挤压，使附近的岩石破碎，形成与断层面大致平行的破碎带，称为断层破碎带，简称断裂带。

fault gouge

【释义】 断层泥

【例句】 New data have been obtained by tracing and observations of the characteristics of faults F7(8) and F201, overburden stripping and fault

fault fracture zone

gouge dating in the Daliushu dam region, which verify that fault F201 is a regional seismogenetic fault that has been very active since the latest Pleistocene and Holocene, and determine the existence of left-lateral strike-slip motion for fault F7(8). 通过对大柳树坝址区 F201、F7(8) 断层带特征的追索观测、工程揭露和断层泥测年，获得 F201、F7(8) 断层活动特征的新资料，证实了 F201 断层是一条晚更新世晚期和全新世以来强烈活动的区域发震断层，同时确定了 F7(8) 断层水平左旋走滑作用的存在。

【中文定义】 指未固结或弱固结的泥状岩石。发育在地壳浅层脆性断层带中，呈各种彩色条带平行断层面展布，带宽几毫米至数十米不等。

fault gouge

fault groove

【释义】 断层擦沟，断层刻槽

fault scarp

【释义】 断层崖

【例句】 A fault scarp is a small step or offset on the ground surface where one side of a fault has moved vertically with respect to the other. 当断层的一盘相对另一盘做垂直向上运动时，便在地面上形成一个小的台阶或偏移，即断层崖。

【中文定义】 由断裂活动造成的陡崖称为断层崖。通常断层崖的走向线平直，在断层崖被侵蚀的过程中随着横贯断层河谷的扩展，完整的断层崖被分割成不连续的断层三角面。

fault sliding deformation monitoring

【释义】 断层错动变形量测

fault striation

【释义】 断层擦痕

【同义词】 slickenside; striation; scratch

【例句】 The stress of fault can be analyzed quantitatively by the attitudes of fault and the orientations of fault striation. 根据断层的产状和断层擦痕方向可定量解析断层应力场。

fault striation

【中文定义】 断层擦痕是指断层两盘相对错动时留下的痕迹，即在断层面上产生的沟槽状细微平行刻痕。组成断层擦痕的擦沟或擦槽往往一端宽且深，另一端窄而浅并逐渐消失，根据它的延长方向及擦槽形状，可以判别断层两盘相对运动的方向。手摸断层擦痕，有光滑感的方向代表断层另一盘滑动的方向。

fault structure

【释义】 断裂构造

【例句】 With the development of economy in Zhujiang delta, the fault structure and its process for earthquake disaster has been regarded again. 对于珠江三角洲地区的断裂构造问题及其在地震

fault structure

灾害中的作用，随着该区经济的迅速发展，又重新引起人们的重视。

【中文定义】 断裂构造又称断裂。断裂或断裂构造是指岩石因受地壳内的动力，沿着一定方向产生机械破裂，失去其连续性和整体性的一种现象。

fault triangular facets

【释义】 断层三角面

【例句】 The fault landforms mainly include: fault triangular facets, fault scarps, ladder-shaped diluvial terraces and buried pluvial fans, bedrock scarps in gully mouth, drop water, and nickpoints in gully bed. 断层构造地貌主要包括：断层三角面、断层陡坎、洪积阶地、埋藏型洪积扇以及冲沟裂点。

【中文定义】 指断层崖经河流或冲沟切割侵蚀后，形成的三角形陡崖，是现代活动断层的标志，常见于山区或山地与盆地、平原的分界处。

fault triangular facets

fault trough valley

【释义】 断层槽谷，断裂槽谷

【中文定义】 断层谷指沿断层线发育的谷。

fault trough valley

fault upper plate

【释义】 上盘

【同义词】 hanging wall; top wall

【例句】 A vertical fault has neither a footwall nor a fault upper plate. 具有直立断层面的断层既无上盘，也无下盘。

【中文定义】 上盘是地质学名词，断层的两盘之一。只有在倾斜断层中才有上下盘之分，断层层面倾斜时，位于断层面上部的称为上盘；位于断层面下部的称为下盘。

faulted basin

【释义】 断陷盆地

【例句】 The pore confined aquifers are mostly distributed in the deep overburden of the mountainous rivers, the quaternary faulted basin, and the thick sediments of piedmont alluvial-proluvial fan. 孔隙承压含水层多分布于山区河流深厚覆盖层、第四纪断陷盆地和山前冲洪积扇的巨厚沉积层内。

【中文定义】 断陷盆地指断块构造中的沉降地块，又称地堑盆地，它的外形受断层线控制，多呈狭长条状。

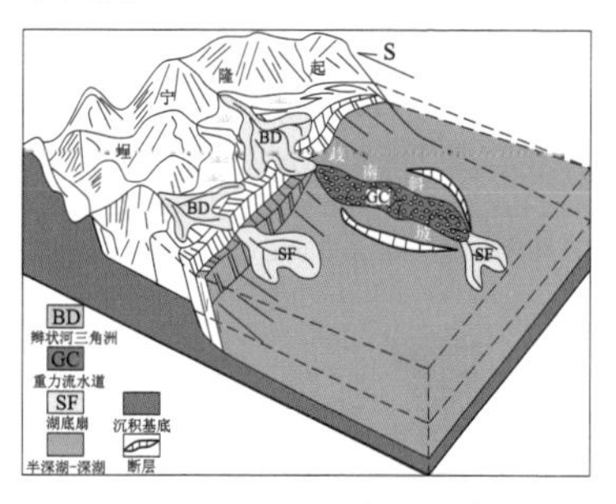

faulted basin

feasibility stage for landslide prevention project

【释义】 滑坡防治工程可行性论证阶段

feather joint

【释义】 羽状节理

【中文定义】 一组呈羽状斜列的节理。

feather joint

feedback information

【释义】 反馈信息

ferruginous rock

【释义】 铁质岩

【例句】 Tianheng mountain with altitude of 72m is connected with Danya mountain in the southeast, where the rock is red and belongs to the Sinian ferruginous rock. 田横山海拔72m，东南与丹崖山相连，岩石亦呈赭红，属震旦纪含铁质岩。

【中文定义】 含大量铁矿物的沉积岩，若其中铁矿物含量很高达到工业品位时，即为沉积铁矿。铁质沉积岩中常见的铁矿物包括赤铁矿、针铁矿、褐铁矿、磁铁矿等。

ferruginous rock

fiber monitoring

【释义】 光纤监测

【例句】 This article introduces the functions structure and design idea of an optical fiber monitoring system, OMS98, which is effective, flexible and reliable. 本文介绍了一种有效、灵活、可靠的OMS98光纤监测系统的功能结构和设计思想。

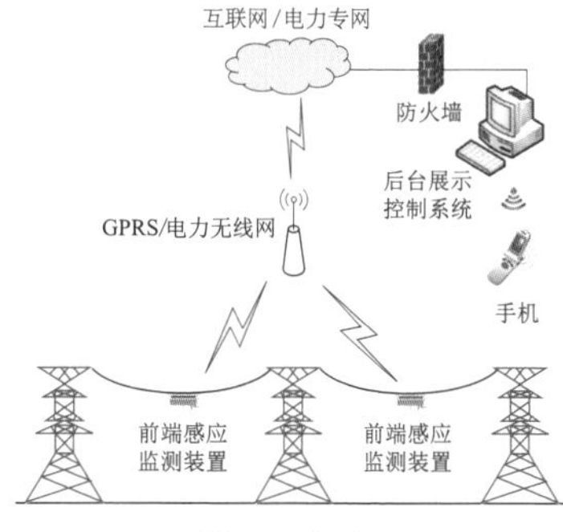

fiber monitoring

field investigation

【释义】 现场调查

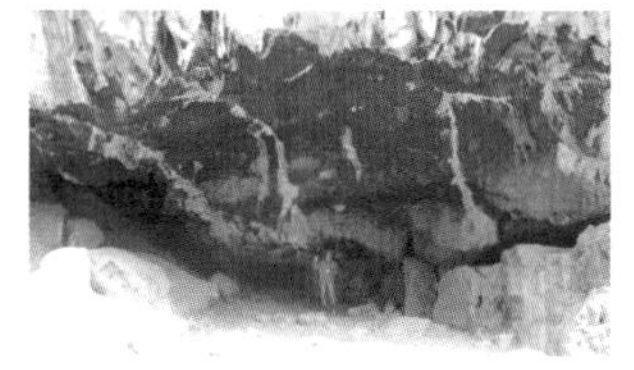

field investigation

field monitoring

【释义】 现场监测

【例句】 In the U. S., we are expanding our field monitoring team and increasing our use of onboard vehicle diagnosis technology. 在美国，我们开始扩大现场监测团队规模、增加对车载诊断技术的使用。

field reconnaissance

【释义】 野外查勘

【例句】 Palm GPS can rapidly and accurately conduct field reconnaissance to get engineering mapping and conduct boat measurement annotation, which is widely used in daily life. 掌上型GPS可以快捷准确地应用于野外查勘来进行工程测绘和船测调绘作业，并在生产与生活中得到了广泛的应用。

field reconnaissance

field work

【释义】 外业

【例句】 Field work data collection is not only the precondition and foundation of mapping projects, but also the most complex and difficult work in mapping projects. 外业数据采集是测绘项目的前提与基础，也是整个测绘项目中最复杂、工作环境最艰苦的工作。

field-acquired geological profile

【释义】 实测地质剖面图

【例句】 Aiming at the problem that it is a time-consuming and laborious work to draw the field-acquired geological profile, a section drawing method which calculates all the geological factors according to the rules of complex calculation was implemented. 针对实测地质剖面图绘制费时费力的问题，提出一种按照复数运算规则计算实测地质剖面要素的绘图方法。

【中文定义】 经实地测绘而编制成的地质剖面图。

filling [ˈfɪlɪŋ]

【释义】 *n*. 充填物，充填

【常用搭配】 argillaceous filling 泥质充填物；debris filling 岩屑充填物

【例句】 The thickness of filling affects the mechanical properties of structural plane. 充填物厚度影响结构面的力学性质。

【中文定义】 指充填于岩体结构面中的各种物质，主要由构造作用或风化作用、地下水作用形成。

filling

film water

【释义】 薄膜水

【例句】 For water drop in macroporous rocks, the surface tension is not large enough to dray them. However, for film water in pore rock, the surface is large enough to retain them. 对于大孔岩石中的水滴而言，表面张力不够大，不能将其托住，但对于小孔岩石中的薄膜水而言，表面张力就足够大可将其保留住。

【中文定义】 指在水蒸气凝结时形成的水或者是滴状液体水（重力水）离去后遗留在岩石中的水，在岩石微粒上围绕吸着水的薄膜形成较厚的薄膜水。

filter ['fɪltə]

【释义】 *n.* 过滤，反滤

【常用搭配】 filter material 反滤料；filter blanket 反滤铺盖

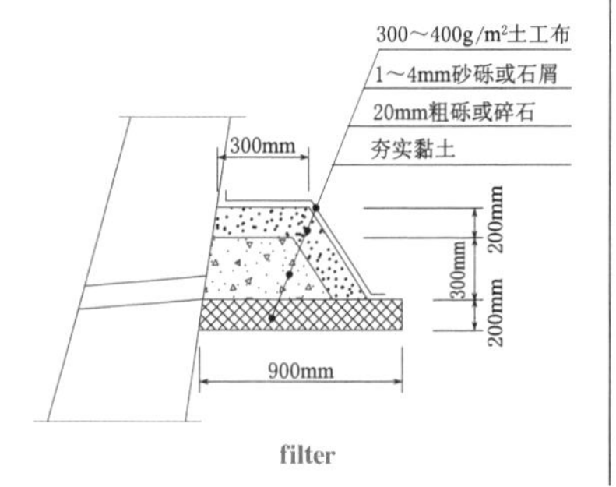

filter

filter material

【释义】 反滤料

【常用搭配】 filter layer 反滤层

【例句】 Filter material means approved sand and gravel or crushed rock from river bed, excavation, or quarries, used as a drainage material. 反滤料是一种排水材料，主要来源于河床、开挖场或石料场的砂、砾石或碎石。

【中文定义】 防止土体在不同级配土层界面或自由表面的渗流逸出处发生渗透变形，而又能排水的材料。

filter material

filtering ['fɪltərɪŋ]

【释义】 *n.* 滤波；*vt.* 过滤

【例句】 Synthetic Aperture Radar (SAR) speckle filtering techniques can be classified into three ca-tegories: multi - look processing, statistical filtering and filtering in transform domain. 合成孔径雷达（SAR）图像相干斑噪声滤波技术分成三类：多视处理、统计滤波和变换域滤波。

fine aggregate

【释义】 细骨料

【例句】 Provision shall be made so that the fine aggregate will bypass the rinsing screen and finish screens. 要有一定预防措施使细骨料不从漂洗筛及二次筛分筛通过。

【中文定义】 指混凝土中起骨架或填充作用的、粒径在 4.75mm 以下的粒状松散材料。

fine aggregate

fine grained texture

【释义】 细粒结构

【中文定义】 颗粒直径 2～0.2mm 的显晶质。

fine gravel

【释义】 细砾，细砾石

【常用搭配】 fine gravel aggregate 细砾骨料

【例句】 Study on dynamic properties of core material and dam-foundation fine gravel of earth-rock dam. 土石坝心墙料及坝基细砂砾料动力特性试验研究。

【中文定义】 指直径在 1～2mm 的砾石。

fine gravel

fine sand

【释义】 细砂

【中文定义】 粒径大于 0.075mm 的颗粒质量超过总质量的 85%，细度模数为 2.2～1.6 的土。

fine sand

fine-grained [fəin'greɪnd]

【释义】 细粒

【常用搭配】 fine-grained soil 细粒土

【例句】 With the rapid development of constructional engineering, especially the development of express highway and reservoir dam, which are beneficial to the people's livelihood, the diseases such as uneven sedimentation of structure, settlement, leakage and frost boiling have a harmful impact on the safety and service life of these major projects. The microshock liquation of saturated fine-grained soil, caused by dynamic load in the large area, generates these diseases. 随着建筑工程特别是关系国计民生的高速公路、水库大坝建设的迅猛发展，大区域饱和细粒土在动荷载作用下出现的微振液化，引发结构物不均匀沉降、沉陷、渗漏及翻浆等病害，影响高速公路等重点工程的安全及使用寿命。

【中文定义】 指小于或等于 0.075mm 的颗粒。

fineness modulus (FM)

【释义】 细度模数 (FM)

【例句】 Compared with natural sand, machine sand has the characteristics of poor grade gradation, bigger fineness modulus, rough surface, sharp particles and so on. 与天然砂相比，机制砂具有级配差、细度模数大、表面粗糙、颗粒尖锐等特点。

【中文定义】 评价砂粗细程度的一项指标，用筛分试验中孔径小于 5mm 各号筛的累积筛余百分率的总和除以 100 来表示。

finished cross-section

【释义】 竣工断面

finite element analysis

【释义】 有限元分析

【例句】 Simplified finite element analysis indicated that the stress release uplift during excavation is probably manageable. 简化有限元分析表明开挖期间应力释放引起的上抬变形或许是可控的。

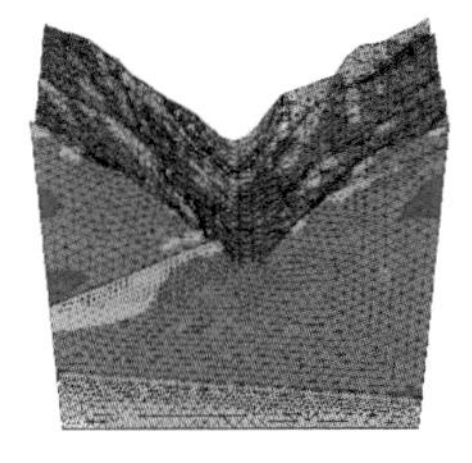

finite element analysis

fissure ['fɪʃə]

【释义】 *n.* 裂隙

【常用搭配】 fissure water 裂隙水；fissure zone 裂隙带

【例句】 Bedrock fissure water is one of the groundwater types, distributed widely in China. 基岩裂隙水是我国分布最为广泛的地下水类型之一。

【中文定义】 通常把岩体中产生的无明显位移的裂缝称为裂隙。

fissure water

【释义】 裂隙水，裂缝水

【例句】 The groundwater is composed of pore water,bedrock fissure water and vein-like fissure water in structural zones. 地下水有孔隙潜水、基岩裂隙水和构造带脉状裂隙水三种类型。

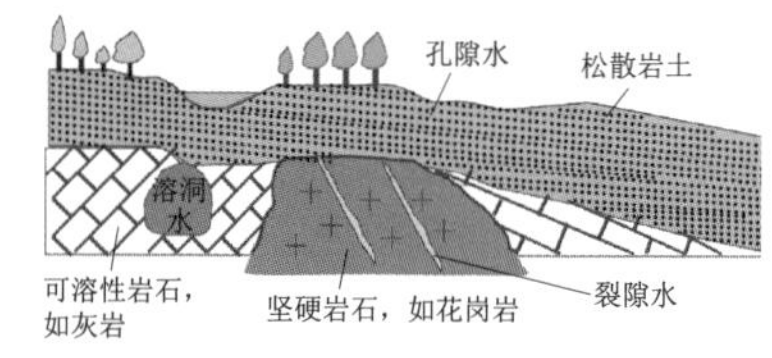

schematic diagram of fissure water

【中文定义】 指存在于岩石裂隙中的地下水；与孔隙水相比较，它分布不均匀，往往无统一的水力联系，是丘陵、山区供水的重要水源，也是矿坑充水的重要来源。

fitting ['fɪtɪŋ]

【释义】 *n.* 拟合

【例句】 Adoption of the above surface fitting and linear interpolation can be other non-test state of the physical point. 通过曲面插值和线性拟合的办法可以得到其他非测试状态点下的物理量。

【中文定义】 指已知某函数的若干离散函数值 $\{f_1, f_2, \cdots, f_n\}$，通过调整该函数中若干待定系数 $f(\lambda_1, \lambda_2, \cdots, \lambda_n)$，使得该函数与已知点集的差别（最小二乘）最小。

fixed offshore exploration platforms

【释义】 海上固定式勘探平台

fixed offshore exploration platforms

flake-shaped particle

【释义】 片状颗粒

【中文定义】 指厚度小于该颗粒所属平均粒径0.4倍的颗粒。

flat jack test

【释义】 扁千斤顶试验

【例句】 The deformation modulus of rock mass can be obtained by flat jack test. 扁千斤顶试验可获取岩体的变形模量。

flexible foundation

【释义】 柔性基础

【例句】 Flexible foundations and wider foundation dimensions should be used as far as possible for nonuniform foundation in mountainous areas. 山区不均匀地基应尽量采用柔性基础和较宽的基础尺寸。

【中文定义】 指刚度小、在竖向荷载作用下无抗弯能力，完全随地基变形的基础。

flexible foundation

flexural toppling

【释义】 弯曲倾倒

【例句】 Flexural toppling failure is one of the most common failure types of counter-tilt slopes of layered rock. 弯曲倾倒破坏是岩质反倾边坡的一种主要失稳破坏模式。

【中文定义】 陡倾层状岩体在重力作用下向临空方向发生弯曲的现象。

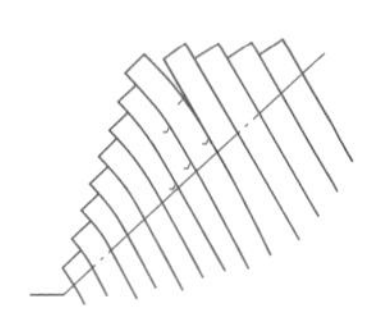

flexural toppling

flexure ['flekʃə]

【释义】 *n.* 挠曲

【常用搭配】 flexure curve 挠曲曲线

【例句】 Laminated composite has a new type structure which is used to improve the fracture toughness and flexure strength. 叠层复合是一种新型的复合构型，它被用来提高材料断裂韧性和挠曲强度。

【中文定义】 在水平或平缓的岩层中，由一般岩层突然变陡而表现出的膝状弯曲，或是由于岩层翘曲或其他和缓变形所形成的弯曲均称挠曲。

flexure

floating offshore exploration platforms

【释义】 海上漂浮式勘探平台

floating offshore exploration platforms

flood plain

【释义】 河漫滩，泛滥平原

【例句】 The majority of floods occur in areas underlain by Quaternary deposits, for it is the recent sediments which form the flood plains and river bottoms. 大多数洪水发生在第四纪沉积物分布的区域，因为河漫滩和河床底部正是由这些近代沉积物组成的。

【中文定义】 指河谷底部在洪水期才被淹没的部分，位于河床和河谷谷坡之间，由河流的横向迁移和漫堤的沉积作用形成。

flood plain

fluvioglacial terrace

【释义】 冰水阶地

【中文定义】 在山谷及山前地带，冰期时冰水形成的大量碎屑堆积，经后期流水的侵蚀切割而造成的阶地，称冰水阶地。

fluvioglacial terrace

flow characteristics

【释义】 流动特性

【例句】 To optimize the design of pump-turbine runner, we should understand the flow characteristics of the pump turbine, including the runner. 要优化水泵水轮机转轮设计，应了解包括转轮在内的水泵水轮机全流道的流动特性。

flow [floʊ]

【释义】 *vi.* 流动，涌流，川流不息，飘扬；*vt.* 淹没，溢过；*n.* 流动，流量，涨潮，泛滥

【常用搭配】 flow chart 流程图；flow control 流量控制

【例句】 The conservation of momentum equations for the compressible, viscous flow case are called the Navier-Stokes equations. 可压缩的黏性流的动量守恒方程被称为纳维-斯托克斯方程。

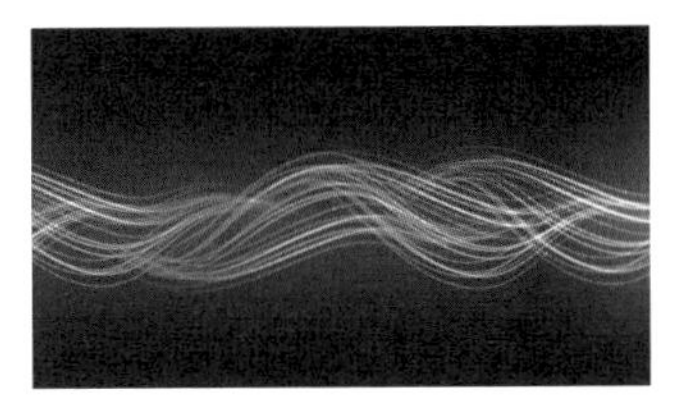

flow

flow cleavage

【释义】 流劈理

【例句】 It's gliding nappe system, which was composed by C－P structure layer, developed gliding tectonics such as synclinorium, bedding recumbent fold, penetrative stretching and flow cleavage, etc. 滑覆系统由C－P构造层组成，发

育有复式向斜、顺层掩卧褶皱、透入性拉伸线理和流劈理等滑动构造。

【中文定义】 流劈理是劈理的一种类型。岩层、岩体或矿体在地应力作用或变质作用下，沿着一定方向产生的裂隙构造。在强烈褶曲的岩层、断层两侧的岩体和变质岩内较为发育。

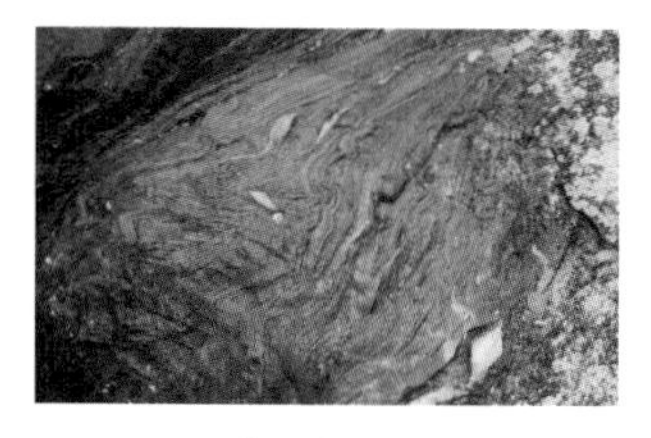

flow cleavage

flow cross-section

【释义】 过水断面

【同义词】 water-carrying section

【中文定义】 某一研究时刻的水面线与河底线包围的面积，称过水断面。

flow direction monitoring

【释义】 流向监测

flow logging

【释义】 流量测井

flow velocity monitoring

【释义】 流速监测

flow velocity monitoring

fluctuation [ˌflʌktʃʊ'eʃən]

【释义】 *n*. 起伏度

【例句】 The fluctuation degree of structural plane has an important influence on shear strength. 结构面起伏度对抗剪强度有重要影响。

【中文定义】 表征结构面在延伸方向的表面起伏程度的一个指标。

fluid ['fluːɪd]

【释义】 *adj*. 流动的，流畅的，不固定的；*n*. 流体，液体

【常用搭配】 brake fluid 制动液；fluid mechanics 流体力学；fluid dynamics 流体动力学

【例句】 Otherwise, fluids are generally viscous, a property that is often most important within a boundary layer near a solid surface, where the flow must match onto the no-slip condition at the solid. 另外，流体通常具有黏性，表现在固体表面形成一个边界层，保证其流动匹配固体上的无滑移条件。

【中文定义】 指一种受任何微小剪切力的作用都会连续变形的物体。

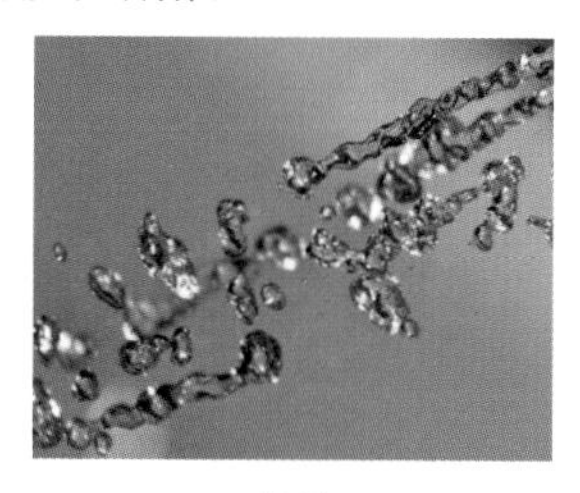

fluid

fluid bowl

【释义】 溶槽

【例句】 The double-arch tunnel passes through a large location of fluid bowl by setting foundation beam method. 双跨连拱隧道以设置地基梁的方式穿过大型溶槽地段。

【中文定义】 地表水沿可溶蚀岩石的节理裂隙进行溶蚀与侵蚀，形成纵横交错的凹槽称为溶沟，深度较大的称溶槽。

fluorite ['flʊəˌraɪt]

【释义】 *n*. 氟石，萤石

【常用搭配】 fluorite structure 萤石型结构；fluorite deposit 萤石矿床

【例句】 The average homogenization temperature of inclusions in fluorite is 250℃. 萤石中包裹体的平均均化温度为 250℃。

【中文定义】 是一种矿物，等轴晶系，其主要成分是氟化钙（CaF_2）。

fluorite

fluvial accumulation

【释义】 河流堆积

【中文定义】 被河流搬运的物质，因河流搬运能力减弱而堆积下来的现象。

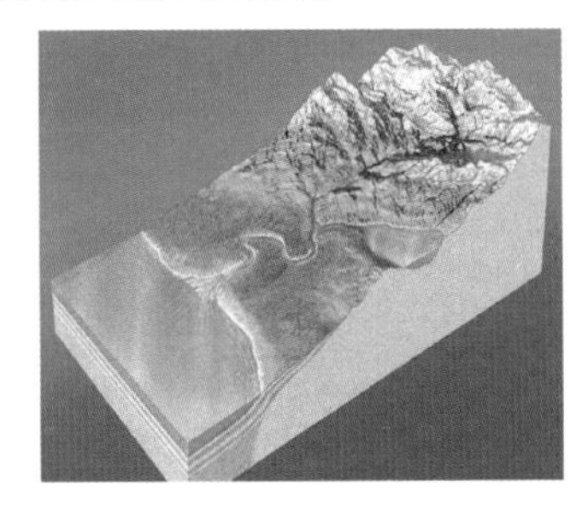

fluvial accumulation

fluvial erosion accumulation

【释义】 河流侵蚀堆积

fluvial erosion accumulation

fluvial facies

【释义】 河相，河流相

【常用搭配】 Fluvial facies coefficient 河相系数；fluvial facies sandstone 河流相砂岩，河流相砂体

【例句】 Plane heterogeneity was resulted from fast variations of fluvial facies. 河流相相变快是造成平面非均质性的重要因素。

【中文定义】 又称冲积相，属陆相沉积类型，是由河流或其他径流作用形成的一套沉积物或沉积岩。它的亚相类型主要有：谷底滞流沉积、边滩沉积、心滩沉积、天然堤沉积、决口扇沉积、河漫滩沉积、废弃河道填谷沉积等等。古代河流沉积的主要标志主要为：岩石由砾砂、粉砂、黏土等碎屑沉积物组成，成分成熟度低，常见底部冲刷面，其上有泥砾；分选性差到中等，粒度分布多双峰态，粒度概率图显两段式；层理类型多样，以反映单向水流的大型槽状和板状交错层理为特征，波痕不对称，可见砾石呈迭瓦状排列；具明显的间断正韵律，砂体呈透镜状，平面上沿水流方向呈弯曲的带状分布。

fly ash

【释义】 粉煤灰

【例句】 The cement-fly ash-gravel pile is widely used in the collapsible loess, clay soil, silt and sandy soil, bue the usability in the artificial filled soil and silty clay isn't recognized. 水泥粉煤灰碎石桩在湿陷性黄土、黏性土、粉土、砂土地区已经被应用和推广，但在人工填土和淤泥质粉质黏土中能否应用，工程界尚未统一认识。

【中文定义】 是从煤燃烧后的烟气中收集下来的细灰，是燃煤电厂排出的主要固体废物。

fly ash

focal depth

【释义】 震源深度

【例句】 After the relocation the maximum epicenter offset is 10.93km, the minimum one is 0.08km, and the mean is 1.87km. The mean focal depth is 6.59km. 重新定位后的地震最大偏移量为 10.93km，最小为 0.08km，平均偏移量为 1.87km，平均震源深度为 6.59km。

【中文定义】 地震波发源的地方，称为震源。震源在地面上的垂直投影，即地面上离震源最近的一点称为震中，它是接受振动最早的部位。震源到地面的垂直距离称为震源深度。

focal mechanism solution

【释义】 震源机制解

【例句】 A host of focal mechanism solutions show that there is a locally small regional stress field with horizontal tensile stress with the dominant horizontal extrusion component in the north of Yuncheng basin. 大量震源机制解结果表明，运城盆地存在以水平拉张为主要特点的局部小区域应力场，盆地北部水平挤压分量占优势。

【中文定义】 震源机制解亦称断层面解，是用地球物理学方法判别断层类型和地震发震机制的一种方法。震源机制解不仅可以了解断层的类型（是正断层、逆断层还是走滑断层），还可以揭示断层在地震前后具体的运动情况。

fold [fould]

【释义】 *n.* 褶皱

【例句】 General direction of the structural line is NW-SE and the main structures consist of folds and faults running NNW and nearly EW, with minor secondary faults running NW and NE. 构造线总体方向为 NW-SE，主要构造为 NNW 向和近 EW 向的褶皱和断层，其次为 NW、NE 向的次级断层。

【中文定义】 褶皱是岩石中的各种面（如层面、面理等）受力发生的弯曲而显示的变形。

fold model

fold axis

【释义】 褶皱轴

【例句】 By statistics and analysis of fold axis, lines and conjugate joints in structures of every stages, the paleotectonic stress fields in Indo-chinese stage, Yanshan stage and Himalayan stage were resumed. 通过对各期构造的褶皱轴迹和共轭节理的统计分析，恢复了印度支那阶段、燕山期和喜马拉雅期的古构造应力场。

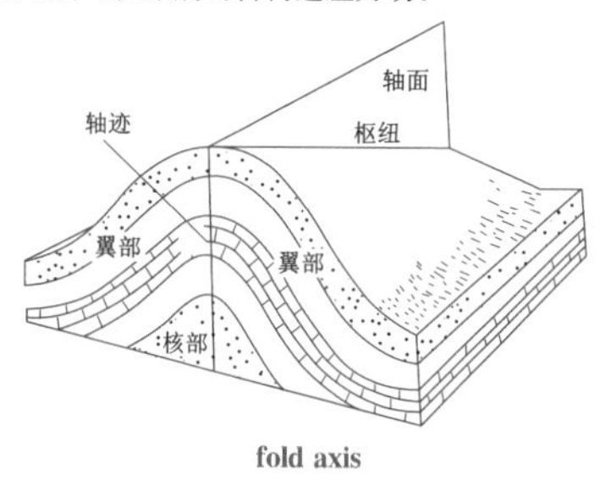

fold axis

【中文定义】 褶皱轴是轴面与水平面的交线，即褶皱在水平面上的轴迹，代表褶皱的延伸方向，常用于地质构造图。

fold belt

【释义】 褶皱带

【例句】 Nd isotopic model ages of basement rocks suggested that the crustal age of Yili plate in the north (1.9Ga) is younger than that of the South Tianshan (STS) fold belt in the south (2.2Ga). 本区北部伊犁地块基底岩石的 Nd 同位素模式年龄约为 1.9Ga，小于部南天山褶皱带的 2.2Ga。

【中文定义】 指地槽中的沉积岩层经过剧烈的地壳运动后，由线型褶皱组合上升形成的强烈构造变形地带。

fold belt

fold system

【释义】 褶皱系

【例句】 In the northern regions as Junggar fold system and Tianshan fold system, Q_0 was also lower (250 - 300) and η was varying between 0.5 and 0.9。在北部的准噶尔褶皱系和天山褶皱系，Q_0 呈现出较低的值，在 250～300 的范围内变化，η 值的变化范围为 0.5～0.9。

【中文定义】 是由若干个褶皱带和夹持其间的中间地块所组成的一个大尺度的、地壳上的一级构造单元。

fold system

folded mountain

【释义】 褶皱山

【中文定义】 指由地壳运动形成的褶皱岩层所组成的山体，是地表岩层受水平方向的构造作用力而形成岩层弯曲的褶皱构造山地。

folded mountain

foreshock [ˈfɔːˌʃɑk]

【释义】 *n.* 前震

【例句】 Although the five foreshocks differed from each other in total magnitude, their waveforms were remarkably similar. 尽管这五次前震的震级不同，但它们的波形却显著相似。

【中文定义】 指在一个地震序列中，地面的前几次震动。

Formation [fɔrˈmeʃən]

【释义】 *n.* 组

【中文定义】 指具有相对一致的岩性、岩相、变质程度和一定结构类型的地层体，是基本的岩石地层单位。

forward intersection method

【释义】 前方交会法

fossil [ˈfɑːsl]

【释义】 *n.* 化石；*adj.* 化石的

【常用搭配】 index fossil 标准化石；living fossil 活化石；fossil fuel 化石燃料

【例句】 The fossil lungfish is a new species, but a name cannot be given because of its uncertain origin. 这种化石的肺鱼是新的物种，但因为它不确定的来源导致至今无法命名。

【中文定义】 指地壳中保存的属于古地质年代的动物或植物的遗体、遗物或生物留下的痕迹。

trilobite fossil

foundation [faʊnˈdeʃən]

【释义】 *n.* 地基

【常用搭配】 reinforced soil foundation 加筋土地基

【例句】 Comparing with no reinforcement the bearing behavior of geogrid reinforced soil foundation are improved, and the effect of two-layer reinforcement is significantly better than one-layer reinforcement. 土工格栅加筋土地基与无筋地基相比，承载性能得到改善，双层加筋明显优于单层加筋。

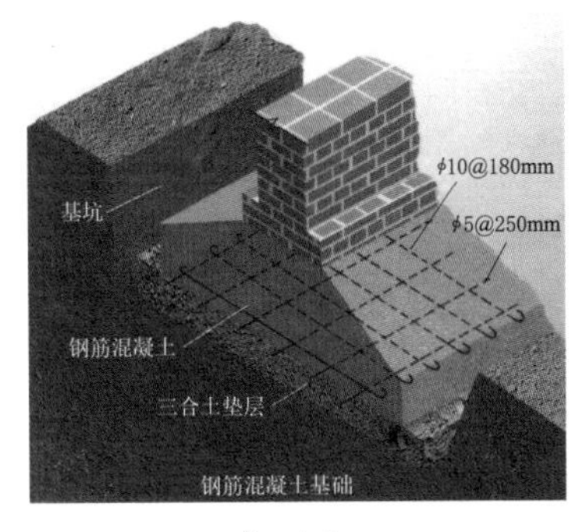

foundation

foundation data model

【释义】 基础数据模型

【中文定义】 基于工程地质勘察活动采集的数据建立的地质测绘模型、勘探模型、物探模型、试验模型、观测模型，并作为地质体分析建模的基础。

foundation failure

【释义】 地基失效

【例句】 When liquefaction happens, the building subsidence and sometimes with serious tilting may be caused by foundation failure. 当砂土液化发生时，地基失效导致建筑下沉，有时还会严重倾斜。

【中文定义】 地基土因载荷过大或承载力降低，发生较大变形或破坏而导致建筑物失稳的现象。

foundation pad

【释义】 基础垫层

【例句】 Based on the character and practical status of salt-affected soil, it is economical, reliable, simple and feasible to adopt replacement combined with deepen foundation pad to treat the foundation. 根据盐渍土的特性及实际情况，采用换土法并结合加深基础垫层等方法对地基进行处理是

经济可靠和简便易行的。
【中文定义】 指设置在基础和地基土之间的、用于隔水、排水、防冻以及改善基础和地基工作条件的低强度等级混凝土、三合土、灰土等铺贴层。

foundation settlement

【释义】 地基沉降
【例句】 It shows that cofferdam dyke and foundation settlements have obvious aging characters, and the time-settlement relationship can be fitted by negative exponent secular character curve. 结果表明：围堰堤坝及地基沉降具有显著的时效特性，沉降与时间的相关关系可用负指数时效特性曲线拟合。
【中文定义】 指地基土层在附加应力作用下压密而引起的地基表面下沉。过大的沉降，特别是不均匀沉降，会使建筑物发生倾斜、开裂而不能正常使用。

foundation treatment

【释义】 地基处理，基础处理
【例句】 The design of replacement underlayer thickness is crucial to the replacement method of foundation treatment. 换土垫层厚度设计是换填法地基处理设计的关键。
【中文定义】 为提高地基强度，或改善其变形性质或渗透性质而采取的工程措施。

foundational fault-block

【释义】 基底断块
【例句】 Within these blocks foundational fault-blocks and superficial fault-blocks may also be differentiated. 在断块内部，基底断块和盖层断块存在着差异。
【中文定义】 基底断块是地壳断块内部更次一级的、被基底断裂切割和围限的断块。它可沿康拉德面滑动，均分布在现在大陆区。

fraction ['frækʃn]

【释义】 *n.* 粒组

fracture ['fræktʃə]

【释义】 *n.* 破裂，断裂，断口
【常用搭配】 conchoidal fracture 贝壳状断口；splintery fracture 参差状断口；step-like burst fracture 梯状破裂断口
【例句】 In the end, microscopic characteristics of rheological failure shaping fracture of rock under different confining pressures were analyzed. 最后，对不同围压作用下岩石流变破裂断口微细观特征进行了分析。
【中文定义】 材料在受到外力作用发生的随机的无方向性的破裂，破裂面呈各种凹凸不平的形状，称为断口。

fracture cleavage

【释义】 破劈理
【例句】 The second episode is characterized by NW-SE direction compression-shortening deformation with formation of two conjugate fracture cleavages or kink bands. 第二期为平行造山带的挤压缩短变形，形成区域透入性共轭破劈理或膝折带。

fracture cleavage

【中文定义】 破劈理是雷思（C. K. Leith，1905年）和克尼尔（J. L. Knill，1960年）曾列为劈理成因分类中的一种，是指岩石中一组与矿物的排列方向无关的、密集的平行破裂面，一般为剪裂面。

fracture fillings

【释义】 裂隙充填物
【中文定义】 充填于裂隙内的物质。

fracture surface

【释义】 折断面
【例句】 The toppling fracture surface then stretches from the middle part of the bank slope to the slope crest and cuts through, shaping major fracture surface. 倾倒折断面沿岸坡中部向坡顶延伸并贯通形成主折断面。

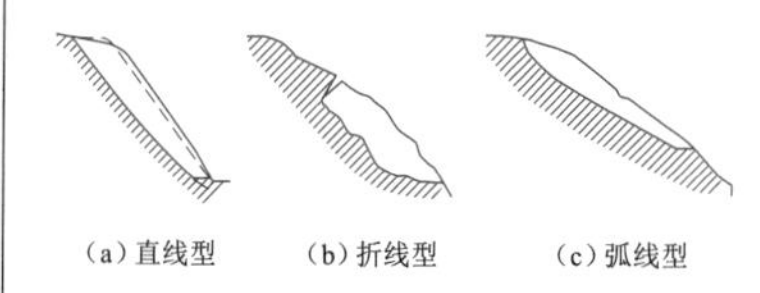

Basic shape of fracture surface

fracturing ['fræktʃərɪŋ]

【释义】 *n.* 龟裂，破碎

【中文定义】 沉积物暴露于水面之上而形成的裂纹，在剖面上呈 V 字形，在平面上形似龟裂纹。

fracturing

fragmental rock

【释义】 碎块岩

fragmental rock

fragmental structure

【释义】 碎屑状结构

【中文定义】 岩石中的颗粒是机械沉积的碎屑物。碎屑物可以是岩石碎屑、矿物碎屑、石化的有机体或其碎片以及火山喷发的固体产物等。

framework [ˈfreɪmwɜːk]

【释义】 *n*. 格构

【常用搭配】 framework frame 格构梁

【例句】 Compound structure of reinforced concrete framework frame and anchor can be applied to reinforce the slope whose stability is bad. 现浇钢筋混凝土格构梁与锚杆（管）复合结构适用于斜坡稳定性较差、有一定下滑力的情况。

framework

free oscillation test

【释义】 自由振荡法试验

【例句】 Free oscillation test is a hydro-geological test which adopts excitation method to make an instantaneous change in borehole water level. 自由振荡法试验是一种通过一定激发手段使钻孔内水位发生瞬时变化的水文地质试验。

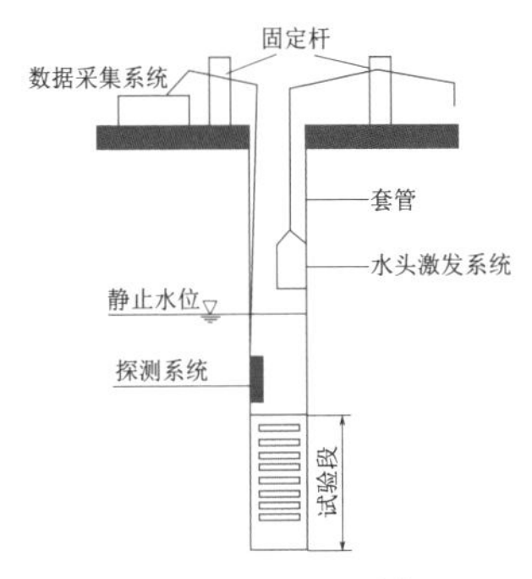

instrument diagram of free oscillation test

free section

【释义】 自由段

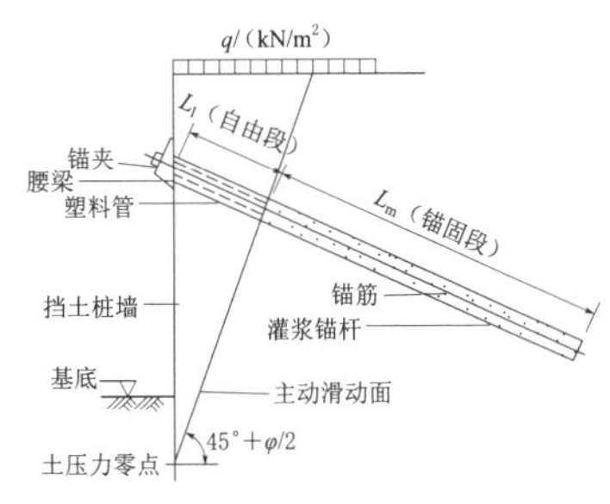

free section

free swelling ratio

【释义】 自由膨胀率

free water

【释义】 自由水，游离水

【中文定义】 岩土中在重力作用下能自由运动的地下水。它能传递静水压力，有溶解能力，易于流动。

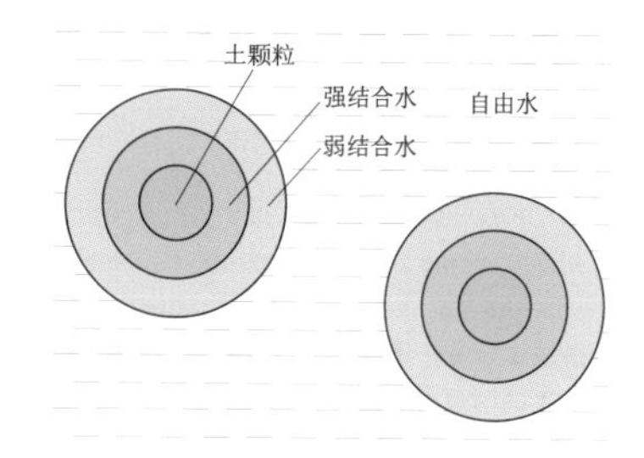

free water

free-flow tunnel

【释义】 无压隧洞

free-flow tunnel

fresh [frɛʃ]

【释义】 *adj*. 新鲜的

【例句】 Generally, the foundation of high arch dam is normally required to be placed on fresh or slightly weathered rock. 通常高拱坝建基面要求开挖至新鲜～微风化基岩。

【中文定义】 裂隙面紧密、完整或焊接状充填，仅个别裂隙面有锈膜浸染或轻微蚀变。

fresh

fresh groundwater

【释义】 地下淡水

【例句】 The effects of tides and the density difference between seawater and fresh groundwater are neglected in the model. 在模型概化过程中，忽略了海潮作用以及海水与地下淡水之间的密度差异等。

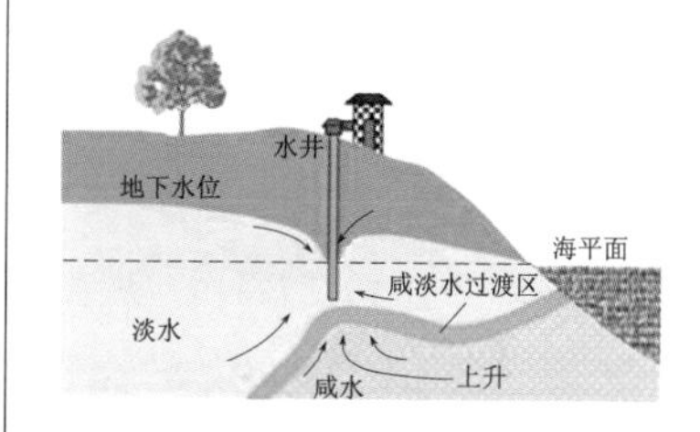

the seawater has infiltrated

fresh groundwater

【中文定义】 含盐量小于 0.5g/L 的地下水。

frictional strength

【释义】 摩擦强度

【例句】 Based on the experimental results, the authors here suggest that the crust composed of sandstone possesses the maximum failure strength and frictional strength at the depth of about 10km. 根据实验结果，笔者推测由砂岩组成的地壳在地下 10km 左右的深度有最大破坏强度和摩擦强度。

frictional strength of structural plane

【释义】 结构面的面摩擦强度

【例句】 Once the shear stresses exceed the frictional strength of structural plane, the fault will be unlocked, and it will slip suddenly. 当剪应力超过断层面的摩擦强度时，断层解锁并突然错动。

friction-resistance ratio

【释义】 摩阻比

front edge of landslide

【释义】 滑坡体前缘

frost heaving

【释义】 冻胀

【例句】 After cold winters, frost heaving and boiling, pits and cracks frequently occur on roads, especially in early April. 经历了一个寒冷的冬季，特别是 4 月初路面被反复地冻融，导致冻胀、翻浆、坑槽、裂缝的情况发生。

frost-heave amount

【释义】 冻胀量

frost-heave force

【释义】 冻胀力

【例句】 The horizontal frost-heave force acting on the retaining wall is one of the primary design loads in seasonal frost regions. 作用于挡墙上的水平冻胀力是季节冻土区挡土墙设计的主要荷载。

frozen soil

【释义】 冻土

【例句】 The research of the failure theory and criterion is the foundation of the frozen soil fracture mechanics. 研究冻土的破坏理论和破坏准则，是冻土力学的基本理论问题。

【中文定义】 冻土是指零摄氏度以下，并含有冰的各种岩石和土壤。一般可分为短时冻土（数小时/数日至半月)、季节冻土（半月至数月）以及多年冻土（又称永久冻土，指的是持续两年或两年以上的冻结不融的土层)。

frozen soil

fry-dry method

【释义】 炒干法

fly-dry method

full face tunnel boring

【释义】 全断面隧道掘进

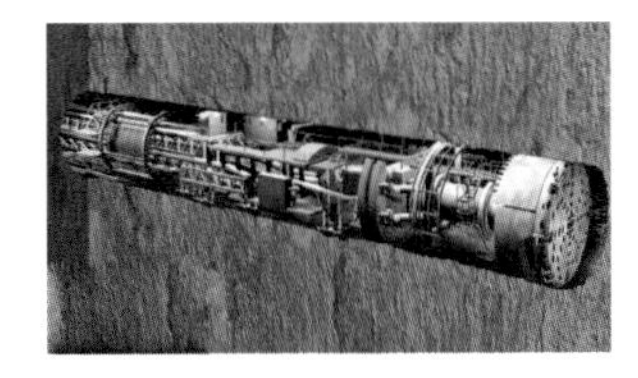

full face tunnel boring

G

gabbro ['gæbroʊ]

【释义】 *n.* 辉长岩

【常用搭配】 alkali gabbro 碱性辉长岩；orthoclase gabbro 正长辉长岩；olivine gabbro 橄榄辉长岩

【例句】 The Haladala mafic pluton in the southwestern Tianshan Mountains mainly consists of troctolite, olivine gabbro and gabbro. 西南天山哈拉达拉杂岩体主要由橄长岩、橄榄辉长岩和辉长岩组成。

gabbro

gamma-ray survey

【释义】 伽马测量

gap gradation

【释义】 间断级配

【例句】 The bigger the size of intermittent division in gap gradation is adopted, the higher the strength becomes. 间断级配间断区间的粒径越大，二灰碎石的强度越高。

【中文定义】 在矿料颗粒分布的整个区间里，从中间剔除一个或连续几个粒级，形成一种不连续的级配，称为间断级配。

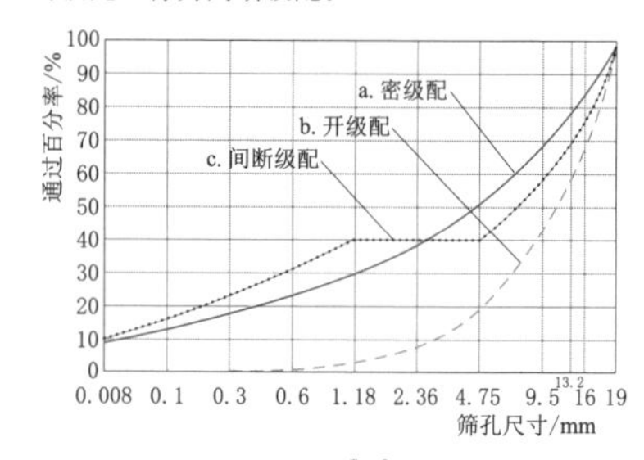

gap gradation

gap model

【释义】 缝隙模型

gap-graded soil

【释义】 不连续级配土

garnet ['gɑrnɪt]

【释义】 *n.* 石榴石

【例句】 A mineral like garnet, for example, has a number of varieties. 比如，石榴石族矿物有多个种类。

garnet

gas

【释义】 瓦斯

Gauss-Krueger plane rectangular coordinate system

【释义】 高斯平面坐标系

【中文定义】 根据高斯-克吕格投影所建立的平面坐标系，简称高斯平面坐标系。

Gauss-Krueger Projection

【释义】 高斯-克吕格投影

【中文定义】 又称为横轴墨卡托投影、切圆柱投影，其投影带中央子午线投影成直线且长度不变，赤道投影也为直线，并与中央子午线正交。

general investigation

【释义】 普查

general map

【释义】 普通地图

【中文定义】 综合反映地表的一般特征，包括主要自然地理和人文地理要素，但不突出表示其中的某一种要素的地图。

genetic analysis

【释义】 成因分析

【例句】 According to the genetic analysis of mountain channels, the treatment and prevention measures were proposed coming with practice. 针对山区渠道滑坡成因分析，结合实际，提出了滑

坡处理、防治措施。

【中文定义】 形成原因的分析。

genetic classification

【释义】 成因分类

genetic type

【释义】 成因类型

【例句】 To determine the genetic type of deposit is one of the important methods for exploration. 确定矿床成因类型是明确找矿方向的主要方法。

gentle slope

【释义】 平缓坡

【中文定义】 水电工程实践中，习惯将地形坡度小于25°的岸坡称为平缓坡。

geocell

【释义】 *n.* 土工格室，土工固格网

【常用搭配】 geocell cushion 土工格室垫层

geocell

【例句】 Geocell is a new geosynthetics which has spatial configuration. 土工格室是一种具有立体结构的新型土工合成材料。

geocentric coordinate system

【释义】 地心坐标系

【例句】 It is an important mission of Geodesy to improve the accuracy of control networks and to establish the geocentric coordinate system via combined satellite and terrestrial network adjustment. 通过地面网与卫星网的联合平差，以改善控制网的精度和建立高精度的地心坐标系统，是当前大地测量的一项重要任务。

【中文定义】 以地球质心为原点建立的空间直角坐标系，或以球心与地球质心重合的地球椭球面为基准面所建立的大地坐标系。

geochronologic unit

【释义】 地质年代单位

【例句】 The boundaries of geochronologic units are defined by GSSP. 地质年代单位的边界是由全球年代地层单位界线层型剖面和点位确定的。

【中文定义】 地质年代单位是指地质时期中的时间划分单位，又称地质时间单位。划分的主要依据是生物演化的不可逆性和阶段性。

代码	符号	名称
	宙	
101	PN	显生宙
102	PT	元古宙
103	AR	太古宙
104	HD	冥古宙
	代	
201	Kz	新生代
202	Mz	中生代
203	Pz	古生代
204	Pz_2	上古生代
205	Pz_1	下古生代
206	Pt_3	上元古代
207	Pt_2	中元古代
208	Pt_1	下元古代
209	Ar_3	上太古代
210	Ar_2	中太古代
211	Ar_1	下太古代
212	An∈	前寒武纪
213	AnZ	前震旦纪
	纪	
301	Q	第四纪
302	R	第三纪
303	N	上第三纪

代码	符号	名称
	世	
401	Qh Q^4	全新世
402	Q_p	更新世
403	Q^3	上更新世
404	Q^2	中更新世
405	Q^1	下更新世
		上第三纪
406	N_2	上新世
407	N_1	中新世
		下第三系
408	E_3	渐新世
409	E_2	始新世
410	E_1	古新世
411	K_2	上白垩世
412	K_1	下白垩世
413	J_3	上侏罗世
414	J_2	中侏罗世
415	J_1	下侏罗世
416	T_3	上三叠世
417	T_2	中三叠世
418	T_1	下三叠世
419	P_2	上二叠世
420	P_1	下二叠世
421	C_3	上石炭世

代码	符号	名称
435	$\in_2$	中寒武世
436	$\in_1$	下寒武世
437	Z_2	上震旦世
438	Z_1	下震旦世
		青白口纪
439	Qn_2	上世
440	Qn_1	下世
		蓟县纪
441	Jx_2	上世
442	Jx_1	下世
		长城纪
443	Ch_2	上世
444	Ch_1	下世
		期
501	K_2m	马斯特里赫特期
502	K_2kf	坎潘期
503	K_2sd	三冬期
504	K_2k	康尼亚克期
505	K_2l	土仑期
506	K_2s	赛诺曼期
507	K_1ar	阿尔比期
508	K_1a	阿普第期
509	K_1bl	巴列第期

geochronologic unit

geodetic azimuth

【释义】 大地方位角

【例句】 In order to transform conveniently the directional angle of a geodesic line into the geodetic azimuth, the solution of the meridianal convergence is expressed by geodesic coordinates of this point. 该解式直接采用该点的测地坐标来求解子午线收敛角，从而能方便地将测地坐标系中任一大地线的方向角转换为大地方位角。

【中文定义】 参考椭球面上一点的大地子午线与该点到目标点大地线之间的夹角，由大地子午线（北向）顺时针量取。

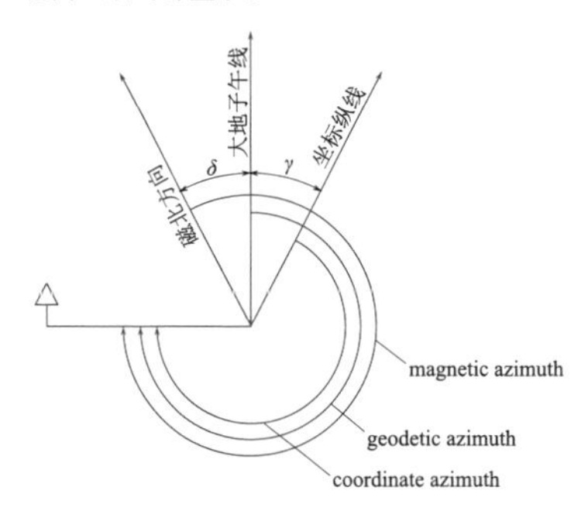

geodetic azimuth, coordinate azimuth, magnetic azimuth

geodetic coordinate

【释义】 大地坐标

【中文定义】 大地测量中以参考椭球面为基准面的坐标，通常以大地经度、大地纬度和大地高表示。

geodetic coordinate system

【释义】 大地坐标系

【例句】 The practical formulas applying it to the 1954' and 1980' geodetic coordinates system of China are given. 给出了适用于我国 1954 年大地坐标系和 1980 年大地坐标系的实用公式。

【中文定义】 以参考椭球面为基准面建立起来的坐标系。

geodetic database

【释义】 大地测量数据库

【例句】 The experiment, which is based on the geodetic database in Hubei province, has been made to analyse the effect of the range, account and precision grade of known control points on the calculating result. 基于湖北省大地测量数据库，试验对已知控制点的范围、计算量和精度等级对计算结果的影响进行了分析。

【中文定义】 利用计算机存储的各种大地测量数据的数据文件及数据管理软件的文件集合。

geodetic height

【释义】 大地高程

【中文定义】 地面点沿法线到参考椭球面的距离。

geodetic latitude

【释义】 大地纬度

【中文定义】 参考椭球面上某点的法线与赤道面的夹角。

geodetic longitude

【释义】 大地经度

【中文定义】 参考椭球面上起始大地子午面与某点的大地子午面的夹角。

geodetic meridian

【释义】 大地子午线

【中文定义】 指地面上连接南北两极的线。

geodetic meridian plane

【释义】 大地子午面

【中文定义】 与地球自转轴平行，或包含地球椭球体短轴的平面；是量度经度的起始面或终止面，大地子午面包含地球椭球体短轴及其法线。

geoenvironmental engineering

【释义】 环境岩土工程

geofabriform

【释义】 *n.* 土工模袋

【例句】 The construction method of geofabriform in the bank protection engineering of Bingjiang Stage of Yongan dyke in Changjiang River is stated. And the action of water control about the geofabriform is also stated. 叙述了用混凝土模袋在长江永安堤滨江段进行护岸的施工方法，并阐述了该工艺在治水方面的重大作用。

geofabriform

geographic grid

【释义】 地理坐标网

【中文定义】 按经纬线划分的坐标格网。

geographic information systems (GIS)

【释义】 地理信息系统

【例句】 While they have historically been drawn by hand, maps are often created today by using a modern geographic contribution, geographic information systems (GIS). 地图过去主要采用手绘，然而现在通常通过现代地理工具——地理信息系统（GIS）来制图。

【中文定义】 在计算机软硬件支持下，把各种地理信息按照空间分布，以一定的格式输入、存储、检索、更新、显示、制图和综合分析的技术系统。

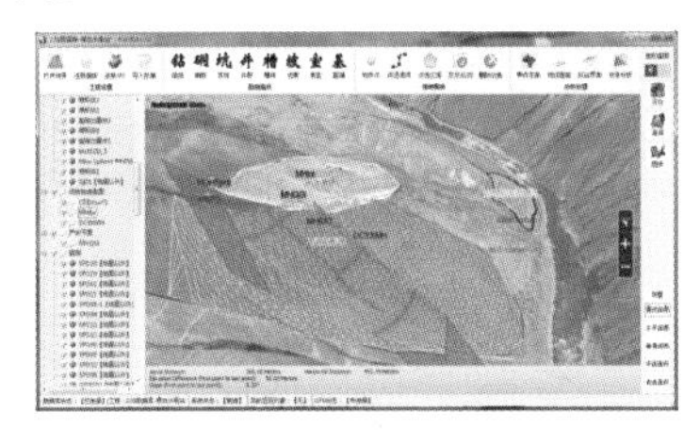

GIS

geographic name database

【释义】 地名数据库

【例句】 In order to extract the boundary of rural habitation, based on geographic name database and basic geographic information data, an extraction method that use polygon aggregation is raised. 为了提取农村居住地的边界，基于地名数据库和基础地理信息数据，列举了一种使用多边形合并的提取方法。

【中文定义】 利用计算机存储的各种地名信息的数据及数据管理软件的文件集合。

geographical name

【释义】 地名，地理名称

【例句】 Geographical Names, as the name implies, refers to the various geographical entities by name. 地名，顾名思义是指各类地理实体的名字。

【中文定义】 具有固定地理位置的特性，用以识别各个地理物体的名称。

geogrid [dʒe'ɪəʊgrɪd]

【释义】 *n.* 土工格栅

【例句】 According to the test results, the calculation model is defined, and theoretical calculation formula is determined forward on uplift bearing capacity of geogrid reinforced sand land of spread foundation. 根据试验结果，确定了加筋风砂土地基扩展基础承受上拔荷载的计算模式和理论计算公式。

geogrid

geoid ['dʒiˌɔɪd]

【释义】 *n.* 大地水准面

【常用搭配】 geoid height 大地水准面高程；oceanic geoid 海洋大地水准面

【例句】 A precise model of Earth's geoid is crucial for deriving accurate measurements of ocean circulation, sea-level change and terrestrial ice dynamics. 一个精确的大地水准面模型对于推导出精确的洋流、海平面的变化和地球冰川运动来说都是非常重要的。

【中文定义】 大地水准面是指与平均海水面重合并延伸到大陆内部的重力等位面。

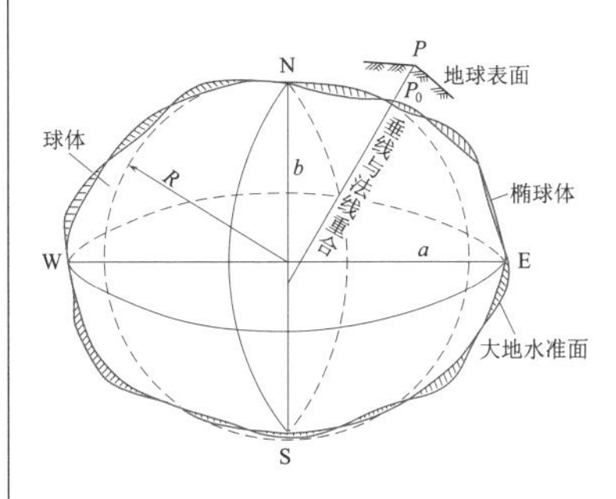

geoid

geologic body model

【释义】 地质体模型

【中文定义】 在CAD软件中，根据基础数据模型进行精确定位和合理推测，按照地质体发育规律创建的几何模型。

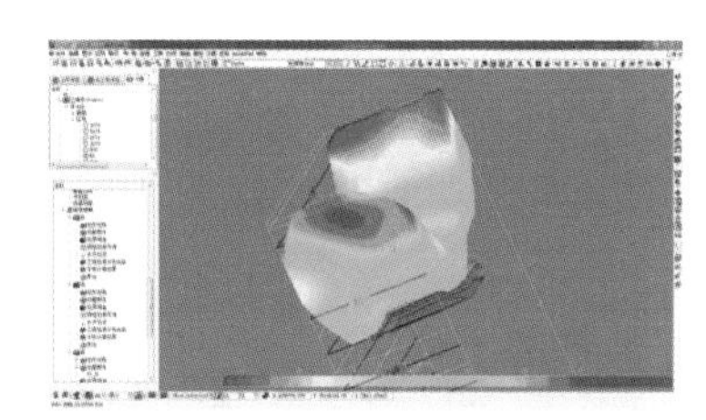

geologic body model

geologic database

【释义】 地质数据库

【例句】 We maintain a geologic database that cater for geologic database. 我们维护着一个服务于地质专业的地质资料库。

【中文定义】 将工程地质勘察数据以一定方式存储在一起、能为多个用户共享、具有尽可能小的冗余度、与应用程序彼此独立的数据集合，通常被当作为三维地质系统的核心组成部分，是三维地质建模、模型分析、二维出图、统计查询等工作的基础。

geologic hazards inquiry

【释义】 地质灾害调查

【例句】 The evaluation of geohazard emergence(its proneness) is an important mission in geologic hazards inquiry and evaluation. 地质灾害易发性评价是地质灾害调查评价的一项重要内容。

geologic hazards inquiry

geologic log of dam foundation

【释义】 坝基编录图

【中文定义】 依一定比例尺和图例，根据开挖揭露的坝基工程地质、水文地质现象编制而成的原始图件。

geologic reserves

【释义】 地质储量

【例句】 Geologic reserves are an important factor to decide oil and gas exploration and exploitation as well as investment sum. 石油、天然气地质储量是指导油气田勘探、开发，确定投资规模的重要依据。

geologic risk

【释义】 地质风险

【常用搭配】 geologic risk evaluation 地质风险评价；geologic risk analysis 地质风险分析

geologic structure

【释义】 地质构造，地质背景

【例句】 Guanyin'ge Dam is located in a limestone area. The geologic structure of the foundation is complex with developed karst. 观音阁水库大坝地处石灰岩地区，地质构造复杂，岩溶发育。

【中文定义】 指地壳中的岩层地壳运动的作用发生变形与变位而遗留下来的形态。地质构造是岩石或岩层在地球内动力的作用下产生的原始面貌。地质构造因此可依其生成时间分为原生构造与次生构造。

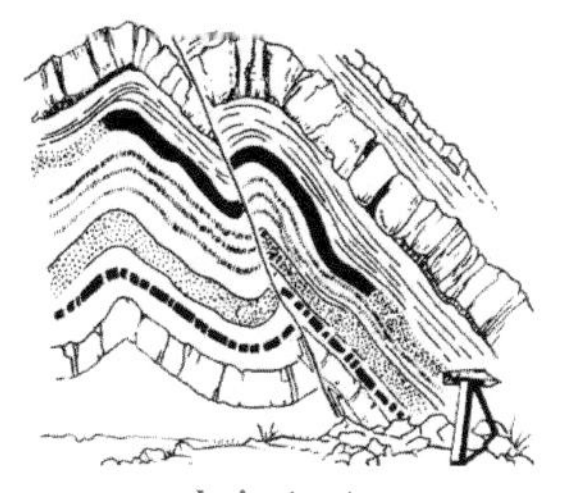

geologic structure

geological attribute model

【释义】 地质属性模型

【中文定义】 在 CAD 软件中，为表示地质体随空间分布的不均一性，通过有限元方法将几何实体模型划分成网格单元，并在网格节点、面片、单元上赋予地质体特征值的模型。

geological body probably resulting in hazard

【释义】 致灾地质体

【中文定义】 可能导致灾害的地质体。

geological body probably resulting in hazard

geological boundary

【释义】 地质界线

【例句】 Geological boundary is used to indicate geological boundary of different geological body and geological phenomenon in geological map. 地质图中用地质界线表示不同地质体和地质现象之间分界线。

【中文定义】 地质图中不同地质体和地质现象之间的界线。

geological briefing

【释义】 地质简报

geological daily record during construction

【释义】 施工地质日志

geological environment

【释义】 地质环境

geological geometry model

【释义】 地质几何模型

【中文定义】 在CAD软件中，以点符、线框、表面、实体等几何图元表示地质体位置、规模、形态、相互关系等空间属性特征的模型。

geological hazard

【释义】 地质灾害

【例句】 Landslide is one of the most serious geological hazard. 滑坡是地质灾害中最为严重的一类。

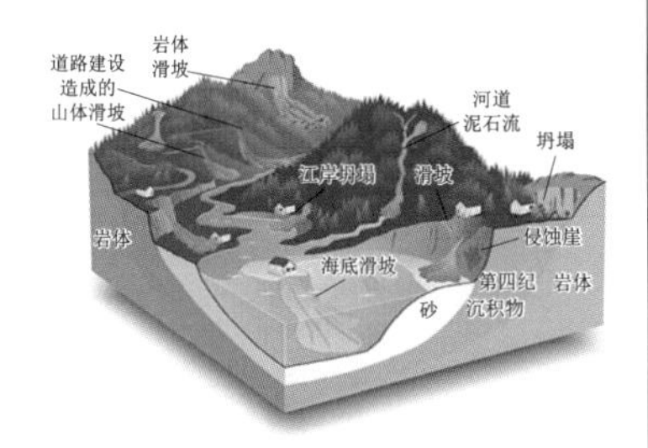

geological hazard

geological hazard potential

【释义】 地质灾害隐患

geological interpretation map of geophysical exploration results

【释义】 物探成果地质解释图

geological log

【释义】 地质日志

geological mapping

【释义】 地质测绘

【例句】 Geological mapping and recording is one of the main tasks in the construction geology of water conservancy projects. 地质测绘与编录是水利水电工程施工地质工作的一项主要内容。

geological mapping

geological memorandum during construction

【释义】 施工地质备忘录

geological model of slope

【释义】 边坡地质模型

【例句】 Based on the rebuilt 3D geological model of slope of the reservoir banks, the mechanism of instability is analyzed by using 3D limit equilibrium method firstly, which provides an environmental background for the visual simulation. 基于重建的库岸三维边坡地质模型，运用三维极限平衡法分析滑坡的失稳诱发机制，为失稳运动的可视化仿真提供环境模拟背景。

geological monitoring

【释义】 地质监测

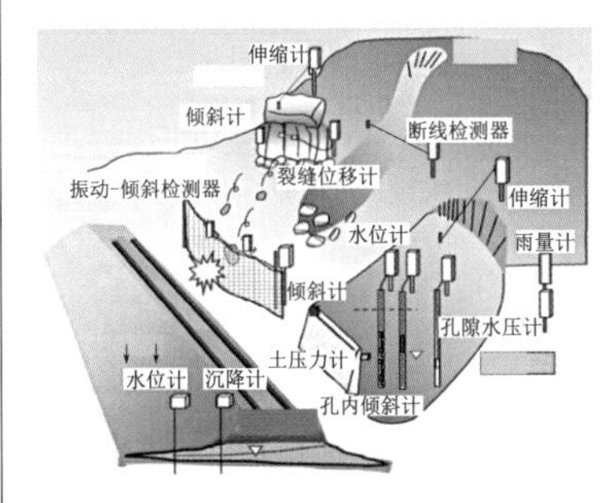

geological monitoring

geological notification during construction

【释义】 施工地质通知

geological plan of immersion area and protection area

【释义】 浸没区及防护地质图

【中文定义】 反映水库浸没区及防护工程地质条

件的地质图件。

geological point survey

【释义】 地质点测量

【同义词】 geological spot survey

【例句】 It abolish control and reduce cost in small-scale geological point survey that locate single geological dots utterly using GPS. 在小比例尺地质点测量中，用GPS单点绝对定位技术测定地质点，减少了控制测量环节，降低了成本。

geological point survey

geological prediction

【释义】 地质预报

geological specification

【释义】 地质说明书

geological surface

【释义】 地质曲面

【例句】 According to geologic surface, the virtual space to display the three dimension geological surface dynamically is constructed using GRID file and Elevation Grid node. 根据地质曲面的实际情况，提出采用网格化数据文件与海拔栅格节点相结合方法建立虚拟空间，动态显示三维地质曲面。

【中文定义】 反映地质体形态的空间几何曲面。

geological works during construction

【释义】 施工地质工作

【例句】 On the basis of geological works during construction, the engineering geology assessment is carried out for the dam base of Baiyekou Reservoir. 在施工地质工作的基础上，对柏叶口水库工程面板堆石坝坝基进行了工程地质评价。

geological works during construction

geological zoning boundary

【释义】 地质分区界线

【例句】 Geological zoning boundary is used to indicate geological boundary of engineering geological zoning interface in geological map. 地质图中地质分区界线用来表示工程地质分区界面。

【中文定义】 地质图中用来表示工程地质分区界面的地质界线。

名称	符号	名称	符号
工程地质区界线	——	工程地质区编号	Ⅰ
工程地质亚区界线	——	工程地质亚区编号	Ⅱ
工程地质地段界线	——	工程地质地段编号	Ⅱ A

geological zoning boundary

geomechanical model

【释义】 地质力学模型

【例句】 Based on the geomechanical model of cavern surrounding rock, the numerical simulation calculation is carried out by using ANSYS finite element analysis software。基于洞室围岩地质力学模型，利用ANSYS有限元分析软件进行了数值模拟计算。

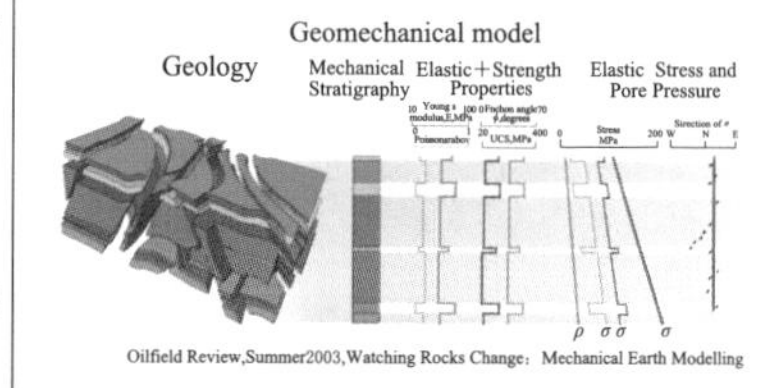

geomechanical model

geomechanical model test

【释义】 地质力学模型试验

geomechanics [ˌdʒiːəumi'kæniks]

【释义】 *n.* 地质力学

【常用搭配】 geomechanics theory 地质力学学说；coupled geomechanics 耦合地质力学

【例句】 The history of regional tectonic study

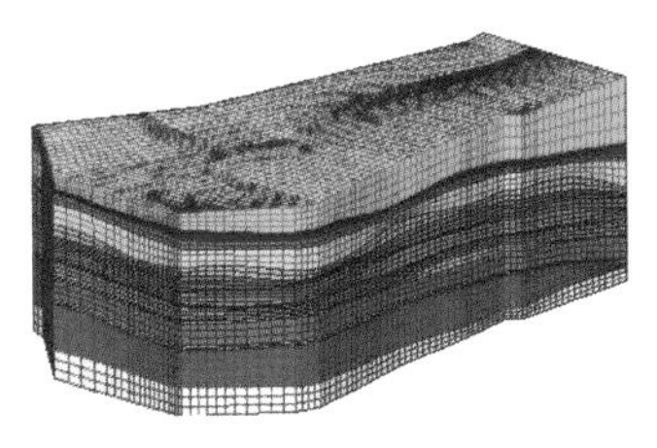

geomechanics model test

can be divided into three stages: geosyncline - platform theory, geomechanics theory and plate tectonics theory. 区域构造研究的历史可分为三个阶段：地槽台地理论、地质力学理论和板块构造理论阶段。

【中文定义】 广义的理解是指地质学和力学结合的边缘学科。在中国地质学界，地质力学是指中国地质学家李四光在研究中国和东亚构造的基础上于 20 世纪 40 年代创立的一种构造地质学说。它主要是用力学的观点研究地质构造现象，研究地壳各部分构造形变的分布及其发生、发展过程，用来揭示不同构造形变间的内在联系。

geomembrane [dʒiːoʊmemb'reɪn]

【释义】 *n.* 土工膜

【例句】 As a new impervious synthetic geotechnical material, PE compound geomembrane has been widely applied to water conservancy construction. PE 复合土工膜作为一种新型结构的防渗型土工合成材料，在水利工程上广泛应用。

geomembrane

geomorphic symbol

【释义】 地貌符号

【例句】 Geomorphic symbol is used to indicate different landforms in geological map. 地质图中

名称	符号	名称	符号
鳍脊(锯齿状陡窄山脊)		剥蚀阶地	
鼓丘		常年积雪地区	
冰斗		终积尾积	
槽装冰川谷		蛇形丘	
悬谷		冰积阜	
悬谷口的阶梯		冰川堆积阶地	10~12 数字为相对高度/m
羊背石		冰水堆积阶地	12~15

geomorphic symbol

用地貌符号表示不同地貌对象。

【中文定义】 地质图中绘制地貌对象所使用的各种图形符号。

Geomorphologic and Quaternary geological map

【释义】 地貌与第四纪地质图

【例句】 In prefeasibility study stage and feasibility study stage of hydropower engineering, Geomorphologic and Quaternary geological map may be submitted as appropriate in report for engineering geological investigation. 水力发电工程预可行性研究阶段和可行性研究阶段工程地质勘察报告中可根据需要附加地貌与第四纪地质图。

【中文定义】 反映地区的地貌形态成因年代、物质组成以及在第四纪堆积形成的地层分布岩性变化构造及其他重大地质事件等地质条件的图件。

geomorphy [dʒiːəʊ'mɔːfɪ]

【释义】 *n.* 地貌，地貌学

【常用搭配】 original geomorphy 原始地貌；paleo-geomorphy 古地貌；karst geomorphy 岩溶地貌

【例句】 Based on the analysis of the monitoring

karst geomorphy

data of an offshore sand mining area in the Pear River estuary, the submarine geomorphy of the mining area can be classified into four types. 根据珠江三角洲一近海采砂区的监测数据分析可知，该区域海底地形可以被分为四类。

【中文定义】 地貌即地球表面各种形态的总称，也称为地形。

geophysical detection and test

【释义】 物探检测与测试

geophysical field

【释义】 地球物理场

【常用搭配】 naturally geophysical field 天然地球物理场；artificially geophysical field 人工地球物理场

【例句】 The geophysical field can be divided into naturally occurring geophysical field and artificially excited geophysical field. 地球物理场可分为天然存在的地球物理场和人工激发的地球物理场。

geophysical plane map

【释义】 物探平面图

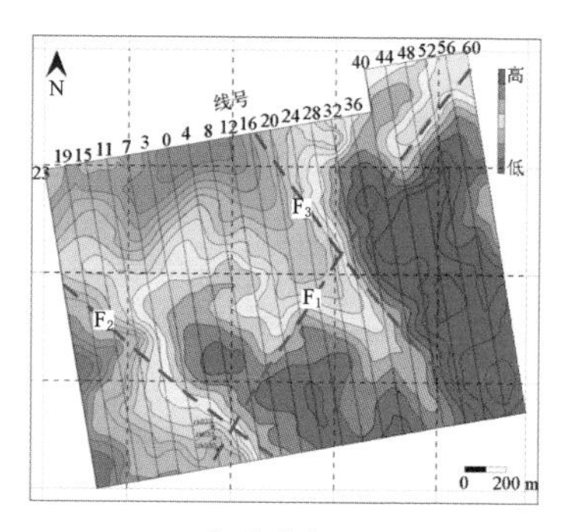

geophysical plane map

geophysical profile

【释义】 物探剖面图

【例句】 There exists lots of geophysical profile mapping work in geophysical survey, and thus a rapid and efficient automation mapping method is highly needed. 在物探扫面工作中，有大量的剖面数据需要绘制成物探剖面图，要求寻找一个高效快速的制图方法。

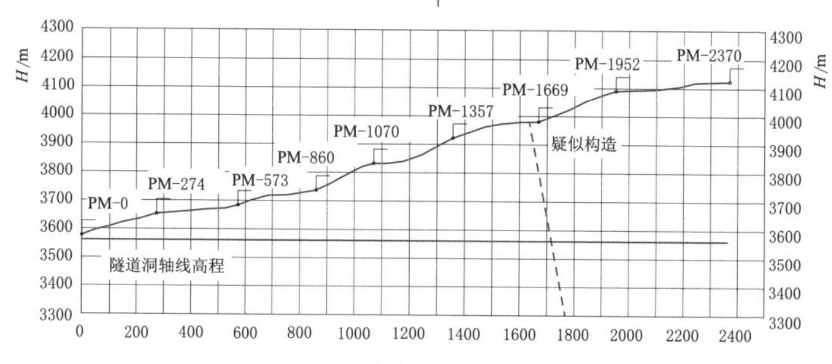

geophysical profile map

geophysical prospecting report

【释义】 物探报告

geophysical result map

【释义】 物探成果图

【例句】 Using the geophysical result map and combined with the analysis and research of surface geological data, it is possible to understand some characteristics of the strata in a certain depth. 利用物探成果图并结合地表地质资料的分析和研究，可以了解在一定深度内岩层的某些特征。

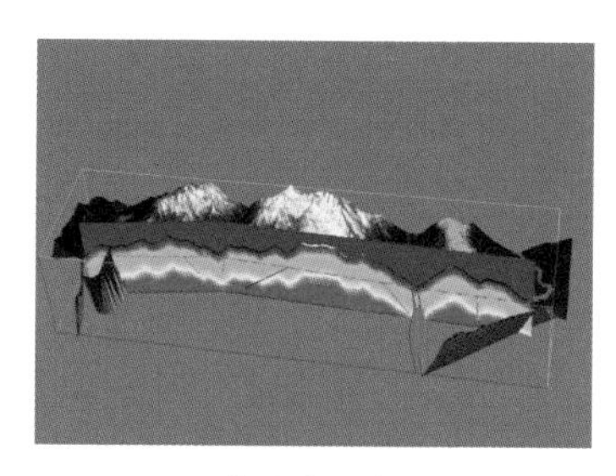

geophysical result map

geophysical station

【释义】 物探测点

geostatic stress

【释义】 自重应力，地静应力

【例句】 It is concluded that the average degree of radial consolidation for a given soil layer increases with its buried depth since the geostatic stress increases with the depth. 结果表明，因自重应力随深度增加，土层径向排水平均固结度随其埋深而增大。

【中文定义】 自重应力是岩土体内由自身重量引起的应力。

geostress field

【释义】 地应力场

【常用搭配】 initial geostress field 初始地应力场

【例句】 The initial geostress field of tunnel area is analyzed by 3D multivariate FE regression based on limited points of in-situ geostress measurement. 采用三维有限元多元回归分析方法，基于有限测点的地应力量测结果，拓展分析隧道区原始宏观地应力场。

【中文定义】 地壳内各点的应力状态在空间分布的总合。

geosynclinal system

【释义】 地槽系

【例句】 Copper-nickel deposit of Hongqiling is located in Huifahe fault which is contacting zone of North China platform and Jihei geosynclinal system. 红旗岭铜镍矿床地处华北地台和吉黑地槽系接触带-辉发河断裂带内.

【中文定义】 地槽系是地壳上特别强烈的活动部分，它包括了成因上有密切联系的、大致同一时期连续发展的所有地槽和其间的中间地块。

geosyncline [ˌdʒioʊ'sɪnklaɪn]

【释义】 *n.* 地槽，地向斜

【常用搭配】 Ouachita Geosyncline 瓦失陶地槽；Andean Geosyncline 安地斯地槽；East Greenland Geosyncline 东格陵兰地槽

【例句】 The geosyncline hypothesis with an obsolete concept involving vertical crustal movement has been replaced by plate tectonics to explain crustal movement and geologic features. 地槽假说是一个被解释地壳运动和地质特征的板块构造学说取代的过时的概念；地槽假说认为地壳以垂直运动为主。

【中文定义】 地槽是地壳上的槽形坳陷，呈长条状分布于大陆边缘或两大陆之间，具有特征性的沉积建造并组成地槽型建造序列。广泛发育强烈的岩浆活动、构造变形和区域变质等。

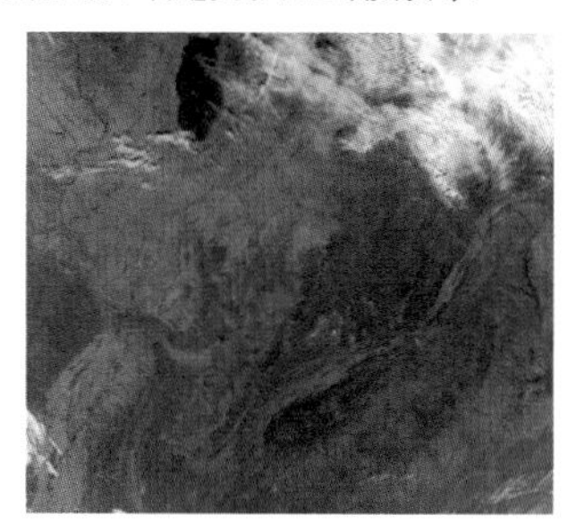

geosyncline

geosyncline-platform theory

【释义】 地槽-地台说

【例句】 The history of regional tectonic study of Guizhou can be divided into three stages, the stage of geosyncline-platform theory, the stage of geomechanics theory and the stage of plate tectonics theory. 贵州地区区域构造研究历史可以划分为三个阶段，地槽-地台说阶段、地质力学说阶段和板块构造说阶段。

【中文定义】 地槽-地台说简称槽台说，其基本论点是：地壳运动主要受垂直运动控制，地壳此升彼降造成振荡运动，而水平运动则是派生或次要的。驱动力主要是地球物质的重力分异作用。物质上升造成隆起，物质下降则造成凹陷。主要的构造单元有地槽和地台两类，地台是由地槽演化而来的。

geotechnical centrifugal model test

【释义】 土工离心模型试验

【例句】 Based on the geotechnical centrifugal model test, the deformation and stability of a certain pit

geotechnical centrifugal model test

in Shanghai is studied. The rule of deformation with time and its effect to the stability are revealed. 通过土工离心模型试验对上海某放坡式开挖基坑边坡的变形特性和稳定性进行了研究，揭示了基坑边坡变形的时间效应及其对边坡稳定性的影响。

geotechnical electrical parameter testing

【释义】 岩土电性参数测试

geotechnical test report

【释义】 岩土试验报告

geotectonic [ˌdʒioʊtek'tɒnɪk]

【释义】 *adj*. 大地构造的，地壳构造的

【常用搭配】 geotectonic cycle 大地构造旋回；geotectonic map 大地构造图；geotectonic system 大地构造体系；geotectonic geology 大地构造地质学

【例句】 To reduce this multiplicity of mechanism, one must be aware of the time spectrum of geotectonic stress relief. 为了减少这种机制的多样性，必须考虑构造应力释放的时间谱。

geotectonics [dʒiːəʊtek'tɒniks]

【释义】 *n*. 大地构造地质学

【中文定义】 大地构造地质学是地球科学的一个分支学科。它主要研究地球的构造、演化及其运动变形和发展规律等问题的学科，是研究地球科学的基础理论之一，不仅对深入认识地球发展史和地壳、岩石圈运动史有重要的理论意义，而且对研究成矿条件、地表成因及预测矿产资源等都具有重要的实际意义。

geotextile ['dʒɪoʊtekstaɪl]

【释义】 *n*. 土工织物，土工布

【常用搭配】 woven geotextile 织造土工织物；nonwoven geotextile 非织造土工织物

【例句】 The characteristics, application principle and construction method of geotextile are mainly elaborated. 重点对土工织物的性质、应用原理、施工方法进行了阐述。

geotextile

geotherm ['dʒiːəuθəm]

【释义】 *n*. 地热，地温

geothermal anomalous area

【释义】 地热异常区

【例句】 The geothermal anomalous area of Yuzhong Basin is on the regional geothermal background and related to deep source geothermal mechanism. 榆中盆地地热异常区，处于区域性地热背景之上，具有深源热机制。

geothermal field

【释义】 地热田

【例句】 Cerro Prieto is the oldest and largest Mexican geothermal field in operation. Cerro Prieto 是墨西哥经营历史最悠久、规模最大的地热田。

geothermal field in Ali Area

geothermal fluid

【释义】 地热流体

【例句】 All geothermal wells of Wumishan have geothermal fluids, which shows that the reservoir has the relative homogeneous characteristics in the study region. 雾迷山组地层所有地热井均能开采出地热流体，反映了该热储层在研究区具有相对均一的发育特征。

【中文定义】 地热流体是地下热水、地热蒸汽以

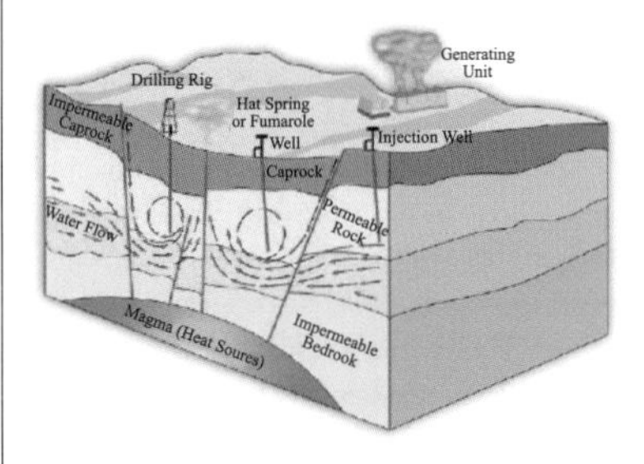

Schematic diagram of geothermal fluid

及载热气体等存于地下、温度高于正常值的各种热流体的总称。

geothermal gradient

【释义】 地热梯度

【例句】 Based on 418 geothermal gradients and 418 heat flow values from South China, both the geothermal map and the heat flow map of South China have been constructed. The present-day distributions of geothermal gradient and heat flow have been analyzed. 根据中国南方418个地温梯度和418个大地热流的数据，编制了中国南方地温梯度图和中国南方大地热流图，研究了中国南方现今地温梯度分布特征和大地热流分布特征。

geothermal reinjection

【释义】 地热回灌

【常用搭配】 geothermal reinjection technique 地热回灌技术

【例句】 It is shown that geothermal reinjection is feasible in Tianjin. 研究表明，在天津实施地热回灌开采是可行的。

geothermal reserves

【释义】 地热储量

【例句】 The regional gravitation was used in the geothermal exploration and analysis of geothermal reserves in Lanzhou fault basin 区域重力被应用于兰州断陷盆地地热勘察和热储分析。

geothermal reservoir engineering

【释义】 热储工程

【常用搭配】 study of geothermal reservoir engineering 热储工程研究

【例句】 The geothermal reservoir engineering in low - middle temperature was studied in this paper. 本文研究了中低温储热工程。

geothermal reservoir model

【释义】 热储模型

【常用搭配】 geothermal reservoir concept model 热储概念模型

【例句】 Through primary analysis of the geothermal reservoir model in Qingdao area, it is regarded that heat reservoir model belongs to deep-circulation convective type. 通过青岛地区热储模型分析认为，其热储概念模型属深循环对流型。

geothermal resources exploration

【释义】 地热资源勘察

【例句】 Studies of the origin and evolution mechanism of groundwater in a type area not only have great significance for guiding the rational utilization and development of geothermal water resources but also provide important information for future geothermal resources exploration and evaluation. 典型地区地下水成因和演化机制的研究不仅对于热水资源的合理利用与开发具有重要的指导意义，而且可以为日后的地热资源勘察评价提供重要信息。

geothermal resources

【释义】 地热资源

【常用搭配】 geothermal anomaly 地热异常；geothermal gradient 地热梯度

【例句】 In exploitation and utilization of geothermal resources, it is of practical significance to study the cyclic pattern of deep Karst groundwater. 在地热资源开发利用中，对深部地下水循环模式进行研究具有实际意义。

geothermal resources

geothermal resources assessment

【释义】 地热资源评价

geothermal system

【释义】 地热系统

【例句】 The enhanced geothermal system, exploited with a three-well system in the granite formation at depth of 5000m, is expected to stabilize its operation in the coming year. 在5000m深花岗岩中，采用3个井开采的增强型地热系统预计将在明年实现稳定运行。

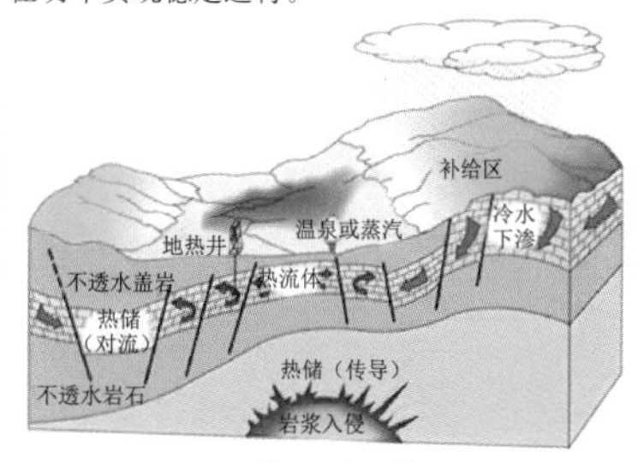

geothermal system

geothermal water

【释义】 地下热水

【例句】 The utilization of medium-low temperature geothermal water has broad and bright prospects. 中低温地下热水有着广阔的开发利用前景。

【中文定义】 地下水在一定地质条件下，因受地球内部热能影响而形成温度不同的地下热水。

geyserite [ˈɡaɪzəˌraɪt]

【释义】 *n.* 硅华

【常用搭配】 geyserite terrace 硅华台地

【例句】 The fact that the geyserite Cs-deposit is discovered only in Tibet has relation with the modern tectogenetic movement of the Tibet plateau by all appearances. 硅华铯矿仅在青藏高原发现，显然跟青藏高原在近期的构造运动有关。

【中文定义】 是指地下水或地表水经化学作用而形成的二氧化硅沉积物，是一种典型的化学成因的硅质岩。

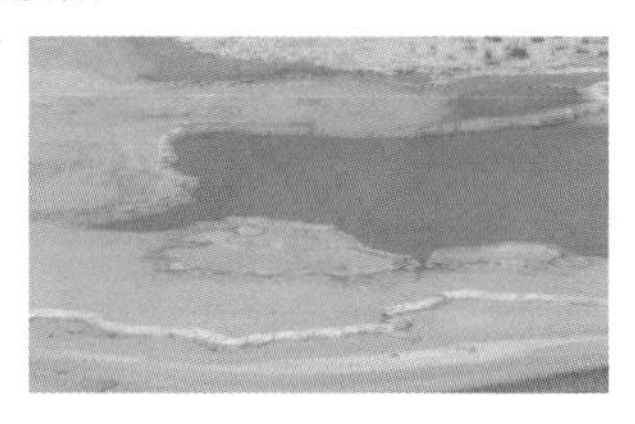

geyserite

giant-thick layer structure

【释义】 巨厚层状结构

【中文定义】 层面厚度大于 100cm 的层状结构。

glacial deposit

【释义】 冰川沉积

【例句】 An important source of fresh water on Lopez, San Juan, Orcas, and Shaw Islands in San Juan County of the northwestern coast of Washington is glacial deposit and bedrock aquifers. 位于华盛顿西北岸圣胡安县的洛佩兹、圣胡安、奥卡斯以及肖尔岛上的淡水主要来源于冰川沉积和基岩含水层。

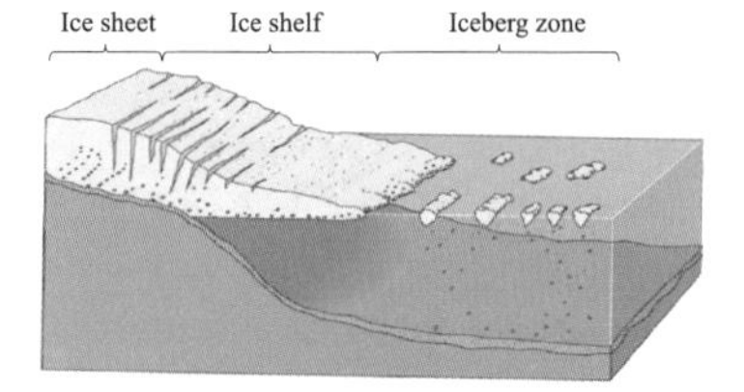

the origin of glacial deposits

【中文定义】 是在冰川运动中或消融时因搬运能力降低，而将其携带的各种岩石碎屑沉积下来的堆积作用。

glacial kar

【释义】 冰蚀凹地

【中文定义】 由于冰川具有强大的挖掘能力，常在冰川的幽谷中挖掘成凹地，称为冰蚀凹地。

glacial kar

glaciation [ˌɡleɪsiˈeɪʃn]

【释义】 *n.* 冰川作用，冻结成冰

【例句】 At least, two major phases of glaciation occurred on the west slope of the Mt. Gongga during Quaternary, one at the late of Middle Pleistocene and another at the Late Pleistocene. 第四纪期间，贡嘎山西坡至少发生过两次明显的冰川作用，一次发生在中更新世晚期，另一次发生在晚更新世。

glaciation

【中文定义】 冰川作用，广义上泛指冰川的生成、运动和后退；狭义上仅指冰川运动对地壳表面的改变作用，包括冰川的侵蚀、搬运和堆积。

glacier [ˈɡleɪʃər]

【释义】 *n.* 冰川，冰河

【常用搭配】 mountain glacier 山岳冰川；continental glacier 大陆冰川；glacier debris flow 冰川泥石流

【例句】 Material which is carried along on the surface or within the glacier will be dropped as the ice melts, a process which is gradual. 当冰融化时，冰川表层或内部携带的物质会逐渐沉积下来。

【中文定义】 冰川是在重力和压力的影响下由雪源地向外缓慢移动着的冰体。

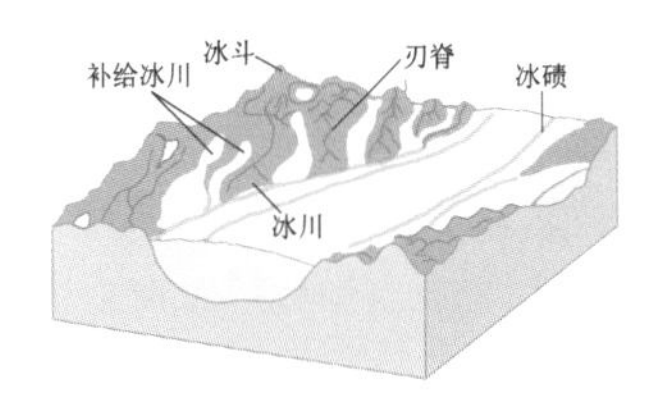

glacier

glacier debris flow

【释义】 冰川泥石流

【例句】 Glacier debris flow is a kind of debris gravitational flow deposits that result from the mix of moraines and ice snow melt water. 冰川泥石流是由冰碛物与冰雪融水混合形成的一种碎屑重力流沉积。

【中文定义】 发育在现代冰川和积雪边缘地带，由冰雪融水或冰湖溃决洪水冲蚀形成的含有大量泥沙石块的特殊洪流。

glacier debris flow

glacier glen

【释义】 幽谷，冰川槽谷

glacier glen

【中文定义】 正U形谷常称幽谷，又叫冰川槽谷，是冰窖中伸出的冰舌在山谷中流动、磨蚀和掘蚀冰床岩石形成的谷地。幽谷底部较宽，两坡陡立，横剖面呈U形，与河流的V形谷有明显的区别。

global navigation satellite system (GNSS)

【释义】 全球导航卫星系统

global navigation satellite system

【例句】 Global navigation satellite system (GNSS) is the trend of development and application of navigation technology in civil aviation. 全球导航卫星系统（GNSS）是民航导航技术发展和应用的趋势。

global positioning system (GPS)

【释义】 全球定位系统（GPS）

【例句】 These handheld instruments provide latitude, longitude and altitude by picking up timing signals from global positioning system satellites. 这些手持式装置通过接收全球定位系统的卫星时间信号，可以提供经度、纬度以及海拔高度。

【中文定义】 利用GPS定位卫星，在全球范围内实时进行定位、导航的系统，称为全球卫星定位系统，简称GPS。

GPS

gneiss [nais]

【释义】 *n.* 片麻岩

【常用搭配】 layered gneiss 层状片麻岩；mica gneiss 云母片麻岩；mylonite gneiss 糜棱片麻岩

【例句】 These granitoids are characterized by a very coarse-grained and huge augen gneiss texture. 这些花岗岩以特有的粗粒、巨大的眼球状片麻结构为特征。

gneiss

【中文定义】 片麻岩是一种变质岩，而且变质程度深，具有片麻状构造或条带状构造，鳞片粒状变晶结构，主要由长石、石英、云母等组成，其中长石和石英含量大于50%，长石多于石英。如果石英多于长石，就称为片岩而不再是片麻岩。

gneissic structure

【释义】 片麻状构造

【例句】 Due to the ductile shearing at the end of Late Carboniferous most rocks are deformed to be granitic mylonites and super-mylonites macroscopically in a gneissic structure. 由于晚石岩世的韧性剪切，大部分岩石在宏观上变形为片麻状构造的花岗质糜棱岩和超糜棱岩。

【中文定义】 变质岩中不同颜色的片状、柱状矿物交替断续定向排列的一种构造。

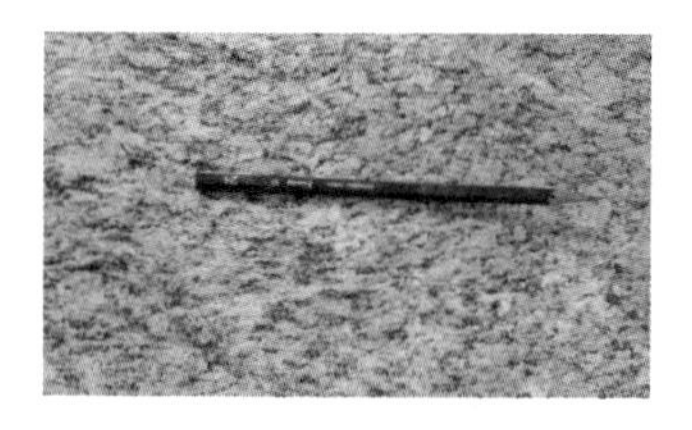

gneissic structure

goat-back stone

【释义】 羊背石

【中文定义】 冰川谷底，冰蚀后残留的石质小丘，呈椭圆形，其长轴方向就是冰川流动方向，两坡不对称，迎坡面为缓坡，较圆滑，有冰川擦痕或磨光面，背冰面为陡坡，坎坷不平。

goat-back stone

gorge [gɔrdʒ]

【释义】 *n.* 峡谷

【常用搭配】 narrow gorge 嶂谷；gorge dam 峡谷坝

【同义词】 canyon

【例句】 Rock mass stress state is considered as one of the significant investigation content for dams built in gorge area. 岩体地应力状态是峡谷区建坝勘察的重要内容之一。

【中文定义】 峡谷是深度大于宽度谷坡陡峻的谷地，V形谷的一种。一般发育在构造运动抬升和谷坡由坚硬岩石组成的地段。

gorge

graben ['grɑbən]

【释义】 *n.* 地堑，裂谷

【常用搭配】 graben fault 地堑断层；dogleg graben 折曲地堑；graben shoulder 地堑肩；volcanic graben 火山地堑

【例句】 A graben is formed when a block that is bounded by normal faults slips downward, usually because of a tensional force, creating a valley like depression. 地堑是由正断层所包围的块体向下滑动而形成，通常是拉应力作用下，形成似凹陷的山谷。

【中文定义】 两侧被陡倾断层围限，中间下降的槽形断块构造。多指大、中型的构造，大者延长可达数百公里。地堑常呈长条形的断陷盆地，其边界可以是平直的，但更常见的是折线状边界，一般由多条陡倾角的正断层联合而成。

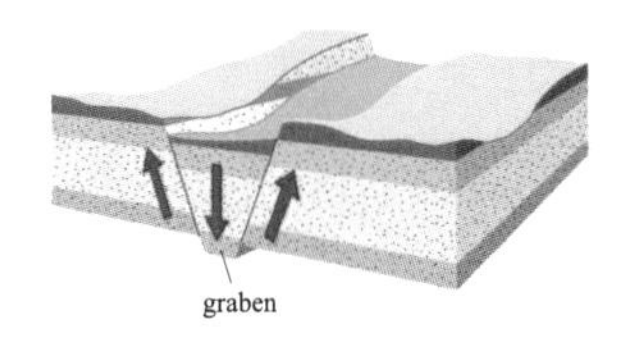

graben

gradation curve

【释义】 颗粒分析曲线，颗粒级配曲线

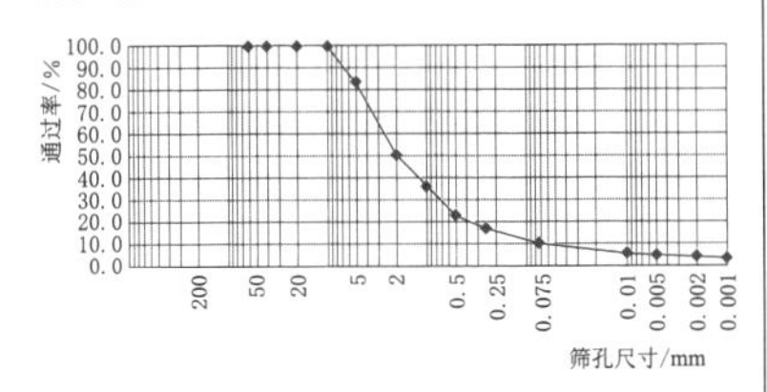

gradation curve

graded bulk density

【释义】 分级堆积密度

【例句】 Graded bulk density is the mass per unit volume of gravel in natural accumulation state. 分级堆积密度是各粒径组砾石（碎石）在自然堆积状态下单位体积的质量。

gradient ['greɪdɪənt]

【释义】 *n*. 坡度，梯度，倾斜度

【中文定义】 斜坡的斜度，多以角度表示。

grading of geological disaster

【释义】 地质灾害分级

【例句】 Furthermore, a coastal erosion disaster degree grading scheme is presented by consulting other gradings of geological disaster. 在参考其他地质灾害灾度分级方案的基础上，给出了海岸侵蚀灾度分级方案。

grading of structural plane

【释义】 结构面分级

【例句】 In the *Code for Hydropower Engineering Geological Investigation* (GB 50287—2016), there are 5 grades of classification of structural plane. 在《水力发电工程地质勘察规范》(GB 50287—2016)中，岩体结构面分5级。

【中文定义】 一般按结构面宽度和长度进行分级，从而区分结构面的规模。

grain modulus (GM)

【释义】 粒度模数

【例句】 The grain modulus is an indicator to evaluate the fineness of gravel, expressed as the sum of the cumulative percentage of the cumulative sieve in the sieve with a pore size greater than 5 mm. 粒度模数是评价砾石粗细程度的一项指标，用筛分试验中孔径大于5mm各号筛的累积筛余百分率的总和除以100来表示。

grain size

【释义】 粒径

【例句】 Before treatment, the soil must be pretreated to make the moisture content, maximum grain size, pH value and clay content suitable for subsequent treatment. 在处理之前，先要将土壤进行预处理，使得土壤的含水率、土壤最大颗粒直径、pH、黏土含量等适宜于之后的处理。

grain size analysis test

【释义】 颗粒分析试验，颗分试验

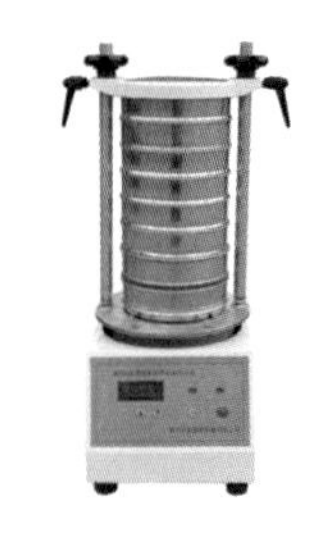

grain size analysis test

grain size distribution curve

【释义】 粒径分布曲线

【例句】 Through random redistribution and form coefficient conversion are further conducted, a continuous and complete grain size distribution curve is obtained. 通过随机再分布和形状系数转换，得出一条连续而完整的粒度分布曲线。

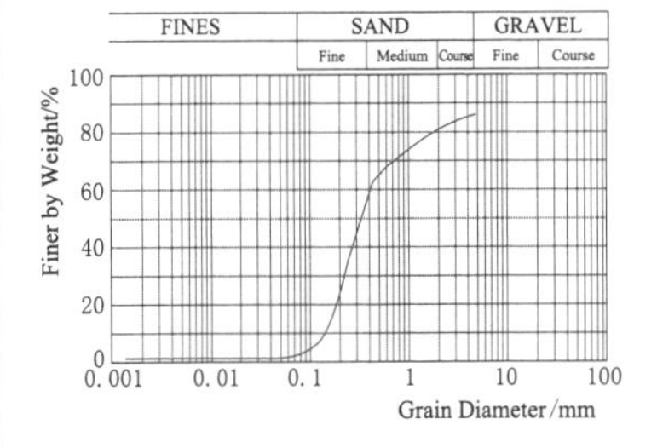

grain size distribution curve

grain size frequency-diagram

【释义】 颗粒大小频率图

grain size gradation

【释义】 颗粒级配

【例句】 It is very important to determine reasonably the maximum grain size and graded sizes gradation of the coarse aggregate. 合理确定混凝土粗骨料的最大粒径和颗粒级配是非常重要的。

granite ['grænɪt]

【释义】 *n*. 花岗岩

【常用搭配】 granite pegmatite 花岗伟晶岩; granite porphyry 花岗斑岩

【例句】 The granite from Beishan failed at the similar values of axial strain under the same confining pressure and failed at the strain about 0.34%, 0.54% and 0.71% under the confining pressure of 2MPa, 10MPa and 30MPa respectively. 北山花岗岩在相同围压下相近的轴向应变值时破坏，并且在 2MPa、10MPa 和 30MPa 围压下破坏的应变值分别为 0.34%、0.54% 和 0.71%。

granite

granite pegmatite

【释义】 花岗伟晶岩

【例句】 This paper reports the SHRIMP U-Pb dating of zircons from granite pegmatite. 本文报道了花岗伟晶岩脉中锆石的 SHRIMP U-Pb 年龄测定。

granite pegmatite

granite porphyry

【释义】 花岗斑岩

【例句】 Tangjiaping granite porphyry is located in the northern part of Dabie orogen. 汤家坪花岗斑岩产于大别造山带北麓。

granite porphyry

granodiorite [ˌgrænəu'daiərait]

【释义】 *n*. 花岗闪长岩

【例句】 A large number of directional microgranular dioritic enclaves occur in Changshannan granodiorite from Kunyushan granitoids in the Jiaodong Peninsula. 胶东半岛昆嵛山花岗质岩石中的长山南花岗闪长岩存在大量定向微粒的闪长质包体。

granodiorite

granularity [grænjʊ'lærɪtɪ]

【释义】 *n*. 粒度

【同义词】 grain size, particle size

【例句】 The average grain diameter and cumulative relative curve of red sandstone are obtained by the granularity size analysis. 对红砂岩进行粒度分析，得出其平均粒径和概率累积曲线。

【中文定义】 粒度是指颗粒的大小。通常球体颗粒的粒度用直径表示，立方体颗粒的粒度用边长表示。对不规则的颗粒，可将与该颗粒有相同行为的某一球体直径作为该颗粒的等效直径。

granulite ['grænjʊlaɪt]

【释义】 *n*. 麻粒岩

【常用搭配】 granulite facies 麻粒岩相

【例句】 Granulite xenoliths are important samples for understanding the formation and evolution of the lower crust. 麻粒岩捕虏体是了解下地壳形成和演化的重要样品。

【中文定义】 又称粒变岩，是一种颗粒比较粗，变质程度较深的岩石。其特点是岩石中没有云母、角闪石等含水矿物，常含有斜长石、铁铝榴石或辉石。具粒状变晶结构，具有不明显的片麻状构造或块状构造。常见的类型有：①长英麻粒岩——由粉砂岩、硅质页岩、中酸性岩浆岩等变质而成，如中国河北宣化庞家堡的石榴二长麻粒岩（又称白粒岩），主要由铁铝榴石、钾长石、斜长石和石英组成；②辉石麻粒岩——常由基性岩浆岩变质而成，如中国河北宣化产有辉石麻粒岩，主要由铁铝榴石、透辉石（部分变为角闪石）和中基性斜长石组成。

granulite

granulitic rock

【释义】 碎粒岩

【例句】 The interformational slip structure occurs between competent and incompetent beds, and the cataclasites and granulitic rocks may be formed. 层滑构造发生于软硬岩层之间，可形成碎裂岩和碎粒岩等。

【中文定义】 碎粒岩是研磨得很细的断层岩，由原岩经碾磨形成的岩粉，或矿物碎粒构成。在偏光显微镜下，岩石具有压碎结构。碎粒岩中如残留一些较大矿物颗粒，则构成碎斑结构，称为碎斑岩。碎粒岩的粒径一般为 0.1～2mm。

graphite ['græfaɪt]

【释义】 *n.* 石墨

【常用搭配】 amorphous graphite 无定形石墨，土状石墨；expanded graphite 膨胀石墨；graphite carbon 石墨碳

【例句】 Graphite is a polymorphous (or allotropic) mineral with diamond, chaoite, and lonsdaleite. 石墨是一种具有多种形态（或同素异形体）的矿物，包括金刚石、蜡石、六方碳等。

【中文定义】 元素碳的一种同素异形体。

graphite

gravel texture

【释义】 砾状结构

【中文定义】 碎屑颗粒粒径大于 2mm 的碎屑结构。

gravelly sand

【释义】 砾砂

【例句】 The pile-soil stress ratio of medium-coarse sand, gravelly sand and gravel cushion on the top of CFG pile is studied by loading test. 利用载荷试验研究了 CFG 桩顶中粗砂、砾砂、砾石垫层的桩土应力比。

【中文定义】 是砂土中砾粒（粒径大于 2mm）含量占总质量 25%～50%的砂。

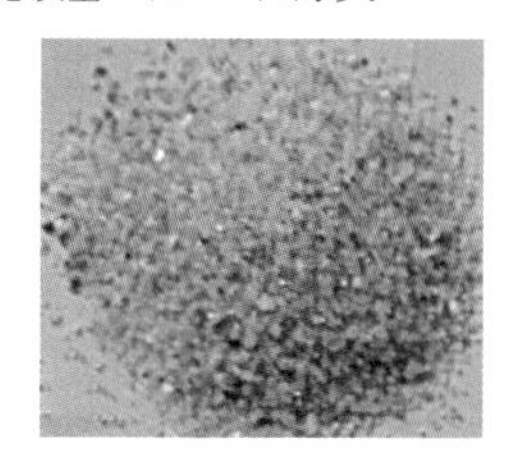

gravelly sand

gravelly soil

【释义】 砾质土

【例句】 The strength parameters of gravelly soils can be estimated from the sand matrix. 砾质土的强度参数可以从砂基质性状估计。

【中文定义】 砾质土是砾粒含量在 10%～15%之间，且砾粒含量少于砂粒、粉粒或黏粒的砾类土。

gravelly soil

gravel-packed screen

【释义】 填砾过滤器

【例句】 The theoretical bases and influential factors to the choice of filter material for gravel-packed screen are discussed. 探讨了填砾过滤器滤料选择的理论依据及影响因素。

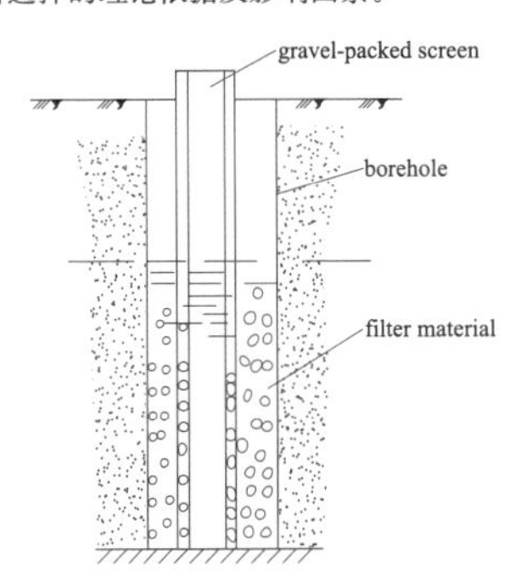

gravel-packed screen

gravitational and magnetic anomaly

【释义】 重磁异常

【例句】 A New method of downward continuation for gravitational and magnetic anomalies is proposed to overcome the unstability and the distortion due to noises. 本文针对重磁异常向下延拓的不稳定性和原始资料中的噪声干扰，提出了一种新方法。

gravity ['grævəti]

【释义】 *n.* 重力，地心引力

【常用搭配】 gravity dam 重力坝；gravity field 重力场，引力场；specific gravity 比重

【例句】 The contractor shall submit for review water-cement ratio, slump, air content, airtemperature and specific gravity. 承包商应提供水灰比、坍落度、掺气量、气温和比重以供复核。

gravity water

【释义】 重力水

【例句】 The research shows that the influx depth of gravity water in soil can indicate the existence of dried soil layer and the degree it developed, which is the key factor to decide the kind of vegetation. 已有的研究表明，土壤中重力水的入渗深度能够指示土壤干层的有无和发育强弱，是决定植被类型的最主要和最关键的因素。

【中文定义】 岩土中在重力作用下能自由运动的

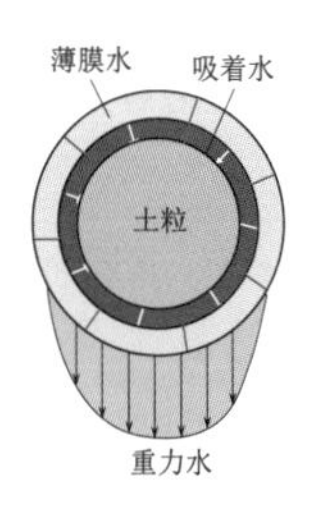

schematic diagram of adsorbed water, film water and gravity water distribution

地下水。泉水、井水和矿坑涌水都是重力水，是水文地质学研究的主要对象。

greasy luster

【释义】 油脂光泽

【例句】 Greasy luster is a kind of luster presented by minerals similar to oily glass such as nephelite, nephrite, soapstone, and quartz. 油脂光泽是类似于油性玻璃的矿物所呈现的一种光泽，例如霞石、软玉、岫玉、石英等。

【中文定义】 透明矿物解理不发育时，在断口上呈油脂光亮。

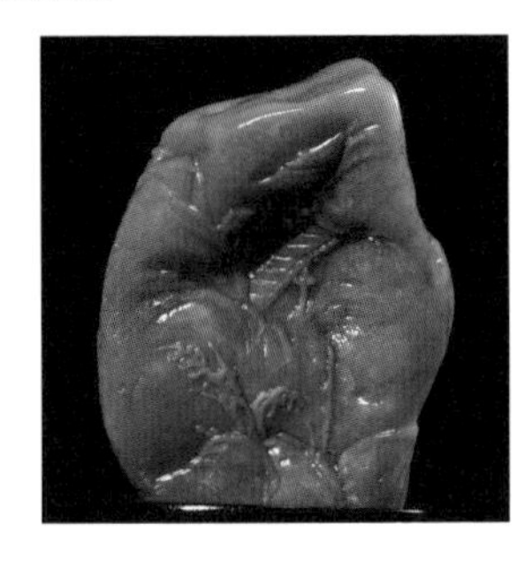

greasy luster

Griffith's strength criterion

【释义】 格里菲斯强度准则

groove-shaped weathering

【释义】 槽状风化，槽形风化

【例句】 The abrupt change of weathering boundary caused by groove-shaped weathering is difficult for engineering treatment. 槽状风化引起风化界限的突变，给工程处理带来困难。

【中文定义】 当岩体中存在规模较大、延展较深的断层碎带或断裂交汇带、不稳定矿物富集带、

岩脉与断裂交叉带及地下水循环交替较强的局部裂隙发育带时，风化的底部界限局部较深的一种现象。

ground basic network

【释义】 地面基本台网

【例句】 Ground basic network technology has matured. 地面基本台网技术已成熟。

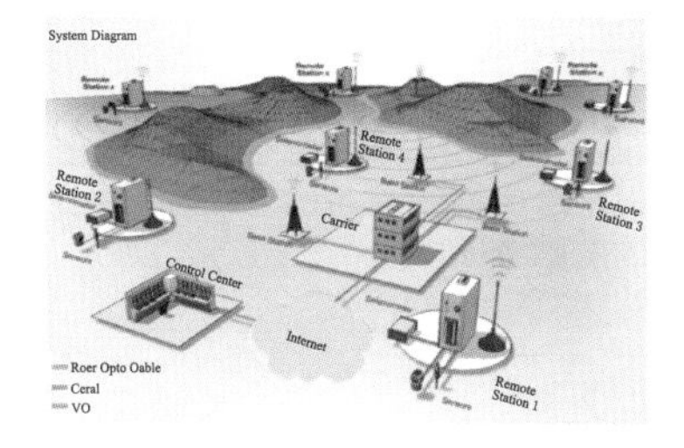

ground basic network

ground collapse

【释义】 地面塌陷

【常用搭配】 karst ground collapse 岩溶地面塌陷

【中文定义】 土体或岩体向下塌落并在地面形成坑、洞和洼地的现象。

ground collapse

ground crack

【释义】 地裂缝

【例句】 For the east section, the hazards are mainly ground subsidence, ground cracks and expansion and shrinkage of expanded soils. 东区段主要的地质灾害是地面沉降、地裂缝和膨胀土的胀缩灾害。

【中文定义】 为地面裂缝的简称，是地表岩层、土体在自然因素（地壳活动、水的作用等）或人为因素（抽水、灌溉、开挖等）作用下产生开裂，并在地面形成一定长度和宽度的裂缝，是一种宏观地表破坏现象。

ground crack

ground fissure site

【释义】 地裂缝场地

【中文定义】 地裂缝通过或可能通过的场地。

ground fissure

【释义】 地裂缝

【中文定义】 区域性的地面开裂现象。

ground fracture

【释义】 地裂

ground fracture

【中文定义】 由干旱、地下水水位下降地面沉降、地震、构造运动或斜坡失稳等原因造成的地面开裂。

ground motion parameter

【释义】 地震动参数

【例句】 The effect of soft soil layer on earthquake ground motion parameter has been taken into account for engineers. 软弱土层对地表地震动参数的影响已引起工程师们的重视。

【中文定义】 地震动参数是表征地震引起地面运动的物理参数，包括峰值、反应谱和持续时间等。

ground penetrating radar (GPR)

【释义】 探地雷达

【例句】 Ground penetrating radar can be used to detect various materials such as rocks, soil, gravel. 探地雷达可用于探测各种材料，如岩石、泥土、砾石。

ground penetrating radar (GPR)

ground stress measurement

【释义】 地应力测试

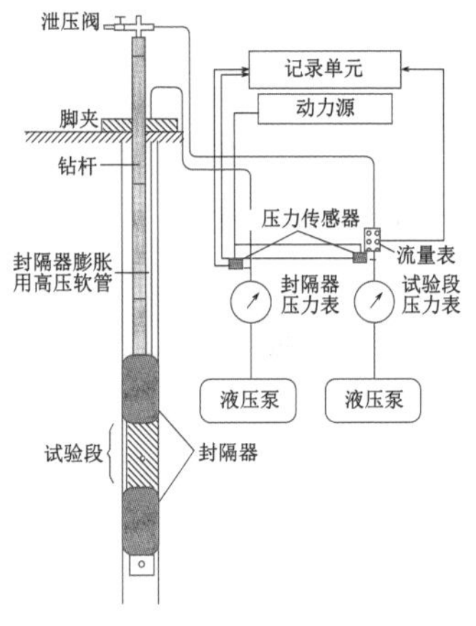

hydraulic fracturing ground stress measurement

【例句】 Ground stress measurements in deep granitic rock were carried out on the project site last week. 上周在工程区进行了深部花岗岩的地应力测试。

ground subsidence

【释义】 地面沉降，地层陷落，地层下陷

【例句】 Ground subsidence inflicts expensive damage on buildings, roads, and pipelines. 地面沉降会对建筑物、道路、管道造成严重破坏。

【中文定义】 是大面积区域性的地面下沉，一般由地下水过量抽吸产生区域性降落漏斗引起。大面积地下采空和黄土自重湿陷也可引起地面沉降。

ground surface exploration

【释义】 地面勘探

【例句】 In the course of prospecting, exploration wells proton magnetometer with ground surface exploration will be able to play an important role. 在勘探过程中，井中质子磁力仪勘探结合地面勘探将能发挥重要作用。

ground surface exploration

ground temperature

【释义】 地温

【例句】 Shallow earth temperature is influenced by surface temperature change everyday and every year, topographic change, elevation change and surface feature change, so it should be corrected. 浅层地温受到地表气温的日际及年际变化、地形变化、高程变化和地貌变化等的影响，应对实测值进行相应校正。

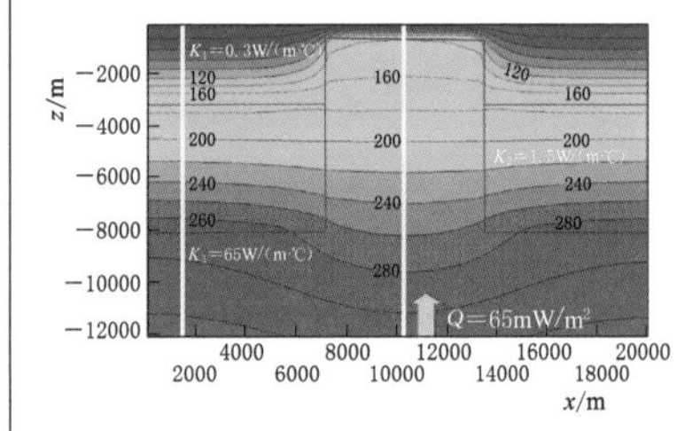

ground temperature

ground treatment

【释义】 地基处理

【常用搭配】 design of ground treatment 地基处

理设计

【例句】 The design of replacement underlayer thickness is crucial to the replacement method of ground treatment. 换土垫层厚度设计是换填法地基处理的关键。

ground treatment

groundwater ['graʊnd'wɔtə]

【释义】 *n.* 地下水

【常用搭配】 groundwater level 地下水水位，潜水面；groundwater recharge 地下水补给；groundwater contamination 地下水污染

【同义词】 ground water

【例句】 The recharge and exploitation of groundwater aquifer is therefore an important component in the sustainable development and management of water resources. 地下水含水层的补给及其开发利用是水资源可持续开发利用与管理的重要组成部分。

【中文定义】 指赋存于地面以下岩石空隙中的水，狭义上是指地下水面以下饱和含水层中的水。在《水文地质术语》(GB/T 14157—93) 中，地下水是指埋藏在地表以下各种形式的重力水。

groundwater connectivity test

【释义】 地下水连通试验

groundwater consumption rate

【释义】 地下水消耗量

【例句】 The city's current groundwater consumption rate has gone 10 million cubic meters over warning levels. 该市目前的地下水消耗量已经达到 1000 万 m^3，超过警戒水平。

groundwater contaminant source

【释义】 地下水污染源

【例句】 The results from investigating and evaluating groundwater contaminant sources indicated that waste landfill was one of main pollutant sources. 地下水污染源调查评价研究表明，垃圾填埋堆放场是造成地下水污染的主要污染源之一。

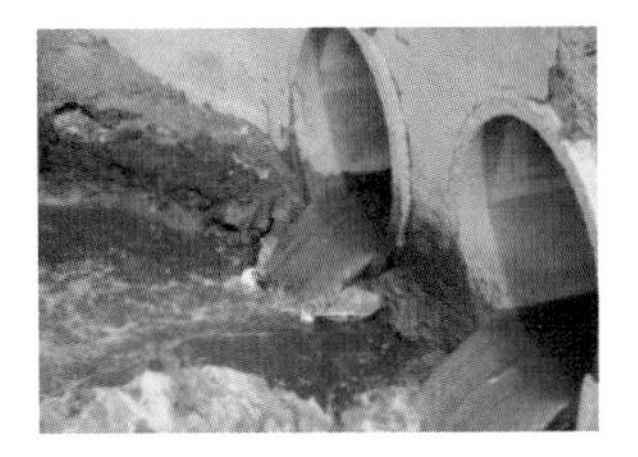

groundwater contaminant source

groundwater contamination

【释义】 地下水污染

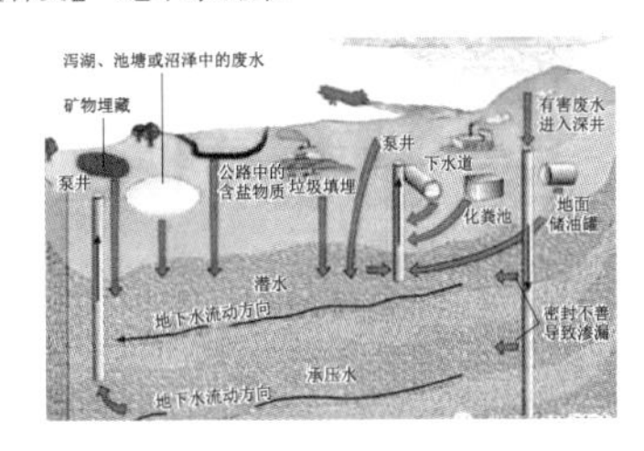

groundwater contamination

groundwater contour chart

【释义】 地下水等水位线图

【例句】 The groundwater contour chart and environment effect of the Yellow River Beach are analyzed. 分析了黄河滩地下水等水位图及环境影响。

【中文定义】 指地下水水位高程相等的各点所连成的曲线。

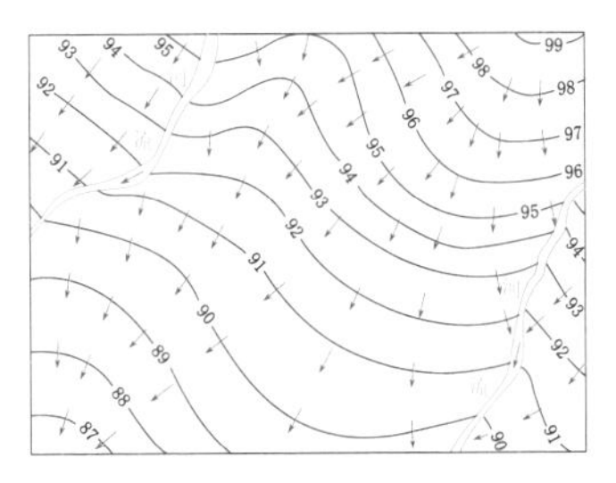

Measured groundwater contour chart

groundwater corrosivity

【释义】 地下水腐蚀性

【中文定义】 指地下水对混凝土的侵蚀破坏能力。

groundwater depletion

【释义】 地下水枯竭

【例句】 However, only over the last 50 or so years have their siting and design significantly improved, and, unlike in earlier times, many wells today are constructed not for drinking or irrigation purposes, but rather for monitoring groundwater depletion and chemical migration. 然而，仅仅在过去50年左右的时间里水井的选址和设计就得到了显著改善，但如今水井建造的目的不再像过去仅用于饮用或灌溉，而更多用于监测地下水枯竭和化学物质迁移。

groundwater depletion

groundwater detection

【释义】 地下水探测

groundwater discharge

【释义】 地下水流量

【例句】 Base flow in surface streams is maintained by groundwater discharge. 地表水的基本流量是靠地下水补给维持的。

groundwater environment quality evaluation

【释义】 地下水环境质量评价

【例句】 The GIS technology, database technology, water quality evaluation technology, etc. are applied in the development of groundwater environment quality evaluation system. GIS技术、数据库技术、水质评价模型技术等可用于开发地下水环境质量评价系统。

groundwater environmental background value

【释义】 地下水环境背景值

【例句】 On the basis of the study on groundwater environmental background value in the area of Poyang lake, a tentative programme of groundwater environment quality standards was put forward. 根据鄱阳湖地区地下水环境背景值研究成果，提出了制定该区地下水环境质量标准的初步方案。

groundwater hardness

【释义】 地下水硬度

【例句】 To solve the problem of groundwater hardness, it is necessary to find out the temporal and spatial development of the hardness in the shallow groundwater. 要解决地下水硬度问题，必须掌握浅层地下水硬度的时空发展趋势。

【中文定义】 指地下水中钙、镁离子的总浓度。

groundwater level

【释义】 地下水水位

【例句】 Environmentalists say that diverting water from the river will lower the groundwater level and dry out wells. 环保主义者称引流河水会导致地下水水位下降和水井干枯。

【中文定义】 指地下水面相对于基准面的高程。

groundwater level observation

【释义】 地下水水位观测

groundwater level observation

groundwater monitoring

【释义】 地下水监测

【例句】 Such procedures are helpful to convert the point data observed by groundwater monitoring wells into continuous surface data. 这类程序有助于把地下水监测井观测的点数据转换为连续的面状数据。

groundwater numerical model

【释义】 地下水数值模型

【例句】 Features and current situation of application of several popular groundwater numerical models in China were introduced in this paper. 本文介绍了国内常用地下水数值模型的特征及应用现状。

【中文定义】 在地下水研究中使用软件等建立的分析模型。

groundwater quality evaluation

【释义】 地下水水质评价

【例句】 Dispersion coefficient is an important hydrogeological parameter for the groundwater quality evaluation. 弥散系数是进行地下水水质评价的重要水文地质参数。

groundwater quality model

【释义】 地下水水质模型

【例句】 A series of groundwater quality models in the typical karst area in Guizhou is demonstrated. 展示了贵州典型岩溶区地下水水质系列模型。

【中文定义】 又称地下水水质数学模型，是水体水质的变化规律的数学描述。它可用于水体水质的预测、研究水体的污染与自净以及排污的控制等。

groundwater recharge

【释义】 地下水补给量

【例句】 These results are essential to quantify infiltration rates and groundwater recharge. 这些结果对于量化渗透率和地下水补给量是必不可少的。

groundwater resources

【释义】 地下水资源

【例句】 Once these regions are identified, precautions can be taken to restrict activities or prevent contamination of the groundwater resources. 一旦这些区域得到确定，可以采取预防措施限制或防止地下水资源的污染。

groundwater salinity and alkalinity

【释义】 地下水盐碱度

【例句】 The characteristics of groundwater salinity and alkalinity and its relationship with vegetation growth were studied. 地下水盐碱度特征及其与植被生长的关系被研究。

【中文定义】 即单位体积地下水中含 NaCl 和 $CaCO_3$ 等可溶性盐类的质量，常用单位为 g/L 或 mg/L。

groundwater source field

【释义】 地下水水源地

【例句】 The results indicate that the main recharge source of riverside groundwater source field in frigid area in winter is lateral runoff of aquifer. 结果表明，高寒地区傍河型含水层地下水水源地的补给来源在冬季主要为含水层侧向径流补给。

groundwater source field of Ertan reservoir area

groundwater steady flow

【释义】 地下水稳定流

【例句】 Groundwater pumping test can be used to determine whether the type of groundwater is groundwater steady flow or unsteady groundwater flow. 地下水抽水试验可以判断地下水的类型是地下水稳定流或是地下水不稳定流。

【中文定义】 指在渗流场内运动过程中各运动要素（水位、流速、流向等）不随时间改变的水流。

groundwater storage

【释义】 地下水储存量，地下水储量

【例句】 The world's largest groundwater storage system is beneath Stockholm's Arlanda Airport. 世界上最大的地下水储藏系统位于斯德哥尔摩阿兰达机场地下。

groundwater tracer test

【释义】 地下水示踪试验

【例句】 The principle of ion-selecting electrode tracer tecnology and its testing process has been used in the study of groundwater tracer test. 离子选择电极示踪技术方法原理及试验过程被用到地下水示踪试验研究中。

groundwater vulnerability evaluation

【释义】 地下水脆弱性评价

【例句】 In order to evaluate the groundwater vulnerability reasonably, entropy weight coefficient method is applied for the first time, which provides a new way to groundwater vulnerability evaluation. 为了合理地对地下水脆弱性进行评价，首次将熵权系数法模型应用于地下水脆弱性评价之中。

group [grup]

【释义】 *n.* 群

【常用搭配】 supergroup 超群；subgroup 亚群

【中文定义】 群是最大的岩石地层单位，通常是岩性相近、成因相关、结构类型相似的组的联合。

group of grottoes

【释义】 洞室群

group of grottoes

grout hole sealing

【释义】 灌浆封孔

【例句】 In the selective examination to observe the grout hole sealing effect of grouted curtain on Geheyan water conservancy project, it was surprisedly found that some paste had not hardened. 隔河岩水电工程帷幕灌浆封孔效果抽检发现部分水泥浆没有硬化。

grout hole sealing

grout leaking

【释义】 串浆

【常用搭配】 treatment device for grout leaking and returning 串浆返浆处理装置

grout leaking

grout oozing out

【释义】 冒浆

grout oozing out

groutability

【释义】 可灌性

【常用搭配】 rockmass groutability 岩体可灌性；groutability ratio 可灌比值

grouting curing material method

【释义】 灌入固化物法

【中文定义】 指利用气压、液压或电化学原理把某些能固化的浆液注入各处介质的裂缝或孔隙，以改善地基的物理力学性质。

grouting test

【释义】 灌浆试验

【例句】 Based on the related grouting test, the rationality of the grouting curtain design is verified and the suitable grouting parameters are obtained as well. 通过灌浆试验，验证了帷幕设计的合理性，并取得了合适的灌浆参数。

grouting test

guide adit

【释义】 导洞

【中文定义】 指隧洞施工中，为探查掌子面前方的地质条件，并为整个隧道作导向而开挖的小断面坑道。

gully [ˈgʌli]

【释义】 *n.* 冲沟，水沟；*vt.* 在……上开沟；*vi.* 形成沟

【常用搭配】 gully slope 沟坡

【中文定义】 一种基本的洪流地形，在洪流作用下，向源侵蚀使沟谷加深，旁蚀作用使沟谷加宽，使得冲沟得以形成与发展。

gully debris flow

【释义】 沟槽型泥石流，沟谷型泥石流

gully debris flow

gypsum ['dʒɪpsəm]

【释义】 *n*. 石膏

【常用搭配】 gypsum board 石膏板；gypsum powder 石膏粉

【例句】 Gypsum is a common white or colorless mineral (hydrated calcium sulphate). 石膏是一种白色或无色的矿物（水合硫酸钙）。

【中文定义】 单斜晶系矿物，主要化学成分为硫酸钙（$CaSO_4$）。

gypsum

H

Hadean ['heidɪən]

【释义】 *n*. 冥古宙（宇）

【中文定义】 冥古宙是太古宙之前的一个宙，开始于地球形成之初，结束于 38 亿年前，依据不同的文献可能有不同的定义，这个时期的岩石几乎没有保存到现在的，所以并没有正式的细分。

halite ['hælaɪt]

【释义】 *n*. 岩盐

【例句】 The drilling methods and well completion techniques of halite exploration were summarized. 岩盐勘探中的钻探施工方法及成井工艺被总结。

halite

【中文定义】 岩盐，化学成分为氯化钠，晶体都属等轴晶系六八面体晶类的卤化物。

hamada ['hæmədə]

【释义】 *n*. 石漠，岩漠

【中文定义】 又称戈壁滩，是指在平坦的或微有起伏的山区、山麓平原等表面布满大小不一的、有棱角的石块地，这种砾石流沙混杂的地面，主要由巨砾和裸露的基岩组成。

The ‘black’ hamada at Tademayt, Algeria

hanging valley

【释义】 悬谷

【例句】 Hanging valleys are also simply the product of varying rates of erosion of the main valley or the tributary valleys. 悬谷是主河道或支流侵蚀比率变化产生的。

【中文定义】 冰川谷的两侧高悬支谷，其距主谷底高差常达数十米，甚至数百米。

A hanging valley in Yosemite National Park

hard rock

【释义】 坚硬岩

【例句】 In general, fresh quartzite, granite and basalt are typical of hard rock. 新鲜的石英岩、花岗岩、玄武岩通常是典型的坚硬岩。

【中文定义】 单轴饱和抗压强度大于 60MPa 的岩石。

hardness ['hɑrdnɪs]

【释义】 *n*. 硬度，硬性

【例句】 With regard to hardness, the diamond is in a class by itself. 讲硬度，金刚钻是独一无二的。

【中文定义】 指材料局部抵抗硬物压入其表面的能力。

harmful gas detecting instrument

【释义】 有害气体检测仪（器）

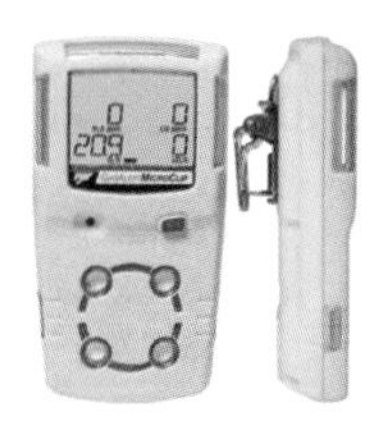

harmful gas detecting instrument

heat conductivity

【释义】 热导率；热传导；导热率；热传导性

【例句】 The heat conductivity of bentonite will increase with the rising of water content under the same compact density. 结果表明，在相同的压实密度条件下，随着膨润土含水率的增加，导热系数也增加。

heat source

【释义】 热源

【例句】 The Anomalous Heat Sources over the Tibetan Plateau has a great influence on the Anomalous Activities of the 1999 East Asian Summer Monsoon. 青藏高原热源异常对 1999 年东亚夏季风异常活动有极大影响。

height measurement

【释义】 高程测量

【同义词】 height surveying

【例句】 Coordinate transformation and height interpolation are the main methods for GPS-RTK height measurement/surveying. GPS-RTK 高程测量的主要方法是坐标转换和高程拟合。

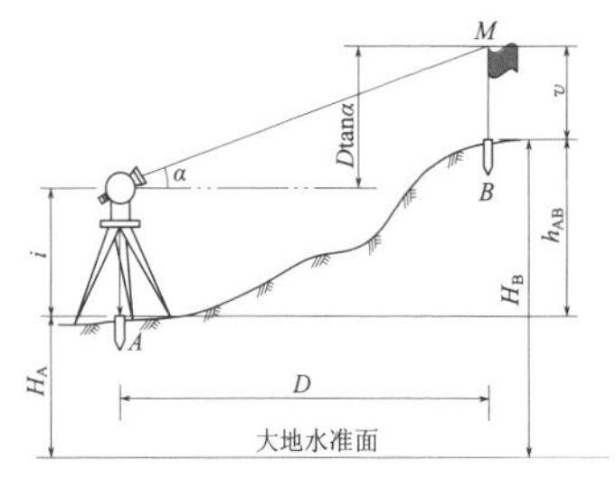

height measurement

hematite ['hiːmətait]

【释义】 *n.* 赤铁矿

【例句】 The results show that Fe in the ore is predominantly as hematite. 结果表明，矿石中铁主要以赤铁矿形式存在。

hematite

hemicrystalline texture

【释义】 半晶质结构

【中文定义】 岩石由部分晶体和部分玻璃质组成。多见于浅成岩和火山岩中。

hidden ground fissure

【释义】 隐伏地裂缝

【中文定义】 未在地表出露的破裂。习惯上把在地表出露的地裂缝和未在地表出露的地裂缝统称为地裂缝。

hidden trouble investigation

【释义】 隐患排查

high frequency debris flow

【释义】 高频泥石流

【中文定义】 发生频率为一年多次至五年一次的泥石流。

high mountain

【释义】 高山

【常用搭配】 medium high mountain 中高山；low high mountain 低高山

【中文定义】 海拔介于 3500～5000m，相对高差介于 200～1000m 的地区。

high mountain

high precision microseism

【释义】 高精度微震

high speed landslide

【释义】 高速滑坡

【例句】 The aerodynamic effect must be taken in account for the analysis of landslide with large-scale and high speed landslide. 对于大型高速滑

坡，必须考虑空气动力效应。

high water pressure test

【释义】 高压压水试验

【例句】 High water pressure test in pressure tunnel design is applied. 高压压水试验在压力隧洞设计中已被应用。

【中文定义】 对于水库大坝、深埋地下工程等水头很高的工程而言，在常规压水试验结果不能反映实际水头压力情况下，采用的高于常规压力的压水试验。

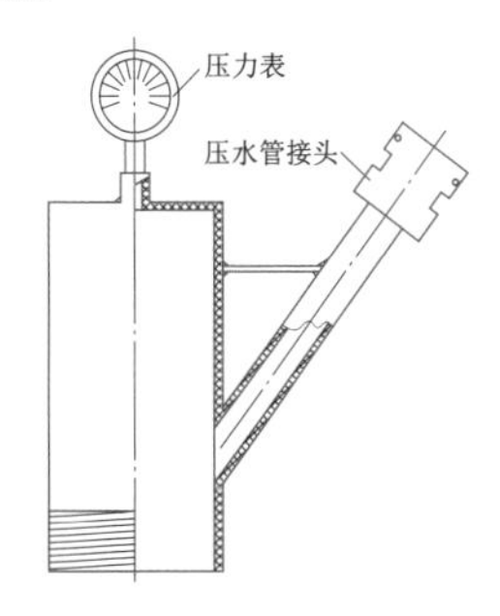

structure of orifice device for high water pressure test

highland ['haɪlənd]

【释义】 *n.* 高地，高原；*adj.* 高地的，高山的

【常用搭配】 highland climate 高山气候；highland barley 青稞；Scottish Highland 苏格兰高地

【例句】 The Mambilla plateau constitutes the western Nigerian boundary of the Adamaoua highlands. 蒙贝拉高原是阿达马瓦高地在尼日利亚的西边界。

Cameron Highlands

highly weathered

【释义】 强风化，强烈风化的

【常用搭配】 highly weathered rock 强风化岩石；highly weathered granite 强风化花岗岩

【例句】 The rock mass quality of the engineering slope is generally poor, and most of rock mass are highly weathered. 工程边坡岩体质量总体较差，大多数岩体强风化。

【中文定义】 岩石的组织结构大部分已破坏；小部分岩石已分解或崩解成土，大部分岩石呈不连续的骨架或心石，风化裂隙发育，有时含大量次生夹泥。

highly weathered

hill [hil]

【释义】 *n.* 丘陵，山丘，凸起地形

【常用搭配】 loess hill 黄土峁；sand hill 沙丘；hillside 山坡

【例句】 The Mambilla Plateau consists of undulating hills with softly rounded slopes, mostly being covered by grassland. 蒙贝拉高原是由起伏的山丘构成的，坡缓且大部分被草原覆盖。

【中文定义】 地表面起伏较缓、冈丘错综连绵、大部分地面的倾斜角在 2°～6°、地面高差在 20～150m 的地区。

hill

Himalayan orogeny

【释义】 喜马拉雅运动

【例句】 Gansser suggested a maximum shortening of the crust of the order of 500km during the Himalayan orogeny. Gansser 认为在喜马拉雅运动期间，地壳最大缩减了约 500km。

【中文定义】 新生代以来的造山运动被黄汲清称

为喜马拉雅运动。这一造山运动因首先在喜马拉雅山区确定而得名。

hole-wall strain over-coring

【释义】 孔壁应变法

Holocene ['hɔləsiːn]

【释义】 *n.* 全新世（统）

【中文定义】 第四纪的第二个世，是地质时代最新阶段，开始于约1.17万年前，持续至今。

holocrystalline texture

【释义】 全晶质结构

【中文定义】 同“显晶质结构”。

homogeneous landslide

【释义】 均质滑坡

honeycomb weathering

【释义】 蜂窝状风化

【例句】 Honeycomb weathering is a type of weathering that is described by morphological characteristics. 蜂窝状风化是一种按形态特征进行描述的风化类型。

honeycomb weathering

【中文定义】 在风化作用下，其中一部分矿物发生氧化分解和淋滤流失，形成空洞，而另一些难风化或难溶解矿物（如石英、石髓和褐铁矿等）被残留下来形成骨架，构成蜂窝状构造。

horizontal acceleration

【释义】 水平加速度

【例句】 The seismic zoning map was worked out separately by use of earthquake intensity and peak value of horizontal acceleration of bedrock. 区划图是分别用地震烈度和基岩水平加速度峰值来编制的。

【中文定义】 地震水平加速度是指地震时地面水平运动的加速度。

horizontal bedding

【释义】 水平层理

【中文定义】 又称水平纹层，其特点是纹层呈直线状互相平行，并且平行于层面。

horizontal bedding

horizontal displacement measurement

【释义】 水平位移测量

【例句】 Horizontal displacement measurement of a dam is of great importance for its safety. 大坝水平位移测量对于大坝安全来说非常重要。

horizontal displacement observation

【释义】 水平位移观测

【例句】 An analysis is carried out on the accuracy of horizontal displacement observation by polar coordinate method. 对极坐标法水平位移观测的精度进行了分析。

horizontal monitoring network

【释义】 平面监测网

horizontal motion

【释义】 水平运动

【例句】 In projectile motion, the horizontal motion and the vertical motion are independent of each other, that is, neither motion affects the other. 在抛弹运动中，水平运动和垂直运动是相互独立的；也就是说，两者都不会影响另一方。

【中文定义】 指沿着地球球面切线方向发生的运动。

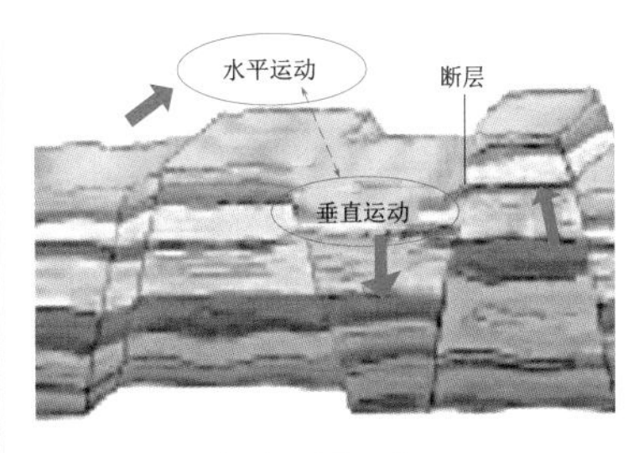

horizontal motion

horizontal runoff zone

【释义】 水平径流带

【中文定义】 地表水或地下水等径流在水平方向

上流动的区域。

hornblende ['hɔrnˌblənd]

【释义】 *n.* 角闪石

【常用搭配】 common hornblende 普通角闪石；hornblende schist 角闪片岩；hornblende granite 角闪花岗岩

【例句】 Hornblende is an important rock-forming mineral of amphibole group of double chain silicates. 角闪石是一种重要的造岩矿物，属于双链状结构硅酸盐矿物中的闪石族。

hornblende

【中文定义】 角闪石族矿物的总称，角闪石属闪石族中一员。与辉石族形态、组成相近。

hornfels ['hɔːnfels]

【释义】 *n.* 角页岩

【常用搭配】 hornfels facies 角页岩相；hornfels texture 角页岩构造

【例句】 Hornfels is a non foliated metamorphic rock. 角页岩是一种无叶理的变质岩。

【中文定义】 区域变质中变质带的产物。是角闪岩相的典型代表，由基性、中基性火成岩或含铁镁多的石灰质白云岩在中至高温区域变质条件下变质而成，主要矿物为角闪石和斜长石，色深绿或浅绿，片理不发育。

hornfels

horst [hɔst]

【释义】 *n.* 地垒

【常用搭配】 step horst 阶状地垒；structural horst 构造地垒

【例句】 In physical geography and geology, a horst is the raised fault block bounded by normal faults or graben. 在自然地理学和地质学中，地垒是以正断层或地堑所界限的被抬起的断块。

【中文定义】 地壳中被两侧倾向相反的正断层偶为逆断层所界限而中间断盘上升的凸起断块构造。纵向延伸可达数百千米。地垒的形成与地壳的水平拉伸作用有关，常与地堑相间出现。

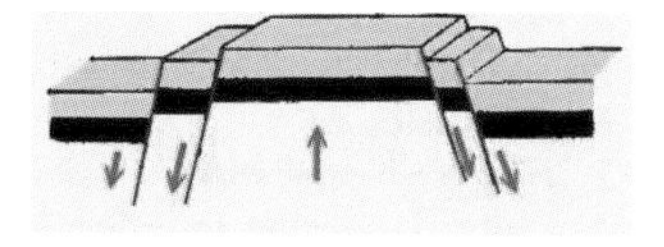

horst

host rock

【释义】 母岩

【例句】 Based on the results of in-situ direct shear tests and laboratory shear tests on interlayer staggered zones at Baihetan hydropower station, the shear behaviors of interlayer material and interlayer material/host rock interface (soil/rock interface) were investigated. 基于白鹤滩水电站层间错动带的室内和现场剪切试验结果，对层间材料、层间材料/母岩接触面（土/岩接触面）的剪切力学特性进行研究。

hot spring

【释义】 热泉

【例句】 Small blue-flowered fringed gentian of western United States (Rocky Mountains) is beautiful, especially around hot springs in Yellowstone National Park. 美国西部（洛矶山脉）一种小型开蓝花的流苏龙胆十分漂亮，尤其是生长在黄石国家公园热泉周围的。

hot spring

【中文定义】 指泉温高于45℃而又低于当地地表水沸点的地下水露头。一般分为海底热泉和陆地热泉两种。

Huanghai elevation system 1956

【释义】 1956年黄海高程系

huge-scale landslide

【释义】 特大型滑坡（巨型滑坡）

【例句】 It shows that in-situ experiment is important to the landslide stability analysis and the optimization of landslide engineering treating, especially to the huge-scale landslide. 试验研究表明巨型滑坡滑带土的大型原位剪切试验对滑坡稳定性分析和工程治理优化具有重要价值。

human activity

【释义】 人类活动

【例句】 Our climate is changing and that change is being caused by human activity. 地球的气候正在改变，而这正是我们人类活动造成的。

【中文定义】 人类为了生存发展和提升生活水平，不断进行了一系列不同规模不同类型的活动，包括农、林、渔、牧、矿、工、商、交通、观光和各种工程建设等。

hummocky moraine

【释义】 冰碛丘陵，基碛丘陵

【例句】 Especially in the Bodo Zangbo valley, there are lots of hummocky moraines like tumulus. 尤其是在博渡藏波河谷，有许多外形像坟墓一样的冰碛丘陵。

【中文定义】 冰碛丘陵属于冰碛地貌的一种，是冰川消融后表碛、中碛、内碛沉落于底碛上形成的起伏不平的地面形态，冰碛厚度由数米至百余米不等。

hummocky moraine

hydraulic fill

【释义】 冲填土，水力充填

【例句】 The hydraulic fill concretion drainage process includes forepart air discharge, penetrate concretion in vacuum seepage field and water gasification discharge under the vacuum condition. 吹填土固结排水过程分为早期空气排出、真空渗流场作用下的渗透固结和真空条件下的水分气化排出三个阶段。

【中文定义】 又称吹填土，是由水力冲填泥沙形成的填土，它是我国沿海一带常见的人工填土之一。

hydraulic fill

hydraulic fracturing

【释义】 水压致裂法

【例句】 Fracturing pressure in formations was the basic data for designing mud weight, well profile and hydraulic fracturing. 地层破裂压力是钻井液密度确定、井身结构和压裂设计施工的基础数据。

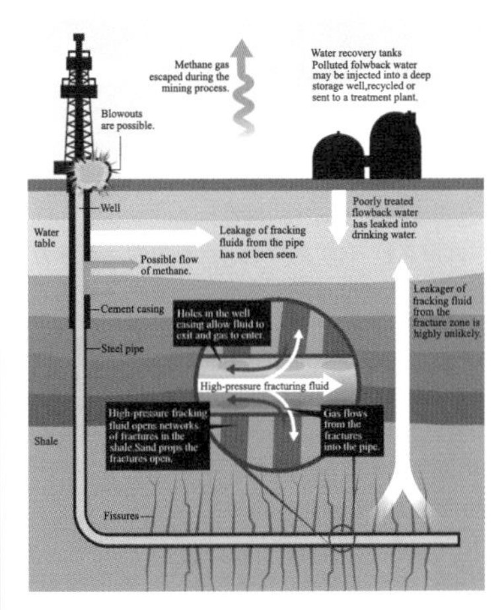

hydraulic fracturing

hydraulic fracturing technique

【释义】 水力劈裂法

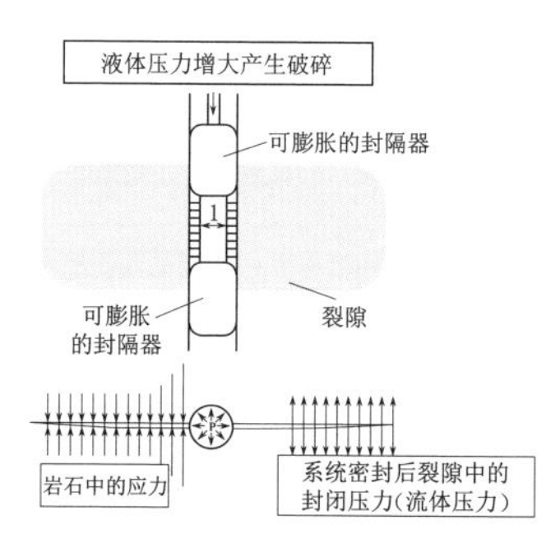

hydraulic fracturing technique

hydraulic gradient

【释义】 水力梯度，水力坡降

【例句】 It is found that both the initiation hydraulic gradient and the failure hydraulic gradient increase with confining pressure. This is mainly due to the increase of friction between particles. 试验结果表明，起始水力梯度和临界水力梯度都随着围压的增大而增大，这是由于增大围压使得颗粒间的摩擦力增大的结果。

【中文定义】 水力梯度，又称水力坡降或者水力坡度，指沿渗透途径水头损失与渗透途径长度的比值。

hydrogeological conditions

【释义】 水文地质条件

【例句】 Therefore, the hydrogeological conditions should be considered as one of the important factors when underground space planning is formulated. 因此在编制地下空间资源开发利用规划时应注意对水文地质条件的研究。

【中文定义】 通常把与地下水有关的问题称为水文地质问题，把与地下水有关的地质条件称为水文地质条件。

hydrogeological map of karst areas

【释义】 岩溶区水文地质图

【中文定义】 针对岩溶区的各种水文地质现象和资料编制的水文地质图件。

hydrogeological parameter

【释义】 水文地质参数

【例句】 Groundwater recharge is an important hydrogeological parameter. 地下水入渗补给量是重要的水文地质参数。

【中文定义】 水文地质参数是反映含水层或透水层水文地质性能的指标，如渗透系数、导水系数、水位传导系数、压力传导系数、给水度、释水系数、越流系数等。

hydrogeological parameters testing

【释义】 水文地质参数测试

【例句】 The methods of hydrogeological parameters testing in the aquifer are mainly involved in pumping test, slug test, water pressure test and numerical inverse analysis. 含水层水文地质参数测试的方法主要涉及抽水试验，冲击试验，水压试验和数值反演分析。

hydrogeological phenomenon

【释义】 水文地质现象

【中文定义】 地下水的数量和质量及随空间和时间变化的自然现象。

hydrogeological test

【释义】 水文地质试验

【例句】 Hydrogeological test is a necessary method to obtain permeability parameters of rock soil mass. 水文地质试验是获取岩土体渗透性参数的必要方法。

【中文定义】 为定量评价水文地质条件和取得水文地质计算参数而进行的各项野外测试工作的总称，包括抽水试验、注水试验、渗水试验、压水试验、地下水连通试验、地下水流速流向测定等。

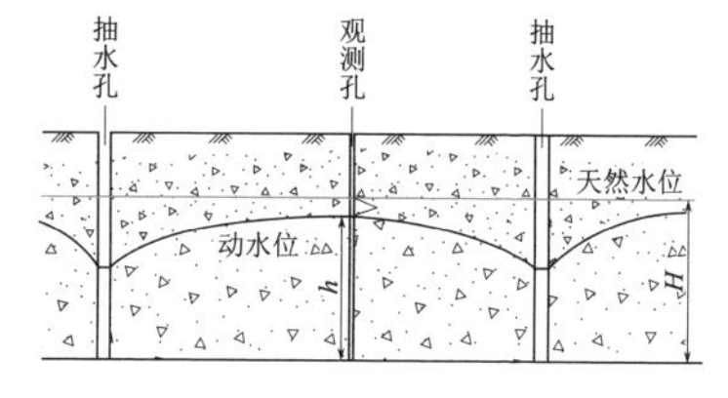

hydrogeological tests

hydrograph cutting method

【释义】 流量过程线切割法

hydrographic surveying and charting

【释义】 海洋测绘

【例句】 In pace with the generation of digital ocean infrastructure, historic changes have taken place in hydrographic surveying and charting in China. 随着数字海洋基础框架的构建，我国海洋测绘发生了历史性的变革。

【中文定义】 以海洋水体和海底为对象所进行的测量。主要包括：海洋大地测量、海底地形测量、海道测量、海洋专题测量和海图编绘等。

hydromica [haɪdrə'maɪkə]

【释义】 *n*. 水云母

【例句】 Hydromica is a typical alteration mineral in granite-type uranium deposit, and also an important indication of uranium. 水云母是花岗岩型铀矿床蚀变带中的一种典型蚀变矿物，它也是寻找铀矿的一个重要标志。

hydromica

【中文定义】 水云母族（也称伊利石族）矿物的总称。其化学成分中的钾含量较云母低而水含量则较之为高，是云母族矿物向蒙脱石族矿物转变的过渡产物。

hydrostatic pressure

【释义】 静水压力，水流静压

【例句】 Below the water table at static equilibrium, hydrostatic pressure potential increases with increasing depth. 当静态平衡时，在水面以下，静水压力随深度而增加。

【中文定义】 静水压力由均质流体作用于一个物体上的压力。

hydrothermal active area

【释义】 水热活动区

【例句】 The hydrothermal active area of the plateau is mainly exposed between the Himalaya and Gangdise-Nyainqentanglha. peratures > 80℃ 高原水热活动带主要出露在喜马拉雅和冈底斯—念青唐古拉之间。

hydrothermal activity

【释义】 水热运动

【常用搭配】 recent hydrothermal activity 现代水热活动

【例句】 The major big rift and two groups of the minor rifts approximately pointing to S－N in the area are the main pathway and the main tectonic crack of heat controller for the hydrothermal activity. 区内近 S－N 向的主要大断裂和两组次一级的断裂是水热活动的主要通道和主要控热构造裂隙。

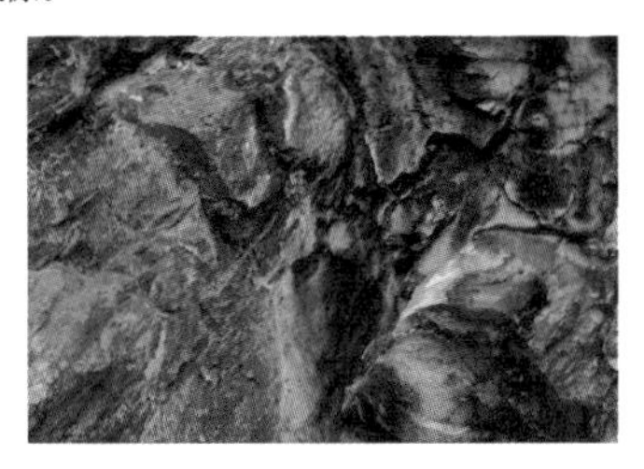

hydrothermal activity

hydrothermal alteration

【释义】 热液蚀变，水热蚀变

【例句】 Hydrothermal alteration affects the physical and mechanical strength of rock and the availability of rock mass. 热液蚀变影响岩石的物理力学强度和岩体的可利用性。

【中文定义】 因热液活动引起岩石矿物成分、化学成分及物理化学特性变化的过程。

hydrothermal system

【释义】 水热系统

【常用搭配】 index of hydrothermal system 水热系统指数

【例句】 The solubility of the hydrothermal system was controlled at 3.891×10^{-6} mol/L. 水热系统溶解度控制在 3.891×10^{-6} mol/L。

hypometamorphic rock

【释义】 深变质岩

【例句】 All the gems that have been discovered occur in the hypometamorphic rock belts. 已发现的宝石均产于深变质岩带中。

I

ice avalanche

【释义】 冰崩

【例句】 A rock avalanche or landslide may be partly an ice avalanche. 岩崩或山崩可能是冰崩的一部分。

ice avalanche

【中文定义】 主要是由于气候变暖而使冰川容易融化、裂开，同时产生移动，继而发生冰崩。

illite [ˈɪlaɪt]

【释义】 *n.* 伊利石

illite

【例句】 From chemical analyses, X-ray diffrection, infrared spectra, differential thermal curves and scanning electrical microscopy we recognized that the mineral deposit is mainly composed of kaolinite with lower crysallinity and little of montmorillonite, illite and their mixlayer minerals. 经化学分析、X射线衍射分析、红外光谱分析、差热和电子显微镜扫描等研究，其矿物组成主要为结晶度差的高岭石、少量蒙脱石、伊利石及其混层矿物。

image map

【释义】 影像地图

【中文定义】 以航空和航天遥感影像为基础，经几何纠正，配合以一定的地图符号和少量注记，将制图对象综合表示在图面上的地图。

image processing

【释义】 图像处理

【例句】 The invention is a device equipped with a camera that captures digital images and is connected to an image processing system. 这一发明配置了一台用来捕捉数字图像的相机，而且已经连接到一个图像处理系统中去。

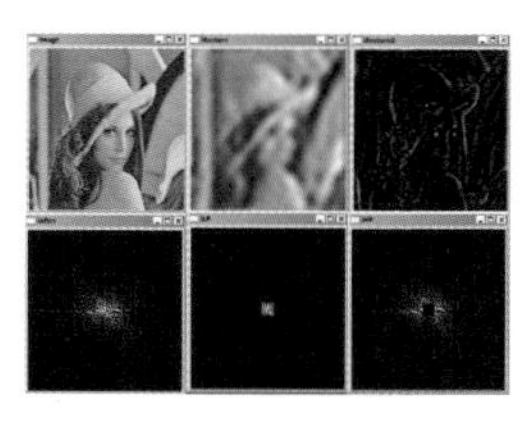

image processing

imbalance thrust force method (ITFM)

【释义】 不平衡推力法

【例句】 Imbalance thrust force method (ITFM) is originally developed approach for slope stability analysis in China, which has been widely used in the field of landslide stability analysis. 不平衡推力法是我国独创的边坡稳定分析方法，在滑坡稳定分析和治理中得到广泛的应用。

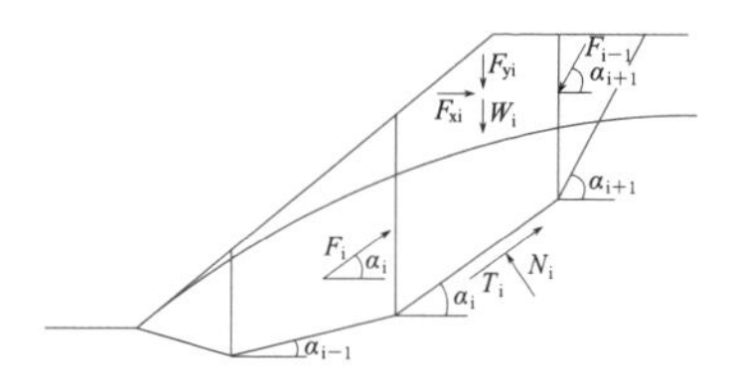

imbalance thrust force method

imbricate fault

【释义】 叠瓦断层

【例句】 These reflections may be caused by imbricate fault slices or by the interbedding of mudstone and limestone. 这些反应可能是由叠瓦断层或泥岩和石灰岩的嵌入引起的。

【中文定义】 叠瓦构造是一系列产状相近、大致等距、近似平行或逆冲断层或滑动所形成的逆断层的组合构造。各断层的上盘依次相对上升，在剖面上呈屋顶的瓦片依次叠置，故名叠瓦构造，又称叠瓦断层。

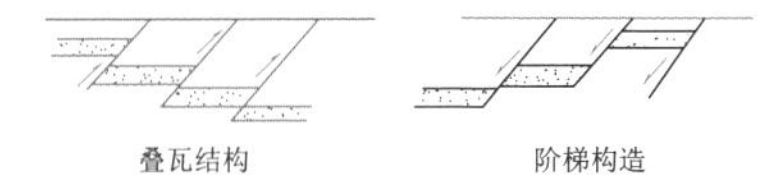

imbricate fault

imbricate structure

【释义】 叠瓦构造

【例句】 The structural types mainly are fault-related folds, imbricate structure and fault delta zone. 其构造类型主要为断层相关褶皱、叠瓦状构造及断层三角带。

【中文定义】 由多个逆断层（冲断层或逆掩断层）平行重叠出现而形成的一种构造。

imbricate structure

immersion [ɪ'mɜrʒən]

【释义】 *n.* 浸没，沉浸

【中文定义】 浸没：水库蓄水使水库周边地区地下水位壅高而引起土壤盐渍化和沼泽化、建筑物地基沉陷或破坏、地下工程和矿井充水或涌水量增加等灾害现象的统称。

impervious blanket

【释义】 防渗铺盖

【例句】 In the meantime we will discuss the influence of impervious blanket's length for earth-rock's seepage on infinite deep pervious foundation. 同时，本文还对无限深透水地基上土石坝水平铺盖长度对渗流的影响进行分析。

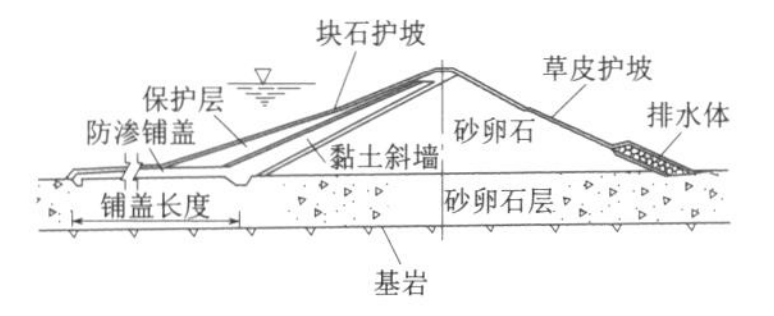

impervious blanket

impervious curtain

【释义】 防渗帷幕

【常用搭配】 suspended impervious curtain 悬挂式防渗帷幕

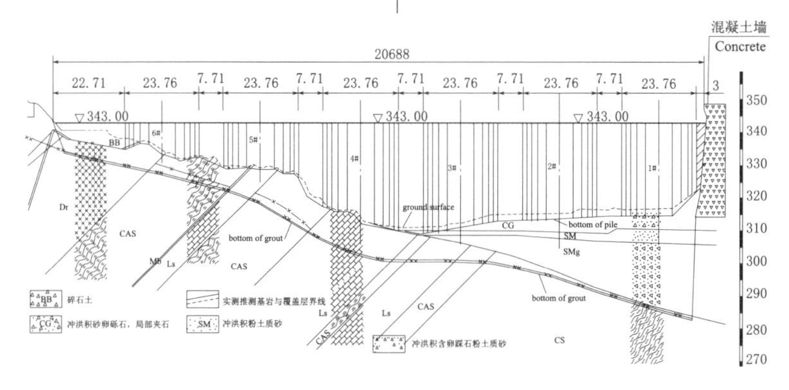

impervious curtain（unit：m）

impounding

【释义】 蓄水

【例句】 The density changes before and after the first water impounding of Three Gorges Reservoir are inversed. 反演了三峡水库第一次蓄水前后地下岩体密度的变化。

in-situ compression curve

【释义】 现场压缩曲线

【中文定义】 根据室内试验所得压缩曲线（孔隙比-压力对数值关系曲线）推测得出的符合土体

天然性状的原始压缩曲线。

in-situ direct shear test

【释义】 原位直接剪切试验

【例句】 With the in-situ direct shear test and indoor direct shear test, shear strength properties of high fill and compaction soil of an airport are studied. 采用原位直剪试验、室内直剪试验测试手段，对机场高填方压实土的抗剪强度特性进行了分析。

in-situ direct shear test

in-situ rock stress test

【释义】 岩体原位应力测试

in-situ stress

【释义】 地应力

【同义词】 geostress, crustal stress

【例句】 Because of limited knowledge, people usually use side pressure coefficient to control the scale of the in-situ stress in the analysis and design. 因为人们对初始地应力的了解有限，分析设计中往往用侧压系数来控制初始地应力的作用范围。

【中文定义】 地应力是在漫长的地质年代里，由于地质构造运动等原因使地壳物质产生了内应力效应，这种应力称为地应力，它是地壳应力的统称。

in-situ test

【释义】 原位测试

【例句】 Plate load test is a kind of in-situ test methods used to measure pressure-deformation characteristic of ground soil. 平板载荷试验是一种用于测量地基土压力变形特征的原位试验方法。

inclination [ˌɪnklɪˈneɪʃn]

【释义】 倾斜

【常用搭配】 inclination angle 倾斜角

【例句】 Geological surfaces that are inclined are said to dip, the angle of dip being the maximum inclination of the surface measured, by convention, from the horizontal. 倾斜地质面的倾角，按惯例是指该面与水平面间的最大倾斜角度。

inclination survey

【释义】 井斜测量

inclined fold

【释义】 倾斜褶皱

【例句】 In profile, the syncline is asymmetrical and called inclined fold. 在横剖面上，这个向斜构造是不对称的，被称为倾斜褶皱。

【中文定义】 指轴面倾斜，两翼倾向相反、倾角不等的褶皱。倾斜褶皱包括两翼陡峻而轴面倾斜的褶皱，但不包括倒转褶皱。轴面倾斜平缓（倾角 10°～30°）的称平缓倾斜褶皱；30°～60°的称中等倾斜褶皱；60°～80°的为陡倾斜褶皱。

inclined fold

inclined shaft

【释义】 斜井

【例句】 Undisturbed samples shall be sampled in borehole as far as possible, and in-situ tests and laboratory tests of samples shall be conducted in representative sections in vertical shafts, inclined shafts and adits. 钻孔中应尽量采取原状样，在竖井、斜井或平洞内代表性地段应开展必要的现场试验和取样进行室内试验。

independent coordinate

【释义】 独立坐标系

【例句】 This paper discussed the method of establishing local independent coordinate system through Ellipsoid transformation. 阐述了应用椭球变换建立区域独立坐标系的方法。

【中文定义】 任意选定原点和坐标轴的直角坐标系。

independent foundation

【释义】 独立基础

independent foundation

【例句】 The strengthening treatment of independent foundation under industrial plant column is discussed. 对工业厂房立柱独立基础的加固处理进行了讨论。

Indian Plate

【释义】 印度洋板块

【例句】 Originally a part of the ancient continent of Gondwana，Indian Plate broke away from the other fragments of Gondwana 100 million years ago and began moving north. 印度洋板块起初属于冈瓦纳古陆的一部分，1 亿年前，印度洋板块从冈瓦纳古陆脱离并开始向北移动。

【中文定义】 印度洋板块为大陆板块，包括印度洋的北部、印度半岛、阿拉伯半岛、大洋洲的大陆、岛屿等。

induced polarization method

【释义】 激发极化法

【例句】 Compared with traditional direct current method，the induced polarization method has strong point of multiple parameters. Integrated analysis using parameters of s，ρ，η，J and D，may determine karstic development status，karstic connectivity and to solve water research in karstic areas. 与常规直流电法相比，激发极化方法具有多参数的显著特点，利用 s、ρ、η、J、D 等参数综合分析，可以确定岩溶发育程度，判别岩溶连通性，解决岩溶区找水问题。

induced stress

【释义】 感生应力

【例句】 Combined with the underground engineering practice in high geo-stress area，the stress restitution testing and lag monitoring are made to determine the induced stress field of surrounding rock mass for underground engineering. 结合高地应力区某地下工程实践，提出了一套岩体地下工程围岩表面二次应力场的应力恢复现场测试与滞后监测方法。

【中文定义】 人类从事工程活动时，在岩体天然应力场内，因挖除部分岩体或增加结构物而引起的应力，称为感生应力。

induction electromagnetic method

【释义】 感应电磁法

inequigranular texture

【释义】 不等粒结构

【中文定义】 同种矿物颗粒的大小不等，多见于侵入体的边缘及浅成侵入岩类。

infiltration test

【释义】 渗水试验

【例句】 These problems are frequently encountered in field test and are common factors influencing the precision of infiltration test. 这些问题在野外试验中常常遇到，是影响渗水试验精度的常见因素。

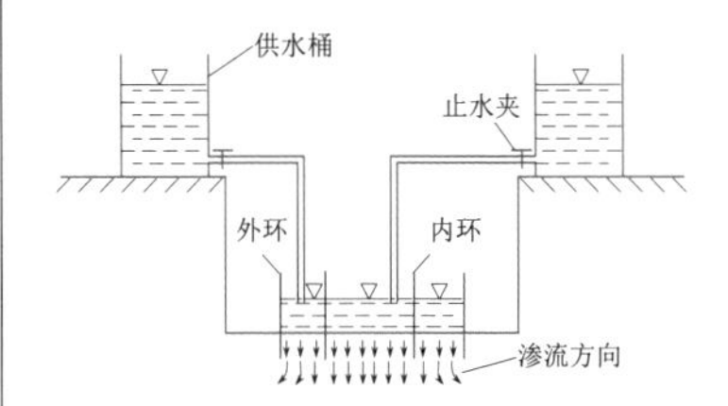

diagram of double ring infiltration test device

【中文定义】 渗水试验一般采用试坑渗水试验，是野外测定包气带松散层和岩层渗透系数的简易方法。试坑渗水试验常采用的是试坑法、单环法和双环法。

infiltration velocity

【释义】 渗透速度，渗流速度，达西流速

【同义词】 seepage velocity

【例句】 Infiltration velocity of groundwater influences its reaction rate with rocks. 地下水的渗透速度影响着其与岩石的反应速度。

【中文定义】 亦称渗流速度、达西流速（Darcy veloeity）或比流量（specifc discharge），是一种假想的充满整个断面的液体的流速。只反映渗流在与渗流方向垂直的断面上的平均流速，不反映多孔介质中任何真实水流的流速。渗透速度 v 和液体的实际平均流速 u 有下列关系：$v=nu$，式中 n 为有效孔隙度。绝大多数情况下，与水力坡度成正比，满足达西定律。

influence area

【释义】 影响区域

【例句】 A building is located within the influence area of the earthquake，so the damage is more serious. 某建筑位于地震影响范围内，因此破坏比较严重。

influence belt of faults

【释义】 断层影响带

【例句】 The book introduces a method used to forcast the attitude character and its position of influence belt of faults. 这本书介绍了一种方法，用于预测断层影响带产状和位置。

influence belt of faults

influence radius

【释义】 影响半径

【例句】 We calculated the surface subsidence, horizontal displacement, incline, and current curvature using the groundwater drawdown, conical depression influence radius of water cone and ground property. 根据地下水降深、降落漏斗影响半径、岩土工程地质性质，计算降水引发的地表沉降值、水平移动、倾斜、曲率现状值。

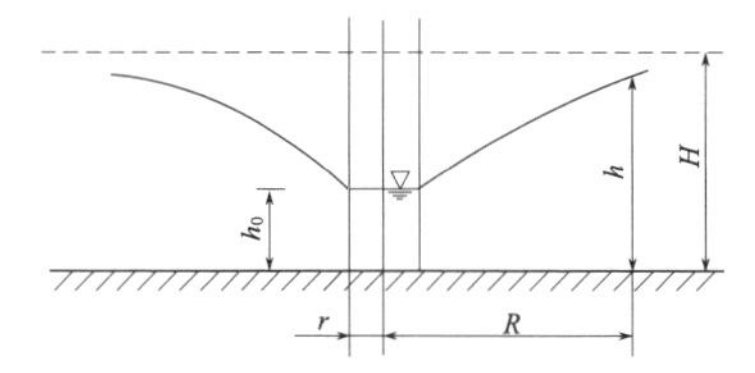

H—含水层的厚度；R—影响半径；r—承压井的半径；h_0—抽水后水位；h—抽水前潜水水位

influence radius

infrared color photography

【释义】 红外摄影，红外彩色照相法

inhomogeneous deformation

【释义】 不均匀变形

【例句】 If the inhomogeneous deformation of subgrade exceeds a certain limit, the cement concrete pavement will get destroyed. 路基的不均匀变形超过一定限度会导致水泥混凝土路面发生破坏。

【中文定义】 变形体在外力作用下产生的塑性变形沿高向、宽向及纵向分布不均匀的现象。

initial collapse pressure

【释义】 湿陷起始压（应）力

initial compression curve

【释义】 初始压缩曲线

initial creep

【释义】 初始蠕变

【例句】 The creep strain of test Ⅲ is less than test Ⅰ and test Ⅱ during initial creep stage and steady creep stage. 在初始蠕变阶段和稳定蠕变阶段，试验Ⅲ的蠕变应变小于试验Ⅰ和试验 Ⅱ。

【中文定义】 指应变随时间延续而增加，但增加的速度逐渐减慢。

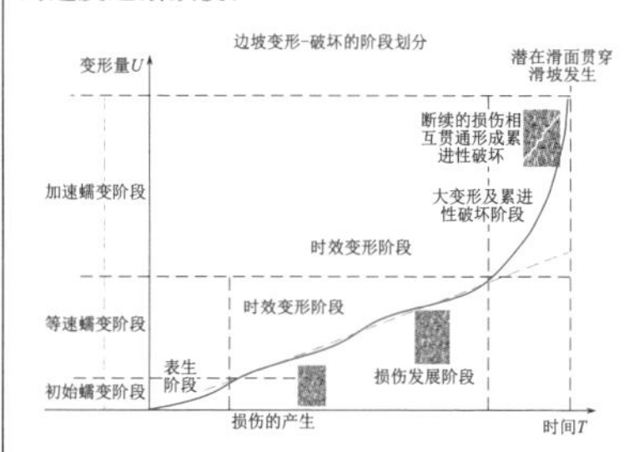

division of deformation and failure stages of slope

initial data

【释义】 原始数据

【中文定义】 用户数据库中的数据，是终端用户所存储使用的各种数据，他构成了物理存在的数据。

initial impoundment

【释义】 初次蓄水

【例句】 Seepage analysis is an important part in initial impoundment analysis. 渗流分析是初次蓄水分析的重要内容之一。

initial pressure

【释义】 初始压力

【例句】 Numerical results show that, the maximum explosion pressure is proportional to initial pressure. 计算结果表明：最大爆炸压力与初始压力成正比。

initial stress

【释义】 初始应力

【例句】 Stress relaxation is dependent upon time, temperature, the initial stress level, and material. 应力松弛由于时间、温度、初始应力水平和材料的不同而不同。

【中文定义】 指在施加所考虑的荷载之前土体中已存在的应力。

initial stress field

【释义】 初始应力场

【常用搭配】 rock mass initial stress field 岩体初始应力场
【例句】 Initial stress field is the basic data for the design and construction of large underground hydropower house. 初始地应力场是大型水电地下厂房设计施工的基础资料。

initial stress of rock mass (natural stress)
【释义】 岩体初始应力（天然应力）

initial stress state
【释义】 初始应力状态
【例句】 Because the structural response differs with differing tunnel shape, wall rock properties, and initial stress state, the safety of the tunnel in different location should be estimated respectively. 由于隧道的形状、岩性及其初始应力状态的不同，结构不同部位的动态响应有较大的差别，应分别评价其危险性。
【中文定义】 岩体在天然状态下所存在的内在应力。

initial support
【释义】 初期支护
【中文定义】 洞室开挖后立即施作的支护。

initial support

initial water level
【释义】 初见水位
【例句】 In general, the initial water level in the permeable soil is basically the same as that of the stable water level. 一般地，强渗透性土中的初见水位与潜水稳定水位基本持平。

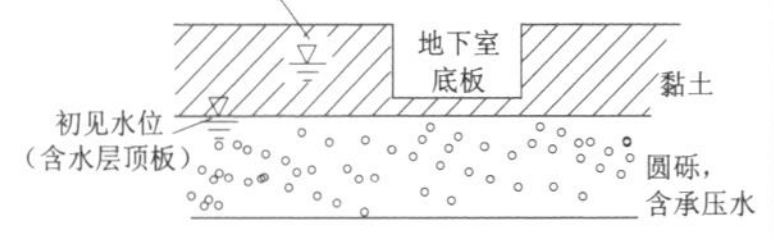

diagram of initial water level and steady water level

inorganic soil
【释义】 无机土，无机土壤
【例句】 After adding inorganic soil amendment to soil, the soil showed a superior water-holding capacity, compared with the control. 无机土壤改良剂加入土壤后，土壤持水能力较对照有所增加。
【中文定义】 指由地表岩石经风化作用而形成的没有胶结或弱胶结的颗粒堆积物。

inorganic soil

insequent valley
【释义】 斜向谷
【中文定义】 河谷延伸方向与岩层走向斜交（30°～60°）。

insequent landslide
【释义】 切层滑坡

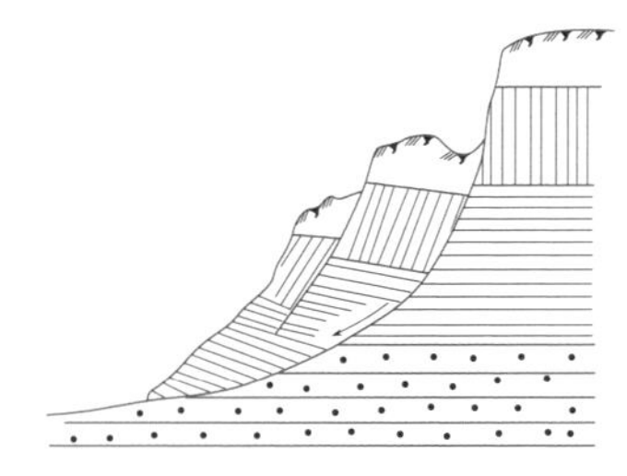

insequent landslide

inshore area
【释义】 近岸区

inshore area

inspection of seismic
【释义】 地震波测试

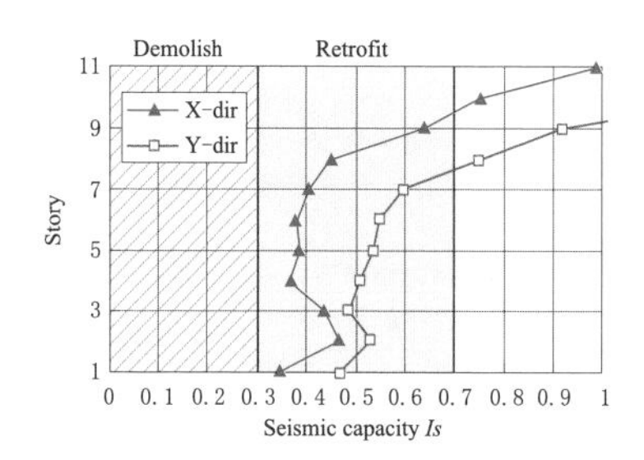

inspection of seismic

inspection of shear wave velocity

【释义】 剪切波测试

【例句】 The inspection of shear wave velocity of soils is an important aspect in dynamic soil mechanics and seismic engineering, and is also a new developing technology which can be used in both seismic exploration and scientific evaluation on the improved foundation in recent years. 土体剪切波测试是土动力学和地震工程学研究中的一项重要内容，也是近年来发展起来的浅层地震勘探和对已加固地基进行科学评价的一项新技术。

instantaneous (immediate) settlement

【释义】 瞬时沉降

【例句】 An immediate settlement analysis that takes soil nonlinearity into account is needed. 瞬时沉降计算有必要考虑土体的非线性特征。

【中文定义】 荷载作用瞬间所产生的地基沉降。

integral coefficient

【释义】 完整性系数

integral structure

【释义】 整体状结构

【中文定义】 岩性单一，构造变形轻微的巨厚层沉积岩、变质岩、火山熔岩和火成侵入岩。

interbedded compressed zone

【释义】 层间挤压带

【例句】 The geologic condition at Jinping-I arch dam site is rather complicated with main defects of f5, f8, and f14 faults, fissures in depth, interbedded compressed zones, lamprophyre dikes and deformed tension cracked rock mass. 锦屏一级拱坝坝区地质条件复杂，主要地质缺陷有 f5、f8、f13 和 f14 断层，深部裂缝，层间挤压带，煌斑岩脉及变形拉裂岩体等。

interfluve ['ɪntərflʊv]

【释义】 *n.* 河间地块，江河分水区

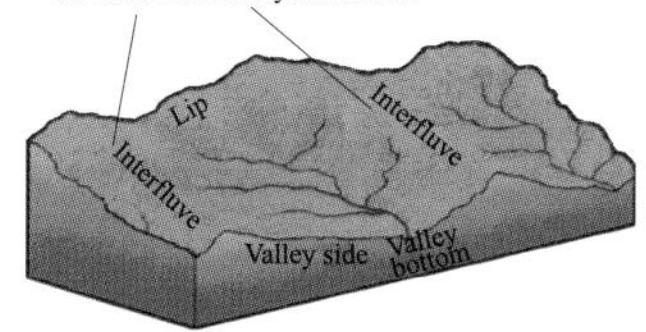

valleys and interfluves

interglacial period

【释义】 间冰期

【常用搭配】 Cromerian interglacial period 克鲁莫间冰期；Mindel-Riss interglacial period 民德里斯间冰期

【例句】 Interglacial period is a useful tool for geological mapping and also for anthropologists, as they can be used as a dating method for hominid fossils. 间冰期对于地质绘图和人类学家非常有用，因为间冰期能用于判定原始人类化石的年份。

【中文定义】 介于两次冰期之间的气候较为温暖的地质时期。是大冰期中相对温暖的时期。间冰期冰川作用相对地变弱，冰盖向高纬度退缩，雪线升高，由于冰体大量消融，冰融水注入海洋，致使海平面上升形成大面积海侵。

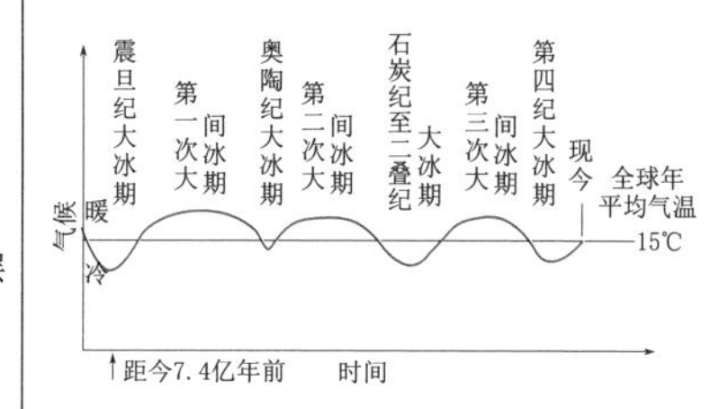

interglacial period

intergrowth texture

【释义】 交生结构

【中文定义】 两种矿物互相穿插有规律地生长在一起。

interlayer ['intəleiə]

【释义】 *n.* 夹层，隔层

【例句】 The depth of a buried soft interlayer will influence the rupture process and the rupture range of the overlaying soil. 软夹层的埋深将影响

intergrowth texture

上覆土层破裂过程和破裂的范围。

【中文定义】 某种岩性为主的岩层中夹有另一种岩层。

interlayer joint

【释义】 层间节理

【例句】 Due to the influence of the fault zone, interlayer joints are mainly developed in a certain range on both sides of the fault zone. 由于受断裂带的影响，在断裂带两侧一定范围内主要发育有层间节理。

【中文定义】 指受不同岩性及层间界面所控制，而局限于某个岩层内的节理。

interlayer joint

interlayer-gliding fault

【释义】 层间错动带

【例句】 The interlayer-gliding fault has a great influence on the stability of surrounding rock mass of underground caverns. 层间错动带对地下洞室群围岩稳定性有很大影响。

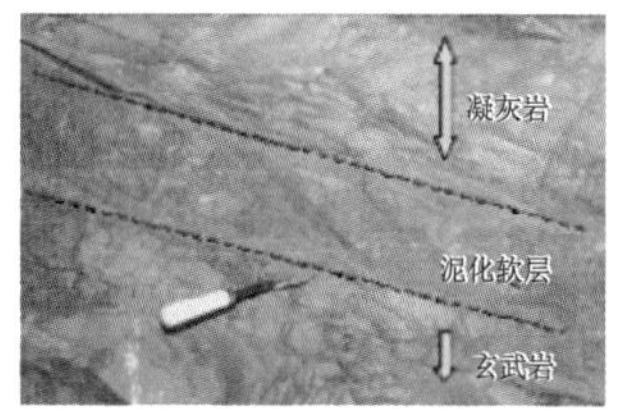

interlayer-gliding fault

【中文定义】 由于构造作用在两个岩层之间发生剪切、错动而形成的软弱带。

interlocked structure

【释义】 镶嵌结构

【中文定义】 沉积岩碎屑和填隙物胶结类型的一种，颗粒之间呈凹凸线状接触，似乎没有胶结物。

interlocked structure

intermediate acidic eruptive rocks

【释义】 中酸性喷发岩

intermediate principal stress

【释义】 中间主应力，中主应力

【例句】 The effect of the intermediate principal stress on the calculated results of the Rankine's earth pressure is a critical factor. 中主应力对朗肯土压力计算结果的影响是一个不可忽视的重要因素。

【中文定义】 中间主应力是某个单元中第二大主应力，即第二主应力。

intermediate rock

【释义】 中性岩

【例句】 The outcropped mesozoic intrusive rock of ShanMen is a composite intrusion which consists of basic, intermediate and acid rocks. 山门出露的中生代侵入岩是一套由基性岩、中性岩和酸性岩组成的复式侵入体。

intermediate speed landslide

【释义】 中速滑坡

intermittent uplift

【释义】 间歇性隆升

【例句】 Since the Tertiary period, the region had gone through several stages of rising and descending with the intermittent uplift of the Himalayas, and gradually formed the modern topography in the mid-Tertiary period. 自第三纪以来，随着喜马拉雅山脉的间歇性隆升，该地区经历了上升和下降的几个阶段，逐渐在第三纪中期形成了现代地貌。

intermontane basin

【释义】 山间盆地，山间坳陷

【例句】 A rational model of agro-ecologic-economic system of intermontane basin must be formed so as to exploit and utilize rationally the natural resources. 必须建立一个合理的山间盆地农业生态经济系统模型，合理利用自然资源。

【中文定义】 指处于造山带之间的盆地。在地槽发展最后阶段表现为普遍回返的隆起过程中，由于上升运动的差别，在地槽转变成为褶皱山系中间所形成的内部陷落。

intermontane basin

intermontane depression

【释义】 山间凹地

【中文定义】 被环绕的山地所包围而形成的堆积盆地，称为山间凹地。

intermontane depression

internal friction angle

【释义】 内摩擦角

【例句】 Along with the increase of the water content, the cohesion and internal friction angle decrease. 随着含水率的增大，内摩擦角和黏聚力呈下降趋势。

【中文定义】 内摩擦角反映了材料的摩擦特性，一般认为包含两个部分：材料颗粒的表面摩擦力，颗粒间的嵌入和连锁作用产生的咬合力。

internal water pressure

【释义】 内水压力

【例句】 The dam displacement and internal water pressure have a great influence on the structural stresses of penstock laid on downstream face of dam. 坝体变位和内水压力对坝后背管结构应力的影响。

【中文定义】 内水压力是指输水隧洞内的水对隧洞衬砌的压力。以单位面积衬砌上所受水压力大小表示，单位为帕。

interpolation [ɪnˌtərpə'leʃən]

【释义】 *n.* 插值，内插，插补

【例句】 The scholar used the riging method and the AASN neural networks as the tools to construct spatial interpolation model. 学者采用克利金法与 AASN 神经网络作为建构空间内插模型的工具。

【中文定义】 在离散数据的基础上补插连续函数，使得这条连续曲线通过全部给定的离散数据点。插值是离散函数逼近的重要方法，利用它可通过函数在有限个点处的取值状况，估算出函数在其他点处的近似值。

interstratified weathering

【释义】 夹层风化

【例句】 Under certain conditions, the weak interlayer formed by interstratified weathering affects the anti-sliding stability of dam foundation. 一定条件下，夹层风化形成的软弱夹层影响坝基抗滑稳定。

【中文定义】 沿块状岩体的断层和节理密集带或变质岩的片理、板理等结构面形成风化程度高于两侧岩体的现象。

interstratified weathering

intrusive rock

【释义】 侵入岩

【例句】 The main intrusive rocks in the mining area are granite pegmatite and diorite dikes. 矿区主要侵入岩为花岗伟晶岩及闪长岩脉。

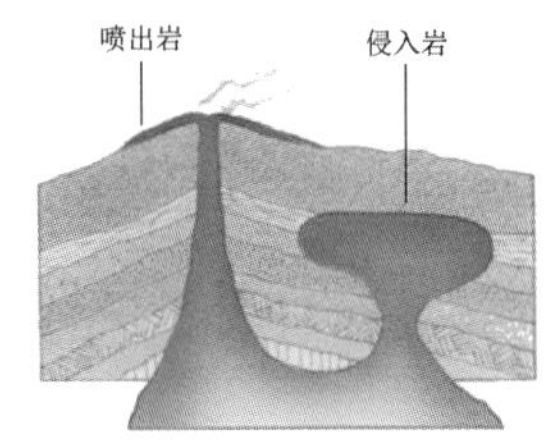

extrusive rock and intrusive rock

inundation [ˌɪnʌn'deɪʃn]

【释义】 *n.* 淹没，泛滥

【常用搭配】 inundation zone 泛滥地区

investigation [ɪnˌvɛstɪ'geɪʃn]

【释义】 *n.* 勘测；*v.* 勘测

【常用搭配】 highway survey 公路勘测；hydropower investigation 水电勘测

【例句】 By site geologic investigation, the report evaluates stability of goaf. 通过现场地质勘测，报告对采空区的稳定性作出评价。

investigation code

【释义】 勘察代号

【例句】 Investigation code is used to express various investigation object categories or name in geotechnical engineering investigation. 岩土工程勘察中用勘察代号表示各种勘察对象的类别或名称。

【中文定义】 用来表示各种勘察对象类别或名称的符号。

代号	名称	代号	名称
钻孔	ZK	裂隙	L
大口径钻孔	ZKd	地质点	D
平洞	PD	综合试样	ZH
竖井	SJ	直剪试样	KJ
探槽	TC	抗压试样	KY
探坑	TK	原状试样	YZ
泥化夹层	NJ	扰动试样	RD
断层	F 或 f	水样	SY
节理	J	物探	WT

investigation code

investigation of prevention and treatment for geologic hazard

【释义】 地质灾害防治工程勘察

investigation of prevention and treatment for geologic hazard

investigation reserves

【释义】 勘察储量

【例句】 According to the specification, investigation reserves shall not be less than 3 times the design requirements. 根据规范，勘察储量不得少于设计需要量的 3 倍。

investigation stage

【释义】 勘测阶段

【例句】 Geology exploration by means of borehole drilling in the job site is one of the necessary and effective measures taken during the investigation stage of municipal underground works. 现场地质勘探钻孔是市政地下工程勘测阶段必要和有效的手段之一。

investigation symbol

【释义】 勘察符号

【例句】 Investigation symbol is used to indicate investigation content in geological map. 地质图中勘察符号用来表示勘察内容。

【中文定义】 地质图中绘制勘察内容所使用的各种图形符号。

名称	符号		说明
	平面	剖面	
钻孔	ZK1 $\frac{831.4}{221}$(10) 1 2	ZK1 $\frac{831.4}{221}$ ZK1 1 ZK1 2	1—已完成的；2—计划的 编号 $\frac{地面高程(m)}{孔深(m)}$ 覆盖层厚度(m)

investigation symbol

iron disseminated

【释义】 铁质浸染

【例句】 The iron disseminated on the structural plane is the result of weathering along the plane. 结构面上的铁质浸染是沿结构面风化作用的结果。

【中文定义】 结构面上含铁氧化物富集的现象。

iron disseminated

iron-manganese disseminated

【释义】 铁锰质浸染

【例句】 The iron-manganese disseminated on the structural plane is the result of weathering along the plane. 结构面上的铁锰质浸染是沿结构面风化作用的结果。

【中文定义】 结构面上含铁锰质氧化物富集的现象。

island arc

【释义】 岛弧，弧形列岛

【常用搭配】 double island arc 双重岛弧；circum pacific island arc 环太平洋岛弧；volcanic island arc 火山岛弧

【例句】 Andesitic volcanism often forms a curved chain of islands, or island arc, that develops between the oceanic trench and the continental landmass. 在安第斯山脉与海沟之间形成一条弯曲的岛屿链，即岛弧。

【中文定义】 岛弧大陆边缘连绵呈弧状的一长串岛屿。与强烈的火山活动、地震活动及造山作用过程相伴随的长形曲线状大洋岛链。

isoseismal [ˌaɪsəʊ'saɪzməl]

【释义】 *n.* 等震线

【例句】 The distortion of isoseismals was specially affected by the main control factor of intensity. 等震线图的扭曲尤其受主要控制因素烈度的影响。

【中文定义】 地震后，在地图上把地面震度相似的各点连接起来的曲线叫等震线，画有等震线的地图称作等震线图。根据等震线图，可以观察一次地震各地区的破坏和地震能量传布的情形。等震线图的型式有呈同心圆的、同心椭圆的或不规则形状的。

isotopic age

【释义】 同位素年龄

【常用搭配】 isotopic age determination 同位素时代测定；uranium isotopic age 铀同位素年龄

【例句】 Zircon also forms multiple crystal layers during metamorphic events, which each may record an isotopic age of the event. 锆石在发生变质作用时形成多晶层结构，每个晶层记录着一次变质事件的同位素年龄。

【中文定义】 同位素地质年龄的简称，指利用放射性同位素衰变定律，测定矿物或岩石在某次地质事件中，从岩浆熔体、流体中结晶或重结晶后，到现在所经历的时间。

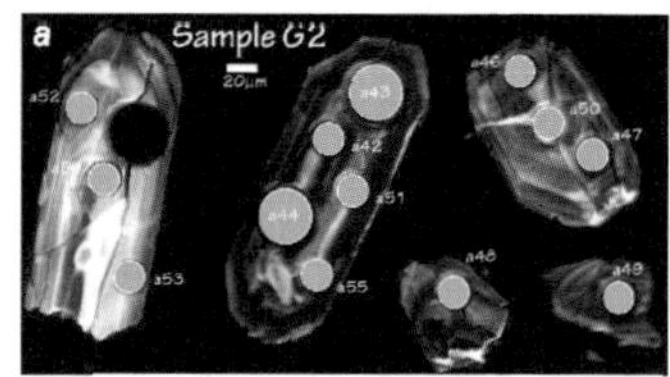

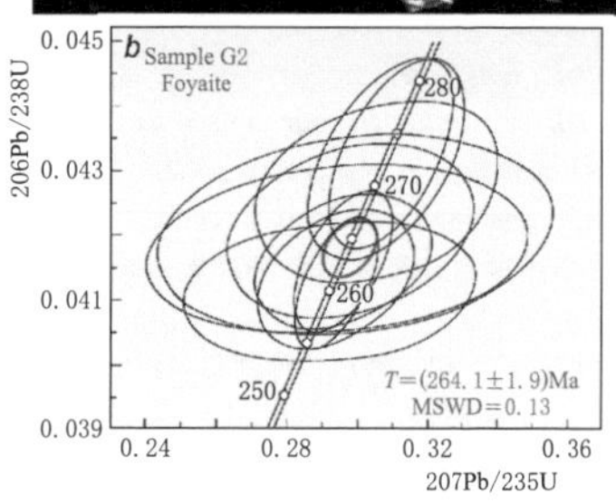

U-Pb isotope ages of the Goryachegorsk foyaites

isotopic geochronology

【释义】 同位素地质年代学

【例句】 The study of isotopic geochronology shows that the ages of the Houhe Groups and Xixiang Group which constitute the metamorphic basement of the northern margin of the Yangtze Craton, are much older than those previously suggested. 同位素地质年代学研究表明，扬子克拉通北缘中段基底岩系后河群和西乡群变质地层的原岩形成时代比早

期推测要早。

【中文定义】 又称同位素年代学，是同位素地质学分支之一。利用自然界放射性衰变规律研究测定各种地质体的形成时代的同位素计时方法。

isotopic tracer method

【释义】 同位素示踪法

【例句】 At present, the isotopic tracer method and turbine flowmeter are universally used for testing annulus well in home oilfield. 同位素示踪法和涡轮流量计法是目前国内各油田普遍采用的环空测井方法。

isotropy [aɪ'sɔtrəpɪ]

【释义】 *n.* 各向同性，无向性，均质性

【例句】 The theory of isotropy damage is based on the hypothesis that the material and damage are symmetrical and isotropic. 各向同性损伤理论是以材料的各向同性以及损伤的均匀分布假设为前提的。

【中文定义】 各向同性指物体的物理、化学等方面的性质不会因方向的不同而有所变化的特性，即某一物体在不同的方向所测得的性能数值完全相同，也称均质性。

J

Janbu's method

【释义】 詹布法

【例句】 What's more, the limiting state hyperplane models of both simplified Bishop's and Janbu's method adaptive to slope project are obtained, and have been applied to the analysis of mine slope stability in Dexing copper open pit. 更重要的是，简化毕肖普法和詹布法都适用于极限状态超平面模型的边坡工程，并已应用于德兴铜矿露天矿边坡稳定性分析。

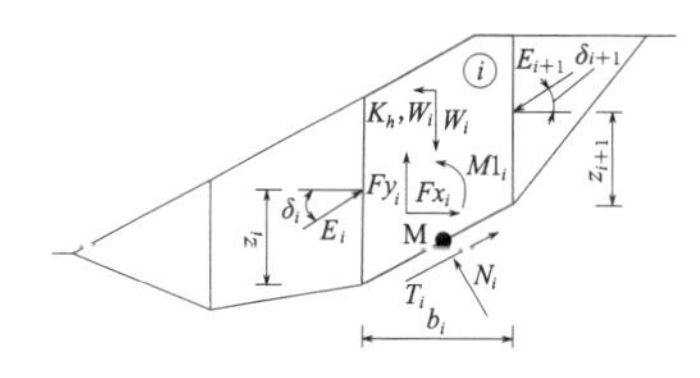

Janbu's method

jet grouting with high pressure

【释义】 高压喷射注浆法

【例句】 Jet grouting with high pressure has unique advantages in strengthening thick alluvion and soft soil layer damaged seriously by earthquake. 对震害严重的厚冲积层或厚软土层，高压喷射注浆法进行加固处理有其独特的优越性。

【中文定义】 用高压水泥浆通过钻杆由水平方向的喷嘴喷出，形成喷射流，以此切割土体并与土拌和形成水泥土加固体的一种地基处理方法。

joint [dʒɔɪnt]

【释义】 *n.* 节理

【常用搭配】 joint plane 节理面；rock joint 岩石节理；bottom joint 水平节理

【例句】 The joint plane is the most important factor for instability of slope. 边坡内侧节理面是构成潜在滑动面的主要因素。

【中文定义】 节理，指岩石在自然条件下形成的裂纹或裂缝。

joint density

【释义】 节理密度

【例句】 The stability of surrounding rock mass is controlled by joint density. 岩体中的节理密度控制着围岩的稳定性。

【中文定义】 是指节理法线方向上单位长度内的节理条数。

joint filling

【释义】 节理充填物

【例句】 Calcite and quartz minerals are common joint fillings. 方解石和石英矿物是常见的节理充填物

【中文定义】 节理往往被充填，形成节理充填物，充填物可以是硅质、铁质、钙质或泥质等。

joint filling

joint frequency

【释义】 节理频数

【例句】 The joint frequency refers to the number of joints on the unit length of the line along the direction of geological mapping. 节理频数是指沿地质测绘的测线方向上单位长度内节理的条数。

joint length

【释义】 节理长度

【例句】 According to the results of joint attitude, joint spacing and joint length, joint probability distribution was deduced. 根据节理裂隙产状、间距及节理长度，推断节理裂隙概率分布特征。

【中文定义】 指节理的延伸长度。

joint persistence ratio

【释义】 节理连通率

【例句】 The joint persistence ratio influence the effective elastic modulus of rock mass. 节理连通率影响岩体的等效弹性模量。

【中文定义】 岩体沿某一剪切方向发生剪切破坏所形成的破坏路径中结构面所占的比例。

joint polar stereonet

【释义】 节理极点投影图

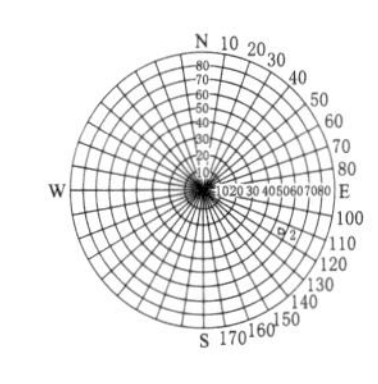
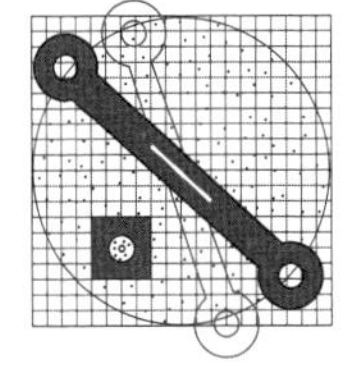

joint polar stereonet

【例句】 Joint polar stereonet is usually compiled on the polar equal area projection network. The circumference direction of the stereonet indicates dip direction from 0° to 360°, and the radius direction indicates dip from 0°to 90° from the center of the circle to the circumference. 节理极点图通常是在极等面积投影网上编制的，网的圆周方位表示倾向，由 0°至 360°，半径方向表示倾角，由圆心到圆周为 0°至 90°。

【中文定义】 节理极点投影图，是一种将节理的法线，以极点投影在乌尔夫网或施密特网上的极点图。

joint set

【释义】 节理组

【例句】 A joint set include all joints with a certain relative occurrence, that formed at the same time, with the same cause. 同期形成，同一成因，具有一定相关产状的所有节理称节理组。

joint spacing

【释义】 节理间距

【例句】 The calculation result shows that the factors that influence the blasting effect are as follows: the speed of lengthwise rock, average joint spacing and the maximal joint spacing. 计算结果表明，影响爆破效果的主要因素依次是：岩石纵波波速、平均节理间距、最大节理间距。

【中文定义】 指不同节理之间的距离。

joint system

【释义】 节理系

【例句】 The dominant joint system strikes roughly east, normal to the axis of La Veta Synclche. 主要节理系大致为东西走向，垂直于罗韦塔向斜轴。

【中文定义】 指在统一应力场中同期形成的力学性质相同的节理群，其产状也大致相同。

joint-concentrated zone

【释义】 节理密集带

【同义词】 densely jointed belt

【例句】 The joint-concentrated zone have great influence on the classification of rock mass structure. 节理密集带对岩体结构分类有重大影响。

【中文定义】 节理密集成带发育，节理间距一般为毫米至厘米级。

Jurassic (J) [dʒʊ'ræsɪk]

【释义】 *n.* 侏罗纪（系）

【中文定义】 侏罗纪介于三叠纪和白垩纪之间，约 1 亿 9960 万年到 1 亿 4550 万年。侏罗纪时发生过一些明显的地质、生物事件，包括最大的海侵事件和环太平洋带的内华达运动。

K

kame [keɪm]

【释义】 *n*. 冰碛阜，冰砺阜

【常用搭配】 kame plateau 冰砺阜高地；kame ridge 冰砺阜脊；kame complex 冰砺阜群

【例句】 Kames are sometimes compared to drumlins, but their formation is distinctively different. 冰砺阜有时被拿来与鼓丘相比，但其成因明显不同。

a kame near Kirriemuir, Scotland

【中文定义】 冰砺阜是指由于冰川消融后冰面上的沉积物沉落到底床上堆积而成的一种圆形的或不规则的小丘。通常由具有层次的粉沙、细沙组成，表面有冰碛物覆盖。

kaolinite ['keəlɪˌnaɪt]

【释义】 *n*. 高岭石

【例句】 Results show the following alkali consumption sequence for those monoclay minerals: montmorillonite(S)> kaolinite(K)≈illite(I). 结果表明黏土矿物中蒙脱石耗碱最大，高岭石的耗碱量与伊利石的接近。

kaolinite

karst [kɑrst]

【释义】 *n*. 喀斯特，岩溶

【常用搭配】 karst process 喀斯特作用；buried karst 埋藏型喀斯特；karst spring 喀斯特泉；karst declogging 喀斯特突水

【例句】 Areas of limestone in which extensive underground drainage has evolved as a result of solution weathering may be described to be 'karst'. 由于溶蚀风化而形成大量地下排水系统的灰岩地区可能被称作为"喀斯特"。

【中文定义】 可溶性岩石长期被水溶蚀以及由此引起各种地质现象和形态的总称。

south china karst landforms

karst aquifer

【释义】 岩溶含水层

【例句】 The residual thicknesses of aquifuge and mining reactivations of the fault are the key factors to control the water inrush when mining is near the large-scale fault with karst aquifer. 在有岩溶含水层的大断层附近开采时，隔水层的剩余厚度和断裂开采活化程度是控制突水的关键因素。

【中文定义】 指含地下水的岩溶地层。其中地下水的富集和运移受各种岩溶形态所控制。其基本特征是其中地下水分布的不均匀性。

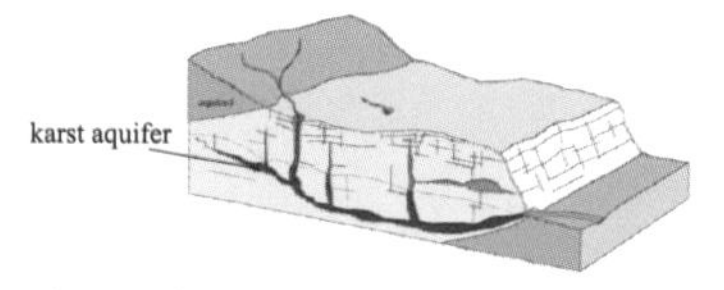

Karst aquifers are carbonate (e.g.limestone) aquifers with caves abd conduit networks formed by dissolution.

karst aquifer

karst base level

【释义】 岩溶基准，喀斯特（侵蚀）基准

【常用搭配】 karst drainage base level 岩溶排水基准面

【例句】 Deep karst was formed in a deep buried condition not controlled by the karst base level. 深层岩溶形成于深埋条件下，不受基准侵蚀面的控制。

【中文定义】 喀斯特作用向地下深处所能达到的下限。一些学者认为，喀斯特地块内的地下水主要向邻近河谷或海洋排泄，在地下水面附近有一个强烈的喀斯特地貌发育带，喀斯特发育受此带控制，因此，河流基准面或海平面为喀斯特基准面。

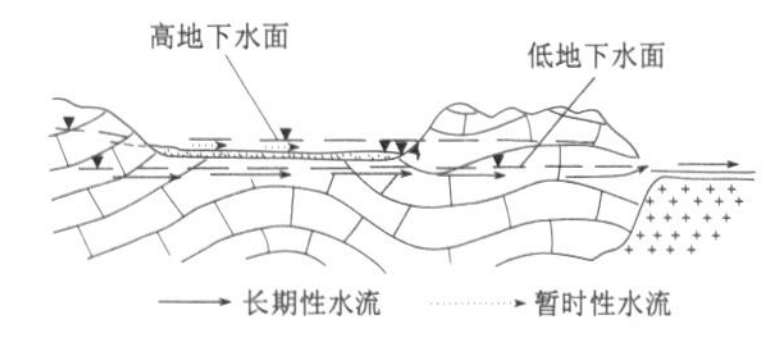

schematic diagram of karst base level

karst basin

【释义】 喀斯特盆地

【例句】 Palaeokarst landform developed in the top of Ordovician weathering crust in mideast Ordos Basin controls the reservoiring of natural gas and is divided into 3 types such as karst highland, karst platform, karst basin and 9 sub types. 鄂尔多斯盆地中东部奥陶系风化壳古岩溶地貌形态对天然气储集起着控制作用，划分为岩溶高地、岩溶台地、岩溶盆地 3 种类型和 9 个亚类型。

【中文定义】 是大型溶蚀洼地，又名坡立谷，常生成在地壳运动长期相对稳定的地区，代表岩溶发育的后期阶段，多在热带气候条件下形成。

karst basin

karst cave

【释义】 溶洞，喀斯特洞穴

【常用搭配】 covered karst cave 隐伏溶洞

【例句】 Karst cave, an important component of karst geomorphology, is formed by geological agent corrosion, erosion and collapse. 喀斯特洞穴是喀斯特地貌的重要组成部分，在地质营力溶蚀，侵蚀和崩塌作用下形成。

【中文定义】 是可溶性岩石中因喀斯特作用所形成的地下空间。溶洞的形成是石灰岩地区地下水长期溶蚀的结果，石灰岩里不溶性的碳酸钙受水和二氧化碳的作用能转化为可溶性的碳酸氢钙。由于石灰岩层各部分含石灰质多少不同，被侵蚀的程度不同，就逐渐被溶解分割成互不相依、千姿百态、陡峭秀丽的山峰和奇异景观的溶洞，由此形成的地貌一般称为喀斯特地貌。

karst cave

karst channel

【释义】 岩溶通道

karst collapse column

【释义】 岩溶陷落柱

【中文定义】 地下溶洞的顶部岩层及覆盖层失去支撑，发生坍塌和剥落产生上小下大的锥状陷落体。

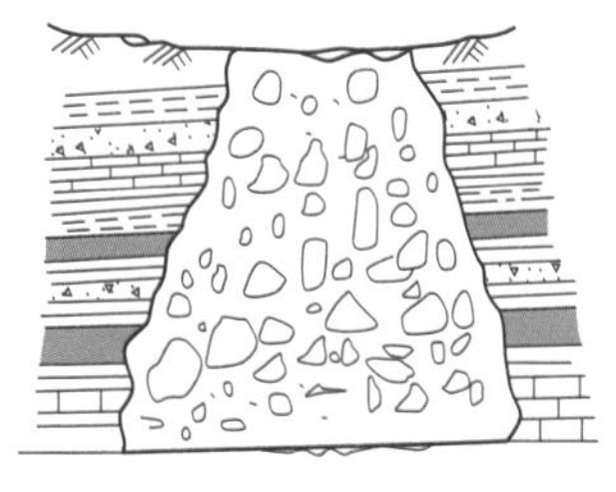

karst collapse column

karst collapse

【释义】 岩溶塌陷

【例句】 The characteristics of fault karst breccia are the combination of fault breccia and karst breccia. It is a product of fault actions and karst collapses. 断溶角砾岩兼具断层角砾岩和岩溶角砾岩的特点，是断裂和岩溶塌陷共同作用的产物。

【中文定义】 指在岩溶地区，下部可溶岩层中的溶洞或上覆土层中的土洞，因自身洞体扩大或在自然与人为因素影响下，顶板失稳产生塌落或沉陷。

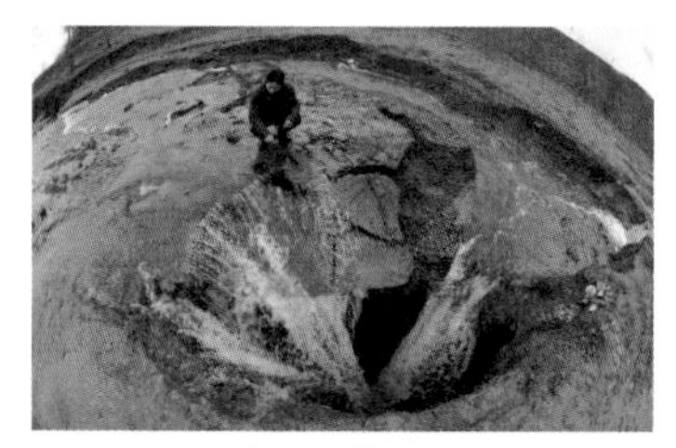

karst collapse

karst depression

【释义】 喀斯特洼地

【例句】 The diurnal observations on atmospheric CO_2 dynamics in the Guilin karst depression ecosystem indicate that: the karst depression has a certain regulation effect on atmospherical CO_2 and may accelerate karst development。对桂林岩溶试验场岩溶洼地生态系统中大气 CO_2 动态的昼夜观测结果说明，岩溶洼地对大气 CO_2 具有一定的调蓄作用，并促进岩溶发育。

【中文定义】 碳酸盐岩地区由于溶蚀作用所形成负地形的总称，又称溶蚀洼地。它包括小至漏斗，大致喀斯特盆地等一类喀斯特地貌。是由喀斯特水（指喀斯特地区流水，包括地表水和地下水）垂直循环作用加强形成，也可由地下洞穴塌陷形成。大洼地底部平坦，有较厚的沉积物；小洼地底部平地很小，沉积物很薄甚至缺乏。洼地的规模主要受集水面积的控制。

karst depression

karst detection

【释义】 岩溶探测

【例句】 The space distribution and physical characteristic of karst had supplied a basic prerequisite for karst detection with Ground Penetrating Radar (GPR). 岩溶的空间分布和物性特性为探地雷达技术进行岩溶探测提供了基本前提条件。

karst erosion

【释义】 岩溶侵蚀

【例句】 It was believed that dilution, eluviation, karst erosion and anthropogenic activity can explain the ion variations, hence this study helps to understand environmental problem in karst. 稀释、淋溶、岩溶侵蚀和人为活动可以解释离子变化的原因，因此本研究有助于了解岩溶中的环境问题。

karst funnel

【释义】 岩溶漏斗

【例句】 The largest doline or karst funnel of Dashiwei is situated at the karst slope zone of the northwest Guangxi in the southern of Yunnan and Guizhou plateau. 大石围最大的岩溶漏斗位于滇黔高原南部桂西北的岩溶斜坡地带。

【中文定义】 又称斗淋，石灰岩地区呈碗碟状或漏斗状的凹地。

karst funnel

karst landform

【释义】 岩溶地貌

【例句】 Karst landform is mainly distributed in yunnan, Guizhou and Guangxi provinces. The area of Karst landform is about 300000km^2, and occupies more than 3% of the total land area of China, in which half of the Karst regions have been desertificated. The rocky desertification area is outspreading by the rate of 2500km^2 per year. 中国岩溶地貌主要分布在云南、贵州和广西等省区，总面积约为 30 万

km^2，占国土面积的3%以上，其中石漠化面积已占1/2，平均每年以2500km^2的速度在扩展。

【中文定义】 是具有溶蚀力的水对可溶性岩石进行溶蚀等作用所形成的地表和地下形态的总称

karst landform

karst landscape

【释义】 岩溶景观

【常用搭配】 karst geological landscape 岩溶地质景观

【例句】 Some suggestions, like establishing geologic park, are put forward mainly based on the background of karst geology, karst landscape and the value of tourist resource as well as geographic location. 主要依据岩溶形成的地质背景、岩溶景观特征以及旅游资源的价值、区位条件等，提出在该区建立地质公园等旅游开发的一些措施。

【中文定义】 具有溶蚀力的水对可溶性岩石进行溶蚀等作用所形成的地表和地下形态的总称。

karst landscape

karst peneplain

【释义】 溶蚀准平原

【中文定义】 岩溶盆地经过长期的溶蚀破坏，形成比较开阔的平原，称为溶蚀准平原。

karst peneplain

karst phenomenon

【释义】 喀斯特现象

【例句】 The soil cavity of karst is a special karst phenomenon. It lies in the shallow ground, developing quickly and distributing concentratedly. 岩溶土洞是一种特殊的岩溶现象，具有埋藏浅、发育快、分布密的特点。

karst phenomenon

karst pit

【释义】 岩溶井

karst pit

karst spring

【释义】 喀斯特泉

karst spring

【例句】 The more prominent is that there are many karst caves in Yishui karst geomorphologic forms besides stone buds, melting groove, melting tank, karst springs on the surface. 沂水喀斯特地貌形态除地表的石芽、溶沟、溶槽、喀斯特泉等

小形态外，较为突出的是有许多的溶洞。

【中文定义】 岩溶水在地表流出的天然露头。

karst topography

【释义】 岩溶地形，喀斯特地形，岩溶地貌

【常用搭配】 karst negative landform 岩溶负地形

【例句】 The Shilin is an impressive example of karst topography. 石林是岩溶地形的典型。

【中文定义】 是具有溶蚀力的水对可溶性岩石（大多为石灰岩）进行溶蚀作用等所形成的地表和地下形态的总称，又称岩溶地貌。

karst topography

karst treatment

【释义】 喀斯特处理

【例句】 Karst treatment in karst areas has been widely used in project constructions. In the current underground construction process, draining, grouting, span and circling around are the commonly used methods at home and abroad to deal with underground rivers. 岩溶处理在喀斯特地区工程建设中已得到广泛应用，目前地下工程施工过程中，引排、堵填、跨越、绕避是国内外处理暗河的常用方法。

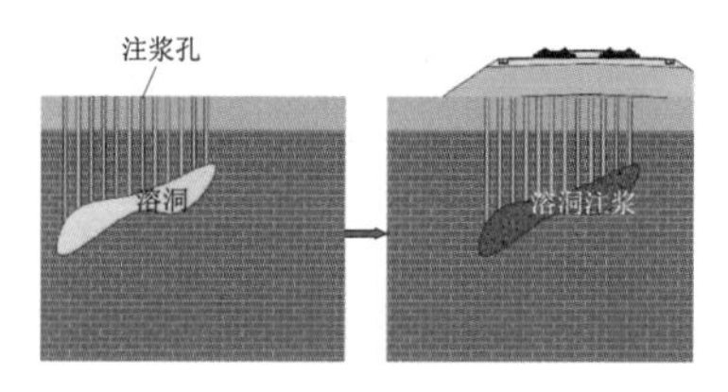

karst treatment

karst valley

【释义】 岩溶槽谷，岩溶谷地

【例句】 Precipitation sinks to the heat storage reservoir via karst valley, then the groundwater flows in the depth along fault structure line from north to south and discharges to the earth surface in maximum deressurization area, such as river valley, to form the geothermal spring. 大气降水通过岩溶槽谷下渗至热储层，在深部沿断裂构造线由北向南径流，并在地表减压最大处，如河谷地段等，排出地表而形成温泉。

【中文定义】 有流水作用参与形成的长条状岩溶洼地。

karst valley

karst water

【释义】 岩溶水

【例句】 Hydrogeological chemistry reflects certain geology environment where karst water has developed. 水化学成分反映了岩溶水形成的一定地质环境。

【中文定义】 赋存于可溶性岩层中溶蚀裂隙和溶洞中的地下水，又称喀斯特水。其最明显特点是分布极不均匀。

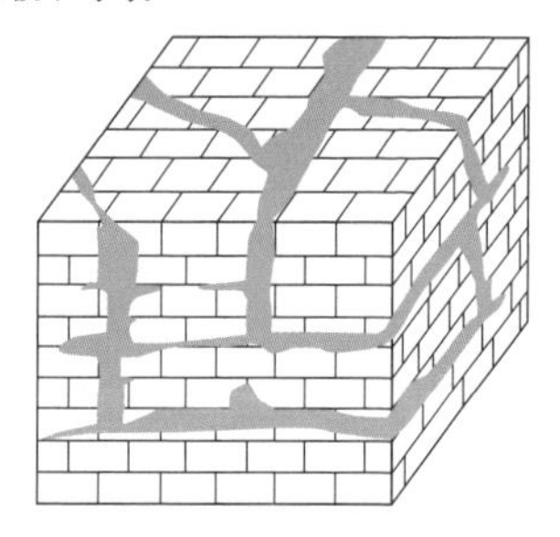

schematic diagram of karst water

karst water inrush

【释义】 岩溶突水

【例句】 The geologic prediction of karst water inrush hazard in the course of baziling tunnel construction of Yiwan railway has been studied by many scholars. 宜万铁路八字岭隧道施工期岩溶突水灾害超前预报被众多学者研究。

【中文定义】 储集和运动于岩溶含水层中的地下水流，被人工揭露或受自然因素影响而骤然产生大量涌水现象。

karst water inrush

karstic network

【释义】 岩溶洞穴网

【中文定义】 在岩溶化岩体中，大小不等、互相连通的洞穴管道系统。

schematic diagram of karstic network

karstification

【释义】 喀斯特作用，岩溶作用

【例句】 The strong karstification along the regional faults makes the reservoir serious leakage that has caused the groundwater polluted. 沿着区域性断层强烈的喀斯特作用使得水库发生严重渗漏，并污染了地下水体。

【中文定义】 水对可溶性岩石（碳酸盐岩、硫酸盐岩、石膏、卤素岩等）以化学溶蚀作用为主，以流水冲蚀、潜蚀和机械崩塌作用为辅的地质作用过程。

karst landform

key bed

【释义】 标志层

【例句】 Coal beds and volcanic ash falls are examples of key beds. 煤层和火山灰堆是典型的标志层。

【中文定义】 一层或一组具有明显特征可作为地层对比标志的岩层。

klippe [klɪp]

【释义】 *n.* 飞来峰；孤残层；构造外露层

【例句】 The long distance thrust system is composed of Ordovician and Silurian and some part of it shows as a klippe. 远距离冲断系统由奥陶系和志留系地层构成，局部地区表现为飞来峰。

【中文定义】 在逆掩断层或辗掩构造中，常见老岩层覆盖在新岩层上，这样的老岩层称推覆体。当推覆体遭受强烈剥蚀，周围地区露出原来的新岩层，而残留的一部分老岩层孤零零地盖在新岩层上，称飞来峰。

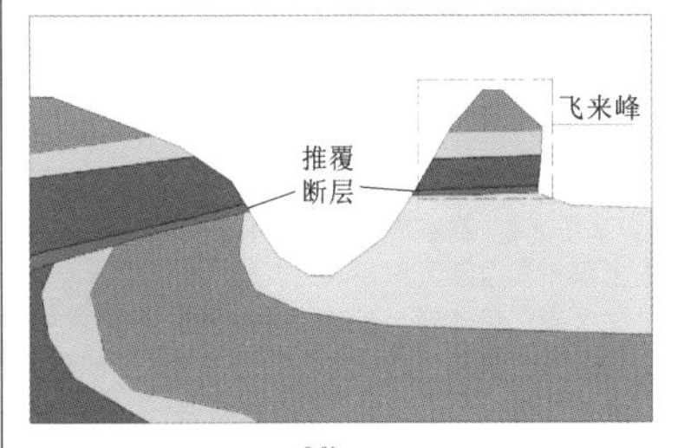

klippe

L

laboratory test

【释义】 室内试验

【例句】 Through the laboratory test, geologists can determine some physical properties of soil. 通过室内试验，地质人员可以确定土层的某些物理性质。

laccolith ['lækəˌlɪθ]

【释义】 *n*. 岩盖

【常用搭配】 composite laccolith 复合岩盖

【例句】 However, unlike sills, the magma that generates laccoliths is believed to be quite viscous. 不同于岩床，形成岩盖的岩浆黏度较大。

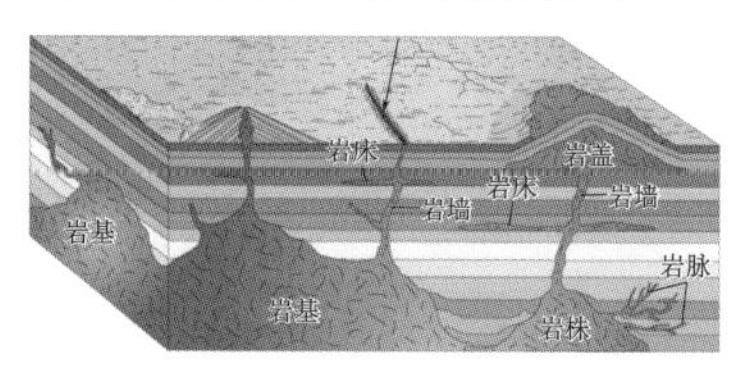

laccolith

lacustrine deposit platform

【释义】 湖积台地

【中文定义】 湖积成因台地，多中央坡度平缓，四周较陡。

lacustrine facies

【释义】 湖相

【例句】 The main sedimentary facies in this area include lacustrine facies and gravity flow facies. 该区域沉积相以湖相和冲积相为主。

【中文定义】 静水湖泊沉积特征。可分为淡水湖泊沉积相和盐湖沉积相两类。淡水湖泊沉积岩石以黏土岩、粉砂岩与砂岩为主，也常有泥灰岩、硅藻土等的沉积。盐湖沉积形成于大陆干旱气候环境，以各种盐岩沉积为主，如石膏、岩盐等，亦有各种细碎屑岩、石灰岩伴生。湖相沉积地层中以水平层理为主，还可有干裂、雨痕和生物搅动构造等。

lacustrine soil

【释义】 湖积土

【例句】 Lacustrine soil samples were collected from the top 20cm of soil layer from Abis area located southeast Alexandria Governorate. 湖积土壤样品是从亚历山大省东南部 Abis 区 20cm 的土壤层顶部收集而来的。

【中文定义】 在湖泊及沼泽等流速极为缓慢的静水条件下沉积下来的土。

lacustrine soil

lacustrine plain

【释义】 湖泊平原，湖积平原

【例句】 The topography of Southern Indiana reflects a system of complex lacustrine plains. 印第安纳南部的地形反映了一个复杂的湖泊平原系统。

【中文定义】 由湖泊沉积物淤积而形成的平原称湖积平原。

Dongting lacustrine plain

lagoon facies

【释义】 潟湖相

【例句】 It can be concluded that clastic facies consisting of tidal facies, lagoon facies, fluvial facies, alluvial fan facies and fan delta and carbonate facies including evaporation platform, restrict platform, open platform and ford of platform margin. 碎屑岩沉积相包括潮坪相、潟湖相、河流相、冲积扇相以及三角洲相，碳酸盐沉积相包括蒸发台地、局限台地、开阔台地和台地边缘浅滩。

【中文定义】 潟湖是海岸带的浅水盆地，由于障壁岛的遮挡，它与广海间的水体循环受到破坏和阻碍。海岸潟湖平行海岸延伸，与广海间一般只以排水口相通。由于蒸发作用或淡水的注入，潟湖水体的含盐度变得极不正常，潟湖按照变异后的含盐度可以划分出淡化和咸化两种潟湖类型。淡化潟湖相的沉积主要为暗色粉砂岩和黏土岩，可见铁锰结核、硫化物和二氧化硅沉积，生物种属单调，体小壳薄，常见畸形及反常纹饰。咸化潟湖相主要沉积细粒碎屑和化学沉积产物，以具膏盐等蒸发岩类沉积为特征，缺少狭盐性生物化石；当盐度超过5%～5.5%时生物灭绝。

lagoon facies

lamprophyre ['læmprəˌfaɪə]

【释义】 *n.* 煌斑岩

【例句】 The lamprophyres of the Jiaobei are collected from Longkou, Yantai and Weihai, and comprise odinites, hornblende lamprophyres and camptovogesites. 胶北煌斑岩分别采自龙口、烟台和威海地区，包括拉辉煌斑岩、斜闪正煌岩和角闪煌斑岩。

lamprophyre

lamprophyre vein

【释义】 煌斑岩脉

【例句】 The geological conditions of the high and steep abutment slope at left bank of Jinping-I Hydropower Project are very complicated, and the faults of f5, f8, f42-9, the lamprophyre veins of X, the release fractures oriented parallel to slope surface and deep fractures are developed. 锦屏一级水电站左岸坝肩高陡边坡地质条件复杂，主要发育有f5、f8、f42-9断层，煌斑岩脉X，顺坡卸荷裂隙及深部裂缝等结构面。

lamprophyre vein

land subsidence hazard

【释义】 地面沉降灾害

【例句】 Land subsidence hazard is an ordinary urban hazard and a typical accumulated geological disaster. 地面沉降是城市常见灾害，是典型的缓慢积累型地质灾害。

land subsidence prevention and cure

【释义】 地面沉降防治

【中文定义】 预防和治理地面沉降灾害的工作。

landmass ['lændˌmæs]

【释义】 *n.* 大陆，陆块

【常用搭配】 stable landmass 稳定地块；Jiangnan landmass 江南陆块；new landmasses 新大陆

【例句】 It seems that the entire Phanerozoic history of New Zealand can be related to a series of Benioff zones dipping westwards under the New Zealand landmass. 看起来，新西兰整个显生宙的历史与新西兰陆块下的一连串向西倾斜的贝尼奥夫地带有关。

【中文定义】 泛指整个地史时期中某个由已固结

land subsidence hazard

陆壳组成的相对稳定地区。陆块范围一般较大，古地理面貌经常发生海陆变迁，可以是隆起剥蚀区也可以是沉积盆地。

landslide ['læn(d)slaɪd]

【释义】 *n.* 滑坡，山崩；*vi.* 发生山崩

【常用搭配】 landslide control 滑坡防治；landslide monitoring 滑坡监测

【例句】 In Japan, 13 people died as an approaching tropical storm triggered floods and landslides in the west of the country. 在日本，步步逼近的热带风暴在日本西部引发了洪水和山体滑坡，共造成 13 人死亡。

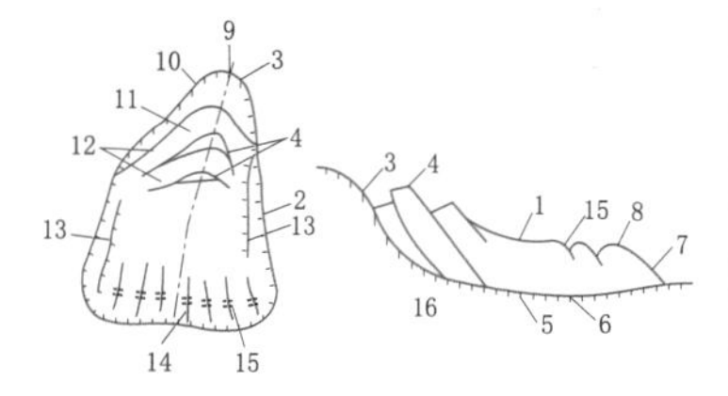

diagram of landslide

1—滑坡体；2—滑坡周界；3—滑坡壁；4—滑坡台阶；5—滑动面；6—滑动带；7—滑坡舌；8—滑动鼓丘；9—滑坡轴；10—破裂缝；11—封闭洼地；12—拉张裂缝；13—剪切裂缝；14—扇形裂缝；15—鼓胀裂缝；16—滑坡床

landslide dam

【释义】 滑坡坝

【例句】 Blasting for excavating a water discharging gap on the body of the landslide dam became an effective method for the excavation. 采用爆破方式炸开缺口泄流成为这一类堰塞湖应急排险的有效手段。

landslide dam

landslide dammed lake

【释义】 滑坡堰塞湖

landslide deposit

【释义】 滑坡堆积

【例句】 According to engineering geological condition of Zhenggang landslide deposit in Gushui Hydropower Station and the fact of slope stability controlled by rainfall conditions, grid control technology and Kriging interpolation method are used to establish 3D model of the slip mass. 结合澜沧江古水水电站争岗滑坡堆积体工程地质条件，依据堆积体稳定性受暴雨工况控制的实际情况，对三维滑面及边界条件采用栅格控制和克里格插值法构建了争岗滑坡体的三维滑块模型。

【中文定义】 是斜坡上的岩石风化产物，在重力作用下沿一定的滑动面整体下滑形成的堆积物。在雨季及多雨的年份容易形成，滑动面上常有擦痕、滑坡泥及滑坡搓碎角砾。

landslide deposit

landslide deposition

【释义】 滑坡沉积

【例句】 Over time the river eroded out the debris, but we still see evidence of the lake in the landslide deposits as well as delta deposits coming out of the tributary valleys. 随着时间的推移，河流侵蚀了残积物，但我们仍然能看到在支流峡谷中出来的滑坡沉积物和三角洲沉积物中湖泊的证据。

landslide drumlin

【释义】 滑坡鼓丘

【中文定义】 滑坡体在滑动过程中，滑坡舌前面常因受阻、挤压而鼓起，称为滑坡鼓丘。

landslide flank

【释义】 滑坡侧缘

landslide fracture

landslide fracture

【释义】 滑坡裂缝

【例句】 Experimental study on the mechanism of landslip crevices is common among landslide analysis. 滑坡裂缝产生机理的实验研究在滑坡研究中十分常用。

landslide graben

【释义】 滑坡洼地

landslide hazard

【释义】 滑坡灾害

【例句】 This paper presents a review on different quantitative models on landslide hazard mapping with emphasis on the use of Geographic Information Systems and Artificial Intelligence in last decades. 该文评述了近几十年来着重应用地理信息系统和人工智能进行滑坡灾害测绘的不同定量模型。

landslide hazard

landslide investigation

【释义】 滑坡勘察

【例句】 The success or failure of a landslide remedial engineering depends on the quality of landslide investigation, remedy engineering design and construction. 滑坡灾害防治工程成败的关键取决于滑坡勘察、设计以及施工质量。

landslide steps

【释义】 滑坡台阶

landslide steps

landslide tongue

【释义】 滑坡舌

【中文定义】 滑坡舌是滑坡地貌的组成部分之一，是在滑坡体前缘形成的舌状突出体。

landslip zone of slope

【释义】 边坡塌滑区

large well drilling

【释义】 大口径钻进

large-scale landslide

【释义】 大型滑坡

【例句】 About 80 percent of large-scale landslides were found in the first slope-descending zone of the mainland topography around the eastern margin of Tibet plateau. 约 80%的大型滑坡发生在环青藏高原东侧的大陆地形第一个坡降带范围内。

Late Paleozoic Era

【释义】 晚古生代

【中文定义】 晚古生代开始于 4.1 亿年前，结束于 2.45 亿年前，持续时间 1.75 亿年，包括泥盆纪、石炭纪和二叠纪三个时期。

lateral discharge

【释义】 侧向排泄量

【例句】 The lateral runoff of piedmont is not only the lateral discharge in hilly area, but also the lateral recharge of groundwater in plain area. It is an important data in the evaluation of groundwater resources. 山前侧向径流量既是山丘区地下水的侧向排泄量，同对又是平原区地下水的侧向补给量，是地下水资源评价中的一个重要的量。

lateral pressure

【释义】 侧压力

【常用搭配】 coefficient of lateral pressure 侧压力系数

【例句】 The content intends to solve the problem of lateral pressure at the walls of large diameter squat silos, especially the problem of overcharge caused by the top pile material. 这部分内容意在解决大直径浅圆仓仓壁上的水平侧压力的计算问题，合理地解决仓壁顶面以上的料堆引起的超压问题。

【中文定义】 是指某种物体的侧向所受的外力大小，有单位面积上所受的力、也有某特定的物体侧面所受的载荷力。

lateral recharge

【释义】 侧向补给量

【例句】 The groundwater lateral recharge can be determined by using single well techniques of a radioactive isotope. 用同位素测井技术可以确定地下水侧向补给量。

lateritic soil

【释义】 红黏土

【例句】 The later period of karst corrosion deposit experienced lateritization, which accomplished the lateritic soil basic characteristic. 岩溶残余堆积物后期又经历红土化作用，造就了红黏土的基本特征。

【中文定义】 红黏土是指碳酸盐类岩石（石灰岩，白云岩，泥质泥岩等），在亚热带温湿气候条件下，经风化而成的残积、坡积或残-坡积的褐红色、棕红色或黄褐色的高塑性黏土 。

lateritic soil

layer wise summation method

【释义】 分层总和法

【中文定义】 将地基受压层划分为若干小层，按无侧限条件分别计算压缩量，而后求和得到地基总沉降量的方法。

leakage around the dam

【释义】 绕坝渗漏

【中文定义】 水库蓄水后由于坝体上下游具有水位差，使库水沿坝体两端以外岩土体之孔隙、裂隙、溶洞、断层破碎带等向下游产生渗漏的现象

leakage coefficient

【释义】 越流系数

【中文定义】 是指表征弱透水层垂直方向上传导越流水量能力的参数。即当抽水含水层（主含水层）与上部（或下部）补给层之间的水位差为一个单位时，垂直渗透水流通过弱透水层与抽水含水层单位界面的流量。

leakage monitoring

【释义】 渗漏量监测

【例句】 The distributed optical fiber temperature sensing technology is a brand-new way for leakage monitoring. 分布式光纤温度传感技术为渗漏监测提供了一个全新的方法。

leakage recharge

【释义】 越流补给

【例句】 The hydrochemistry verified the existence of leakage recharge. 水化学分析证实了越流补给的存在。

【中文定义】 当含水层之间存在较大的水位差时，水头高的含水层中的水通过弱透水层向水头低的含水层中排泄的现象称为越流。含水层通过相邻含水层的越流作用而得到补给称为越流补给。越流补给有时还包括含水层顶底板半透水层的弹性释放量。

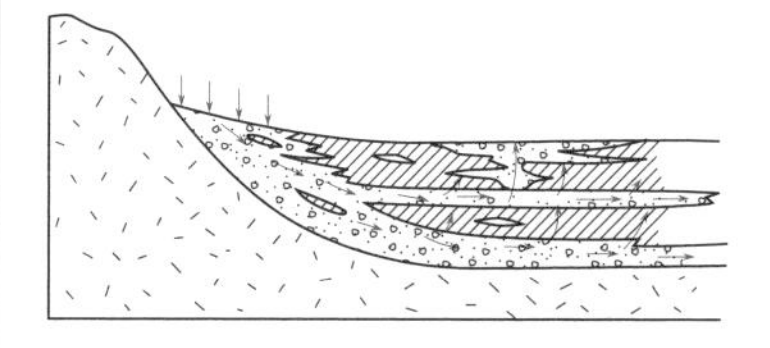

schematic diagram of leakage recharge

leakage through interfluve

【释义】 河间地块渗漏

leakage to adjacent valley

【释义】 邻谷渗漏

legend of geological map

【释义】 地质图图例

【中文定义】 地质图中所用各种符号的说明。它是地质图的附属部分，为读图的工具，放置在图的一侧或其他适合的部位；图例内容有不同的颜色、图形、花纹、字母、数字代号等。

图 例

$Pt_1\gamma o$ 古元古界片麻状石英闪长岩 | 辉长岩层状辉长岩 | 走滑断层 | 角度不整合界线
砾岩 | 辉绿岩 | 性质不明断层推测断层 | 岩体脉动界线
大理岩 | 玄武岩、枕状玄武岩 | 糜棱岩带 | 岩体超动关系 岩体脉动关系（谱系图）
硅质岩 | 辉长岩脉/闪长岩脉 | 40 片麻理及产状 | 第四纪火山口
震旦-寒武纪其曼于特蛇绿（混杂）岩 | 龙岗岩脉/伟晶岩脉 | 90 片理及产状 | 300Ma K-Ar 同位素采集点 年龄值 测试方法
石炭-中二叠世苏巴什蛇绿（混杂）岩 | 区域性断层 一般断层 | 60 层理及产状 | 化石采集点
Σ 超基性岩 | 剥离断层 推覆断裂 | 65 叶理及产状 | 不明地质体
$\psi\iota$ 辉石岩 | 70° 55° 正断层及产状逆断层及产状 | 地层界线、岩体超动及侵入界线

legend of geological map

legend of geomorphic map

【释义】 地貌图例

【中文定义】 地形图或地质图中描绘地貌对象所使用的各种图形符号的说明。

length of test section

【释义】 试段长度

【例句】 According to the test pressure, the length of test section and the steady infiltration water in water pressure test, the permeability of rock mass can be determined. 在压水试验中根据试验压力、试段长度和稳定渗入水量等参数，可以判定岩体透水性的强弱。

lens [lɛnz]

【释义】 *n.* 透镜体

【中文定义】 具有中间厚周边薄特点的固体都叫透镜体，通常形容在压性或压扭性构造破碎带中的不连续块体。

lens

leveling survey

【释义】 水准测量

【同义词】 level survey

【例句】 Third is good operation management: carefully monitoring coupling of geophone to the ground, leveling survey, orientation of geophone and depth and position of shot, which reduce errors to minimum. 第三点是良好的施工管理：细致地监视检波器耦合、水准测量、方位及震源的深度和位置，最大限度减小误差。

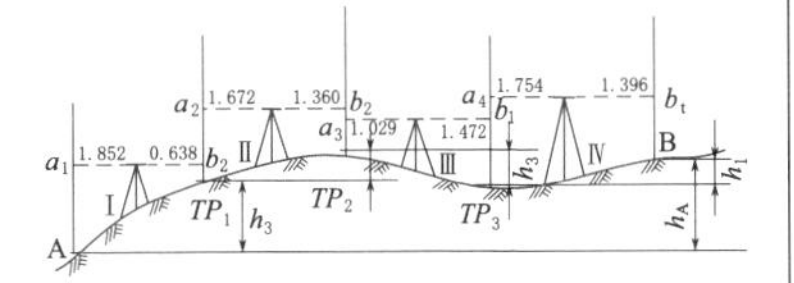

leveling survey

light mineral

【释义】 亮色矿物、浅色矿物

【例句】 Very widespread, biotite gneisses are normally fine-grained with pronounced foliation caused by the parallel arrangement of alternating dark melanocratic ferromagnesian minerals and light leucocratic quartzo-feldspathic minerals. 黑云母片麻岩普遍明显具有的叶理结构（片麻理结构），一般是由暗色铁镁矿物和亮色长英质矿物交替平行排列形成的。

【中文定义】 岩浆岩矿物中的石英和长石颜色较浅，称为浅色矿物。

light mineral quartz crystal cluster

limb of fold

【释义】 褶皱翼

【例句】 All the pre-Shamvaian rocks in the Selukwe area lie on the inverted limb of this fold. 塞卢奎地区的所有前沙姆凡岩石均分布于此推覆体的倒转褶皱翼中。

【中文定义】 泛指褶皱核两侧的岩层，翼与核是相对而言的，无明确划分界线。在同一褶皱面上两个相邻转折端之间的比较平直部分。翼的倾角称为翼角。如两翼不等长，分别称短翼和长翼。

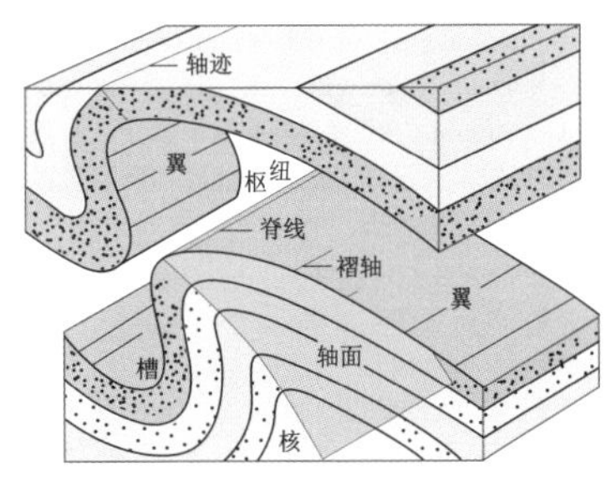

The basic elements of fold

limestone ['laɪmstoʊn]

【释义】 *n.* 石灰岩，石灰石

【常用搭配】 algal limestone 藻灰岩；argillaceous limestone 泥质灰岩；autochthonous limestone 原地灰岩；dolomitic limestone 白云灰岩

【例句】 India's major mineral resources include coal, iron, manganese, mica, bauxite, titanium, chromite, limestone and thorium. 印度的主要矿产资源有煤、铁、锰、云母、铝土矿、钛、铬铁矿、石灰岩和钍。

【中文定义】 一种碳酸盐岩，浅灰至黑色，致密，贝壳状断口，遇稀盐酸剧烈起泡，性脆，小刀能刻动。

limestone

limit ['lɪmɪt]

【释义】 *n.* 限度，界限；*vt.* 限制，限定

【例句】 The soil exhibits different characteristics under different limit. 土体在不同的界限下表现出不同的特性。

limit equilibrium method (LEM)

【释义】 极限平衡法

【常用搭配】 3D rigid limited equilibrium method 三维刚体极限平衡法

【例句】 At present, the rigid-body limit equilibrium method adopted in the design of crane beam on rock wall can not reflect the actual mechanical behavior of crane. 目前，岩壁吊车梁设计所采用的刚体极限平衡设计方法还不能确切反映岩锚吊车梁的实际受力状态。

limitation water ratio test

【释义】 界限含水率试验

limonite ['laɪməˌnaɪt]

【释义】 *n.* 褐铁矿

【常用搭配】 natural limonite 天然褐铁矿；limonite rock 褐铁岩

【例句】 Limonite is a group of brown to yellowish-brown iron oxide minerals, which is a matrix material, in which turquoise often developed. 褐铁矿是一组棕色至黄棕色的铁氧化物矿物，作为一种基质材料，多发育有绿松石。

【中文定义】 主要的铁矿物之一，它是以含水氧化铁为主要成分的、褐色的天然多矿物混合物。

limonite

line of intersection

【释义】 交线

【例句】 The direction of the rock layer is the line of intersection, indicating the horizontal extension of the rock layer in the space. 岩层的走向，也就是交线，表示岩层在空间的水平延伸方向。

【中文定义】 岩层层面与任一假想水平面的相交线，也就是同一层面上等高两点的连线

line type of geological map

【释义】 地质图线型

【中文定义】 地质图中描绘各种地质现象所采用线条的样式。

linear flow structure

【释义】 流线构造

【中文定义】 岩浆在流动过程中所产生的构造。

linear flowing water

【释义】 线状流水

【例句】 Ordovician carbonate weathering crust reservoirs in the central gas field in the Ordos Basin are characterized by evident palaeo karst topography. Unique groove karst landscape has been formed due to the erosive action of surface linear flowing water. 鄂尔多斯盆地中部气田奥陶系碳酸盐岩风化壳储层具有显著的喀斯特地貌特征。在地表线状流水冲蚀作用下形成了独特的沟槽地貌景观。

【中文定义】 地下水活动状态之一。根据《水力发电工程地质勘察规范》（GB 50287—2016）的规定，当地下洞室内地下水每 10m 洞长水量 25L/min$<q\leqslant$125L/min 或压力水头 10m$<H\leqslant$100m 时，地下水状态为线状流水。

linear interpolation

【释义】 线性插值

【例句】 Adoption of the above surface fitting and linear interpolation can be other non-test state of the physical point. 通过曲面插值和线性拟合的办法可以得到其他非测试状态点下的物理量。

【中文定义】 以线性函数为插值函数的插值方法。

linear shrinking rate

【释义】 线缩率

【例句】 Linear shrinking rate-time curve is almost linear in logarithmic relationship. 在双对数坐标下，线缩率和时间几乎呈线性关系。

lineation [ˌlɪnɪ'eɪʃən]

【释义】 *n.* 线理

【常用搭配】 parting lineation 流痕线理；crenulation lineation 纹线理；striation lineation 擦痕线理；flow lineation 流动线理

【例句】 The metamorphic core complex is considered to be symmetric type metamorphic core complex in the light of deformation schistosity, lineation, and the indication of the shear movement. 变质核杂岩的变形片理、线理及运动指向说明该变质核杂岩为对称型变质核杂岩。

【中文定义】 线理是岩石中发育的一般具有透入性的线状构造。根据成因可分为原生线理和次生线理。

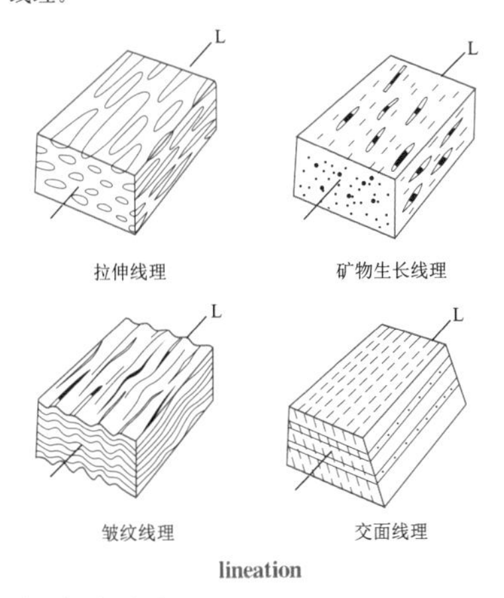

lineation

liquefaction index

【释义】 液化指数

【例句】 New liquefaction index is presented on the basis of analyzing the liquefaction index of *Code for Seismic Design of Buildings* (GBJ 50011). 在对《建筑抗震设计规范》(GBJ 50011)液化指数分析基础上，提出了新的液化指数。

liquid limit

【释义】 液限

liquid state

【释义】 液态

liquidity index

【释义】 液性指数

【例句】 It can appraise the state of the clay stickability by liquidity index. 可通过液性指数来评价黏性土稠度状态。

lithofacies [ˌlɪθə'feɪʃɪz]

【释义】 *n.* 岩相

【常用搭配】 lithofacies map 岩相图

【例句】 The unconformity, river rejuvenation surface and lithology and lithofacies transition surface are the main types of sequence boundaries. 层序界面类型主要有不整合面、河流冲刷侵蚀作用面和岩性岩相转换面等。

【中文定义】 岩相是一定沉积环境中形成的岩石或岩石组合，它是沉积相的主要组成部分。

lithology [lɪ'θɑlədʒi]

【释义】 *n.* 岩性，岩石学

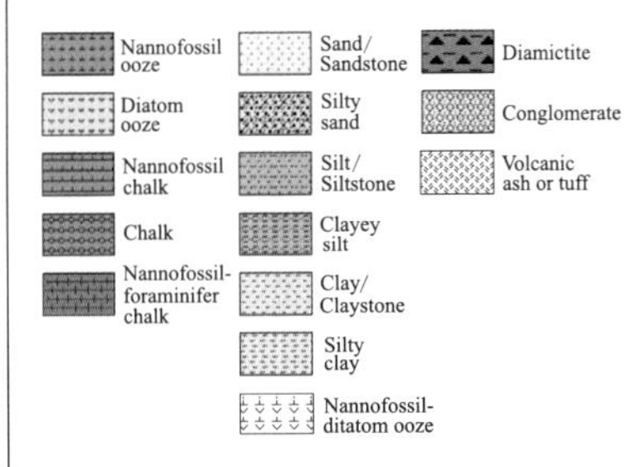

lithology

【常用搭配】 stratigraphic lithology 地层岩性；岩性征貌 lithology feature；lithology distribution 岩性分布

【例句】 These surfaces are defined usually by changes in lithology between one layer and that underlying or overlying it. 这些界面是基于某一地层与其上覆（或下伏）地层的岩性变化得出来的。

【中文定义】 是反映岩石特征的一些属性，如颜色、成分、结构、胶结物及胶结类型、特殊矿物等。

lithospheric fault-block

【释义】 岩石圈断块

【例句】 From the viewpoint of fault-block tectonics, our lithospheric fault-blocks confined by lithospheric fractures are just equivalent to the plates of "plate tectonics". 从断块构造的角度来看，岩石圈断裂所限制的岩石圈断块刚好等同于"板块构造学说"的板块。

【中文定义】 是地球表部（地壳和上地幔顶部，软流圈以上）被岩石圈断裂切割和围限的最大一级的断块。可沿软流圈滑动，相当于板块学说的岩石圈板块。

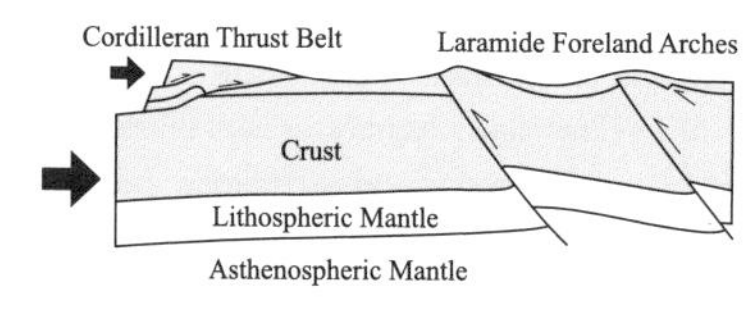

lithospheric fault-block

littoral facies

【释义】 滨海相

【常用搭配】 littoral neritic facies 滨浅海相

【例句】 Elevation and subsidence rates of southeast coast crust obtained by littoral facies sediment samples. 由滨海相沉积样品反映的东南沿海海岸升降速率。

【中文定义】 发育于低潮线和高潮线之间及其邻近地区的狭长滨海地区的沉积相，潮汐作用和波浪作用占主要地位。常常发育一些交错层理，以表示其动荡的沉积环境。

lixiviation [lɪksɪvɪ'eɪʃən]

【释义】 *n.* 淋滤（溶滤）

【例句】 The fixed sulphur baking produce the water-soluble sulfuric acid calcium, which make against the cyaniding lixiviation. 固硫焙烧法则产生了水溶性的硫酸钙，不利于氰化浸出。

【中文定义】 岩石中某些组分进入水中的过程，称为溶滤作用。

loam [loʊm]

【释义】 *n.* （尤指适合植物生长的）壤土，沃土；肥土（含有黏土、沙和有机物质的土地）；亚黏土 *vt.* 用壤土填

【常用搭配】 sandy loam 沙壤土，沙质壤土；silt loam：粉沙壤土

【例句】 The left bank and riverbed water retaining dam of Kangyang hydropower station is a dam with inclined loam core. 康扬水电站左岸及河床挡水坝为壤土斜墙坝。

【中文定义】 指土壤颗粒组成中黏粒、粉粒、沙粒含量适中的土壤。质地介于黏土和沙土之间，兼有黏土和沙土的优点，通气透水、保水保温性能都较好，是较理想的农业土壤。

loam

local coordinate system

【释义】 地方坐标系，局部坐标系

【中文定义】 是因建设、城市规划和科学研究需要而在局部地区建立的相对独立的平面坐标系统。

local inclination

【释义】 局部倾斜

【中文定义】 砌体承重结构沿纵墙 6～10m 内基础两点下地基沉降差与其间距的比值。

locally unstable

【释义】 局部不稳定

【中文定义】 根据《水力发电工程地质勘察规范》（GB 50287—2016），Ⅲ类围岩稳定性多为局部不稳定，表现为围岩强度不足，局部会产生塑性变形，不支护可能产生塌方或变形破坏。完整的较软岩，可能暂时稳定。

loess ['loʊɪs]

【释义】 *n.* 黄土

【常用搭配】 aeolian deposit loess 风积黄土；loess ridge 黄土梁；loess plain 黄土平原

【例句】 Ansai is the most concentrated and representative area for the preservation and development of Chinese loess culture, and it has a special position. 安塞是中国黄土文化保持和发展的最集中、最有代表性的地域，具有特殊的地位。

【中文定义】 指在地质时代中的第四纪期间，以风力搬运的黄色粉土沉积物。它是原生的、成厚层连续分布，掩覆在低分水岭、山坡、

loess

丘陵，常与基岩不整合接触，无层理，常含有古土壤层及钙质结核层，垂直节理发育，常形成陡壁。

loess ridge

【释义】 黄土梁

【中文定义】 脉状的黄土丘陵。两侧为平行的沟谷，地下水埋藏深，在基岩和黄土接触处有时可见到地下水流出。类型有平梁、斜梁、峁梁。

loess ridge

loess tableland

【释义】 黄土塬

【中文定义】 黄土覆盖的面积较大的平坦高地。黄土塬具有巨厚的黄土堆积、完整的地层序列和连续的自然剖面，地形上表现为平坦高地，高地边缘陡峻，为沟谷所环蚀，由于溯源侵蚀而参差不齐。

long term strength

【释义】 长期强度

【例句】 According to the result of shear creep test, the rheologic behavior and long term strength of gravel have been studied in this paper. 在室内剪切蠕变试验研究的基础上，对成都地区卵石土的流变特性及长期强度特征作一初步探讨。

loess tableland

long tunnel

【释义】 长隧洞

long tunnel

longitudinal fault of river bed

【释义】 顺河断层

【例句】 Drilling practice of river-crossing inclined holes in longitudinal fault of river bed at dam site of a hydropower station. 某水电站坝址河床顺河断层穿江斜孔钻探实践。

longitudinal geological section

【释义】 工程地质纵剖面图

【常用搭配】 longitudinal geological sections of main structures 主要建筑物工程地质纵剖面图

【例句】 In prefeasibility and feasibility study stages of hydropower engineering, longitudinal geological sections of alternative dam sites, water way routes and other structures shall be submitted in report for engineering geological investigation. 水力发电工程预可行性研究阶段和可行性研究阶段工程地质勘察报告中应附各比较坝址、引水线路或其他建筑物纵剖面图。

【中文定义】 ①沿拟建场地的延长方向或结构物长轴线的工程地质剖面图。②沿岩层走向或构造线的工程地质剖面图。

longitudinal joint

【释义】 纵节理

【例句】 In general, the rock mass structure planes are divided into three major types, including the primary structural plane, tectonic structural plane and secondary structural plane. longitudinal joint, cross joint, bedded joint, diagonal joint and columnar joint belong to the primary structural plane. 一般而言，岩体结构面分为原生结构面、构造结构面和次生结构面三大类，纵节理、横节理、层节理、斜节理和柱状节理都属于原生结构面。

【中文定义】 指走向节理的几何关系与褶皱轴或区域构造线走向大致平行的节理。

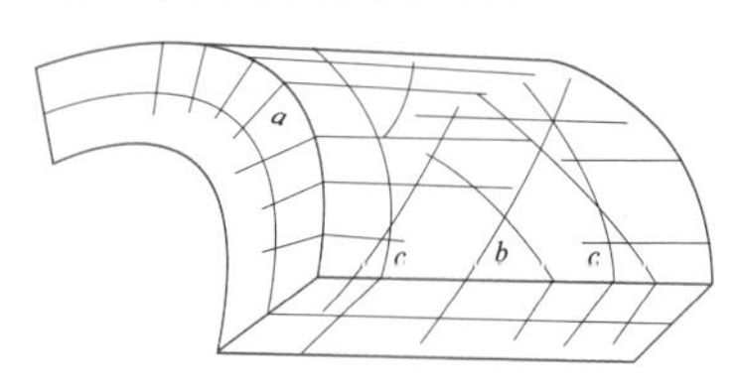

a—纵节理；*b*—斜节理；*c*—横节理

longitudinal joint (see a)

longitudinal profile

【释义】 纵剖面图

【例句】 Knickpoints in fluvial channel longitudinal profiles and channel steepness index values derived from digital elevation data can be used to detect tectonic structures and infer spatial patterns of uplift. 从高程数据提取的河道纵剖面的陡降点和河道陡峭指数可以用来发现构造，并推断构造的空间展布形态。

【中文定义】 ①垂直于拟建场地的延长方向或结构物长轴线的地质剖面图；②垂直于岩层走向或构造线的地质剖面图。

longitudinal valley

【释义】 纵向谷

【中文定义】 指河谷延伸方向与岩层走向基本一致（0°～30°）。

longitudinal wave velocity

【释义】 纵波波速

【中文定义】 纵波在岩土体中传播的速度。纵波是质点的振动方向与传播方向同轴的波。

long-term modulus

【释义】 长期模量

【例句】 The analysis and research of long-term shear modulus of soft soils. 软土长期剪切模量的分析与研究。

long-term stability

【释义】 长期稳定性

【中文定义】 洞室、边坡岩土体，经二次应力长期调整后的稳定性。

loose [lus]

【释义】 *adj*. 松的，宽松的，散漫的

loose granular structure

【释义】 散体结构

【中文定义】 指岩体节理、裂隙很发育，岩体十分破碎，岩石手捏即碎。

loose zone

【释义】 松动区

【例句】 In order to have a profound understand of the internal force and deformation rules in loose zones of the loess tunnel caused by the construction process of tunnel, the internal force and surrounding rock displacement change rules at loose zone are analyzed and summarized. 为了掌握黄土隧道施工过程对围岩松动区内力和位移变化的影响，对黄土隧道施工各阶段围岩松动区内力和位移变化规律进行了分析和总结。

【中文定义】 边坡、隧道等工程活动开挖后所产生的具有松动迹象的区域。

loosen rock mass

【释义】 松散岩体

【例句】 In the excavation of the tunnel, steel arch support forms are often used to deal with the loosen rock mass structure. 在隧洞开挖中，钢拱架支撑的形式往往被用来处理松散岩体结构

【中文定义】 指松弛或解体的岩体，结构破碎，整体体积增大。

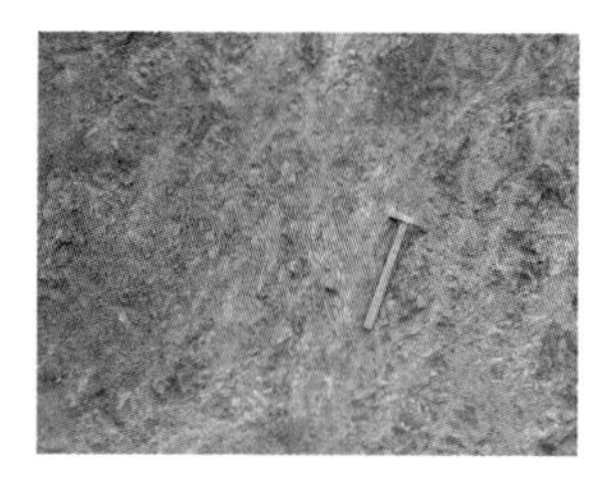

loosen rock mass

lopolith ['lɔpəliθ]

【释义】 *n*. 岩盆

【例句】 After the formation of the above-mentioned structural type, the magmatic withdrawal in the magma vent leads to the subsidence of the cyclic structure, thus forming the lopolith. 薄层状多旋回构造形成后，由于岩浆管道中岩浆冷却收缩下降，导致旋回构造相应下沉而形成岩盆。

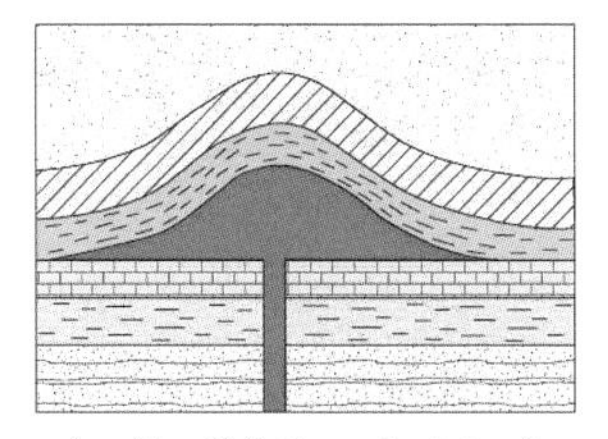

laccoliths with flat floor and arched roof

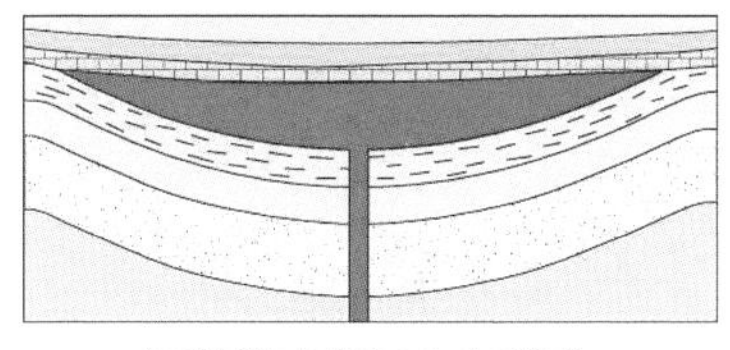

lopolith intrudeol into a structural basin

igneous structures and field relationships lopolith

Love wave

【释义】 勒夫波

【例句】 From the images of Love wave, we can know that there are differences of velocity among the major tectonic units in this area. 从勒夫波的成像结果我们可以看出各主要构造单元的速度差异。

【中文定义】 又称Q波（或地滚波）。在半无限介质之上出现低速层的情况下，一种垂直于传播方向的在水平面内振动的波。在垂直面上，粒子呈逆时针椭圆形振动，震动振幅一样会随深度增加而减少。

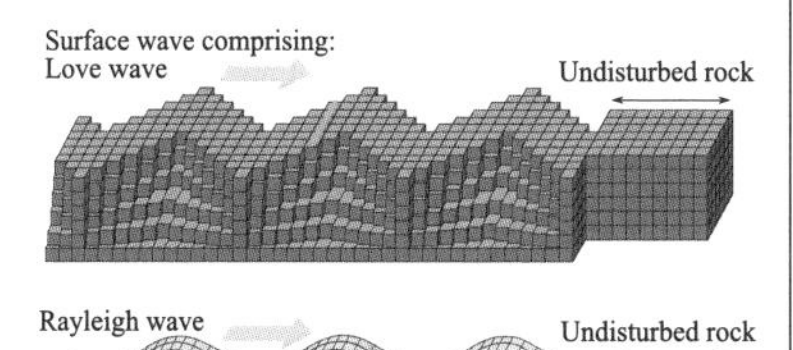

Love wave

low frequency debris flow

【释义】 低频泥石流

【中文定义】 5～20年发生一次的泥石流。

low mountain

【释义】 低山

【常用搭配】 medium low mountain 中低山

【中文定义】 高程500～1000m，相对高差200～1000m的地区。

low mountain

low speed landslide

【释义】 慢速滑坡

low velocity layer

【释义】 低速带

Lower [louər]

【释义】 下统（早世）

【常用搭配】 Lower Jurassic 侏罗系下统（早侏罗世）

lower bound solution

【释义】 下限解

【常用搭配】 lower bound solution of limit analysis 极限分析下限解

【例句】 The analysis starts with the lower bound solution proposed by Hoffman and Sachs and includes the effects of strain, strain rate and pressure on the deformation behaviour. 该分析从霍夫曼和萨克斯提出的下界解方案入手，包括应变、应变率和压力对变形行为的影响。

lower explosion limit (LEL)

【释义】 爆炸下限

【例句】 Detection of explosive atmospheres relies on the accurate measurement of combustible gases below the LEL concentration. 对易爆大气的测量依赖于对可燃气体低于LEL浓度的精确测量。

【中文定义】 指可燃气体爆炸下限浓度（$V\%$）值。

lower hemisphere projection

【释义】 下半球投影

【例句】 Lower hemisphere projection is one kind of stereographic projection. 下半球投影是赤平极射投影的一种。

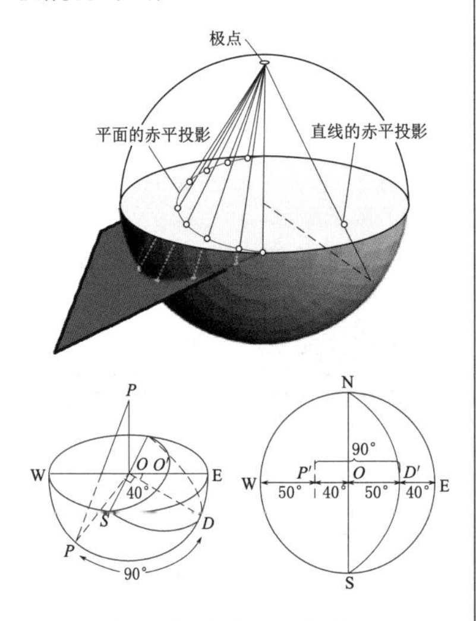

lower hemisphere projection

【中文定义】 投影球上两极的发射点，分上极射点（P）和下极射点（F）。由上极射点（P）把下半球的几何要素投影到赤平面上的投影称为下半球投影。

lower plate

【释义】 下盘

【同义词】 heading wall; footwall

【例句】 When the fault plane is inclined, the block above the fault plane is called upper plate, and the block below it is called lower plate. 当断层面倾斜时，位于断层面上方的称为上盘，下方的称为下盘。

【中文定义】 断层的两盘之一，断层形成后，一般形成有倾斜角的断层面，在倾斜面一侧下方的岩石称为下盘。

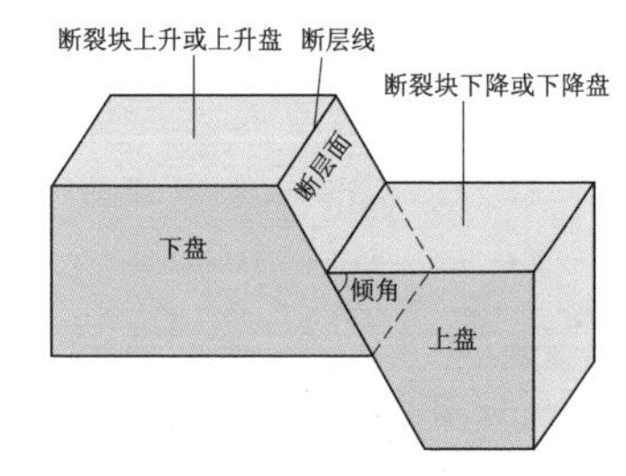

fault element

lower-limit of earthquake magnitude

【释义】 起算震级

【例句】 For the facilities with specific safety requirement, such as nuclear power plant, a substitution of lower-limit of earthquake magnitude 4 for the minimum magnitude may be unsafe. 对核电厂这类具特殊安全性要求的设施，以起算震级4.0代替最小震级，是偏于不安全的。

M

magma ['mægmə]

【释义】 *n.* 岩浆

【常用搭配】 magma chamber 岩浆房；parental magma 母岩浆；magma cooling 岩浆冷却；anatectic magma 深熔岩浆

【例句】 The modes of the fractional crystallization in magma chamber were discussed for many years. 关于岩浆房中岩浆的结晶分异方式，多年来一直存在着争议。

magma

magnetic azimuth

【释义】 磁方位角

【中文定义】 由通过某点磁子午线北端起算，顺时针方向至某一直线间的夹角（即地球南北极上的磁南、磁北两点间的连线）。

magnetic conductivity

【释义】 磁导率

magnetic meridian

【释义】 磁子午线

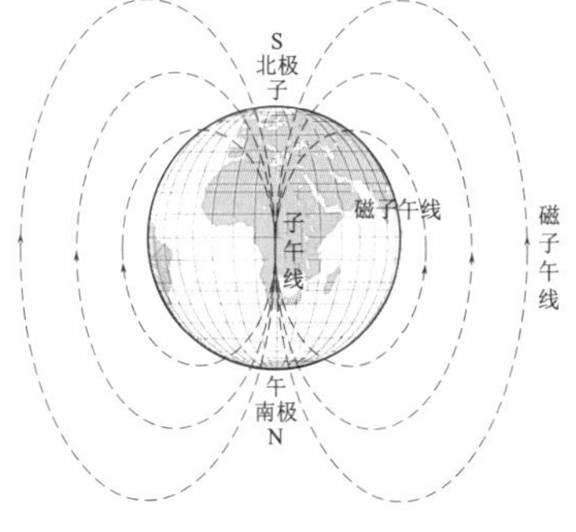

magnetic meridian

【中文定义】 通过地球南北磁极所作的平面与地球表面的交线。

magnetite ['mægnəˌtaɪt]

【释义】 *n.* 磁铁矿

【例句】 Opaque phase includes hematite, magnetite, rutile and chalcopyrite. 暗色子矿物包括磁铁矿、赤铁矿、金红石和黄铜矿。

magnetite

magnitude-frequency relation

【释义】 震级-频度关系

【例句】 Based on the proposed magnitude-frequency relation and the result of extremal method, a earthquake of 6.5～6.8 which will occur probably in the Yang-Dong seismic zone in the future is predicted. 根据扬-铜带修改的震级-频度关系，以及极值方法的结果，预测扬-铜带未来可能发生一次 6.5～6.8 级地震。

magnitude interval

【释义】 震级档

【例句】 Several seismological and geological characteristics have been selected and quantized to describe the seismicity features in time and space of every magnitude interval. 本文选取并量化了多个地震、地质特征，以描述各震级档地震活动在时间上和空间上的性质。

【中文定义】 指地震危险性概率分析中的震级分档间隔（一般取 0.5 级）。

main ground fissure

【释义】 主地裂缝

【中文定义】 错断勘探标志层且有显著最大垂直位移的地裂缝。

main shock

【释义】 主震

【例句】 These happen as the seismic waves from a main shock pass through the earth's crust, which results in transient changes. 动态应力考虑的是地震波从主震经过地壳传递的瞬间时段中，应力所发生的变化。

【中文定义】 一个地震序列中最强的地震称为主震

major exploration profile

【释义】 主勘探剖面

man-made isotope

【释义】 人工同位素

【例句】 Man-made isotope of Cesium-137 can be as environment tracer to research the sediments on river floodplains. 人工同位素 137Cs 可以作为环境示踪因子研究河漫滩沉积物。

【中文定义】 人工制造的具有相同质子数、不同中子数的同一元素的不同核素互为同位素。

map border

【释义】 图廓

【中文定义】 分幅地图的实际和整饰范围线。

map compilation

【释义】 地图编绘

【中文定义】 利用已有地图及有关资料，根据成图要求编制地图的过程。

map digitizing

【释义】 地图数字化

【例句】 The contemporary way of map digitizing has already achieved the transition from manual tracking digitizing to scanning digitizing. 地图数字化方式已经由手扶跟踪数字化逐渐过渡到扫描数字化。

【中文定义】 实现从线画地图到数字信息转换的过程。

map element

【释义】 地图要素

【例句】 Because every map element solely correlates a second characteristic value, the map element can be expressed as a generalized isoline or isoplane of second characteristic value. 由于每一种地图要素唯一关联着一个第二特征值，因而地图要素可以表示为第二特征值的泛等值线或泛等值面。

【中文定义】 构成地图的基本内容。分数学要素、地理要素、整饰要素。

map projection

【释义】 地图投影，地图投影法

【例句】 Map projection is the basis of cartography. 地图投影是地图学的数学基础。

【中文定义】 是利用一定数学法则把地球表面的经、纬线转换到平面上的理论和方法。

map scale

【释义】 地图比例尺

【中文定义】 地图上某一线段的长度和地面上相应线段水平距离之比。

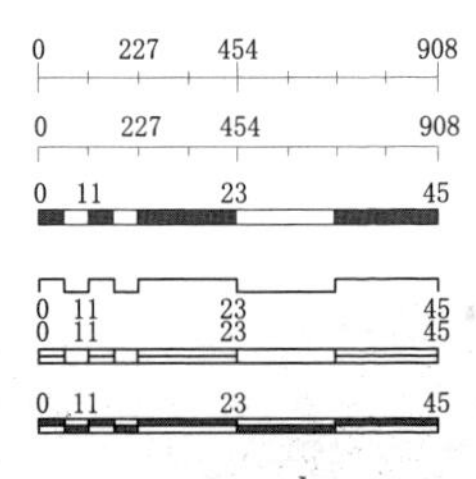

map scale

map symbol

【释义】 地图图式

【中文定义】 对地图上地物、地貌符号的样式、规格、颜色、使用以及地图注记和图廓整饰等所作的统一规定，是测绘标准之一。

map title

【释义】 图名

【中文定义】 赋予每幅地图的名称。

marble ['mɑrb(ə)l]

【释义】 *n.* 大理岩，大理石

【例句】 Conventional triaxial test and pre-and post-peak unloading confining pressure tests of marble have been performed. 对大理岩进行了三轴试验和峰前、峰后卸围压试验。

marble

【中文定义】 变质岩的一种，因在中国由于云南省大理县盛产这种岩石而得名。是由碳酸盐类矿物如方解石、白云石含量大于50%的区域变质或热接触变质的岩石。

marine [mə'rin]

【释义】 *adj*. 海相，海成，海生的

【常用搭配】 marine sediment 海相沉积物；marine terrace 海积阶地；marine erosion 海蚀

【例句】 Along the East, North and West coasts there are sediments of Mesozoic and Tertiary age, deposited in marginal marine basins. 沿着非洲大陆的东、北及西海岸，即海洋盆地的边缘，分布有中生代及第三纪时期的沉积物。

【中文定义】 指海洋环境中形成的沉积相的总称。

marine terrace

marine clastics

【释义】 海相碎屑岩

【例句】 The Late Triassic transition from marine carbonate rock to clastics in the western Sichuan Basin. 四川盆地西部晚三叠世海相碳酸盐岩到碎屑岩的转换过程。

【中文定义】 指海洋环境下，经海洋动力过程产生的碎屑岩。

marine erosion

【释义】 海蚀，海蚀作用

【常用搭配】 marine erosion coast 海蚀海岸；marine erosion relief 海蚀地貌

【同义词】 marine abrasion，sea erosion

marine erosion

【中文定义】 当海浪冲到基岩海岸时，往往会形成破坏力很强大的击岸浪，拍打着海岸带的岩石，对海岸有强烈的破坏作用，像这种由海水而引起破坏海岸的作用，称为海蚀作用。

marine facies

【释义】 海相

【常用搭配】 marine facies formation 海相地层

【例句】 This deposit is a copper deposit of marine facies's volcano rock type. 该矿为海相火山岩型铜矿。

【中文定义】 在海洋环境中沉积的沉积物、岩石或岩层。

marshland ['mɑrʃlænd]

【释义】 *n*. 沼泽地

【例句】 Increasing air temperature would have enhanced grain yields and stimulated marshlands conversion into croplands. 气温升高会提高粮食产量并刺激沼泽地转化为农田。

【中文定义】 指长期受积水浸泡，水草茂密的泥泞地区。

marshland in Ruoergai Plateau

matrix ['metrɪks]

【释义】 *n*. 基质

【例句】 The compositions show a large variation between chondrules, matrix and mesostasis, which indicate the different material source of these components. 组成矿物成分极不均一，在矿物晶体内部，球粒内部及球粒与基质间均有明显变化。

maximum allowable concentration (MAC)

【释义】 最高容许浓度

【中文定义】 指工作地点在一个工作日内、任何时间均不应超过的有毒化学物质的浓度。

maximum credible earthquake (MCE)

【释义】 最大可信地震

【例句】 The dam could be subjected to upgraded maximum credible earthquake (MCE) or probable maximum flood(PMF) loadings. 该坝能经受住强化的最大可信地震荷载或者可能最大洪水荷载。

【中文定义】 指在给定区域或当前构造格架下断层所能产生的最大地震。

maximum design earthquake（MDE）

【释义】 最大设计地震

【中文定义】 是结构设计或评估的最大地震动水平。在该地震作用下，虽然引发严重的经济损失和破坏，但是水工结构的相关性能要求不至于遭到毁灭性的破坏。

maximum dry density

【释义】 最大干密度

maximum principal stress

【释义】 最大主应力

【例句】 Direction of previous main faults that intersect the maximum principal stress by minor angle also probably becomes the major migration direction of oil and gas. 与最大主应力小角度相交的先期主要断裂的方向也可能成为油气运移的主要方向。

【中文定义】 某个单元中最大的主应力，即第一主应力。

maximum water inflow

【释义】 最大涌水量

【例句】 In headrace tunnel, different surrounding rock mass own diverse maximum water inflow. 引水隧洞中各类围岩的最大涌水量都不相同。

【中文定义】 指工作面和两壁的涌水量之和。

maximum water yield of borehole

【释义】 钻孔最大出水量

mean value

【释义】 平均值

【例句】 The key thing to remember about random variables is that actual observations are centered around an average, or mean value. 对于随机的变化关键要记住的是，实际的观测资料都在平均数或者中间值周围。

meandering channel

【释义】 蜿蜒河道

meandering channel

【中文定义】 在冲积平原上，河流在大范围的冲积层中自由摆动，其外形蜿蜒弯曲，又称曲流型河道、自由曲流或蜿曲。

measured boundary

【释义】 实测界线

【常用搭配】 measured boundary of stratum 实测地层界线

【例句】 Measured boundary of stratum should be drawn with solid line on geological section. 地质剖面图上实测地层界线应用实线表示。

【中文定义】 现场实际测量的各种地质界线，如实测地层界线、实测地层不整合界线等。

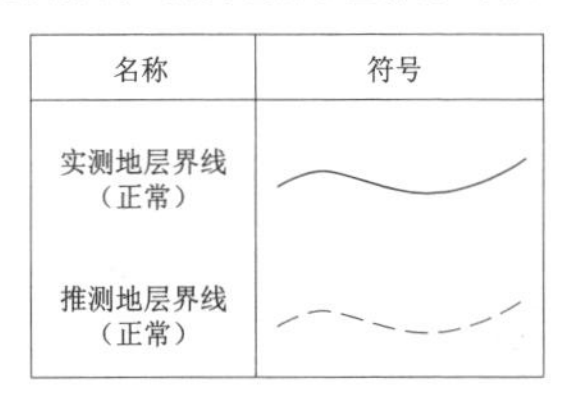

名称	符号
实测地层界线（正常）	
推测地层界线（正常）	

measured boundary and conjectural boundary

measuring method

【释义】 测量方法

【例句】 The experimental results and simulation on SGI workstation prove the correctness and effectiveness of the presented measuring method. 实验结果和 SGI 工作站上的仿真证明了该测量方法的正确性和有效性。

mechanical piping effect

【释义】 机械管涌

【例句】 The piping effect are consist of mechanical piping effect and chemical piping effect. 管涌主要有机械管涌和化学管涌两类。

【中文定义】 指在渗透压力作用下，地下水将土层中细小颗粒由地裂缝、管道、洞穴带走的现象。

medium dense

【释义】 中密

medium gravel

【释义】 中砾，中砾石

medium gravel

medium grained texture

【释义】 中粒结构

【中文定义】 粒径为中等粒度的结构，颗粒直径2～5mm。

medium mountain

【释义】 中山

【常用搭配】 high medium mountain 高中山；low medium mountain 低中山

【中文定义】 高程1000～3500m，相对高差200～1000m的地区。

medium mountain

medium sand

【释义】 中砂

【例句】 When cement content and water cement ratio are constant, the compressive strength of solid mass with medium sand is highest and elastic modulus value maximum, orderly, fine sand and clayey soil. 在水泥含量和水灰比一定时，中砂抗压强度高，弹模大，粉细砂次之，黏性土最小。

【中文定义】 粒径大于0.25mm的颗粒质量超过总质量的50%的砂，细度模数在3.0～2.3的细骨料。

medium sand

medium-scale landslide

【释义】 中型滑坡

【例句】 The Naqin landside that is medium-scale landslide has developed on the granite area. 那勤滑坡发育在花岗岩区，属中型山体滑坡。

meizoseismal area

【释义】 极震区

【例句】 In the meizoseismal area, there appeared a large number of fissures and fractures, consisting of types of tension, compression and shear. 极震区产生了规模较大的地面破裂带，其中有张性、压性和剪切等多种形式裂缝。

【中文定义】 震中附近振动最强烈，破坏比也最严重的地区称为极震区。

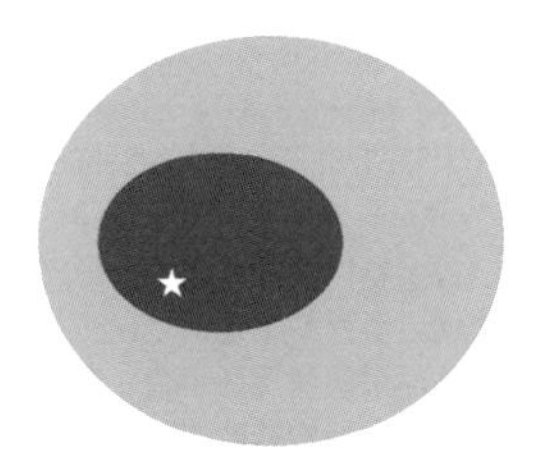

meizoseismal area

Member [ˈmɛmbə]

【释义】 *n.* 段

【中文定义】 为第三级岩石地层单位，是根据组内岩性、结构和地层成因等不同而进行的组的再分。

Mesozoic (Mz) [ˌmesəuˈzəuik]

【释义】 *n.* 中生代（界）

【中文定义】 是显生宙的第二个地质时代，包括三叠纪、侏罗纪和白垩纪三个时期。中生代中、晚期，各板块的漂移加速，在具有俯冲带的洋、陆壳的接触带上俯冲、挤压，导致著名的燕山运动（或称太平洋运动）。由于这时期的优势动物是爬行动物，尤其是恐龙，因此又称为爬行动物时代。

metallic luster

【释义】 金属光泽

metallic luster

【例句】 Metallic luster is a degree of luster that can be seen by certain opaque and compact state minerals such as gold. 金属光泽是指某些不透明的、致密的矿物，比如金，所呈现出的光泽。
【中文定义】 矿物表面对光的反射能力强，呈明显的金属光亮。

metamorphic conglomerate
【释义】 变质砾岩
【例句】 Based on the study of petrology, mineralogy, microelements, palaeontology and geological occurrence characteristics, the authors hold that the thick bedded metamorphic conglomerate in Sinian strata in Laoshuxia, Luoding area of western Guangdong province, is a metamorphic basal conglomerate with simple component suffered structural deformation. 根据岩石学、矿物学、微量元素、古生物以及变质砾岩的形态特征等多方面研究表明，粤西罗定地区老鼠峡震旦纪变质地层中出露的厚层状变质砾岩为一套成分比较单一，属沉积成因并经构造改造的变质底砾岩。
【中文定义】 由砾岩经过变质作用形成的岩石。

metamorphic conglomerate

metamorphic mudstone
【释义】 变质泥岩

metamorphic mudstone

【例句】 Rock debris complex components, mainly igneous rock (39.7%), metamorphic rocks (11.7%), in addition to trace siltstone and metamorphic mudstone has almost no sedimentary rock debris. 岩屑成分复杂，主要有火成岩(39.7%)、变质岩(11.7%)，除微量粉砂岩和变质泥岩外几乎没有沉积岩岩屑。
【中文定义】 由泥岩经过变质作用形成的岩石。

metamorphic rock
【释义】 变质岩
【常用搭配】 contact metamorphic rock 接触变质岩；ultrahigh pressure metamorphic rock 超高压变质岩
【例句】 Hard metamorphic rock consists essentially of interlocking quartz crystals. 坚硬的变质岩，主要由嵌合的石英晶体组成。
【中文定义】 三大岩类的一种，是指受到地球内部力量（温度、压力、应力的变化、化学成分等）改造而成的新型岩石。固态的岩石在地球内部的压力和温度作用下，发生物质成分的迁移和重结晶，形成新的矿物组合。

metamorphic rock

metamorphism [metə'mɔːfɪzəm]
【释义】 *n.* 变质，变质作用，变态，变形
【常用搭配】 regional metamorphism 区域变质；selective metamorphism 选择变质；static metamorphism 静力变质
【例句】 Types of metamorphism include regional metamorphism, contact metamorphism and dynamic metamorphism, etc. 变质作用类型有：区域变质、接触变质和动力变质等。
【中文定义】 指先已存在的岩石受物理条件和化学条件变化的影响，在基本保持固态的情况下，改造其结构、构造和矿物成分，成为一种新的岩石的转变过程。

metasandstone [me'teɪzndstoʊn]
【释义】 *n.* 变质砂岩
【例句】 The study show that the main source of ore is granite, partly from the surrounding rock mass, which consists chiefly of the calcium magnesium carbonatite and metasandstone. 研究表明，矿源主要源于花岗岩岩体，其次来源于钙镁

质碳酸盐岩、变质砂岩等围岩。
【中文定义】 由砂岩经过变质作用形成的岩石。

metasandstone

method of minimum flow rate of springs
【释义】 泉群最小流量比拟法
method of slices
【释义】 条块法
【中文定义】 边坡稳定性分析常用的方法。将边坡滑动体划分成若干垂直条块，对每个条块进行受力分析，计算安全系数。
method of tension wire alignment
【释义】 引张线法
【例句】 Method of tension wire alignment is a method to measure horizontal displacement in deformation observation. 引张线法是变形观测中测定水平位移的一种方法。
"mise-a-la-masse" method
【释义】 充电法
【例句】 Applying ground penetrating radar (GPR), electromagnetic method, galvanic"mise a la masse"method and self potential method ascertains the exact position of water inrush channel and the seepage range in some areas of Hanjiang River embankment. 运用探地雷达、电磁法、直流充电法以及自然电场法，查明了汉江某区段堤防中涌水通道的确切位置、渗水的范围。
mica content
【释义】 云母含量
【例句】 Mica isn't suit to boost up the mould plaster because it will decrease the intensity and sop quotiety. 云母粉不适合作为模型石膏的增强材料，是因为云母粉降低模型石膏的强度和吸水率。
micro opening
【释义】 微张
【例句】 In research area, sandstone and mudstone is on interbed distribution, rock joints and fissures develops early or fully, with micro or full opening. 研究区内砂岩、泥岩互层分布，岩体节理裂隙较发育或发育，微张或张开。

schematic diagram of micro opening structural plane

microbiological index
【释义】 微生物学指标
microcrack ['maɪkrəʊkræk]
【释义】 *n.* 微裂纹

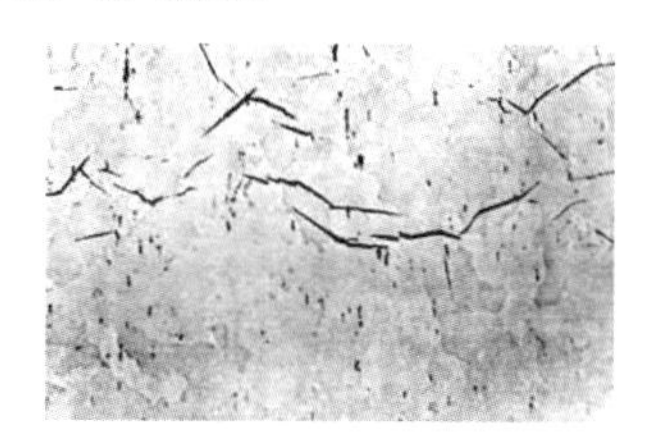

microcrack

microfissure [ˌmaikrəu'fiʃə]
【释义】 *n.* 微裂缝，微裂隙
【例句】 As a result, the ultimate failure resulting from the microfissure propagation and coalescence can be obtained by this model. 岩石的最终宏观断裂破坏与其内部微裂隙的分布、扩展和聚集密切相关。

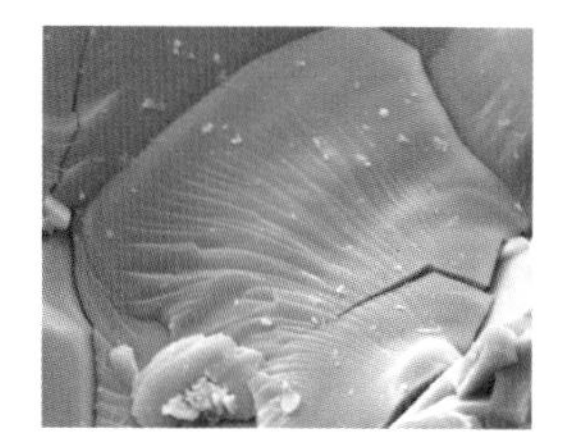

microfissure

【中文定义】 是按裂缝大小（主要依据宽度）而作的一种分类。一般认为裂缝宽度在 100μm 或

150μm 以下的称微裂缝。

microgranular texture

【释义】 微粒结构

【中文定义】 颗粒直径小于 0.2mm 的显晶质。

microquake ['maikrəkweik]

【释义】 *n.* 微震

【例句】 It is the first attempt in China to set up a digital seismic network for microquake. 数字化微震台网在国内是第一次尝试。

【中文定义】 震级大于等于 1 级、小于 3 级的称为弱震或微震

microseismic monitoring

【释义】 微震监测

【例句】 Microseismic monitoring technology is widely used as an effective means of warning in all countries in the world. 世界各国普遍将微震监测技术作为一种有效的预警手段。

microseismograph [ˌmaikrəu'saizməgrɑːf]

【释义】 *n.* 微震仪

【常用搭配】 calibration of microseismograph 微震仪标定

microseismograph

Middle

【释义】 *adj.* 中统（中世）

【常用搭配】 Middle Triassic 三叠系中统（中三叠世）

middle-deep seated landslide

【释义】 中层滑坡

migmatite ['mɪgmətaɪt]

【释义】 *n.* 混合岩

【常用搭配】 banded migmatite 带状混合岩

migmatite

【例句】 The eastern Mountain Ailao consists of gneiss, micaschists, amphibolites, marbles, and migmatite. 哀牢山东部由片麻岩、云母片岩、斜长角闪岩、大理岩和混合岩构成。

【中文定义】 混合岩化作用形成的各种岩石。

mineral ['mɪnərəl]

【释义】 *n.* 矿物

【常用搭配】 primary mineral 原生矿物；dark mineral 暗色矿物；light mineral 亮色矿物；associated mineral 伴生矿物；secondary mineral 次生矿物；clay mineral 黏土矿物；altered mineral 蚀变矿物；mineral aggregates 矿物集合体

【例句】 These schistose rocks are predominantly composed of chlorite, actinolite and epidote, but lack the above described mineral aggregates. 这些片状岩主要由绿泥石、阳起石、绿帘石等矿物组成，但缺乏上述矿物集合体。

【中文定义】 指由地质作用形成的天然单质或化合物，是组成岩石和矿石的基本单元。

fluorite mineral

mineral resources

【释义】 矿产资源

【例句】 The country possesses rich mineral resources. 这个国家有丰富的矿产资源。

【中文定义】 指经过地质成矿作用而形成的，天然赋存于地壳内部或地表埋藏于地下或出露于地表，呈固态、液态或气态的，并具有开发利用价值的矿物或有用元素的集合体。

mineralization analysis

【释义】 矿化分析

mineralization of water

【释义】 矿化度

【同义词】 salinity, mineralization degree

【例句】 Water quality of bedrock fissure water in low mountain area is relatively bad in the south part of a city, and it leads to high mineralization of water in the deep underground water zone. 某城市

南部低山区基岩裂隙水水质较差，导致该市南部地下水深埋带水质矿化度偏高。

【中文定义】 指水中含有钙、镁、铝和锰等金属的碳酸盐、重碳酸盐、氯化物、硫酸盐、硝酸盐以及各种钠盐等的总和。

mineralized water

【释义】 矿化水

【例句】 Electrical resistivity is widely used to measure the saturation of mineralized water in rocks. 电阻率被广泛地用于测量岩石中的矿化水饱和度。

【中文定义】 含有多种微元素和矿物质的水。

minimum mining thickness

【释义】 最小可采厚度

minimum mining width

【释义】 最小可采宽度

minimum principal stress

【释义】 最小主应力

【例句】 The Mohr-Coulomb model is based on plotting Mohr's circle for states of stress at failure in the plane of the maximum and minimum principal stresses. 莫尔-库仑模式基于莫尔圆来判断最大和最小主应力平面上的破坏应力状态。

【中文定义】 某个单元中最小的主应力，即第三主应力。

mining collapse

【释义】 采空塌陷

【例句】 It is introduced that the BP neural network model of prediction of mining collapse is established based on the survey's data and main factors at a mine field where the collapse had happen. 依据某煤炭开采区的勘察资料，综合考虑影响采空塌陷的主要因素，建立了预测采空塌陷的 BP 神经网络模型。

【中文定义】 由于地下挖掘形成空间，造成上部岩土层在自重作用下失稳而引起的地面塌陷现象。

mining intensity method

【释义】 开采强度法

mining modulus analogy

【释义】 开采模数比拟法

mining pumping method

【释义】 开采抽水法

【例句】 Mining pumping method is a method to determine the recharge capacity of the calculated area and to evaluate the groundwater resources. 开采抽水法是确定计算地段补给能力，进行地下水资源评价的一种方法。

mining collapse

Miocene ['maiəusi:n]

【释义】 *n*. 中新世（统）

【中文定义】 第三纪的第四个世。

mirror ['mɪrər]

【释义】 *adj*. 镜面的

【例句】 According to the structure roughness guide proposed by Barton and Bandis, these structures can be defined as mirror structure. 根据 Barton 和 Bandis 提出的结构面粗糙度指南，这些结构面可以定义为镜面的结构面。

【中文定义】 是表示结构面粗糙程度的一种描述，形容结构面表面像镜子面一样光滑。

mirror

miscellaneous fill

【释义】 杂填土

【例句】 On the upper part of foundation in the ancient city, there is a large number of miscellaneous fill which is mainly composed of stove ashes and is variable in thickness. 在古城区建筑地基上部，广泛赋存以炉灰渣为主要成分的杂填土，其厚度不一。

【中文定义】 是由于人类长期的生活和生产活动而形成的地面填土层，其填筑物随着地区的生产和生活水平的不同而异。杂填土的厚度一般变化较大。在大多数情况下，这类土由于填料物质不一，其颗粒尺寸相差较为悬殊，颗粒之间的孔隙大小不一，因此往往都比较疏松，抗剪强度低，压缩性较高，一般还具有浸水湿陷性。

miscellaneous fill

mobile earthquake monitoring

【释义】 流动地震监测

【例句】 The technology of mobile earthquake monitoring in China is developing. 我国的流动地震监测技术正在发展。

mobile earthquake monitoring

mode of slope failure

【释义】 边坡破坏模式

【中文定义】 边坡产生变形破坏可能的方式，包括滑移、倾倒、溃屈等。

moderately developed

【释义】 中等发育

【中文定义】 根据《水力发电工程地质勘察规范》(GB 50287) 的规定，当岩体内结构面发育组数 2～3 组，间距介于 30～50cm 时，其岩体结构面发育程度为中等发育。

moderately hard rock

【释义】 中硬岩

【例句】 In general, fresh marble, limestone and dolomite are typical of moderately hard rock. 通常，新鲜的大理岩、石灰岩、白云岩是典型的中硬岩。

【中文定义】 单轴饱和抗压强度介于 30～60MPa 的岩石。

moderately soft rock

【释义】 较软岩

【例句】 In general, fresh tuff, phyllite and marl are typical of moderately soft rock. 通常，新鲜的凝灰岩、千枚岩、泥灰岩是典型的较软岩。

【中文定义】 单轴饱和抗压强度介于 15～30MPa 的岩石。

moderately thick layer structure

【释义】 中厚层状结构

【中文定义】 层面厚度在 10～30cm 的层状结构。

moderately weathered

【释义】 弱/中风化

【例句】 If we use MMA for bedrock consolidation grouting, the parameters of moderately weathered can be lifted to the level of slightly weathered or fresh rock. 如果 MMA 用于基岩固结灌浆，能使弱/中风化岩体的参数提高到微风化或新鲜岩体的水平。

moderately weathered

【中文定义】 岩石原始组织结构清楚完整，但风化裂隙发育，裂隙壁风化剧烈。

modified Griffith's criterion

【释义】 修正的格里菲斯准则

modulus [ˈmɒdʒələs]

【释义】 *n.* 系数，模数；*pl.* moduli

【常用搭配】 elasticity modulus 弹性模量；deformation modulus 变形模量

【例句】 Soil has different modulus in tension and compression. 土体具有不同的拉压模量和剪胀特性。

【中文定义】 材料在受力状态下应力与应变之比。

modulus of compression

【释义】 压缩模量

【例句】 The foundation settlement is calculated by soil modulus of compression, using layer dividing summation method. 地基的沉降变形依据不同地基土的压缩模量，采用分层总和法来确定。

modulus of pressuremeter

【释义】 旁压模量

【例句】 According to the pressuremeter curve, the relevant soil mechanics indexes of the foundation soil layer at the depth can be obtained, such as the initial pressure, the plastic pressure, the ultimate pressure and the modulus of pressuremeter. 根据旁压曲线可以得到试验深度处地基土层的有关土力学指标，如初始压力、临塑压力、极限压力，以及旁压模量。

modulus of resilience

【释义】 回弹模量

【例句】 At the same time, permanent deformation and affecting factors to the modulus of resilience of the graded broken stones have been analyzed. 同时，对级配碎石永久变形和回弹模量影响因素进行了分析。

Mohr-Coulomb criteria

【释义】 莫尔-库仑准则

【例句】 Minimum in-situ stress could be obtained from normalized Mohr-Coulomb failure criteria through rock mechanical property testing. These properties include cohesion, frictional angle and uniaxial strength along failure plane. 通过岩石力学实验，测定岩样的内聚力、摩擦系数和单轴抗压强度等岩石力学参数，可以用常规的莫尔-库仑破坏准则确定最小主应力。

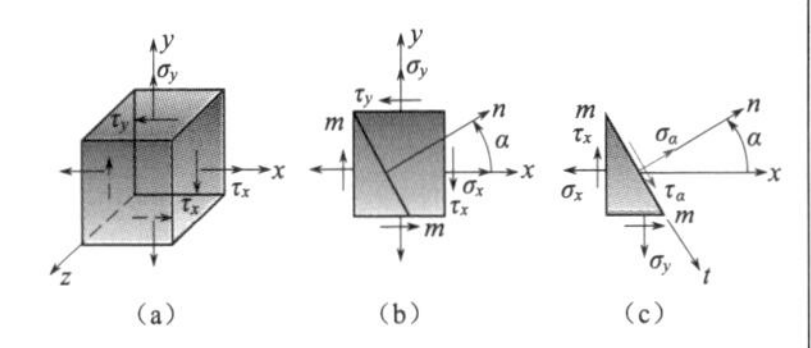

Mohr's Circle criteria

Mohr's rupture envelope

【释义】 莫尔破裂包络线

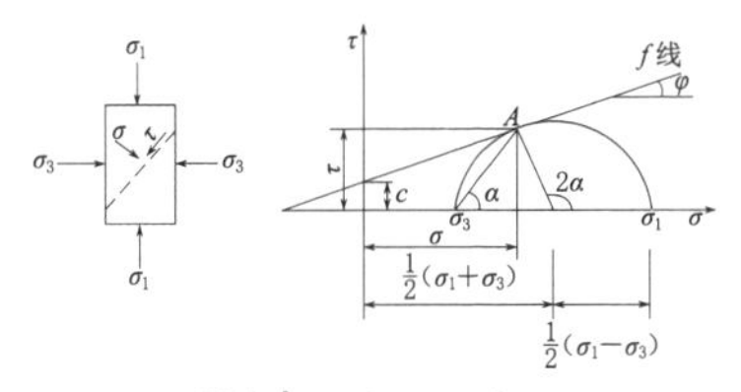

Mohr's rupture envelope

moisture content (MC)

【释义】 含水率，含水量

【常用搭配】 relative moisture content 相对含水率；critical moisture content 临界含水率

【同义词】 water content

【例句】 The major characteristics of soil are texture, acidity, nutrient content, gaseous content, and moisture content. 土壤的主要特性有质地、酸度、养分含量、气体含量和含水量。

monitoring and early warning value

【释义】 监测预警值

monitoring frequency

【释义】 监测频率，监测频次

【例句】 On the basis of error analysis for settlement monitoring of advanced highway soft-base, this dissertation summarized precision indicators, monitoring frequency, monitoring principles and quality guarantee measures that are fit for the settlement monitoring of advanced highway soft-base. 基于对高等级公路沉降监测的误差分析，总结适合于高等级公路软基沉降监测的精度指标、监测频率、监测原则和质量保证措施。

monitoring period

【释义】 监测周期

【例句】 Monitoring period could be divided into preparing stage, construction stage and early stage of running, and a determing method of stages was derived. 监测时段分为施工前期、施工建设期和运营初期，给出了各时段的确定方法。

monitoring section

【释义】 监测断面

【例句】 Monitoring sections have been set in Wuhu Section of Yangtze River to measure the environment index factors in this region. 通过在长江芜湖段设置监测断面进行实验测量，得出地表水的环境指标因子。

monoclinal structure

【释义】 单斜构造

【例句】 In the ore district, the tectonic is simple,

monoclinal structure

which is a monoclinal structure, northwest trend, southeast dip. 矿区构造简单，为一总体走向北西、倾向南西的单斜构造。
【中文定义】 如果岩层在一定范围内其倾斜方向和倾角大体是一致的，则称为单斜构造。

monocline ['mɒnə(ʊ)klaɪn]
【释义】 *n*. 单斜
【例句】 Block L35 oil reservoir, which is monocline lithologic deposit, locates in Tiandong depressed area of the eastern of Baise Basin. L35 块油藏位于百色盆地东部田东凹陷地区，属于单斜岩性油藏。
【中文定义】 是地层受到弯曲或折曲的张性地体。单斜与正断层有密切关系，单斜可能直达深处或者沿移动线深入。

monolithic structure
【释义】 整体结构
【中文定义】 是岩体结构类型的一种。岩性坚硬单一，结构面仅发育 1～2 组，不足以把岩体切割成结构体，结构面延展性差、紧闭、面间连接力强的岩体结构类型。这类岩体中地下水活动微弱。多见于构造变动轻微的巨厚层与大型岩体中。

montmorillonite [ˌmɑntməˈrɪlənaɪt]
【释义】 *n*. 蒙脱石
【例句】 Results show the following alkali consumption sequence for those clay minerals: montmorillonite(*S*)> kaolinite(*K*), illite(*I*). 结果表明三种黏土矿物中蒙脱石耗碱最大，高岭石的耗碱量与伊利石的接近。

montmorillonite

monzonite ['mɑnzəˌnaɪt]
【释义】 *n*. 二长岩
【例句】 The Haigou granitoids, located in the northeastern margin of the North China Craton, consists of monzonite and monzogranite. 海沟岩体位于华北克拉通北缘东段，由二长岩和二长花岗岩组成。

monzonite

moraine plain
【释义】 冰碛平原
【中文定义】 指当冰河进入较暖和的地区而消融后，这些被冰河搬运的碎屑物便产生堆积，形成的冰积地形。

moraine plain in front of Yulong Mountain

moraine soil
【释义】 冰碛土
【例句】 At home and abroad, the native soil, the eolian soil, the moraine soil, the diluvial soil, the age-old river sediments and the recent river shoal sediments were researched as material core for ECRD. 国内外实践中曾经用过残积土、风成土、冰渍土、洪积土、古河川沉积及近代河滩沉积土作为土石坝防渗体的心墙土料。
【中文定义】 冰碛地貌的一种，是土壤在冰期被冰裹挟在其中，在间冰期由于气温升高而形成的一种土壤。主要分布在东欧平原和北美五大湖地区。

morphotectonic analysis
【释义】 构造地貌分析法
【例句】 Four main analytical methods are identified in this paper: morphotectonic framework analysis, morphotectonic figure analysis, related sediment analysis and morphotectonic chronology analysis. 文章中归纳了四种主要的构造地貌分析方法：构造地貌格局分析法、构造地貌形态分析法、构造地貌相关沉积分析法和构造地貌年代

分析法。

【中文定义】 一种研究构造地貌的分析方法，可归纳为构造地貌格局分析法、构造地貌形态分析法、构造地貌相关沉积分析法及构造地貌年代分析法等。

mortar rock

【释义】 碎斑岩

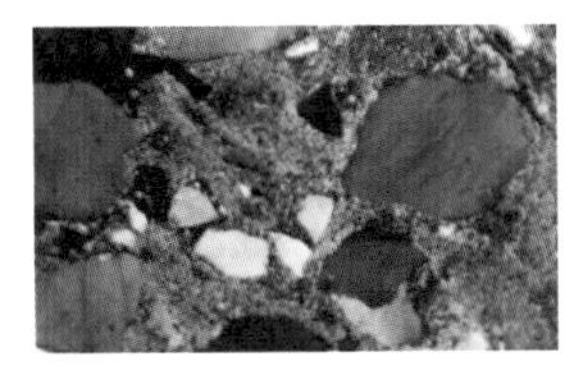

mortar rock

mountain pass

【释义】 山口

【常用搭配】 Tanggula Mountain Pass 唐古拉山口；Kunlun Mountain Pass 昆仑山口

【中文定义】 指高大山脊的相对低凹部分，又称“垭口”或“山鞍”，经常成为通过高大山岭的交通孔道。

mountain pass

mountain range

【释义】 山脉

Tianshan mountain range

【例句】 Many earthquake prone areas in the world are associated with high mountain ranges near to the sea. 世界上许多地震易发区都与近海高山脉有关。

【中文定义】 指沿一定方向延伸、包括若干条山岭和山谷组成的山体，因像脉状且有某种整体性质可以一起称呼，而称之为山脉。主要是由于地壳运动中的内营力作用而形成的。

mountain slope

【释义】 山坡

mountain slope

mountain system

【释义】 山系

【常用搭配】 transverse mountain system 横向山系；longitudinal mountain system 纵向山系

【例句】 Mountain system recharge is the dominant source of recharge to alluvial aquifer system in arid and semiarid mountainous regions. 山系补给是干旱和半干旱山区冲积含水层系统补给的主要来源。

【中文定义】 有成因联系并按一定延伸方向，规模巨大的一组山脉的综合体，多分布于构造带、火山、地震带上，是大地构造作用的产物。

Himalayas mountain system

mountain-along tunnel

【释义】 傍山隧洞

mountain-along tunnel

mountain-cross tunnel

【释义】 越岭隧洞

mountain-cross tunnel

mountainous area

【释义】 山地，山区

【例句】 Most major earthquakes in mountainous areas are associated with widespread landslides arising from an imbalance of forces. 在山区，大多数主地震都与因应力失衡导致的大面积山体滑坡有关系。

【中文定义】 地表面起伏显著、群山连绵交错、大部分地面的倾斜角在 5°～25°之间、一般地面高差在 200m 以上的地区。

mountainous area

moving region of debris flow

【释义】 流通区

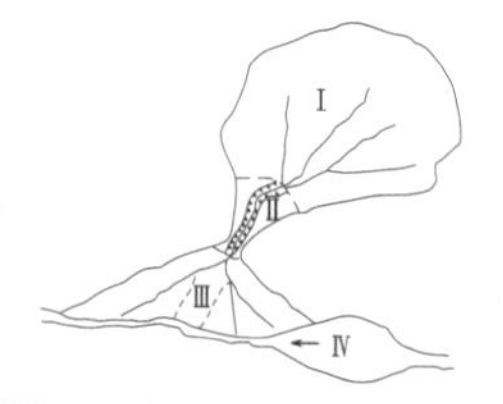

Ⅰ—物源区；Ⅱ—流通区；Ⅲ—沟口堆积区；Ⅳ—主河道

moving region of debris flow (see Ⅱ)

【例句】 Catastrophic process of debris flow is also a mountainous evolution process, which directly affect the changes of Mountain Environment from the origin area, moving region of debris flow to accumulative area. 泥石流的灾变过程也是一种山地演变过程，从源区、流通区到堆积区，直接影响着山地环境的变化。

【中文定义】 泥石流形成后，向下游集中流经的地区。泥石流流通区的地形多为沟谷，有时为山坡，位于泥石流沟谷的中游地段。

mucky soil

【释义】 淤泥质土

【例句】 The investigation into the mechanical properties of field soil-cement mixed with silt or mucky soil can directly provide useful data for the design and construction of soil-cement mixing piles in the Delta of the Pearl River. 研究珠三角地区淤泥、淤泥质黏土形成的水泥土的力学性质可为水泥土搅拌桩的设计和施工提供直接依据。

【中文定义】 天然含水率大于液限、天然孔隙比小于 1.5 但大于或等于 1 的黏性土。

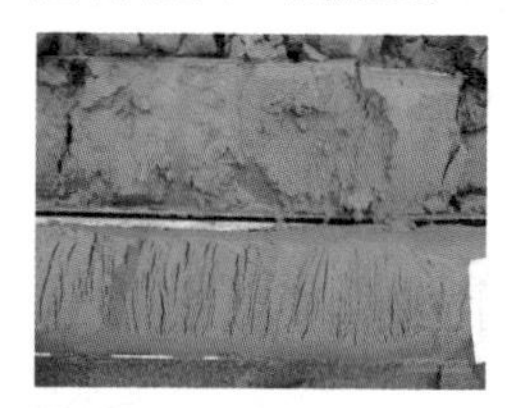

mucky soil

mud [mʌd]

【释义】 *n.* 泥浆

【常用搭配】 mud pit 泥浆池

【例句】 Waste soil and drilling mud should be cleared promptly. 钻探中的废土和钻孔泥浆应及时清理。

mud crack

【释义】 泥裂，干裂

【例句】 Mud cracks are the cracks formed due to

mud crack

unconsolidated sediments exposed to sunlight or dehydration. 龟裂是由于未固结的沉积物暴露于阳光下或脱水收缩形成的裂纹。

【中文定义】 沉积物暴露于水面之上而形成的裂纹，在剖面上呈“V”字形，在平面上形似龟裂纹。

mud desert

【释义】 泥漠

【中文定义】 荒漠的一种，主要由细粒黏土、粉沙等泥质沉积物组成，分布于荒漠中的低洼处，多由湖泊干涸和湖积地面裸露而成。

mud desert

mud inrush

【释义】 突泥

mud inrush

muddy texture

【释义】 泥状结构

【中文定义】 碎屑颗粒大小小于 0.005mm 的碎屑结构。

muddy texture

mudstone [ˈmʌdstəun]

【释义】 *n.* 泥岩

【例句】 The main rocks are sandstone, mudstoneand muddy siltstone or silty mudstone. 岩石类型主要有砂岩、泥岩及泥-粉砂过渡岩类。

【中文定义】 成分复杂、层节理不明显的块状黏土岩。一般以复杂成分泥岩及水云母泥岩为主。

mudstone

multi-hole pumping test

【释义】 多孔抽水试验

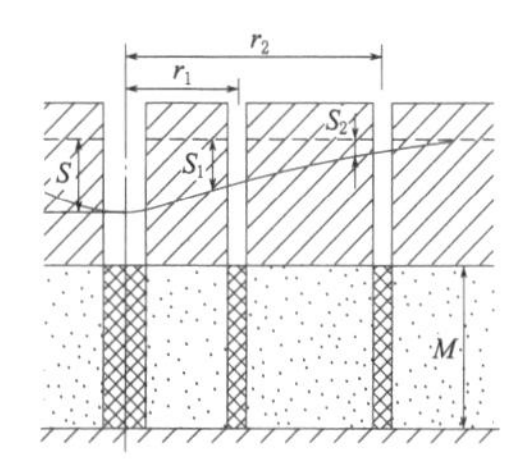

multi-hole pumping test in confined aquifer

multispectral photography

【释义】 多光谱摄影

【例句】 Our proposed scheme overcomes the shortages of traditional multispectral photography, including the system's complexity, high cost, and low feasibility of the “multi wavelength multi-channel” method. 我方推荐的方法克服了传统多光谱摄影中采用“多波多路”方法的系统复杂、造价高和实施困难的缺点。

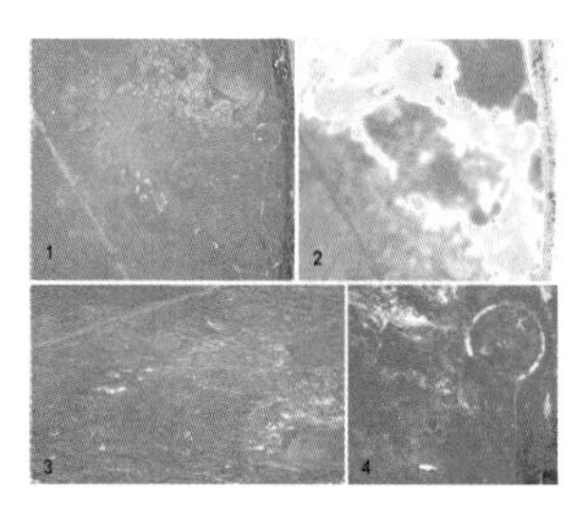

multispectral photography

muscovite [ˈmʌskəˌvaɪt]

【释义】 *n*. 白云母

【常用搭配】 muscovite schist 白云母片岩

【例句】 Other common constituents of granite are muscovite and the dark silicates, particularly biotite and amphibole. 花岗岩其他常见的矿物成分有白云母和暗色硅酸盐矿物，主要为黑云母和角闪石。

muscovite

mushroom rock

【释义】 蘑菇石，风蚀蘑菇

【例句】 Mushroom rocks are deformed in a number of different ways: by erosion and weathering, glacial action, or from a sudden disturbance. 蘑菇石可以通过多种不同的作用形成，包括侵蚀作用、风化作用、冰川作用或者突然的岩体扰动。

【中文定义】 岩石遭受风化之后，因为风、水、生物等对岩石的破坏提供了物质条件，各种外力在运动状态下对地面岩石及风化产物的破坏作用，就称为剥蚀作用。剥蚀作用在破坏山岩外表的同时，也在不断改变着山岩表面的形态，最终形成蘑菇石。

mushroom rock in Kansas

N

natural aggregate
【释义】 天然骨料
【例句】 The difference between recycled aggregate concrete and natural aggregate concrete lies in the application of recycled aggregate in the production of concrete. 再生骨料混凝土与天然骨料混凝土区别就是混凝土生产过程中再生骨料的使用与否。
【中文定义】 采集大自然产生的砂砾石，经筛选分级后制成的混凝土骨料。

natural aggregate

national elevation system 1985
【释义】 1985 国家高程基准

national geodetic coordinate system 1980
【释义】 1980 年国家大地坐标系
【例句】 The construction and application of Chinese astro-geodetic net and national geodetic coordinate system 1980 are of great significance of chinese swrveying industry. 我国天文大地网与 1980 年大地坐标系的建立及应用对我国测量事业具有重大意义。

national geodetic net
【释义】 国家大地测量网
【同义词】 national geodetic network
【例句】 GPS technology, as a new means of measurement, has been universally applied in national geodetic net, city control network, and engineering control network establishment and transformation. GPS 技术作为一种全新的测量手段，在国家大地网、城市控制网、工程控制网的建立与改造等方面已经得到普遍应用。

natural admixture
【释义】 天然掺合料
【中文定义】 指用于混凝土中以改善混凝土性能或减少水泥用量的天然的具有活性的材料。

natural angle of repose
【释义】 天然休止角
【例句】 At last, linear fitting is conducted for four factors, which affect the natural repose angle of sand-sliding slope, like sand grain size, water content, internal repose angle and cohesion by using linear regression theory. The obtained formula can be applied to the estimation of the natural repose angle of sand-sliding slope. 运用线性回归理论对沙粒粒径、含水量、内摩擦角以及黏聚力等 4 个影响沙坡天然休止角的因素进行线性拟合，所得到的公式可用于天然休止角大小的估算。
【中文定义】 无黏性土松散或自然堆积时，其坡面与水平面形成的最大夹角。

natural construction materials
【释义】 天然建筑材料
【中文定义】 指天然产出的可用于水电水利工程建设的砂砾料、土料和石料等。

natural density
【释义】 天然密度
【中文定义】 岩土在天然状态时单位体积的质量。

natural disaster
【释义】 自然灾害
【同义词】 natural hazard
【例句】 The five most prominent natural disasters are floods, windstorms, volcanic eruptions, earthquakes and mass movements(landslides). 五个最突出的自然灾害是洪水、风暴、火山爆发、地震和滑坡。
【中文定义】 指给人类生存带来危害或损害人类生活环境的自然现象。

natural electric field
【释义】 自然电场
【例句】 The analysis through natural electric field and primary field characteristics has obvious effect on working face floor destructive extent monitoring. 利用自然电场和一次场特征进行分析，对于检测工作面底板破坏程度具有明显效果。

natural environmental hydrogeology
【释义】 天然环境水文地质

natural ground

【释义】 原地面线

【中文定义】 指没有被破坏的原始地面在剖面上的投影线。在工程施工中泛指施工之前的原地貌线。

natural isotope

【释义】 天然同位素

【例句】 Scientists have shown for the first time that food enriched with natural isotopes builds bodily components that are more resistant to the processes of ageing. 科学家首次发现富含天然同位素的食品可以构建更抗衰老的机体成分。

【中文定义】 天然的具有相同质子数、不同中子数的同一元素的不同核素互为同位素。

natural sandy gravel material

【释义】 天然砂砾料

【例句】 The first issues are permeability coefficient and permeability stability of natural sandy gravel material and the feasibility of using it as bedding material. 首要的问题是天然砂砾料的渗透系数和渗透稳定性及其作为基础材料的可行性。

natural slope

【释义】 自然边坡

【例句】 Building placement and architecture shall take natural slope and drainage into consideration. 建筑布局和结构应考虑到自然坡度和排水。

natural slope

natural source magnetotellurics

【释义】 天然源大地电磁测深法

natural source surface wave method

【释义】 天然源面波法

nautical chart

【释义】 海图

【中文定义】 以海洋为主要描绘对象的地图。

near-field region

【释义】 近场区

【例句】 Three months before the earthquakes frequency increased and strain release obviously accelerated in the source region, and short-term precursory anomalies was also protruding in the near-field region. 震前 3 个月开始，震源区的频度增加，应变释放加速显著，近场区的短期前兆异常也较突出。

needle-shaped particle

【释义】 针状颗粒

【中文定义】 针状颗粒长度大于该颗粒所属平均粒径 2.4 倍的。

Neoarchean Era

【释义】 新太古代

Neogene (N) [ˈni(ː)əudʒiːn]

【释义】 *n.* 新近纪（系）

【中文定义】 旧称新第三纪，是新生代的第二个纪，生物界的总面貌与现代更为接近，开始于距今 2300 万年，一直延续了 2140 万年。它包括中新世和上新世。新近纪是地史上最新的一个纪，也是地史上发生过大规模冰川活动的少数几个纪之一，又是哺乳动物和被子植物高度发展的时代，人类的出现是这个时代最突出的事件。

neotectonics trace

【释义】 新构造形迹

【例句】 The Liddar Valley of Kashmir is located tectonically at the west of Lesser Himalaya Zone where neotectonics trace is well developed in the Pleistocene sediment of the Karewa Group. 克什米尔利达山谷中的更新世克来瓦群地层发育良好的新构造形迹。

【中文定义】 指新构造运动引起地层发生永久变形而造成各种地质构造形体和地块、岩块相对位移的踪迹。

neotectonics [nioʊtekˈtɒnɪks]

【释义】 *n.* 新构造，新构造学，新构造运动

【常用搭配】 neotectonic element 新构造单元

【例句】 Such data are of paramount importance in the study of neotectonics relating recent plate motions to deformation within continents. 那些数据对于研究近期板块运动和大陆变形的新构造运动至关重要。

【中文定义】 主要指喜马拉雅运动（特别是上新世到更新世喜马拉雅运动的第三幕）中的垂直升降。一般来说，新构造运动隆起区现在是山地或高原，沉降区是盆地或平原。地质学中一般把新近纪和第四纪时期内发生的构造运动称为新构造

运动。

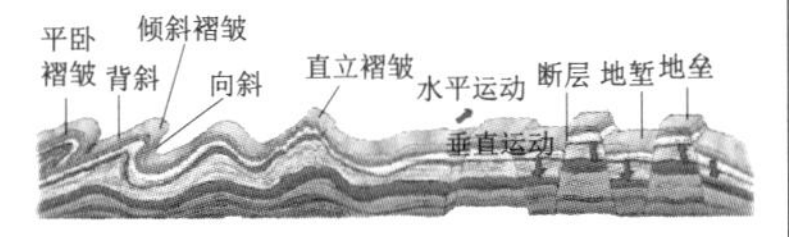

neotectonics

neritic facies

【释义】 浅海相

【常用搭配】 epicontinental neritic facies 陆表浅海相

【例句】 It is suggested that the sandy lenticular bodies of alluvial flat and neritic facies are favourable reservoir zones for ultra shallow biogas exploration. 河漫滩相和浅海相砂质透镜体为超浅层生物气最有利和较有利的储集带。

【中文定义】 海相沉积的一种。海面到海面下200m左右的浅海地区的沉积相。大约相当于海洋大陆架部分。

neritic region

【释义】 浅海区

New Austrian Tunnelling Method

【释义】 新奥法

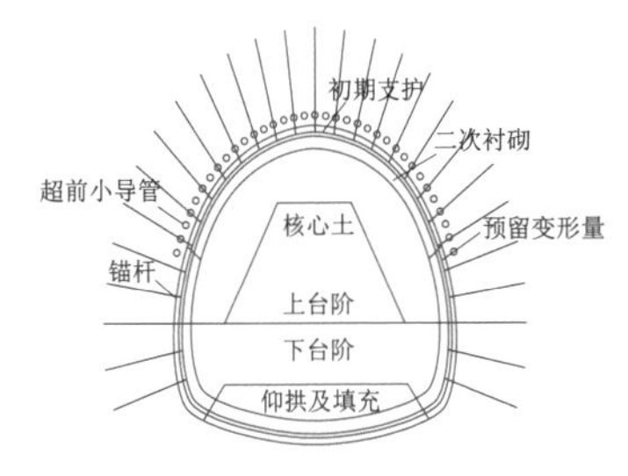

New Austrian Tunnelling Method

non-topographic photogrammetry

【释义】 非地形摄影测量

【中文定义】 不以测制地图为目的的摄影测量。

nonwoven geotextile

【释义】 非织造土工织物

【例句】 Geotextile should divided into woven geotextile, machine woven geotextile, nonwoven geotextile and composite geotextile, etc. considering their fabrication. 土工织物按加工方式分为编织土工织物、机织土工织物、非织造土工织物、复合土工织物等。

nonwoven geotextile

normal consolidated soil

【释义】 正常固结土

normal fault

【释义】 正断层，下落断层

【例句】 In the period 135 - 52Ma, the length of the fault zone increased considerably to 3500km, the depth of penetration reached 30 - 40km and the movement was as a normal fault with a slight component of dextral strike slip. 135～52Ma 时期，该断裂带大幅度地扩展了长度达 3500km，切割深度在 30～40km 之间，为略具右行平移活动的正断层，属地壳断裂。

【中文定义】 地质构造中断层的一种。是根据断层的两盘相对位移划分的。断层形成后，上盘相对下降，下盘相对上升的断层称正断层。它主要是受到拉张力和重力作用形成的。

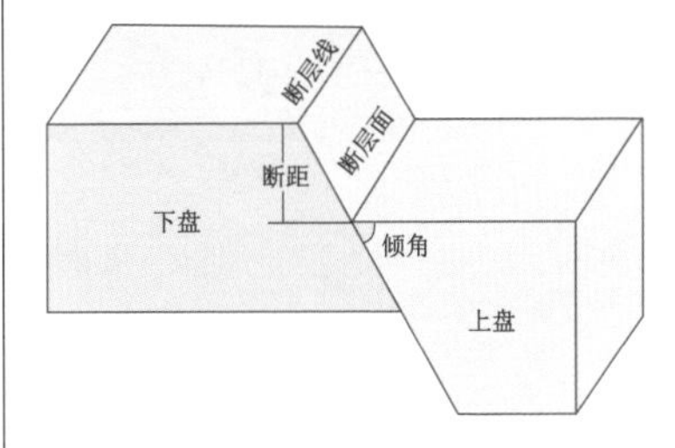

normal fault element diagram

normal water level

【释义】 正常蓄水位

【同义词】 normal storage level; normal water elevation

【例句】 Normal water level is the main characteristic of hydraulic engineering, directly affecting the entire project scale. 正常蓄水位是水电工程的主要特征值，直接影响到整个工程的规模。

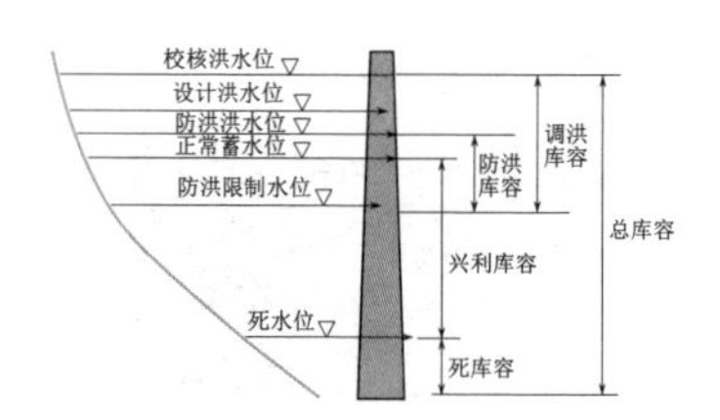

characteristic water level of reservoir

【中文定义】 指水库在正常运行情况下，挡水建筑物允许长期保持的最高库水位，又称为设计蓄水位、正常高水位、最高兴利水位。

numerical method of groundwater resources evaluation

【释义】 地下水资源评价数值法

O

objective function

【释义】 目标函数

【例句】 Characterize the system, identifying important subsystem models and interactions, selecting important design variables, the objective function and constraints. 表现该系统的特性，确定主要子系统模型及其相互作用，选取重要的设计变量，目标函数和约束。

oblique bedding

【释义】 斜层理

【例句】 Sedimentary structures are mainly massive bedding, large trough cross bedding and oblique bedding with few parallel bedding, wave bedding and horizontal bedding. 沉积构造以块状层理、大型槽状交错层理和斜层理为主，见平行层理、波状层理和极少量的水平层理。

【中文定义】 层理基本类型之一，由一系列倾斜层系重叠组成，层系之间界面较平直。

oblique bedding

oblique distance

【释义】 斜距

【例句】 An oblique distance must be greater than a flat distance. 斜距必然是大于平距的。

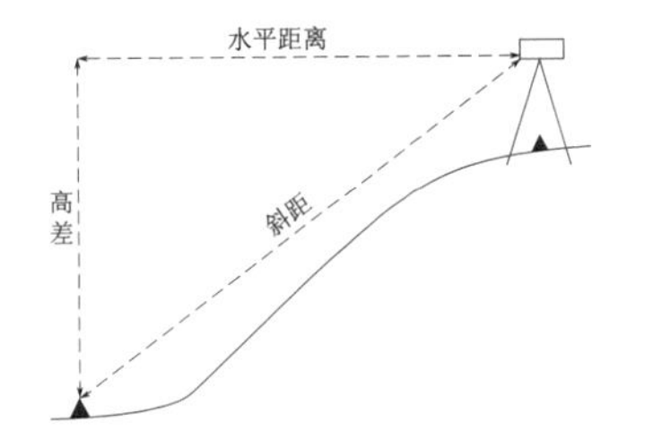

oblique distance

【中文定义】 不在同一高度上的两点之间的距离即为斜距。

oblique fault

【释义】 斜向断层

【例句】 Sudden rupture and displacement occurs with normal, reverse, strike-slip, or oblique-slip faulting. 突然破裂和位移伴随正、逆、走向滑动或斜向滑动断层作用而发生。

【中文定义】 与褶皱轴向或区域构造线走向斜交的断层。

observation array

【释义】 观测台阵

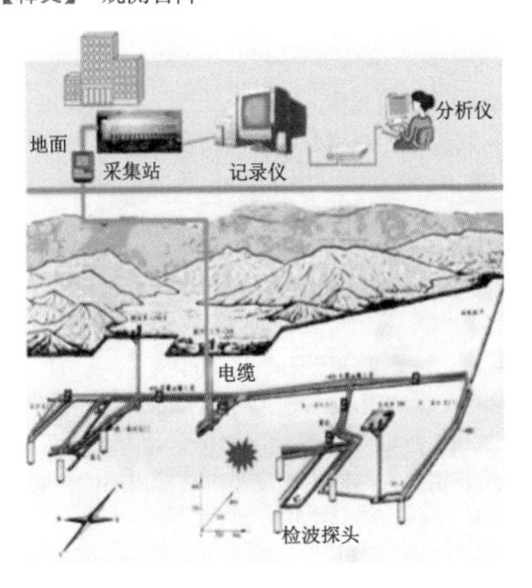

observation array

observation borehole (well)

【释义】 观测孔

【例句】 Observation borehole can be installed at necessary locations where other resources do not exist. 观察孔的位置有必要设置在没有其他资源的地方。

observation point

【释义】 观测点

【常用搭配】 settlement observation point 沉降观测点

【同义词】 observing point

【例句】 During the observation of settlement to the sanitary landfilling which is in high filling of 32 meters, several typical observation points are given and discussed. 在 32m“高填方”沉降观测中，分别选取有代表性的观测点进行讨论，并得出结论。

observation point

observation target

【释义】 测量觇标

【中文定义】 观测照准目标及安置仪器用的测量标架。

office work

【释义】 内业

【同义词】 indoor work

【例句】 In the office work of building GPS engineering control network, the influence to the coordinate of distortion of Guass projection must be deleted. 在 GPS 工程控制网建立的内业处理中，必须设法消去高斯投影变形对坐标成果的影响。

old deposited soil

【释义】 老沉积土

old deposited soil till

【例句】 The mean shaft friction of steel tube pile under compressive load in deep old deposited soil area can reach 76.79 MPa and is about 27.5% to 30%, 6% higher than ordinary clay. 深埋老沉积土区钢管桩在压缩荷载作用下的平均轴摩阻力可达 76.79MPa，约 27.5%～30%，比普通黏土高出 6%。

【中文定义】 是第四纪晚更新世 Q3 及其以前沉积的土层，一般具有较高的强度和较低的压缩性。

Oligocene [ˈɑlɪgoˌsin]

【释义】 *n.* 渐新世（统）

【中文定义】 第三纪的第三个世。

one-dimensional consolidation

【释义】 单向固结

【常用搭配】 Terzaghi's one-dimensional consolidation theory 太沙基-单向固结理论

【例句】 The current calculation methods based on one-dimensional consolidation theory have not met the project's needs. 现有设计理论中的基于单向固结理论的计算方法已无法满足工程要求。

oolitic texture

【释义】 鲕状结构

【例句】 The average content of Gallium in honeycomb texture is higher than that in pisolitic, oolitic texture and massive texture. 蜂窝状矿石中镓的含量明显高于豆状结构、鲕状结构和块状结构的镓含量。

【中文定义】 沉积岩的一种结构，由球形或椭球形颗粒组成，颗粒外形、大小像鱼卵。由鲕体与成分相同的胶结物组成，一般粒径小于 2mm。

opal [ˈopl]

【释义】 *n.* 蛋白石

【常用搭配】 noble opal 贵蛋白石；opalization 蛋白石化；colloform opal 胶状蛋白石；grannular opal 粒状蛋白石

opal

【例句】 The study indicates that the early stage contains a lot of quartz, but the late stage only consists of opal. 研究表明成矿早期，矿石中存在大量的石英，晚期全为蛋白石。

open caisson foundation

【释义】 沉井基础

【例句】 Open caisson foundation is one of built method in deep foundation of soft groundwork. 沉井基础是软土地基中深基础施工方法之一。

open caisson foundation

open well calculation method

【释义】 大井计算法

opening ['oʊpnɪŋ]

【释义】 *v.* 张开，打开，公开，开放

【常用搭配】 opening angle 开度角，孔径角，张角；opening joint rate 张开裂隙率

schematic diagram of tensile crack opening

【例句】 On the basis of analyzing the method for classifying the unloading zones of Emeishan basalts in the project site of a hydroelectric power station in southwestern China, the authors adopt three indexes, that is, opening joint quantity, joint opening degree and wave velocity that have been exercised by many projects as quantitative indexes for classifying the unloading zones. 在分析西南水电工程中峨眉山玄武岩岩体卸荷划分方法的基础上，采用已被大多数工程运用的岩体的张开裂隙条数、裂隙张开度和纵波速度作为卸荷带划分量化指标。

operating basis earthquake (OBE)

【释义】 运行基准地震

【中文定义】 指在工程的服务生命周期中可能合理预期发生的地震，即在工程生命周期中超过50%的发生概率。在该地震作用下，工程的相关性能要求几乎没有或没有破坏，工程的相关功能没有中断。

optical image processing

【释义】 光学图像处理

【例句】 In the field of image processing, there are two completely different methods. One method is optical image processing, and the other is the computer digital image processing. 在图像处理领域，存在着两类完全不同的方法，一类是光学图像处理方法，另一类是计算机数字图像处理方法。

optical resolution

【释义】 光学分辨率

【例句】 The instrument has the advantages of high optical resolution, fast detection speed and high degree of computer control automation. 该仪器具有光学分辨率高、检测速度快、微机控制自动化程度高的优点。

【中文定义】 是用来描述光学成像系统解析物体细节的能力。

optimum moisture content

【释义】 最优含水率

【例句】 The result shows that the wall has a high pore water pressure which is closely related with filling progress and the soil moisture content, and the earth moisture controlled around the optimum moisture content during construction will be benefical to the filling of the wall. 结果表明：心墙在施工期产生的高孔隙水压力与填筑进度以及土料含水量大小密切相关，施工期将土料的含水量控制在最优含水量附近对心墙填筑有利。

【中文定义】 指击实试验所得的干密度与含水率关系曲线上峰值点所对应的含水率。

optimum slope method

【释义】 优定斜率法

【例句】 With the in-situ and indoor direct shear test, the least square method, optimum slope method and other mathematic statistics, shear strength properties of high fill and compaction soil of an airport are studied. 机场高填方压实土的研究中用到了原位直剪试验、室内直剪试验等测试手段以及最小二乘法、优定斜率法等数理统计方法。

【中文定义】 是将同一级别的成果点绘制在关系图上，根据图上散点的总体趋势优定关系曲线斜率的方法。

ordinary microseism

【释义】 普通微震

Ordovician (O) [ˌɔːdəu'viʃən]

【释义】 *n.* 奥陶纪（系）

【中文定义】 古生代的第二个纪，距今约4.8亿～4.4亿年，延续了4200万年。奥陶纪是地史上大陆地区遭受广泛海侵的时代，是火山活动和地壳运动比较剧烈的时代，也是气候分异、冰川发育的时代。

organic soil

【释义】 有机土，有机质土

【例句】 The organic soil must be stored outside the yard in the selected sites, to be used in the stabilization of area, slope protection, or other required activities. 有机土壤必须存放于选定位置的场地，以用于场区稳定、边坡防护、或其他要求的活动。

【中文定义】 当黏土中有机质含量超过5%以及砂土中有机质含量超过3%时，则成为含有有机质的土。含有机质臭味、黑色暗色的有机土。

organic soil

organic texture

【释义】 生物结构

【例句】 The indicator texture of direct biomineralization is biological organic colloid texture and the indicator texture of biomineralization is "mineralized biological". 生物直接成矿作用的标志性结构为生物有机胶体结构，生物间接成矿作用的标志性结构为“矿交代生物”结构。

【中文定义】 生物化学岩的一种结构。具此种结构的岩石，其内部所含的生物骨骼需达30%以上。

origin of control network

【释义】 控制网原点

orogen ['ɔrədʒɪn]

【释义】 *n.* 造山带

【例句】 Formation of an orogen is accomplished in part by the tectonic processes of subduction (where a continent rides forcefully over an oceanic plate(non collisional orogens)) or convergence of two or more continents(collisional orogens). 造山带形成的部分原因是构造过程中两个或多个板块之间发生潜没（指地壳的板块沉到另一板块之下）或者聚合。

【中文定义】 是地球上部由岩石圈剧烈构造变和其物质与结构的重新组建，使地壳挤压收缩所造成的狭长强烈构造变形带，往往在地表形成线状相对隆起的山脉。一般与褶皱带、构造活动带等同义或近乎同义。

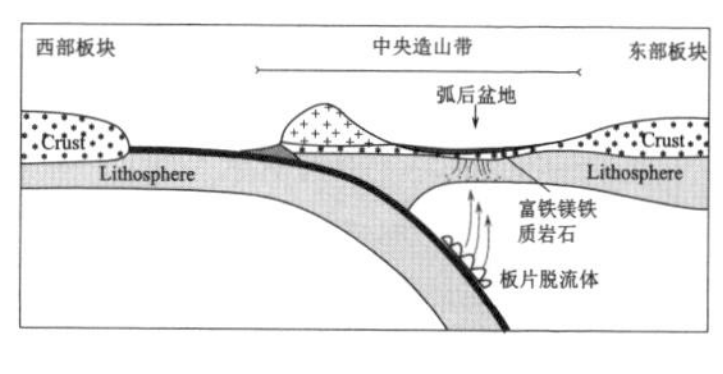

orogen

orogeny [ɔ'rɑdʒəni]

【释义】 *n.* 造山运动

【常用搭配】 Grenville Orogeny 格林威尔造山运动；Wichita Orogeny 威奇塔造山运动；Cadomian Orogeny 卡多姆造山运动

【例句】 In addition to orogeny, the orogen(once formed) is subject to other processes, such as sedimentation and erosion. 除了造山运动，造山带的形成（一次形成）还要受到其他进程的影响，比如沉积和侵蚀。

orogeny

【中文定义】 指地壳局部受力、岩石急剧变形而大规模隆起形成山脉的运动，仅影响地壳局部的狭长地带。

orthoclase [ˈɔːθəkleɪz]

【释义】 *n.* 正长石

【常用搭配】 natron orthoclase 钠正长石

【例句】 Rock-forming mineral include olivine, pyroxene, hornblende, orthoclase feldspar, plagioclase, mica, quartz, biotite, calcite and other common metal and non-metal minerals. 造岩矿物包括橄榄石、长石、角闪石、辉石、云母、石英、方解石和其他常见金属和非金属矿物。

orthoclase

orthometamorphite [ɔːθɒmɪtəˈmɔfaɪt]

【释义】 *n.* 正变质岩

【例句】 Orthometamorphite is formed by metamorphism of igneous rock. 正变质岩是由火成岩受变质作用而形成的。

【中文定义】 由火成岩遭受变质作用形成的变质岩称为正变质岩

orthometamorphite

outcrop of groundwater

【释义】 地下水露头

【例句】 The Jadeite Spring is a typical acidic karst water cold spring. The flow velocity of the water in the Jadeite Spring aquifer is slower than that in the Pearl spring, and its water hence has relatively long detained time. It contains a certain amount of CO_2 and is an outcrop of groundwater of deep aquifer. 矿泉水厂沟的翡翠泉是典型的酸性岩溶冷泉，含水层水的径流较珍珠泉慢且滞留时间相对长些，含有一定的 CO_2，系一深部含水层地下水露头。

【中文定义】 指地下水流出地表的部分。

outline drawing

【释义】 轮廓图

【常用搭配】 outline drawing of building 建筑轮廓图

outline of investigation program

【释义】 勘测大纲

outline of structure

【释义】 建筑物轮廓线

【例句】 Aiming to deal with the problem that manual extraction of contours of buildings on oblique view maps are expensive and ineffective with low accuracy and coarse detail, we present a method of automatic extraction of outline of structure on oblique view maps which based on 3D city models. 针对城市侧视地图上建筑物轮廓线主要依靠人工交互提取，成本高昂、效率低下且精细度和准确度有限的问题，提出了一种任意视角侧视地图中建筑物轮廓线的自动提取方法。

【中文定义】 指建筑外墙面水平投影的外轮廓线。

outwash drip

【释义】 冰水沉积

【例句】 Sometimes the animal multitudes are horizontal, as on Salisbury Plain, a glacial outwash drip delta densely colonized by king penguins, fur and elephant seals, and kelp gulls. 有些地方动物种群是水平分布的，比如在索尔兹伯里平原，这块冰水沉积三角洲密集分布着王企鹅、软毛海豹、象海豹和黑背海鸥。

【中文定义】 又称冰水停积，是以冰川融水为主要营力，由砾石和沙粒组成的沉积物。

outwash drip

oven drying method

【释义】 烘干法

【例句】 The test results from oven drying method show the maximum error is less than 2%, which can meet the need of project. 烘干法的试验结果表明最大误差小于2%，满足工程需要。

overall stability

【释义】 整体稳定性

【中文定义】 自然或工程边坡整体、宏观的稳定性状，一般受控于坡体岩性、结构面和临空面等性状。

overburden landslide

【释义】 覆盖层滑坡

overburden landslide

overburden layer

【释义】 *n.* 覆盖层

【常用搭配】 overburden pressure 积土压力

【例句】 The deep overburden layer will also appear horizontal displacement to upper and down streams after dam construction. 受坝体的影响，坝基深厚覆盖层也会向上、下游发生水平位移。

【中文定义】 泛指基岩上较新的、松散的沉积物，一般指第四纪未固结的沉积物。其厚度随地形和成因而异。

overburden layer

overconsolidated soil

【释义】 超固结土

【例句】 An approach to calculate the in-situ one-dimensional consolidation of slightly overconsolidated soils is proposed in this paper. 本文给出了一种计算轻度超固结土的一维现场固结的方法。

overconsolidation ratio

【释义】 超固结比

overlaid minerals

【释义】 压覆矿产

【中文定义】 指在当前技术经济条件下，因建设项目或规划项目实施后，导致已查明的矿产资源不能开发利用。

oversize aggregate

【释义】 超径骨料

【例句】 A small amount of oversize aggregate is allowed to be mixed in the concrete aggregate. 混凝土骨料中允许有少量超径骨料的混入。

【中文定义】 超径骨料是粒径超过标准范围的骨料。

overturned anticline

【释义】 倒转背斜

【例句】 The Houjia-Songshan overturned anticline and NNE-trending compresso-shear fault belts(F_1, F_2 etc.) are the main rock and ore-controlling structures in the Yongping copper ore field. 侯家-嵩山倒转背斜和北北东向压扭性断裂带（F_1、F_2等），是永平铜矿田的主要控岩控矿构造。

overturned fold

【释义】 倒转褶皱

【例句】 While the first episode of Zunhua age was appressed linear overturned fold stretching for NE, owing to the limit of boundary condition, the fold superimposed on Qianxi Group turned their axes to near SN. 遵化期第一幕为北东向摆动的紧密线形倒转褶皱。由于边界条件的限制，叠加在迁西群上的褶皱，轴向转为近南北向。

【中文定义】 倒转褶皱是轴面倾斜、两翼倾向相同的褶皱。对两翼岩层而言，一翼产状正常，地层顶面向上，称正常翼；另一翼产状倒转，地层顶面向下，称为倒转翼。

overturned fold

overturning stability

【释义】 倾覆稳定性

【中文定义】 是指在自重和外载荷作用下抵抗倾覆的能力。

overthrust

【释义】 逆掩断层

【同义词】overthrust fault; low-angle thrust

【例句】 By the analysis and study of a special kind of overthrust fault occurred in the lower layer B2 of B coal group in Furong Mine, a new method for quantitative prediction, the method of elevation difference value of relative bed interval, is presented. 通过对芙蓉煤矿B煤组下分层B2煤层中出现的一种特殊逆掩断层的分析和研究，提出了一种新的定量预测预报方法——相对层间距高差值法。

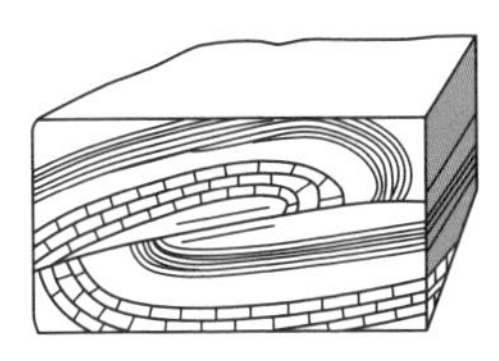

overthrust

【中文定义】 逆掩断层是地质构造名称。断层有多种，随命名的依据而异。按断层两盘相对滑动方向的差异可分为正断层与逆断层。而在逆断层中断层面倾角平缓，断层面倾角小于25°者，称为逆掩断层。

oxbow lake

【释义】 牛轭湖，牛角湖，河迹湖

【例句】 Oxbow lakes are common landforms in meandering river floodplains. 牛轭湖是蜿蜒河流漫滩的常见地形。

【中文定义】 河流在平原地区河曲发育，随着流水对河面的冲刷与侵蚀，愈来愈曲，最后导致河流自然裁弯取直，原来弯曲的河道被废弃，形成湖泊。因这种湖泊的形状恰似牛轭，故称之为牛轭湖。

oxbow lake

P

Pacific Plate
【释义】 太平洋板块
【常用搭配】 East Pacific plate 东太平洋板块
【例句】 The Pacific Plate contains an interior hot spot forming the Hawaiian Islands. 太平洋板块包含一个内部热点形成夏威夷群岛。
【中文定义】 是一块大部分位于太平洋海面下的海洋地壳板块。它是法国地质学家勒皮雄 1968 年首次提出的六大板块之一，自提出以来，其范围基本没有大的变动。

Paleoarchean Era
【释义】 古太古代

Paleocene ['peliəˌsin]
【释义】 *n.* 古新世（统）
【中文定义】 第三纪的第一个世。

paleochannel
【释义】 古河槽，古河道，埋藏河道
【例句】 The paleochannel sandstone type uranium deposit which can be exploited by low cost in situ leach technology is one of the most economic and competitive uranium resources in recent uranium industry. 可采用低成本、原地浸出工艺开采的古河道砂岩型铀矿是当今铀工业中最具经济价值和竞争力的铀资源之一。
【中文定义】 地质历史时期河流改道或因其他原因被废弃后遗留的河床冲刷槽。

paleochannel

paleo-earthquake
【释义】 古地震
【例句】 All the results show that microscopic analysis is an effective method for identifying obscured fault and paleo-earthquake event in soft sediments. 上述结果表明，微观分析是探讨松散沉积物中形迹不明断层和古地震事件的有效手段。
【中文定义】 人类历史记载以前所发生的地震。广义上，古地震应包括整个地质历史时期中所发生的地震，但因向前追溯的历史越长，与现今地震活动的关系越小，所以更有现实意义的是第四纪以来，特别是全新世以来所发生的地震。因此也有人将古地震理解为第四纪以来发生的史前地震。

paleoenvironment [ˌpeiliːəuin'vaiərənment]
【释义】 *n.* 史前环境，古环境
【常用搭配】 paleoenvironment reconstruction 古环境重建；paleoenvironment interpretation 古环境解释
【例句】 Evolution of desert is one of important aspect of Quaternary paleoclimate and paleoenvironment research. 沙漠形成演化是第四纪古气候、古环境研究的重要内容之一。
【中文定义】 随着时间的推移，地球上的大陆在漂移，环境也在变化，地球曾有过的环境，就叫古环境。

Paleogene (E) ['peiliːəudʒiːn]
【释义】 *n.* 古近纪（系）
【中文定义】 旧称早第三纪，是新生代的第一个纪，开始于约距今 6500 万年，结束于 2300 万年，延续了约 4247 万年，包括古新世、始新世和渐新世。

paleokarst
【释义】 古岩溶，古喀斯特
【常用搭配】 paleokarst unconformity 古岩溶不整合面；paleokarst breccia 古岩溶角砾岩；paleokarst reservoir 古岩溶储层

paleokarst

【同义词】 ancient karst
【例句】 The secondary dolomite and paleokarst of the Lower Permian Series in Sichuan Basin is one of the important atectonic reserviors in the basin. 四川盆地下二叠统中次生白云岩及古岩溶是下二叠统主要的非构造圈闭储层之一。
【中文定义】 非现代营力环境下形成的岩溶。如中国热带岩溶分布范围远远超过现在的热带，它们便是古岩溶的现象。

Paleozoic (Pz) [ˌpæliːəu'zəuik]
【释义】 *n.* 古生代（界）
【中文定义】 是显生宙的第一个代，包括寒武纪、奥陶纪、志留纪、泥盆纪、石炭纪、二叠纪，持续约 3 亿年。其中寒武纪、奥陶纪、志留纪又合称早古生代，泥盆纪、石炭纪、二叠纪又合称晚古生代。

palimpsest texture
【释义】 变余结构
【中文定义】 又称残留结构，指变质岩中由于重结晶作用不完全，仍然保留的原岩结构。

palimpsest texture

pangaea [pæn'dʒiə]
【释义】 *n.* 泛大陆，盘古大陆
【常用搭配】 pangaea continent 原始大陆；pangaea cycle 联合古陆时期

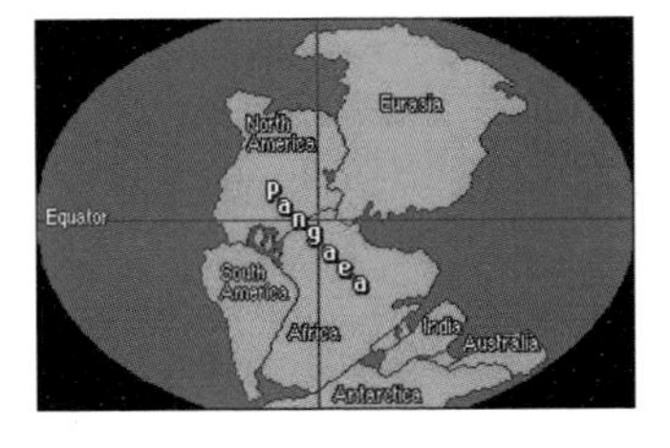

pangaea

【例句】 He theorized that pangaea split apart during Mesozoic time and that the continents then moved steadily in all directions to their present positions. 他的理论认为，泛大陆分裂开始于中生代，各个板块沿一定的方向运动至当前位置。
【中文定义】 大陆漂移说认为，晚古生代时期全球所有大陆连成一体的超级大陆。中生代以来逐步解体，形成现今的大陆、大洋。

panorama of geological drilling
【释义】 钻孔全景数字成像

panthalassa [pænθə'læsə]
【释义】 *n.* 泛大洋，原始大洋
【例句】 Panthalassa had a western prolongation, called Thetys, inserted like a wedge between Eurasia to the north, and Africa, India and Australia to the south. 泛大洋西段的延伸被称为特提斯洋，楔形的在北部嵌入欧亚大陆，在南部嵌入非洲、印度和澳大利亚之间。
【中文定义】 又称泛古洋、盘古大洋，在希腊文中意为"所有的海洋"，是个史前巨型海洋，存在于古生代到中生代早期，环绕着盘古大陆。

parallel section method
【释义】 平行断面法
【例句】 In the process of engineering exploration, parallel section method is suitable for the borrow area of terrain fluctuation and uesful material layer thickness change. 在工程勘探过程中，平行断面法适用于地形有起伏、有用层厚度有变化的料场。
【中文定义】 指相邻两断面面积平均值乘以断面间平均距离，求出两断面间的分段储量，然后总和各分段的储量的方法。

parametamorphic rock
【释义】 副变质岩
【例句】 The Pengjiakuang gold deposit is a new type discovered in east Shandong, which occurs in interstratified glide breccia of the Jinshan Group parametamorphic rock series in the Mesozoic Jiaolai Basin. 蓬家夼金矿是山东乳山地区新发现的金矿类型，受荆山群中的构造角砾岩系控制，沿胶莱盆地边缘莱阳组砾岩与荆山群副变质岩的构造接触带分布。
【中文定义】 沉积岩遭受变质作用后形成的变质岩称为副变质岩。

partially penetrating borehole (well)
【释义】 非完整孔（井）
【例句】 A mathematical model is developed for 3D unsteady flow in confined aquifer subjected simultaneously to pumping and recharge by a sin-

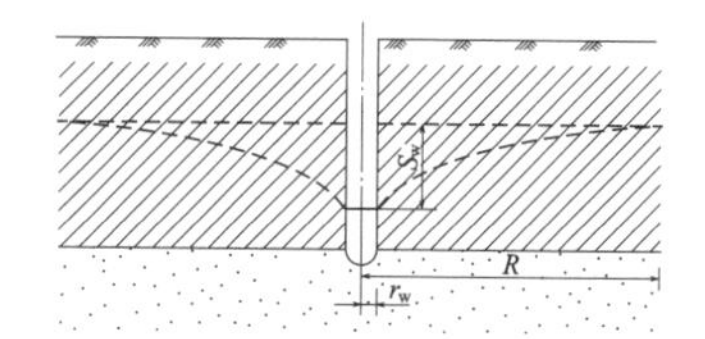

partially penetrating well in confined water

gle partially penetrating well. 对承压含水层在抽-灌同轴非完整井的抽、灌水作用下产生的地下水三维非稳定流建立了数学模型。

particle shape

【释义】 颗粒形状

【例句】 Because the particle shape of MCC is irregular, it causes bad flowability, the variation of tablets weight, and the poor disintegration properties. 然而由于微晶纤维素颗粒形状的不规则，使得其粉末的流动性不佳，造成锭片重量上的差异。

【中文定义】 一个顺粒的轮脚边界或表面上各点的图像及表面的细微结构。

particle shape

particle size analysis

【释义】 颗粒分析试验

particle size analysis

【例句】 Applying the least square method and parabolic interpolation to particle size analysis of soil, this paper gives the method and procedure of computer processing of experimental data. 本文将最小二乘法和抛物线插值法应用于土壤颗粒分析试验，给出了试验数据的计算机处理方法和步骤。

particle size distribution curve

【释义】 颗粒大小分布曲线

particle texture

【释义】 颗状结构

【中文定义】 按颗粒粒径划分的结构。

passive net

【释义】 被动网

【中文定义】 由钢丝绳网、环形网、固定系统、减压环和钢柱四个主要部分构成。钢柱和钢丝绳网连接组合构成一个整体，对所防护的区域形成面防护，从而阻止崩塌岩石土体的下坠，起到边坡防护作用。

passive net

peak acceleration

【释义】 地震峰值加速度

【例句】 The equivalent viscous-elastic model was used to calculate and analyze 2D seismic responses of the Mengyejiang Concrete Faced Rockfill Dam under the conditions of earthquake peak acceleration of 0.146g. 采用等价黏弹性模型，对勐野江面板堆石坝进行了地震峰值加速度为 0.146g 的二维地震响应计算分析。

peak cluster

【释义】 峰丛

【例句】 Status and evaluation of water quality in Karst peak cluster areas of Guangxi. 广西峰丛岩溶区水质现状与评价。

【中文定义】 是一种连座峰林，基部完全相连，顶部为圆锥状或尖锥状的山峰。峰丛多分布于碳酸盐岩山区的中部，或靠近高原、山地的边缘部分。

peak cluster

peak forest

【释义】 峰林

【例句】 The Wulingyuan peak forest is a newly recognized type of peak-forest landform. 武陵源峰林是一种新被认识的峰林地貌类型。

【中文定义】 高耸林立的碳酸盐岩石峰，分散立和丛聚两类。从聚的连座峰林又称峰丛。峰林主要发育在湿润热带、亚热带，它是热带喀斯特的代表。

peak forest

peak forest landform

【释义】 峰林地形

【中文定义】 峰丛、峰林、孤峰及溶丘总称峰林地形，它们是岩溶地区的正地形，都是在高温多雨的湿热气候条件下，长期岩溶作用的产物。

peak forest landform in China

peak ground acceleration (PGA)

【释义】 地震动峰值加速度

【例句】 As the intensity index of ground motion, PGA(peak ground acceleration) is often employed and analyzed. 作为地震动强度指标，峰值加速度经常被运用和分析。

peak strength

【释义】 峰值强度

【例句】 In the biaxial test, peak strength and shape coefficient can be fitted with linear functions well, and the internal friction angle increases with the decrease of shape coefficient. 在双轴试验中，材料的峰值强度与形状系数的变化规律可用线性函数很好地进行拟合，内摩擦角随形状系数的减小而增大。

【中文定义】 是岩石（体）抗剪断试验的剪应力-剪位移关系曲线上最大的剪应力值。

pearly luster

【释义】 珍珠光泽

【例句】 Minerals such as talc, chlorite, biotite, and muscovite, have pearly luster. 滑石、绿泥石、黑云母和白云母等矿物呈珍珠光泽。

【中文定义】 具有最完全解理的透明矿物，由于光线通过几层解理面的连续反射和互相干涉，呈现与珍珠相似的光泽

peat [pit]

【释义】 *n.* 泥煤，泥炭

【例句】 Thick accumulations of peat can form only if the swamp basin slowly subsides. 只有当沼泽盆地慢慢下陷时，厚厚的泥炭堆积才能形成。

peat

【中文定义】 是一种经过几千年所形成的天然沼泽地产物（又称为草炭或是泥煤），是煤化程度最低的煤，同时也是煤最原始的状态，无菌、无毒、无污染，通气性能好。质轻、持水、保肥、有利于微生物活动，增强生物性能，营养丰富，既是栽培基质，又是良好的土壤调解剂，并含有很高的有机质，腐殖酸及营养成分。但是泥炭属于不可再生资源，开采行为对环境破坏很大。

pegmatite vein

【释义】 伟晶岩脉

【例句】 This paper reports in situ U-Pb ages and Hf isotopic data on detrital zircons from muscovite quartz schist, garnet amphibolite, garnet-bearing paragneiss and pegmatite vein of the Dunhuang Group, which belong to metamorphic basement of NE marginal Tarim basin by the LA-MC-ICP-MS. 本文采用 LA-MC-ICP-MS 手段对敦煌地块中敦煌群的白云母石英片岩、石榴斜长角闪岩、石榴黑云斜长片麻岩和长英质伟晶岩脉中的锆石进行了 U-Pb 和 Lu-Hf 同位素分析。

pegmatite vein

pegmatite ['pɛgməˌtaɪt]

【释义】 *n.* 伟晶岩

【常用搭配】 granite pegmatite 花岗伟晶岩; pegmatite type 伟晶岩型

【例句】 On the basis of the achievements, we hold that pegmatite formed from post magmatic hydrothermal solution after the formation of the Linglong granitoid. 综合研究成果，本文认为伟晶岩形成于玲珑花岗岩岩浆期后热液。

pegmatite

peneplain ['pinəˌplen]

【释义】 *n.* 准平原

【常用搭配】 dissected peneplain 切割准平原; uplifted peneplain 上升准平原

【中文定义】 隆起地面经长期侵蚀、剥蚀而成的地面起伏平缓的平原。

penetration resistance

【释义】 贯入阻力

【常用搭配】 specific penetration resistance 比贯入阻力; standard penetration resistance 标准贯入阻力

【例句】 A new kind of miniature cone penetrometer was developed in order to acquire automatically the cone penetration resistance and depth in MCPT. 研制了一种新型的微型触探仪，该仪器能对贯入阻力与贯入深度数据进行自动采集。

penetration test

【释义】 渗透试验

【常用搭配】 penetration test in laboratory 室内渗透试验

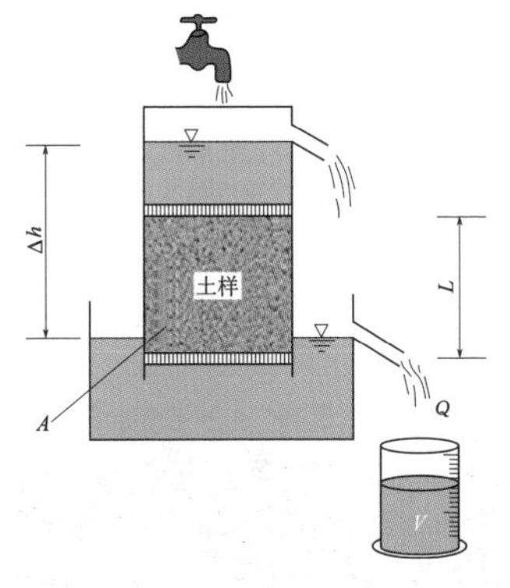

penetrating test

【例句】 On the basis of the penetration test, the penetrating condition of Puhe Reservoir is studied in detail by means of an analytical method of the engineering geology. 本文在渗透试验的基础上，用工程地质分析法，对瀑河水库的渗漏条件进行了详细的分析。

percent passing

【释义】 过筛百分率

【例句】 The percent passing of the material directly reflects the screening of the material in the sieve. 物料过筛率直接反映了物料在筛筒内的筛分情况。

【中文定义】 在给定的时间内物料通过筛子的量，以质量单位表示或装料量的百分比表示。

percentage of open area of screen

【释义】 过滤器骨架管孔隙率

perched water

【释义】 上层滞水，栖留水

【例句】 It shall reckon in tilting movement resulting from perched water in the stability analysis of tilting failure of bedded rock slope, thus it can only explain many irregular phenomena of topping mass. 层状岩石边坡倾倒破坏稳定分析中，应计入上层滞水产生的倾倒力矩，这样才能解释倾倒体的许多不规则现象。

【中文定义】 简称上滞水，是指包气带内局部隔水层之上积聚的具有自由水面的重力水，由雨水、融雪水等渗入时被局部隔水层阻滞而形成，消耗于蒸发及沿隔水层边缘下渗。

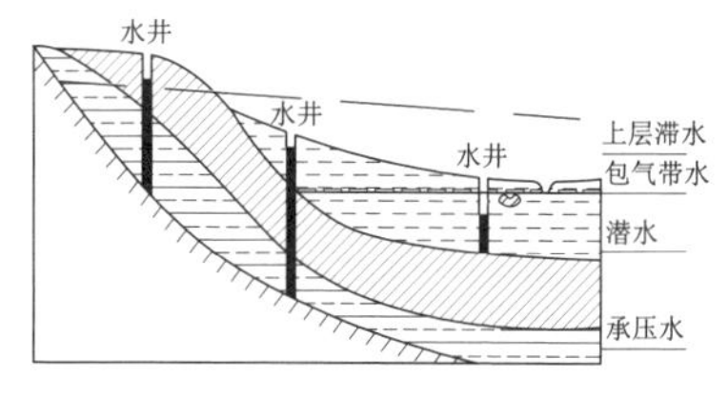

perched water

percussion drilling

【释义】 冲击钻进

【例句】 The percussion drilling, as the important part in enlarging the service field of the current exploration engineering, is taken as the main method in water well drilling. 冲击钻进作为当前工程勘探拓宽服务领域的重要组成部分，被作为水井钻探的主要手段。

percussion-rotary drilling

【释义】 冲击回转钻进

【例句】 In recent years, percussion-rotary drilling technique has gained a rapid development. 近年来，冲击回转钻进技术得到了较快的发展。

peridotite [ˌperɪ'dotaɪt]

【释义】 *n.* 橄榄岩

【常用搭配】 mica peridotite 云母橄榄岩

peridotite

【例句】 More and more attention has been paid to peridotite Re-Os isotopic system as this system behaves as an important method to probe the formation age of the lithospheric mantle. 橄榄岩的Re-Os同位素体系作为探讨岩石圈地幔形成年龄的重要手段已受到人们越来越多的重视。

Period ['pɪrɪəd]

【释义】 *n.* 纪

【常用搭配】 Cambrian Period 寒武纪；Devonian Period 泥盆纪

【中文定义】 第三级地质年代单位，是根据全球生物界新生的最高级生物门类的初次大量发展为依据进行划分的。

peripheral stress

【释义】 周边应力，周界应力

【例句】 The method of photo elasticity is discussed in measuring peripheral stress concentration of rectangular ports on the shape shell under the axial compression. 在轴向压缩下，台形壳体上方形孔周边的应力集中问题，用光测弹性方法测量。

【中文定义】 开挖空间外侧四周的应力。

permafrost ['pɜːrməfrɔːst]

【释义】 *n.* 多年冻土，永冻土

【例句】 The Qinghai-Tibet railway is the first railway in permafrost region which built on Qinghai-Tibet plateau in China. 青藏铁路是中国在青藏高原修建的第一条多年冻土区铁路。

【中文定义】 又称永久冻土或多年冻土，是指持续多年冻结的土石层。可分为上下两层：上层每年夏季融化，冬季冻结，称活动层，又称冰融层；下层长年处在冻结状态，称永冻层或多年冻层。土层的冻融变化是土木工程建设中必须考虑的重要因素，处置不当将带来严重后果。

permafrost

permanent leakage

【释义】 永久渗漏

【中文定义】 指库水沿库边和库盘的透水岩带渗漏至低邻谷、洼地和下游，如果渗漏严重将影响正常蓄水位选择和水库效益，并可引起浸没、边坡塌滑等环境工程地质问题。通常提到水库渗漏时，均指永久性渗漏。

permanent support

【释义】 永久支护

permanent support

permanent works

【释义】 永久工程

permanganate index

【释义】 高锰酸盐指数

【例句】 The quality of the water from Beijing to Tianjin is worse than Grade V. The key pollution indicators are ammonia nitrogen, petroleum, potassium permanganate index and volatile phenol. 北京入天津境内水质为劣Ⅴ类。主要污染指标是氨氮、石油类、高锰酸盐指数、挥发酚。

【中文定义】 指在一定条件下，以高锰酸钾（$KMnO_4$）为氧化剂，处理水样时所消耗的氧化剂的量。表示单位氧的毫克/升（O_2，mg/L）。

permeability [ˌpə-miəˈbɪlɪti]

【释义】 *n.* 渗透性，渗透率，磁导率

【常用搭配】 permeability anisotropy 渗透率各向异性；hydraulic permeability 水力渗透性；permeability coefficient 渗透系数

【例句】 It is feasible to determine permeability of ordinary concrete by measuring electric conduction of the concrete by alternating current. 用交流电测量普通混凝土的电导并以此反映混凝土的渗透性是可行的。

【中文定义】 一种材料在不损坏介质构造情况下，能使流体通过的能力。

permeability coefficient

【释义】 渗透系数，水力传导系数

【同义词】 hydraulic conductivity, filtration coefficient

【例句】 The water inflow increases with the change of permeability coefficient of surrounding rock mass or water depth above surrounding rock. 隧道洞内涌水量随着围岩渗透系数或围岩上覆海水深度的增大呈线性增大。

【中文定义】 地基土的防渗标准多采用渗透系数（*K*，单位 cm/s）表述，防渗标准一般要求渗透系数降低到 10^{-4}cm/s 量级以下。在各向同性介质中，它定义为单位水力梯度下的单位流量，表示流体通过孔隙骨架的难易程度，其值与介质和液体的性质有关。影响渗透系数大小的因素很多，主要取决于土体颗粒的形状、大小、不均匀系数和水的黏滞性等，要建立计算渗透系数 *K* 的精确理论公式比较困难，通常可通过试验方法（包括实验室测定法和现场测定法）或经验估算法来确定 *K* 值。

permeability index

【释义】 渗透性指标

【例句】 Permeability coefficient is an important permeability index of rock and soil. 渗透系数是岩土体一个重要的渗透性指标。

【中文定义】 反映岩（土）体渗透性的相关指标。

permeability of rock and soil

【释义】 岩土渗透性

【例句】 The permeability of rock and soil is graded according to the permeability coefficient or water permeability. 岩土渗透性按渗透系数或透水率进行分级。

【中文定义】 水在岩石或土体中渗透流动的性能。

permeability parameter

【释义】 渗透性参数

【中文定义】 用于评价地下水渗透性的各种变量。

permeable linking

【释义】 渗透联系

permeable rate (*q*)

【释义】 透水率

【例句】 The grouting effect has met the design requirement that the permeable rate is less than 0.044 Lu. 采用这种工艺后的灌浆效果完全满足透水率 $q<0.044$Lu 的设计要求。

【中文定义】 岩石地基的防渗标准采用钻孔压水试验成果来表示，压水试验成果又以透水率 *q* 来表示，单位为吕荣（Lu）。定义为：压水 *p* 为 1MPa 时，每米试段长度 *L*（m）每分钟注入水量 *Q*（L/min）为 1L 时，称为 1Lu。计算公式为：$q=Q/(p\cdot L)$。国内外岩石地基工程防渗标准一般在 1～5Lu，特殊情况高标准可达 0.5Lu，低标准为 10Lu。

permeable stratum

【释义】 透水层

【例句】 The amplitude of the head fluctuation decreases with the storativity and leakage of both semi-permeable stratum and increases with the leakance of the silt-layer. 分析表明，承压含水层中地下水水头波幅是上、下弱透水层贮水率和越流系数的减函数，是淤泥层相对透水系数的增函数。

【中文定义】 是指动水流能够透过的土层或地层。

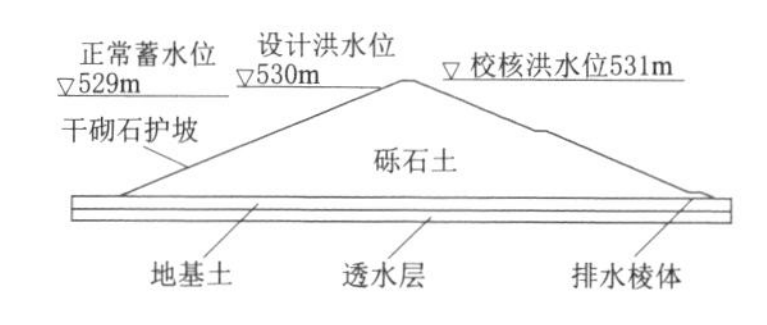

permeable stratum under the foundation of dam

permeation [ˌpɜːmɪ'eɪʃn]

【释义】 *n.* 渗透作用，渗入，透过

【常用搭配】 permeation pressure 渗透压；permeation limit 渗透极限

【例句】 The drug release and permeation were observed to be dependent on some factors. 药物的体外释放和渗透受到多种处方因素的调节。

【中文定义】 水分子从水势高的系统向水势低的系统移动的现象。

Permian (P) ['pɜːmiən]

【释义】 *n.* 二叠纪（系）

【中文定义】 古生代的最后一个纪，开始于距今约 2.99 亿年前，延至 2.5 亿年前，共经历了 4500 万年，也是重要的成煤期。二叠纪的地壳运动比较活跃，古板块间的相对运动加剧，世界范围内的许多地槽封闭并陆续形成褶皱山系，古板块间逐渐拼接形成联合古大陆（泛大陆）。

permissible concentration-short term exposure limit (PC-STEL)

【释义】 短时间接触容许浓度

【中文定义】 指一个工作日内，任何一次接触不得超过的 15min 时间加权平均的容许接触浓度。

permissible concentration-time weighted average (PC-TWA)

【释义】 时间加权平均容许浓度

【中文定义】 指以时向为权数规定的 8h 工作日的平均容许接触水平。

permissible yield

【释义】 容许开采量

petrology [pə'trɑlədʒi]

【释义】 *n.* 岩石学

【常用搭配】 structural petrology 构造岩石学；metamorphic petrology 变质岩石学

【例句】 In many respects, the theories and methods of modern petrology can be utilized in promoting the steel manufacture. 现代岩石学的理论与方法可以用以促进炼钢工业。

【中文定义】 是研究岩石的成分、结构构造、产状、分布、成因、演化历史和它与成矿作用的关系等的学科，是地质学的主要分支学科。

phanerocrystalline texture

【释义】 显晶质结构

【中文定义】 又称全晶质结构，是指火成岩中的矿物晶体颗粒肉眼能分辨的结构。具有显晶质结构的岩石叫显晶岩，如花岗岩、辉长岩等。

Phanerozoic [ˌfænərə'zoɪk]

【释义】 *n.* 显生宙（宇）

【中文定义】 开始出现大量较高等动物以来的阶段，包括古生代、中生代和新生代，从距今大约 5.7 亿年前延续至今。5.4 亿年前，寒武纪始，生物逐渐演化出较高级的动物，动物已具有外壳和清晰的骨骼结构，故称显生宙。

phonolite ['fonəˌlaɪt]

【释义】 *n.* 响岩

【常用搭配】 leucite phonolite 白榴响岩

【例句】 Sodium analcime (nepheline) phonolites in Mibale area of Tibetan Plateau contain abundant clinopyroxene phenocrysts which show the normal, reverse or oscillatory zoning patterns. 西藏高原米巴勒地区产出的中新世钠质方沸石（霞石）响岩中的单斜辉石发育大量环带结构，环带结构包括正环带、反环带和韵律环带。

phonolite

photogrammetry [ˌfotə'græmɪtri]

【释义】 摄影测量

【中文定义】 利用摄影影像信息测定目标物的形状、大小、空间位置、性质和相互关系的科学

技术。

photographic resolution

【释义】 摄影分辨率

photoplan

【释义】 相片平面图

【中文定义】 用经投影变换的相片编制的带有公里格网、图廓内外整饰和注记的平面图。

phreatic water

【释义】 潜水

【例句】 The experimentation results are significant for the transformation relationship between soil moisture and phreatic water, the conservation and protection groundwater resources. 研究成果对于认识土壤水与潜水的转化关系、涵养和保护地下水资源具有重要意义。

【中文定义】 地表以下第一个稳定隔水层以上具有自由水面的地下水。

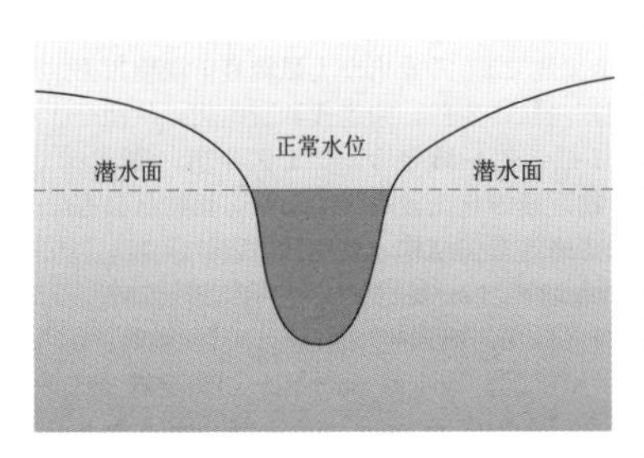

schematic diagram of mutual replenishment between river water and phreatic water

phreatic water table

【释义】 潜水面，浅层地下水水位

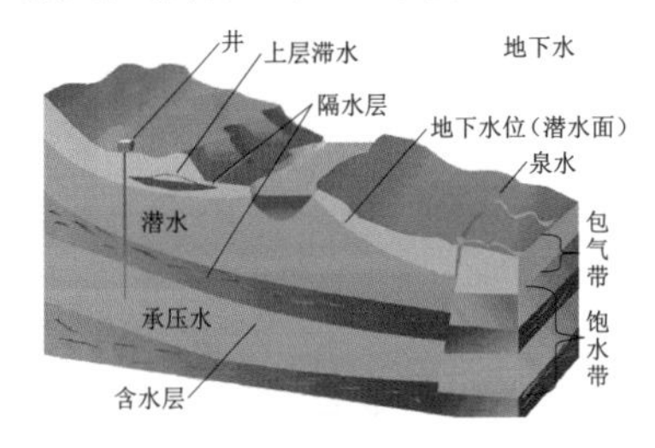

phreatic water table

【例句】 From experiments and analysis show that the daily periodic variation of temperature is one of the main reasons causing the daily periodic variation of phreatic water table. 通过实验和分析证实，温度的日变化是引起浅层地下水水位日变化的主要原因之一。

【中文定义】 潜水的自由表面称潜水面。潜水的补给来源主要是大气降水和河、湖水补给。

phyllite [ˈfɪlˌaɪt]

【释义】 *n.* 千枚岩，硬绿泥岩

【例句】 It was an effective way to control big deformation of surrounding rock mass by adopting strong rigid support of long anchor and taking initial support measure of net grouting steel fibre concrete to the phyllite stratum with big deformation of surrounding rock. 对于围岩变形大的千枚岩地层，采用长锚杆大刚度型钢支架，结合网喷钢纤维混凝土等初期支护措施，是控制围岩产生大变形的有一个效方法。

phyllite

【中文定义】 是具有千枚状构造的低级变质岩石。原岩通常为泥质岩石（或含硅质、钙质、炭质的泥质岩）、粉砂岩及中、酸性凝灰岩等，经区域低温动力变质作用或区域动力热流变质作用的底绿片岩相阶段形成。显微变晶片理发育面上呈绢丝光泽。变质程度介于板岩和片岩之间。

phyllitic structure

【释义】 千枚状构造

【中文定义】 岩石中的矿物颗粒肉眼不能分辨，鳞片状矿物定向排列成细小片理、片理面上具有丝绢光泽的一种构造。

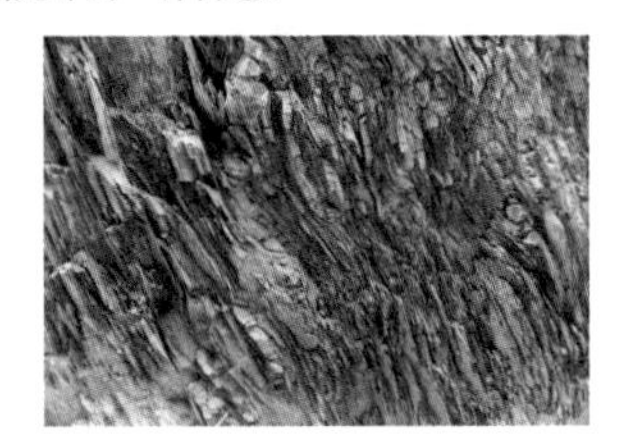

phyllitic structure

phyllonite ['fɪlənaɪt]

【释义】 *n.* 千枚糜棱岩，千糜岩

【例句】 Except the mylonite, the ductile deformed rocks in Yunnan region include the phyllonite and schist with structural origin. 在云南地区韧性变形岩石除糜棱岩类外，还有千糜岩类和构造片岩类。

【中文定义】 又称千枚糜棱岩，是一种原岩遭受强烈挤压破碎后，经明显重结晶作用形成的动力变质岩石。

phyllonite

physical alteration

【释义】 物理蚀变

【例句】 In general, physical alteration and chemical alteration are associated with each other. 一般而言，物理蚀变和化学蚀变是相互伴生的。

【中文定义】 蚀变过程中发生的物理变化。

physical weathering

【释义】 物理风化，物理风化作用

【例句】 The main factors affecting the development of fractures are tectonic, physical weathering and chemical dissolution. 影响裂缝发育的主要因素是构造、物理风化和化学淋溶作用。

【中文定义】 地表岩石发生机械破碎而不改变其化学成分，也不形成新矿物的作用称为物理风化。

physical weathering

physicogeological phenomenon

【释义】 物理地质现象

【常用搭配】 bad physicogeological phenomenon 不良物理地质现象

【例句】 Landslide and mudflow are main physicogeological phenomena in the mountainous region. 滑坡和泥石流是山地区域的主要物理地质现象。

【中文定义】 是指大量的地质变迁现象。有滑坡、泥石流、岩崩、岩溶、岩堆（坡积层）、软弱土、膨胀土、湿陷性黄土、冻土、水害、采空区以及强震区等。

piedmont ['pidmɑnt]

【释义】 *n.* 山麓，山麓地带

【常用搭配】 piedmont plain 山麓冲积平原；piedmont glacier 山麓冰川；piedmont depression 山麓坳陷

【例句】 The geotectonic position of study area lies in the southern margins of the Zhungeer depression and Dalongkou depression, which belong to the Urumchi piedmont depression. 研究区大地构造位置位于乌鲁木齐山麓坳陷之准噶尔南缘坳陷和大龙口坳陷内。

【中文定义】 山麓是指山坡和周围平地相接的部分。

piedmont

piedmont plain

【释义】 山前平原，山麓平原

【常用搭配】 piedmont denudation plain 山麓剥蚀平原

【中文定义】 是指位于山区至平原的过渡地带，由一系列洪积扇或冲洪积扇发展形成的平原。

piedmont plain

piezocone penetration test（CPTU）

【释义】 孔压静力触探试验（CPTU）

【例句】 A soil engineering classification derived from the piezocone penetration test（CPTU）involves the uncertainty of correlation between soil composition and soil mechanical behavior. 采用孔压静力触探（CPTU）测试资料进行土的工程分类时，在土成分和力学性状等相关性方面会产生很多不确定性。

piezometer [ˌpaɪɪ'zɒmɪtə]

【释义】 *n.* 测压管

【同义词】 piezometric tube

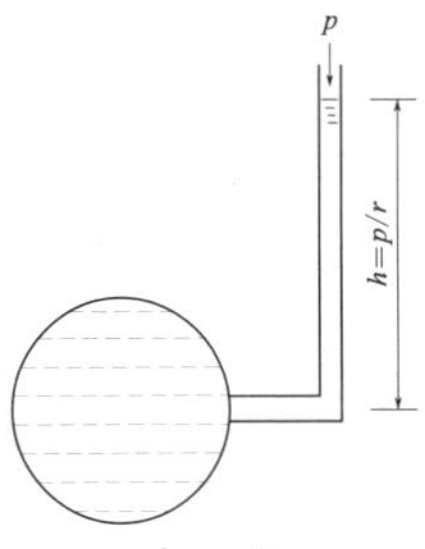

piezometer

【例句】 Based on the analyses of hydrological conditions and piezometer response patterns around the reservoir, the mechanism of the sinkholes and the tendency of their development are involved in the paper. 该文从水文地质条件分析出发，依据坝后地下水动态监测资料，阐述了地表塌陷的成因机制及其可能的发展趋势。

【中文定义】 是测量液体相对压强的一种细管状仪器，上端开口与大气相通，下端连接于容器侧壁上与被测液体连通。管内液体沿管上升至某一高度，据此可算出管下端壁孔处的液体相对压强。可用来监测坝体浸润线、渗压压力、地下水水位及绕坝渗流等。

piezometric height

【释义】 测压高度

【中文定义】 含水层中某点至地下水面的垂直距离，它等于该点的静水压强与水的质量之比。

piling density

【释义】 堆积密度

【中文定义】 把粉尘或者粉料自由填充于某一容器中，在刚填充完成后所测得的单位体积质量。

pillow structure

【释义】 枕状构造

【中文定义】 海底喷发的基性熔岩所形成的一种特殊构造。

pillow structure

pilot tunnel

【释义】 导洞

【例句】 Steel tube column is one of the major bearing structures in the construction method of "Cave-pile-beam" for undercutting in city metro works. Due to the limitation of space in the pilot tunnel, the working procedures of the steel tube columns are different to those on the ground. 钢管柱是城市暗挖地铁“洞桩梁”法施工的主要承重结构之一，因导洞空间的限制，洞内钢管柱各道工序与地面施工工艺不同。

pilot tunnel

pinnate shear joint

【释义】 羽状剪节理

【例句】 The angles between the micro fracture planes of pinnate shear joints and principal fracture plane are generally 5°～15°. 羽状微裂面与主裂面交角一般为 5°～15°。

【中文定义】 主剪裂面由羽状微裂面组成的节理。

pinnate shear joint

pinnate tension joint

【释义】 羽状张节理

【例句】 Pinnate tension joints are a set of joints formed on both sides of the fault plane by the fault movement. 羽状张节理是由断层运动而在断层面两侧形成的一组张节理。
【中文定义】 由断层运动而在断层面两侧形成的一组张节理，它们与断层面斜交而呈羽状排列。张节理与断层面所夹锐角指向本盘相对运动方向。

pinnate tension joint

piping ['paɪpɪŋ]
【释义】 *n*. 管涌
【中文定义】 指土体中的细颗粒在渗流作用下，由骨架孔隙通道流失的现象，主要发生在砂砾石地基中。

piping on contact surface
【释义】 ①：接触冲刷
【例句】 The earth seepage failure includes four types, they are mass flow, piping effect, contact soil flow and piping on contact surface. 土的渗漏破坏包括四种形式，它们是流土、管涌、接触流土和接触冲刷。
【中文定义】 渗流沿着两种不同介质的接触面流动并带走细颗粒的现象称为接触冲刷。
【释义】 ②：接触管涌
【中文定义】 指在两种不同性质介质（其中至少有一种是颗粒材料）的接触面上，在渗透水作用下所发生的管涌现象。

pisolitic texture
【释义】 豆状结构
【中文定义】 结构直径大于 2mm 的沉积岩的一种结构。

plagioclase ['pledʒɪəklez]
【释义】 *n*. 斜长石

plagioclase

【例句】 Plagioclase is usually white-colored with perfect cleavage and albite twins. 斜长石通常呈白色，具有完全解理和聚片双晶。

plain [plen]
【释义】 *n*. 平原
【常用搭配】 pediplain 山麓侵蚀平原；tectonic plain 构造平原；alluvial plain 冲积平原
【例句】 The river and stream beds which traverse flood plains have a limited water carrying capacity. 河漫滩地区的河床持水能力有限。
【中文定义】 指地表面平坦宽广、地面高差不超过 20cm、大部分地面的倾斜角在 2°以下，一般海拔在 200m 以下的地区。

Chengdu Plain

plain fill
【释义】 素填土
【例句】 The results show that the plain fill reinforced by dynamic compacting forms a kind of two-layer structure with harder top whose strength increasing with time. 研究结果表明：强夯加固素填土，强度随时间增长，在竖向形成上硬下软的 2 层结构。
【中文定义】 由天然土经人工扰动和搬运堆填而成，不含杂质或含杂质很少，一般由碎石，沙或粉土，黏性土等一种或几种材料组成。按主要组成物质分为：碎石素填土、沙性素填土、粉性素填土、黏性素填土等，可在素填土的前面冠以其主要组成物质的定名，对素填土进一步分类。

plain fill

plan [plæn]
【释义】 *n*. 平面图

【中文定义】 只表示地形要素的平面位置，不表示起伏形态的地图。

planar ['plenər]

【释义】 *adj*. 平直的

【例句】 The shear strength is different because of the smoothness of the structural plane. There are mainly three kinds of undulating structures, including planar, undulating and stepped. 结构面的平整程度不同，抗剪强度也不同。结构面的形态有平直的、起伏的和有台坎的三种。

【中文定义】 是结构面一种起伏形态特征的描述。结构面的平整程度不同，抗剪强度也不同。

planar slide

【释义】 平面滑动

【例句】 For the rock slope with planar slide, the structural plane plays a controlling role in the stability of slope. 对于平面滑动型的岩质边坡，结构面对边坡的稳定性起着控制作用。

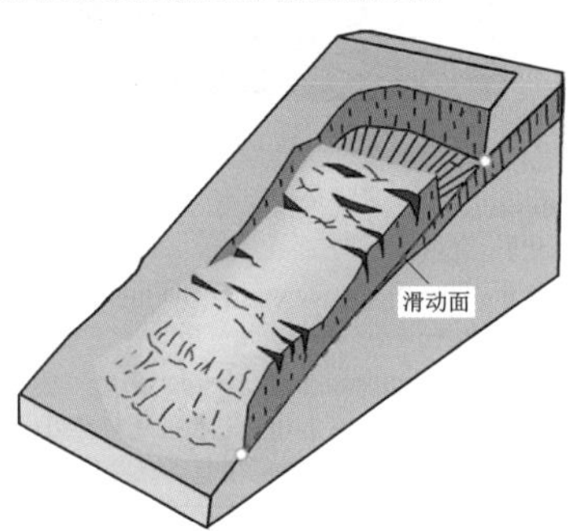

planar slide

planar structural plane

【释义】 平直结构面

【中文定义】 结构面的形态是平直的，反映了结构面的平整程度。

planation surface

【释义】 夷平面

【常用搭配】 denudation-planation surface 剥夷面

【例句】 Planation surfaces have been given various other names, based on various formation hypotheses, such as a peneplane in William Morris Davis's "cycle of erosion". 基于不同的地层假说，夷平面又被冠以不同的名字，比如在威廉·莫里斯·戴维斯"侵蚀循环"里叫作准平原。

【中文定义】 指各种夷平作用形成的陆地表面。是一种陆地抬升或侵蚀基面下降，侵蚀作用重新活跃，经过一个时期后所残留的地表形态。

a planation surface on top of a small plateau in the Bighorn Basin

planting soil

【释义】 耕植土

【例句】 The nature of planting soil is intrinsically related to the natural environment and human activities. 耕植土的性质与自然环境和人类活动存在错综复杂的内在联系。

planting soil

plastic deformation

【释义】 塑性变形

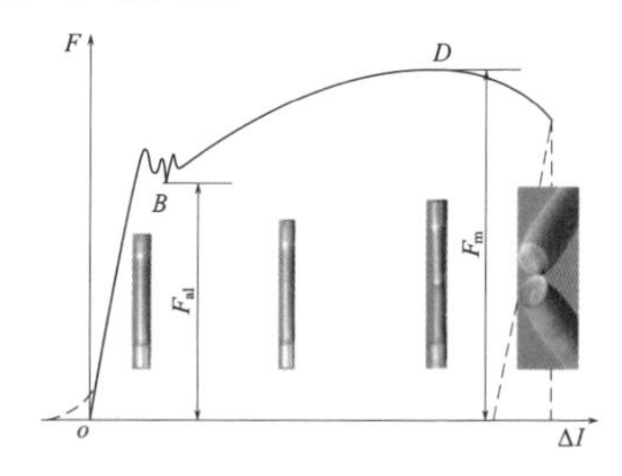

elastic deformation and plastic deformation

【例句】 The petrological study shows that banded ore belongs to mylonite, which has abundant microstructure feature of plastic deformation. 岩石学研究表明条带状矿石属于糜棱岩，具有丰富的塑性变形显微构造特征。

【中文定义】 材料在外力作用下产生形变，而在外力去除后，弹性变形部分消失，不能恢复而保留下来的那部分变形即为塑性变形。

plastic failure

【释义】 塑性破坏

【例句】 Generally, only the plastic failure mode is considered in system reliability analysis of structures. 在一般结构系统可靠性分析中往往只考虑构件塑性破坏的单一模式。

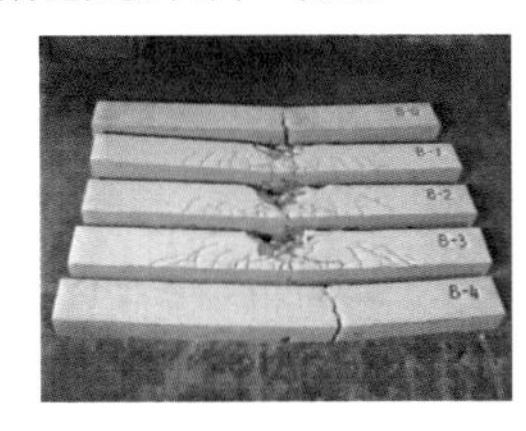

plastic failure

plastic limit

【释义】 塑限

【例句】 The plastic limit is an important property of fine-grained soil, which reflects the interaction between soil and water. 塑限是细粒土的一个重要属性，它反映了土粒和水相互作用时的性态。

plasticity chart

【释义】 塑性图

plasticity index

【释义】 塑性指数

plate loading test

【释义】 平板荷载试验

【例句】 The bearing capacity of pile tip soil is reflected intuitively by deep plate loading test. 通过深层平板载荷试验，直观地反映了桩端土的承载能力。

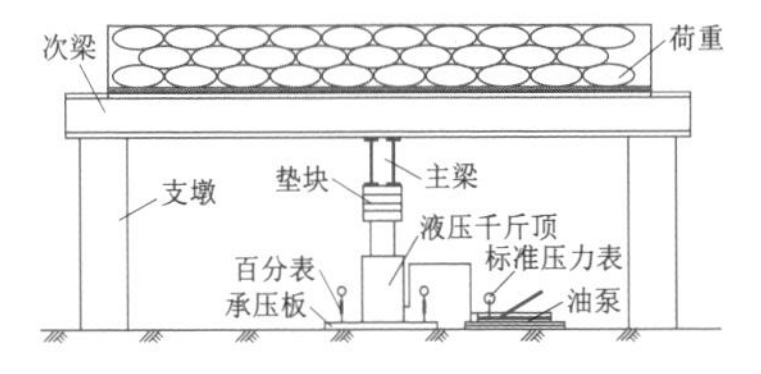

plate loading test

plate tectonics

【释义】 *n*. 板块构造论，大地构造学

【常用搭配】 lithosphere plate tectonics 岩石层板块大地构造；evolution of the plate tectonics 板块构造演化；plate tectonics theory 板块构造学说，板块运动学说

【例句】 Plate tectonics is a unifying concept which provides a coherent model of the outer part of the earth and how it works. 板块构造学说是一个系统的概念，它构建了一个描述地球表部及其运动的连贯模型。

【中文定义】 板块构造学说是在大陆漂移学说和海底扩张学说的基础上提出的。地球表面覆盖着不变形且坚固的板块（岩石圈），这些板块确实在以每年 1～10cm 的速度在移动。

plateau [plæ'to]

【释义】 *n*. 高原，高原环境

【常用搭配】 loess plateau 黄土高原；lava plateau 熔岩高原；Qinghai-Tibet Plateau 青藏高原

【例句】 The Mambilla plateau is a high level peneplain, with an altitudinal range between El 1300 and El 1936, incised by deep valleys. 蒙贝拉高原是一个被深谷切割的准平原，高度介于 1300m 到 1936m 之间。

New Mexico Colorado Plateau

【中文定义】 高度一般在 1000m 以上，面积广大，地形开阔，周边以明显的陡坡为边界，比较完整的大面积隆起地区称为高原。

platform ['plætfɔrm]

【释义】 *n*. 地台

【常用搭配】 activation of platform 地台活化

【例句】 Platforms, shields and the basement rocks together constitute cratons. 地台、地盾和基底岩石一起构成大陆稳定地块（克拉通）。

【中文定义】 指大陆上自形成以后未再遭受强烈褶皱的稳定地区，曾称陆台，1885 年由 E. 修斯提出。在地槽地台学说中，地台是与地槽相对应的地壳稳定构造单元，以含有未变质的沉积盖层区别于地盾。

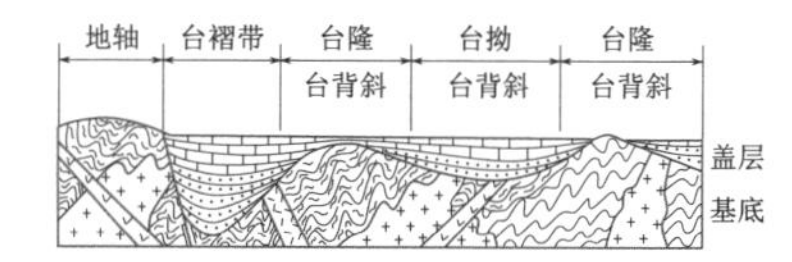

platform structure and internal unit division

platy structure

【释义】 板状构造

【例句】 Steep and antithetic platy structure of rock mass and geomorphology of the longitudinal valley in the dam site area cause wide-range toppling deformation on the cross-strait slope. 对该水电站坝址区陡立反倾板状结构岩体而言，纵向河谷的地形特征，使坝址两岸边坡大范围发育倾倒变形。

【中文定义】 岩石中存在的一组密集互相平行的劈裂面，将岩石劈开成板状的一种构造。

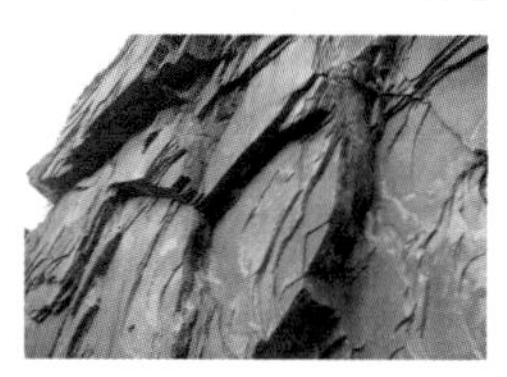

platy structure

Pleistocene ['plaɪstosin]

【释义】 *n*. 更新世（统）

【中文定义】 更新世也称洪积世，是第四纪的第一个世，介于约258.8万年前～约1.17万年前。

Pliocene ['plaɪəˌsin]

【释义】 *n*. 上新世（统）

【中文定义】 第三纪的第五个世。

plot of great circles

【释义】 极点大圆图

plough horizon

【释义】 耕作层

【中文定义】 指经耕种熟化的表土层。除去表土的土壤，一般厚15～20cm，养分含量比较丰富，作物根系最为密集，粒状、团粒状或碎块状结构。

plunge angle

【释义】 倾伏角

【常用搭配】 apparent plunge angle 视倾伏角

【例句】 The plunge angle is the plunging angle of the linear structure. 倾伏角是指线状构造倾伏的角度。

【中文定义】 倾伏角是指线状构造倾伏的角度，即在包含线状构造的铅垂参考面上此斜线与水平线之间的夹角。

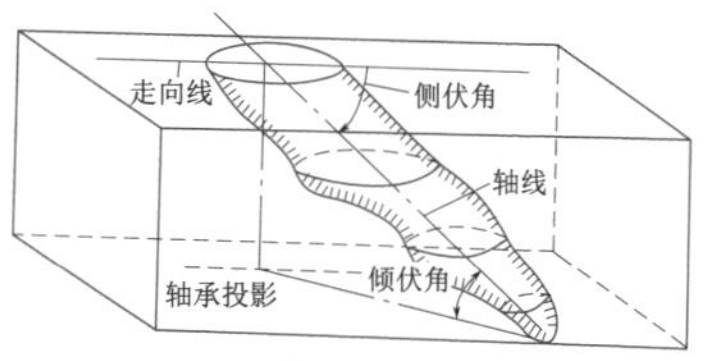

plunge angle

plunge direction

【释义】 倾伏方向

【例句】 The plunge direction is the direction of the straight line which is indicated by the projection line of an oblique line on the horizontal plane. 倾伏方向是某一直线在空间的延伸方向，即某一倾斜直线在水平面上的投影线所指示的该直线向下倾斜的方位。

【中文定义】 某一直线在空间的延伸方向，即某一倾斜直线在水平面上的投影线所指示的该直线向下倾斜的方位，用方位角或象限角表示。

plunging fold

【释义】 倾覆褶皱，倾伏褶皱

【例句】 This paper has recounted the technique for correcting the attitude of plunging fold strata in the application of paleomagnetism to the study of geological structure, and derived the conversion formula for experts' reference. 本文介绍了用古地磁研究地质构造问题时倾伏褶皱地层产状的校正方法，推导了换算公式，以提醒同行们的关注。

【中文定义】 枢纽倾伏的褶皱，两翼岩层的走向不平行，在倾伏端交汇成封闭的弯曲线，为一倾伏背斜。

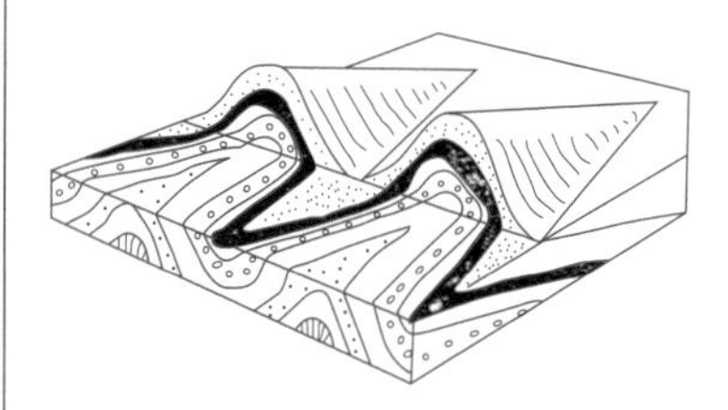

plunging fold

plutonic rock

【释义】 深成岩

【例句】 The Cenozoic magmatic rocks of shoshonitic series in eastern Qinghai-Tibet plateau include potassic alkaline plutonic rocks, volcanic rocks, lamprophyres and acidic porphyries. 青藏高原东部及邻区新生代钾玄岩系列岩石包含钾质碱性深成岩、火山岩、煌斑岩和偏酸性的斑岩。

【中文定义】 也称火成岩，是岩浆侵入地壳深层3km以下，缓慢冷却形成的火成岩，一般为全晶质粗粒结构。

point diagram

【释义】 极点图

【常用搭配】 joint point diagram 节理极点图

【例句】 In this paper, by statistic analyses of fracture rockmass structural plane of FAST, the fractal character of the field structural plane point diagram is known. 本文通过对FAST台址区裂隙岩体结构面的统计分析，得出场地结构面极点图的分形特征。

【中文定义】 是一种将某类面状构造（如片理、节理、轴面、双晶面等）的法线或某类线状构造（如线理、晶体光轴或结晶轴等）的产状测量数据，以极点投影在乌尔夫网或施密特网上的岩石组构图。

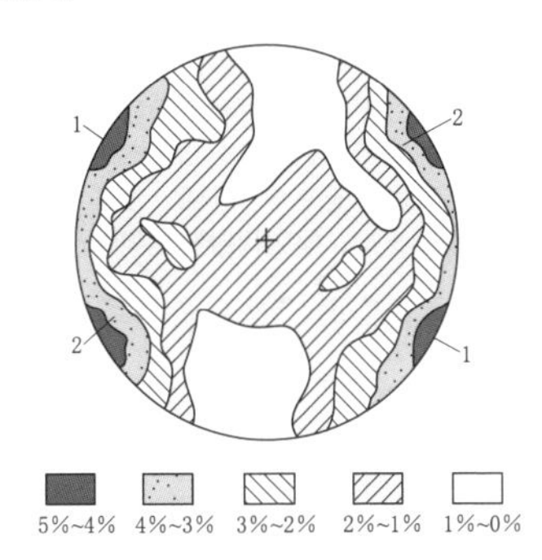

joint point diagram

point load strength

【释义】 点荷载强度

【例句】 The point load strength tests can determine the uniaxial compressive strength of rock. 点荷载试验能够确定岩石的单轴抗压强度。

【中文定义】 指用点荷载试验测试得到的材料强度。

point-loading test

【释义】 点荷载试验

【例句】 Concerning the stipulation and statement for empirical formula and revised index in point-loading test there are still some uncertainties in related specifications and manuals. 关于点荷载强度试验的经验公式及其修正指数的规定和论述，在相关的规范和手册中还有一些有待明确的方面。

point-loading test

points model

【释义】 点符模型

【中文定义】 在CAD软件中，由单个或一组点符按照一定的空间位置放置并赋予特定的模型样式和模型属性代表实物整体或表面上的一点的模型。

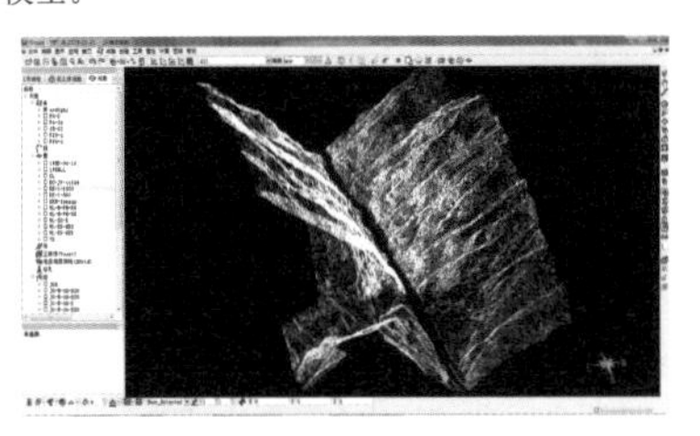

points model

poisonous gases

【释义】 有毒有害气体

pole [pol]

【释义】 *n.* 极点，电极

【例句】 Pole is one of projection elements in stereographic projection. 极点是极射赤平投影的投影要素之一。

【中文定义】 通过球心的直线与球面的交点称为极点，一条直线有两个极点。铅直线交球面上、下两个点（也就是极射点）；水平直线交基圆上两点；倾斜直线交球面上两点。

pole plot of discontinuities

【释义】 断裂极点图

polje ['pəulje]

【释义】 坡立谷

【例句】 According to the study on the hydrogeological conditions and the leakage of the reservoir, the conclusions have been drawn as follows: The characteristics of limestone block bearing water in the Niukouyu reservoir area is basically controlled by the evolution process of Niukouyu polje. 根据水文地质条件和对水库渗漏的研究，可得出如下结论：牛口峪水库地区石灰岩含水体的特征主要受到牛口峪坡立谷演化过程的控制。

【中文定义】 又称槽谷，谷地两侧多被峰林夹峙，谷坡急陡，但谷底平坦，横剖面如槽形。

polje

poor surrounding rock mass

【释义】 差围岩

【中文定义】 地下硐室在采用 RMR（Rock Mass Rating system）分类时，当 RMR 评分值介于 21～40 时，为Ⅳ类围岩，定性评价为差围岩。

poorly integral

【释义】 完整性差

【中文定义】 根据《水力发电工程地质勘察规范》(GB 50287—2016）的规定，当岩体完整性系数 $0.35<K_v\leqslant0.55$ 时，则该岩体的完整程度为完整性差。或根据岩体中结构面发育程度判断，完整性差岩体中结构面多发育 2～3 组，结构面间距平均值多小于 0.3m。

poorly-graded soil

【释义】 不良级配土

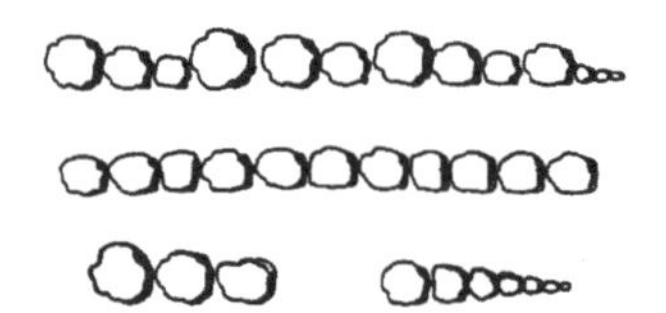

poorly graded soil

polycyclicity theory

【释义】 多旋回构造学说

【例句】 Prof. Huang Jiqing (T. K. Huang) put forward the concept of polycyclic tectonic movement which was developed into the polycyclicity theory that is influential all over the world. 黄汲清教授提出的多旋回构造运动发展成为多旋回构造学说，该学说影响了世界。

【中文定义】 简称多旋回说，由黄汲清于 1945 年提出，一个褶皱带（造山带）的形成往往经历了多旋回造山运动，即多旋回的发展。

pore pressure coefficient

【释义】 孔隙水压力系数

【例句】 Finally, the curves of axial strain and dynamic pore pressure coefficient were figured and analyzed. 最后，绘制了试样轴向应变与动孔隙水压力系数的关系曲线，并分析了曲线的变化规律。

pore water

【释义】 孔隙水

【常用搭配】 pore water pressure 孔隙水压力

【例句】 Pore water pressure distribution dependent on location of grout and drainage curtain. 孔隙水压力分布取决于灌浆、帷幕的位置。

【中文定义】 指主要赋存在松散沉积物颗粒间孔隙中的地下水。在堆积平原和山间盆地内的第四纪地层中分布广泛，是工农业和生活用水的重要供水水源。孔隙水的分布、补给、径流和排泄决定于沉积物的类型、地质构造和地貌等。

pore water

pore water pressure gauge

【释义】 孔隙水压力计

【例句】 The invention discloses a pore water pressure gauge comprising a water pressure display unit. 本发明公开了一种包括水压显示单元

的孔隙水压力计。

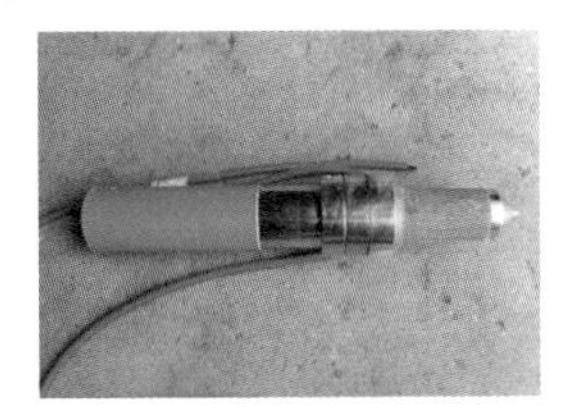

pore water pressure gauge

porosity [pə'rɑsəti]

【释义】 *n*. 有孔性，多孔性，孔隙率

【例句】 The porosity and permeability of rocks are directly related to the particle size and sortability of the rock. 颗粒大小和分选性对岩石的孔隙率和渗透率有直接关系。

porphyaceous texture

【释义】 似斑状结构

【中文定义】 与斑状结构同为颗粒较大的“斑晶”分布于颗粒较小的“基质”上，但斑状结构的基质为隐晶质或玻璃质；而似斑状结构的基质为显晶质，是比斑晶颗粒小的晶体。和斑状结构不同，似斑状结构斑晶和基质是同时形成的，只是一种物质过剩而形成斑晶，另一种物质较少形成基质。

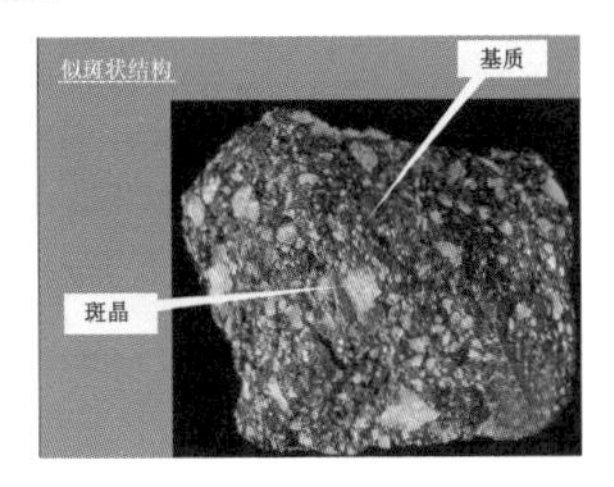

porphyaceous texture

porphyrite ['pɔːfirait]

【释义】 *n*. 玢（发音同“宾”）岩，火成岩的一种

porphyrite

【例句】 The dioritic porphyrite vein existing in the fault is closely related with orebody and usually associates with the gold deposit. 断层内的闪长玢岩脉与矿体关系密切，常与金矿伴生。

porphyritic texture

【释义】 斑状结构

【中文定义】 岩石中矿物颗粒分为大小截然不同的两群，大的称为斑晶，小的及不结晶的玻璃质称为基质。其间没有中等大小的颗粒，可与连续不等粒结构相区别。

porphyritic texture

porphyry ['pɔrfəri]

【释义】 *n*. 斑岩

【例句】 The Central Asian Metallogenic Domain with many large and super-large porphyry Cu deposits is an important metallogenic domain of porphyry Cu deposits in the world. 中亚成矿域发育许多大型和超大型斑岩铜矿床，是世界上重要的斑岩铜矿成矿域。

porphyry

potassium magnesium salt rock

【释义】 钾镁盐岩

【中文定义】 一种纯化学成因的岩石，由蒸发海水或湖泊作用沉淀而成。主要由钾、镁的卤化物及硫酸盐矿物组成。

potassium magnesium salt rock

potential debris flow gully

【释义】 潜在泥石流沟

【中文定义】 经调查无近期泥石流活动史，但存在可能暴发泥石流部分条件的沟谷。

potential seismic source zone

【释义】 潜在震源区

【例句】 Area source in engineering seismology is such a type of potential seismic source zone which the positions and strikes of genetic faults are uncertain. 工程地震学中所指的“面源”属于这样一类潜在震源区——区内发震断层的位置和走向都是不确定的。

【中文定义】 潜在震源区是未来可能发生破坏性地震的地区。

potential unstable rock mass

【释义】 潜在不稳定体

【例句】 The residual loose body and potential unstable rock mass of the mountain posed a serious threat to the China Giant Panda Protection Research Center below. 危岩带上方滞留的松散块石和潜在不稳定块体，对其下方中国大熊猫保护研究中心构成严重威胁。

powdery rock

【释义】 碎粉岩

Precambrian [priːˈkæmbriən]

【释义】 *n.* 前寒武纪（系）

【中文定义】 是地质年代中，对于显生宙之前数个宙所使用的非正式名称。原本正式的名称是隐生宙，后来被拆分成冥古宙、太古宙与元古宙三个时代。前寒武纪开始于大约 45 亿年前的地球形成时期，结束于约 5 亿 4200 万年前大量肉眼可见的硬壳动物诞生之时。

precise leveling

【释义】 精密水准测量

【例句】 It would expend more labor and material in the traditional precise leveling. 传统的方法进行精密水准测量，需消耗较多的人力物力。

preconsolidation pressure

【释义】 先期固结压力

【例句】 The results of laboratory tests showed that the apparent preconsolidation pressure of Sichuan clay is higher than the effective overburden pressure. 室内试验成果表明，四川地区黏土的准先期固结压力较天然有效压力大得多。

pre-drilling pressuremeter

【释义】 预钻式旁压仪

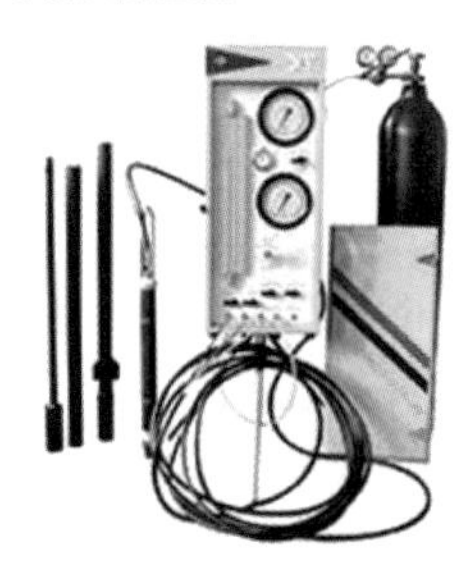

pre-drilling pressuremeter

prefeasibility study stage

【释义】 预可行性研究阶段

【例句】 In every large company you will see projects that have been running for years in prefeasibility study stage. 在每个大型公司，你都可以看到很多项目的预可行性研究阶段就持续数年以上。

preloading method

【释义】 预压法

【常用搭配】 piled-load preloading method

【例句】 Application of piled-load preloading method in silty foundation treatment. 堆载预压法在处理淤泥质地基土中的应用。

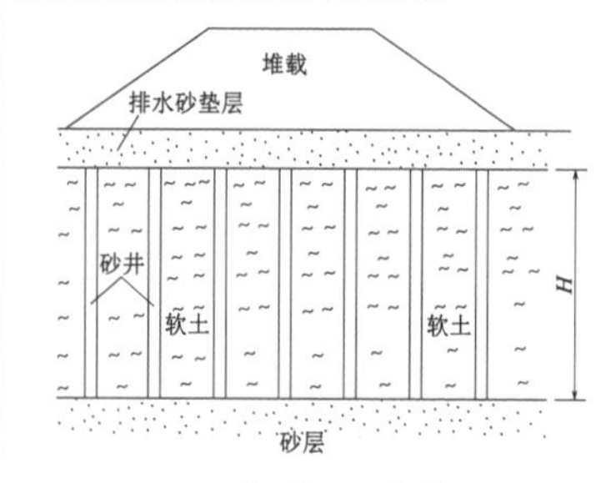

preloading method

pre-reinforcement

【释义】 预压加固

【常用搭配】 vacuum pre-reinforcement 真空预压加固

【例句】 These problems are to be solver in order to achieve rational design for the strengthening of soft clay subgrade by vacuum pre-reinforcement. 这是进行合理的真空预压加固软土地基设计所必须解决的问题。

pre-reinforcement

presplitting blasting

【释义】 预裂爆破

【例句】 The rock property difference of bedded composite rock often does not lead to ideal presplitting blasting effect. 层状复合岩体中岩层性质的差异常常导致预裂爆破效果不理想。

presplitting blasting of diversion canal slope

pressure calibration

【释义】 压力标定

【例句】 The experimental system calibration includes three aspects: displacement calibration, temperature calibration and pressure calibration. 实验系统的标定主要包括三个方面的内容：位移标定、温度标定和压力标定。

pressure conduction coefficient

【释义】 水力扩散系数，压力传导系数

【同义词】 coefficient of pressure conductivity

【中文定义】 压力传导系数又称导压系数，表示弹性液体在弹性多孔介质中不稳定渗流时，压力变化传递快慢的一个参数。

pressure drop

【释义】 压降、压力差

【常用搭配】 initial pressure drawdown 初始压力降

【例句】 This increased pressure drop improves the material flow rate through the flowmeter. 这个增大的压力降增大了通过流量计的材料流率。

pressure head

【释义】 压力水头，压位差

【例句】 Based on the calculation results of unsteady seepage in levees the empirical formulas for maximum value of important parameters in the process of flood including seepage discharge per unit length, height of saturation line and pressure head at the toe are deduced. 根据一般双层地基上堤防非稳定渗流有限元计算成果，分析出单宽渗流量、堤背脚下承压水头和堤身浸润线高度等渗流关键值在洪峰过程中的峰值经验公式，供设计管理堤防者参考。

pressure loss of tube

【释义】 管路压力损失

pressure meter test (PMT)

【释义】 旁压测试，旁压试验

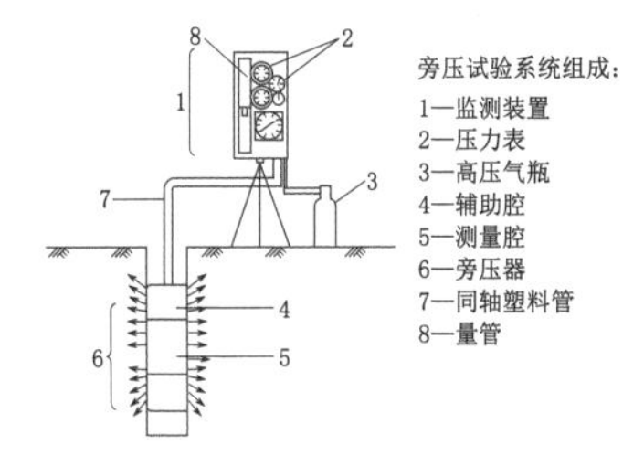

pressure meter test (PMT)

pressure-shear test

【释义】 压剪试验

【例句】 A numerical simulation of pressure-shear test on rock with preexisted cracks under compression is presented by using a rock failure process analysis code, RFPA 2D. 采用岩石破裂过程分析软件 RFPA 2D 对预制裂隙大理岩试样的压剪试验进行数值模拟。

prestressed anchorage

【释义】 预应力锚固

【常用搭配】 Unbonded Prestressing Anchor System 无黏结预应力锚固系统

【例句】 On the premise of not reducing the original mechanical parameters of the anchor rod, the high-strength prestressed anchorage operation of the anchor rod is realized. 在不降低锚杆原有力学参数的前提下，实现锚杆高强预应力锚固。

prestressed anchorage

prevention and treatment of geologic hazard

【释义】 地质灾害防治工程

【例句】 The third phase of the project to prevention and treatment of geological hazard in the Three Gorges Reservoir area was carried out smoothly. 三峡库区三期地质灾害防治工程顺利实施。

prevention and treatment of geologic hazard

primary consolidation settlement

【释义】 主固结沉降

【例句】 In the absence of compressibility data from laboratory tests, the total primary consolidation settlement of a structure founded on clay can be estimated from settlement measurements taken over a period of time. 如果缺乏试验压缩数据，建筑物的主固结沉降可以从一个时期内的沉降值估算出来。

【中文定义】 饱水地基土层的压缩变形速率受孔隙水排出速率控制的地基沉降。

primary gneissic structure

【释义】 原生片麻构造

【中文定义】 岩石中暗色矿物连续定向排列，其间被浅色粒状矿物分开，与变质岩的片麻状构造相类似，但它是岩浆成因的，是侵入体形成过程中流动的岩浆对围岩强烈挤压而产生的，是岩浆流动的遗迹，主要见于中酸性侵入岩。

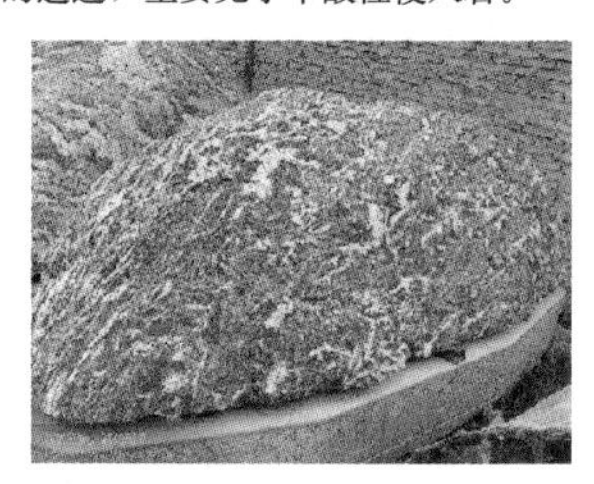

primary gneissic structure

primary joint

【释义】 原生节理

【常用搭配】 primary joint plane 原生节理面

【例句】 The failure of jointed rock mass was usually caused by the initiation, propagation and coalescence of new wing cracks derived from primary joint. 节理岩体的失效破坏往往是由于赋存于其中的原生节理在荷载作用下产生新生裂纹，并且逐渐扩展，使岩桥贯通造成的。

【中文定义】 岩石在成岩过程中所形成的节理，即原生节理，一般由岩浆冷却并收缩而引起。

primary joint

primary mineral

【释义】 原生矿物

【例句】 Thus primary minerals are wholly or partly changed to secondary minerals including serpentine and montmorillonite(from olivine), calcite, chlorite (from biotite), kaolin (from feldspars), zeolites and other clays. 原生矿物全部或部分转变为次生矿物，包括蛇纹石、蒙脱石（自橄榄石转变）、方解石、绿泥石（自黑云母转变）、高岭土（自长石转变）、沸石和其他黏土矿物。

【中文定义】 指在内生条件下的造岩作用和成矿作用过程中，同时期形成的岩石或同时期形成的矿物。

primary mineral malachite

primary wave

【释义】 纵波

【例句】 The method for theoretical calculation of parametric transient signal array, in which the nonlinear absorption of primary wave has been taking into consideration, is presented. 提出考虑到非线性吸收时计算暂态信号参量阵的计算方法。

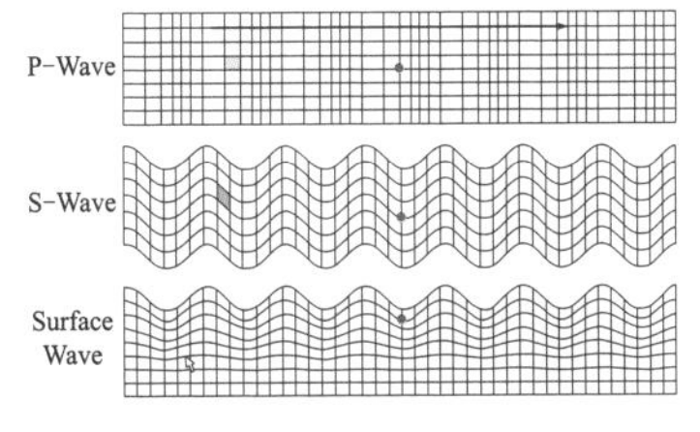

seismic wave

【中文定义】 地震纵波是推进波，地壳中传播速度为5.5～7km/s，最先到达震中，又称P波，它使地面发生上下振动，破坏性较弱。

primitive data map

【释义】 实际材料图

【例句】 In prefeasibility study stage and feasibility study stage and bidding design stage of hydropower engineering, primitive data map shall be submitted as appropriate in report for engineering geological investigation. 水力发电工程预可行性研究阶段、可行性研究阶段和招标设计阶段，工程地质勘察报告中可根据需要附加实际材料图。

【中文定义】 地质工作实际材料图简称实际材料图，是以一定的符号反映野外地质工作中的所获实际资料的图件。

principal stress

【释义】 主应力

【例句】 In order to study the influence of principal stress axis cyclic rotation at low shear stress level on clay's behavior, a series of experiments of Hangzhou typical intact soft clay were conducted. 为研究低剪应力水平主应力轴循环路径可能对黏土性状产生的影响，以杭州地区典型原状软黏土为对象开展了试验研究。

【中文定义】 主应力指的是物体内某一点以法向量为$n=(n_1, n_2, n_3)$的微面积元上，剪应力为零时的正应力。

Probabilistic Seismic Hazard Analysis (PSHA)

【释义】 地震危险性概率分析法、概率地震危险性分析法

【中文定义】 指对特定区域确定其在未来一定设计基准期内地震动参数超过某一给定概率的方法。

probability curve

【释义】 概率曲线

【例句】 Research results: Different variable probability curve with different characteristics, the limestone, the fault breccia, the paste dissolve the breccia relative dispersion to be big, the possibility of large broken rocks to the construction influential. 研究结果：不同的岩性具有不同的概率曲线特征，灰岩、断层角砾岩、膏溶角砾岩离散度大，岩石破碎可能性大，对施工影响大。

probability of exceedance

【释义】 超越概率

【例句】 Occurrence and probability of exceedance are all important indicators of seismic hazard for a given site. 地震烈度发生概率和地震烈度超越概率都是衡量场地地震危险性的重要指标。

【中文定义】 指在一定时期内，工程场地可能遭遇大于或等于给定的地震烈度值或地震动参数值的概率。

process simulation

【释义】 过程模拟

【例句】 The whole process simulation analysis of Sanglang RCC arch dam is done with 3D finite element method(FEM). 利用三维有限元法，对桑郎碾压混凝土拱坝进行了全过程仿真分析。

proctor compaction test

【释义】 普氏击实试验

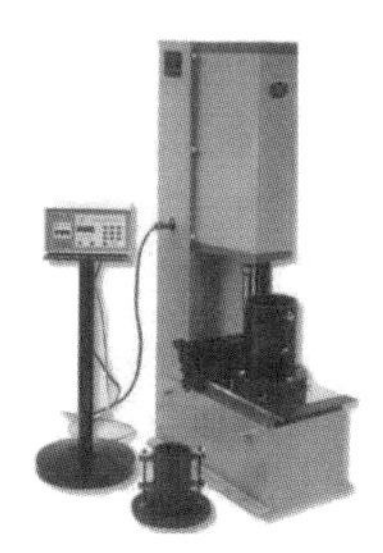

proctor compaction test

profile method

【释义】 剖面法

【例句】 Profile method is important with regard to the description of the shape of a stratum. 剖面法是表达地层形态的重要方法。

proglacial fan

【释义】 冰前扇地

【中文定义】 从冰川末端往外由冰川中融化所成的冰下水挟带了大量的泥砾和冰川研磨所形成的细泥，堆积在终碛堤的边缘，形成了向外扩展的、坡度愈向外愈平缓的扇形地，称为冰前扇地。

proglacial fan

progress [ˈprɑgrɛs]

【释义】 *n.* 进度

【常用搭配】 progress: progress reports 进度报告; actual progress 实际进度; planned progress 计划进度

progressive failure

【释义】 渐进破坏

progressive landslide

【释义】 推移式滑坡

【例句】 The type of landslide can be divided into progressive landslide and retrogressive landslide. 滑坡类型可分为推移式滑坡和牵引式滑坡。

project area

【释义】 工程场区

proluvial fan

【释义】 洪积扇

【常用搭配】 alluvial-proluvial fan 冲洪积扇

【例句】 The pore confined aquifers are mostly distributed in the deep overburden of the mountainous rivers, the quaternary faulted basin and the thick sediments of piedmont alluvial proluvial fan. 孔隙承压含水层多分布于山区河流深厚覆盖层、第四纪断陷盆地和山前冲洪积扇的巨厚沉积层内。

【中文定义】 从山谷出口向山外缓慢倾斜的由洪积作用形成的扇形地貌。

proluvial fan

prospective view

【释义】 透视图

【常用搭配】 prospective view of powerhouse 厂房透视图

prospective view

Proterozoic (Pt) [ˌprɑtərəˈzoɪk]

【释义】 *n.* 元古宙（宇）

【中文定义】 元古宙是一个重要成矿期，同位素年龄从25亿～6亿（或5.7亿）年前，共经历19亿年的悠久时间。元古宙划分为3个代，其中新元古代的后半段，即8亿～6亿（或5.7亿）年单划分称震旦纪。元古宙与太古宙相比，岩石变质程度较浅，并有一部分未经变质的沉积岩。

Protogiakonov's coefficient

【释义】 普氏坚固系数

【例句】 Protogiakonov's coefficient can be used to reflect the ability to resist deformation of rock mass. 普氏坚固系数可以用来反映岩体抵抗变形破坏的能力。

【中文定义】 岩石的坚固性是指岩石在几种变形方式的组合作用下抵抗破坏的能力。坚固系数表征的是岩石抵抗破碎的相对值，又称普氏坚固系数。

pseudosection map

【释义】 拟断面图

pseudo-static method

【释义】 拟静力法

【例句】 Not all kinds of gangue dams can use the pseudo-static method to analyse their seismic stability or great departure may result. 拟静力法并非适合于所有的尾矿堆积坝抗震稳定性分析，即便采用其所得结果也可能会出现较大偏差。

【中文定义】 也称为等效荷载法，即通过反应谱理论将地震对建筑物的作用以等效荷载的方法来表示，然后根据这一等效荷载用静力分析的方法对结构进行内力和位移计算，以验算结构的抗震承载力和变形。

pump discharge

【释义】 泵量

【例句】 During the process of trenchless construction, it is often inconvenient due to the bit water way matching the pump discharge inappropriate. 在实际非开挖施工过程中，由于扩孔钻头水眼大小与泵量不能够很好地匹配，给正常施工带来很大的不便。

pumping test

【释义】 抽水试验

【例句】 Through pumping tests, the hydrogeological parameters were confirmed. 通过抽水试验，确定了该区的水文地质参数。

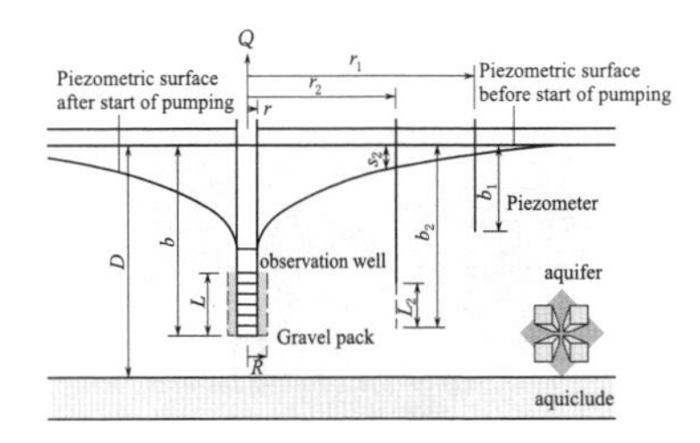

pumping test

【中文定义】 包括自抽水井抽取一定水量而在某距离之间观测井测定不同时间地下水水位的变化，观测数据利用各种地下水流理论式或其图解法分析抽水试验的结果。

pumping test of well group

【释义】 群孔抽水试验

【例句】 The pumping test of well group is done in suitable area, which can obtain the interrelationship between wells and the aquifer productivity. 通过在合适的地方做群孔抽水试验可以得出井群与含水层产水量之间的相互关系。

pumping well (borehole)

【释义】 抽水井（孔）

【例句】 The system consisted of an infiltration gallery, interdiction wells, pumping wells, and monitoring wells. 该系统包括渗渠，阻断井，抽水井和监测井。

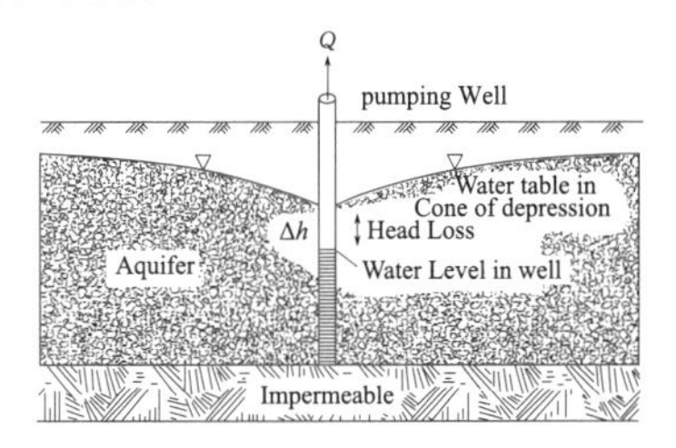

flow to a pumping well with head loss

pure water

【释义】 纯水

【例句】 Pure water is neutral with a pH of 7. 纯水是中性的，酸碱度为 7。

【中文定义】 指中性，酸碱度为 7 的液态水。

pycnometer method

【释义】 比重瓶法

【例句】 The test results show that the pycnometer method has poor repeatability and low quality reliability. 试验结果表明，比重瓶法在使用过程中存在重复使用性差、质量可靠性低等现象。

pyrite ['paɪraɪt]

【释义】 *n.* 黄铁矿

【常用搭配】 magnetic pyrite 磁黄铁矿；pyrite type structure 黄铁矿型构造

【例句】 Crystalline dolostone contains little associated minerals, only pyrite observed occasionally. 晶粒白云岩伴生矿物少，偶见黄铁矿。

pyrite

pyroclastic rock

【释义】 火山碎屑岩

【例句】 The volcanic rock series in the Rucheng basin in southeastern Hunan is composed of the basalt, diabase and basaltic pyroclastic rock. 湘东南汝城盆地火山岩系由辉绿岩、玄武岩和玄武质火山碎屑岩组成，属于低钾拉斑玄武岩系。

【中文定义】 是介于岩浆熔岩和沉积岩之间的过渡类型的岩石，其中50%以上的成分是由火山碎屑流喷出的物质组成，这些火山碎屑主要是火山上早期凝固的熔岩、通道周围在火山喷发时被炸裂的岩石形成的。

pyroclastic rock

pyroclastic texture

【释义】 火山碎屑结构

pyroclastic texture

【中文定义】 正常火山碎屑岩的特有结构。其特点是岩石中火山碎屑物的含量达到90%以上。

pyrophyllite [ˌpaɪˌroˈfɪlˌaɪt]

【释义】 *n.* 叶蜡石

【常用搭配】 active pyrophyllite 活性叶蜡石

【例句】 With the increase of heating temperature, phase of pyrophyllite will undergo a series of trans-formations. 随着加热温度的升高，叶蜡石将产生一系列的相变。

【中文定义】 一种非常软的硅酸盐矿物。

pyrophyllite

pyroxenite [paɪˈrɑksənaɪt]

【释义】 *n.* 辉岩

【例句】 Xiaozhangjiakou ultra basic rocks is the host rock of gold deposit, mainly consisting of pyroxenite rocks, including pyroxenite, pyroxene-peridotite, hornblende-pyroxenite and diopsidite, etc. 小张家口超基性岩体是华北板块北缘超基性岩带的一个典型岩体，岩相分带明显，由透辉岩、闪辉岩、二辉岩和纯橄岩组成，主要矿物组成包括透辉石、贵橄榄石、角闪石等。

pyroxenite

Q

quantity of exploration works
【释义】 勘探工作量
【例句】 Without increase quantity of exploration works, we try to make the calculation of oil and gas field recoverable reserve more closed to the objective and actual condition. 在不增加勘探工作量的前提下，尽可能使油气田可采储量计算更加接近客观实际。

quarry ['kwɔri]
【释义】 *n.* 采石场，料场
【常用搭配】 subterraneous quarry 地下采石场
【例句】 The quarry and dumping area of excavated rock from structures shall be planned according to the layout of project, construction planning and scheduling and quality requirements for embankment materials. 应根据工程枢纽布置、施工组织设计及对坝料的质量要求，做好采石场及建筑物开挖石料堆填区的料场规划。
【中文定义】 指开采石料的场地。

quarry

quartz [kwɔrts]
【释义】 *n.* 石英
【常用搭配】 high quartz 高温石英；low quartz 低温石英；quartz andesite 石英安山岩；quartz basalt 石英玄武岩；quartz diorite 石英闪长岩；quartz schist 石英片岩
【例句】 On the surface of the basal unconformity, there are erosion pits, paleo-karst(including solution groove, cave and breccias), crust of paleo-weathering(including limonite, clay and quartz), and basal conglomerate(such as fine chert conglomerate). 该底部不整合面位于晚三叠系马鞍塘组与中三叠系雷口坡组之间，显示为平行不整合面或角度不整合面，在接触面上发育冲蚀坑、古喀斯特溶沟、溶洞、溶岩角砾、古风化壳的褐铁矿、黏土层及石英、燧石细砾岩等底砾岩。

quartz

quartz vein
【释义】 石英脉
【例句】 The Huangsha tungsten deposit is an important large quartz vein type W-polymetallic deposit in southern Jiangxi Province. 黄沙钨矿床是赣南地区一大型石英脉型钨多金属矿床。

quartz vein

quartz vein infilled
【释义】 石英脉充填
【例句】 The mechanical properties of the quartz veins infilled structures are relatively good. 石英脉充填的结构面力学性质相对较好。

【中文定义】 由地下岩浆分泌出来的 SiO_2 的热溶液填充沉淀在岩石裂缝中形成。

quartz vein infilled

quartzite [ˈkwɔːtsaɪt]

【释义】 *n*. 石英岩，硅岩

quartzite

【例句】 Wuqiangxi Hydropower Station quartzite aggregate processing plant is the largest aggregate processing system in China. 五强溪水电站石英岩人工砂石骨料加工厂是我国最大的人工骨料加工系统。

【中文定义】 石英含量大于 85％的变质岩石。由砂岩和硅质岩经区域变质作用重结晶形成。

Quaternary (Q) [kwəˈtənəri]

【释义】 *n*. 第四纪（系）

【中文定义】 是新生代最新的一个纪，包括更新世和全新世，开始于距今约 260 万年。第四纪期间生物界已进化到现代面貌，其中，在灵长目中完成了从猿到人的进化。第四纪的构造运动属于新构造运动，陆地上新的造山带是新构造运动最剧烈的地区，如阿尔卑斯山、喜马拉雅山等。地震和火山是新构造运动的表现形式。

quick shear test

【释义】 快剪试验

【常用搭配】 consolidated quick shear test 固结快剪试验

【例句】 Besides, correlation between the strength indexes of direct shear test and consolidated quick shear tests were studied. 此外，还对典型土层直剪试验和固结快剪强度指标间的相关性进行了研究。

R

radioactivity survey
【释义】 放射性测量
【例句】 Radioactivity survey is one of the non-seismic exploration techniques. 放射性测量是非地震勘探技术之一。

radioisotope [ˌredɪoˈaɪsətop]
【释义】 *n*. 放射性同位素
【常用搭配】 artificial radioisotope 人造放射性同位素；radioisotope logging 放射性同位素测井
【例句】 Today many radioisotopes are produced using the particle accelerator called a cyclotron. 今天许多放射性同位素是用被称为回旋加速器的粒子加速器来制造的。
【中文定义】 如果两种原子质子数目相同，但中子数目不同，则他们仍有相同的原子序，在元素周期表中是同一位置的元素，所以两者就叫同位素。

radioisotope logging
【释义】 同位素示踪测井
【例句】 Analyzed is economic and social benefit from the application of the radioisotope tracer logging in oilfield development. 分析了放射性同位素示踪测井资料在油田开发应用中所带来的经济效益和社会效益。
【中文定义】 是利用放射性核素或稀有稳定核素作为示踪剂对研究对象进行标记分析的井。

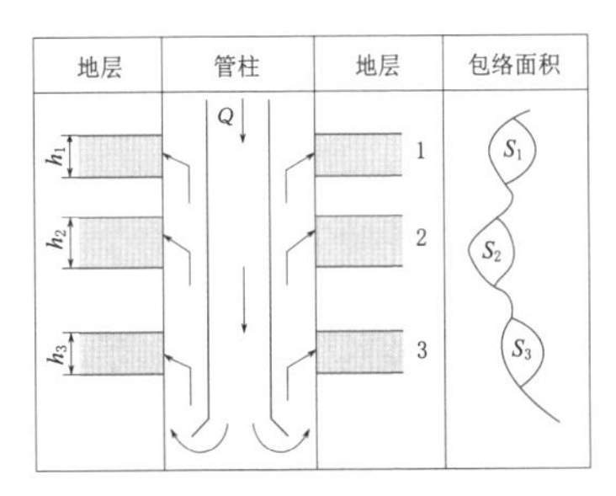

schematic diagram of radioisotope logging principle

radius of influence
【释义】 影响半径
【例句】 The misunderstood Dupuit stable well flow model(model of radius of influence) is the main theoretical basis of the unreasonable selection of exploited horizon of groundwater. 选择地下水开采层位不合理的理论依据主要是裘布依稳定井流模型（影响半径模型）。
【中文定义】 即降落漏斗的周边在平面上投影的半径。影响半径的大小与含水层的透水性、抽水延续时间、水位降深等因素有关。

radon survey
【释义】 氡气测量
【例句】 The results of the radon survey are affected by many factors, for instance, the humidity of the soil, the temperature in the detecting hole, climate, etc. 氡气测量结果受地表各种因素的影响，如土壤湿度、测孔中的温度、气候等。

raindrop imprint
【释义】 雨痕
【中文定义】 指雨点降落在未固结的泥质、砂质沉积物表面，所产生圆形或椭圆形的凹穴。

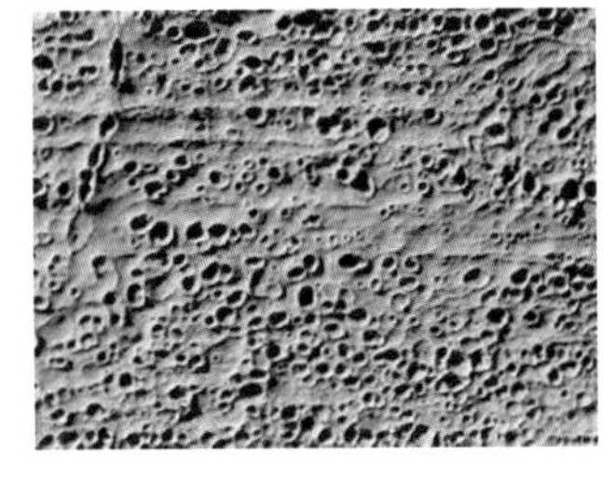

raindrop imprints

rainfall-triggered debris flow
【释义】 降雨型泥石流
【中文定义】 指暴雨将含有沙石且松软的土质山体经饱和稀释后形成的洪流。

rate of karst filling
【释义】 岩溶充填率
【中文定义】 填充物体积与孔洞体积的百分比。

rate of karstification
【释义】 岩溶率
【中文定义】 又称喀斯特率，是反映可溶岩（主要是碳酸盐岩）分布区在一定地段内岩溶发育程度的指标。根据统计方法不同，岩溶率可分为线岩溶率、面积岩溶率和体积岩溶率三种。

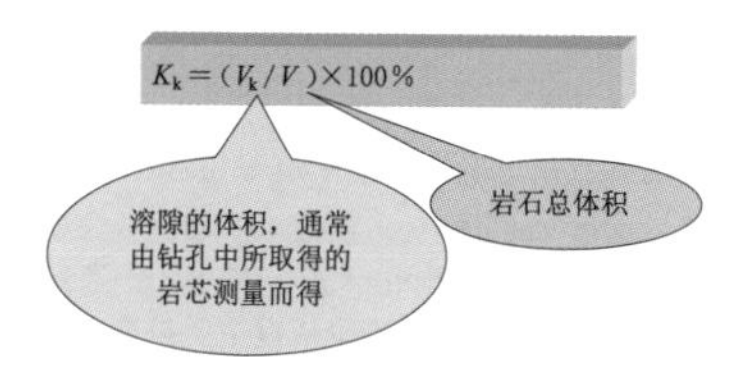

schematic diagram for
calculation rate of karstification

rate of swelling

【释义】 膨胀速率

【例句】 Compared with the higher suction specimens, the development rate of swelling pressure of specimens with the lower suction is inferior at early stage, but it becomes superior after a period of time. 相同干密度下，尽管初始阶段高吸力试样的膨胀力发展更快，但一段时间后低吸力试样的膨胀力变化速率会比高吸力试样的大。

rational utilization of groundwater

【释义】 地下水合理利用

【例句】 In order to prevent karst collapse the rational utilization of the karst groundwater should be solved and the environment stability of the groundwater should be come back to make the sustainable development of the resource and environment. 要防治岩溶塌陷，就要从根本上要解决岩溶地下水的合理利用问题，恢复地下水环境的生态稳定，促进资源、环境的可持续发展。

Rayleigh wave

【释义】 瑞利波

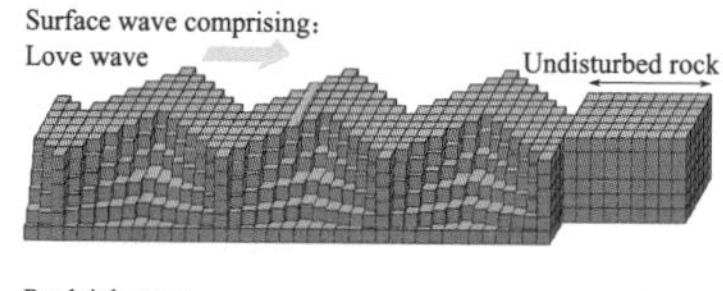

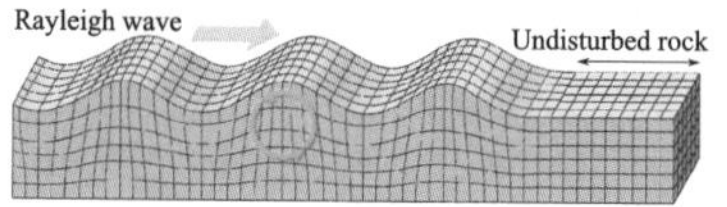

the propagation of Rayleigh wave

【例句】 Low-frequency pseudo Rayleigh wave (high velocity) can be used to estimate shear wave attenuation, then permeability in original state formation. 低频伪瑞利波（高速）能很好地估算原状地层的横波衰减量，进而估算地层的渗透率。

【中文定义】 一种常见的界面弹性波，是沿半无限弹性介质自由表面传播的偏振波 。由 L. 瑞利于 1887 年首先指出其存在而得名，地震学中称其为 R 波或 L 波。在震中附近，不出现瑞利波。瑞利波沿二维自由表面扩展，在距波源较远处，其摧毁力比沿空间各方向扩展的纵波和横波大得多，因而它是地震学中的主要研究对象。

reaction rim texture

【释义】 反应边结构

【中文定义】 岩浆中早期析出的矿物或捕虏晶，因物理化学条件改变，与周围岩浆发生反应，在其外围产生新的反应矿物，即一种矿物颗粒周围生长有另一种矿物镶边，这种现象称为反应边结构。

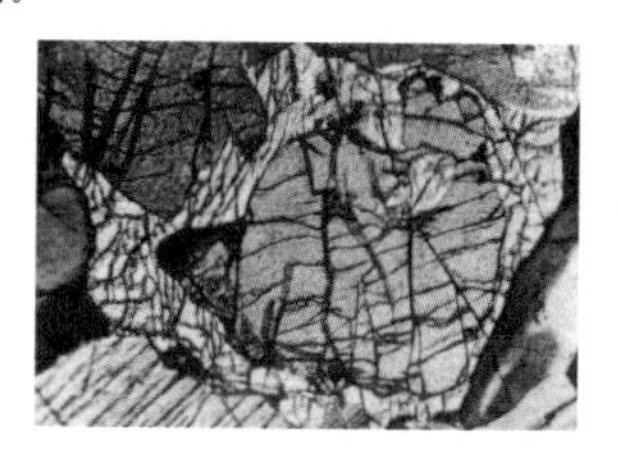

reaction rim texture

rebound of foundation

【释义】 地基回弹

【例句】 The variation rate of subsidence will become a negative value when the groundwater causes the rebound of foundation. 地下水回升引起地基回弹，其沉降变化速率为负值。

【中文定义】 基础施工中，因开挖基坑卸载而引起的地基土表面向上隆起的现象。

rebound survey of foundation pit

【释义】 基坑回弹测量

rebound survey of foundation pit

rebound test hammer

【释义】 回弹仪测试

【例句】 Rebound test hammer is the fastest, easiest and most economical test method for detecting concrete strength. 回弹仪测试是检测混凝土强度的最快速、最简单和最经济的测试方法。

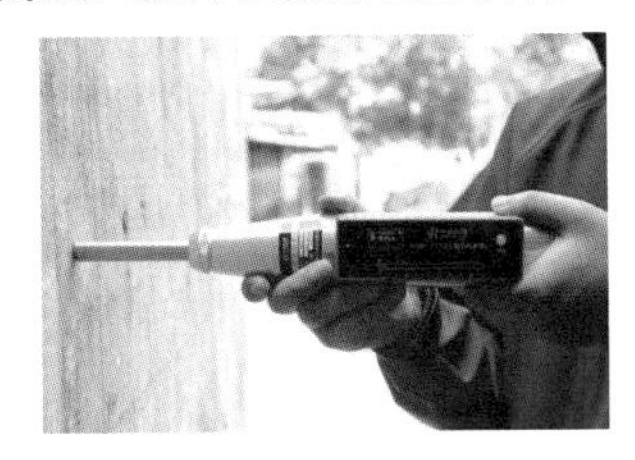

rebound test hammer

recently deposited soils

【释义】 新近沉积土

【例句】 These findings would provide more knowledge about dynamic characteristics of recently deposited soils in this area, and could be used as reference for seismic safety evaluation of general engineering sites. 该成果有利于加深对该地区新近沉积土动力特性的认识，并对该地区一般工程建设场地的地震安全性评价工作具有借鉴和参考价值。

【中文定义】 第四纪全新世中近期沉积的土，定为新近沉积土。

recently deposited soil

recharge area

【释义】 补给区

【中文定义】 指含水层接受大气降水、地表水、回渗（归）水以及其他含水层等入渗补给的地区。

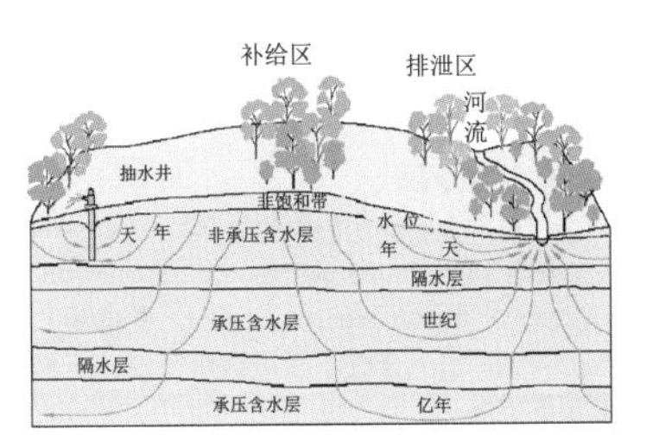

schematic diagram of recharge area

recharge rate

【释义】 补给率

【例句】 Because of large spatial and temporal variation, the direct measurement of groundwater recharge rate is hardly possible, so it is very difficult to reconstruct the groundwater recharge history and estimate the changes of recharge rate. 由于地下水时空变化巨大，直接测量地下水补给率较为困难，很难重建地下水补给史，估计补给率变化。

recoverable reserves

【释义】 可开采储量

【例句】 Shale gas, according to the committee, accounts for two-thirds of America's technically recoverable reserves, enough to supply the country for 90 years. 根据委员会的数据，页岩层天然气在技术上占美国可开采储量的 2/3，足够供应美国达 90 年之久。

recumbent fold

【释义】 平卧褶皱

recumbent fold

【例句】 Without late tectonic deformation, the features of recumbent fold on the geological map would be the same as those of horizontal strata. 平卧褶皱若未经后来构造变动，其图面特征与水平地层区完全一样。

【中文定义】 又称横卧褶皱，是轴面和枢纽都近于水平（倾角和倾伏角均为 0°～10°）的倒转褶皱。

reduction factor

【释义】 折减系数

【例句】 In consideration of long-term degradation, geosynthetic products should respond to minimum requirements of default reduction factor. 考虑到长期降解的需要，土工合成产品应该满足默认折减系数的最低要求。

【中文定义】 是排水工程中的一个概念，其值随长细比而变化且是一个小于或等于 1 的数。

reexamine [ˌriːɪg'zæmɪn]

【释义】 *vt*. 复核

【例句】 The influence of historical earthquakes is evaluated on the basis of actual records and the epicenter location and magnitude of destructive earthquakes is reexamined. 由实际记载资料进行历史地震影响状况评价，并复核确认破坏性地震的震中位置和强度。

reference ellipsoid

【释义】 参考椭球面

【例句】 The determination of projective coordinates in different reference ellipsoid, which is with the transform between ellipsoid elements in different reference ellipsoid, gets the coordinates in corresponding new ellipsoid, and finally projects to horizontal coordinates. 位于不同参考椭球面下投影面坐标的求解，是通过不同参考椭球间椭球元素的转换，求得对应的新椭球体上的大地坐标，最后投影至平面坐标。

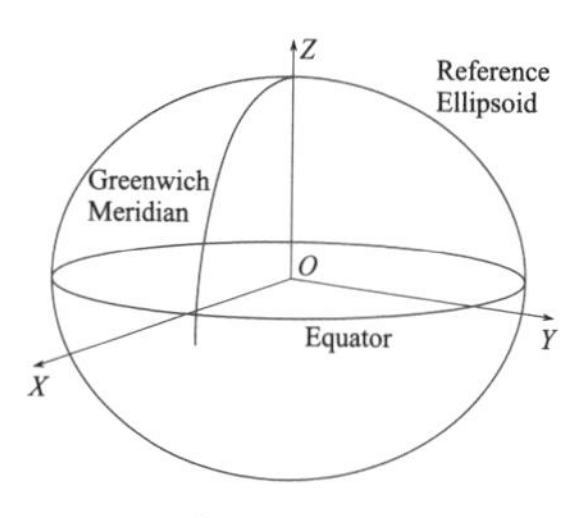

reference ellipsoid

reference supply radius

【释义】 引用补给半径

regional comprehensive geological map

【释义】 区域综合地质图

【例句】 In regional geological investigation, a regional comprehensive geological map of the planned river or river section shall be prepared on the basis of collection and analysis of the available latest geological data in the region. 区域地质勘察工作应在收集和分析已有的各类最新区域地质资料的基础上，编绘规划河流或河段的区域综合地质图。

【中文定义】 反映一定研究区域的地形地貌、地层岩性、地质年代、地质构造、岩浆活动、地壳运动和发展历史、地震分布等内容的综合性图件，通常包括区域地质平面图、地层柱状图、地质剖面图、构造纲要图、地震震中分布图等。

regional metamorphite

【释义】 区域变质岩

【例句】 What kinds of regional metamorphite rocks are present in the area? 该区域有哪些区域变质岩？

【中文定义】 由区域变质作用所形成的岩石。

regional metamorphic rocks schist

regional seismotectonics

【释义】 区域地震构造

【中文定义】 指与地震孕育和发生有关的地质构造。通常按照研究内容和角度的不同，将地震构造分为全球地震构造、区域地震构造、震源构造及工程地震构造。

regional tectonic stability

【释义】 区域构造稳定性

【例句】 Based on finite element analysis, the authors studied the regional tectonic stability of the Hutiaoxia(region) systematically. 通过二维有限元对滇西北虎跳峡地区区域构造稳定性进行了系统研究。

【中文定义】 工程建设术语，是指工程建设地区地壳的稳定程度，即该地区有无活断层和地震等现代构造活动的迹象、它们活动的强度及其对工程的可能影响等情况。

regional tectonic outline map

【释义】 区域构造纲要图

【例句】 In prefeasibility study stage of hydropower engineering, the regional tectonic outline map shall be submitted in report for engineering geological investigation. 水力发电工程预可行性研究阶段工程地质勘察报告中应附区域构造纲要图。
【中文定义】 用不同的线条、符号、色调来表示一个研究区域主要地质构造特征的图件。

regional tectonic stress fields

【释义】 区域构造应力场
【例句】 These results are consistent with NE-NEE pinching action of the regional tectonic stress field. 这与区域构造应力场 NE-NEE 向挤压作用的结果相符。
【中文定义】 通常指导致构造运动的地应力场，或者由于构造运动而产生的地应力场。在地质力学中，构造应力场是指形成构造体系和构造型式的地应力场，包括构造体系和构造型式所展布的地区，连同它内部在形成这些构造体系和构造型式时的应力分布状况，它是确定构造体系或构造形式的必要步骤之一。

relative density

【释义】 相对密度
【例句】 On the basis of Terzaghi formula of flowing soil, the probability density function of critical hydraulic gradient is derived with soil relative density and void ratio as random variables. 基于太沙基公式，将土粒相对密度和孔隙比作为随机变量，推导了临界水力比降的概率密度函数。

relative elevation

【释义】 相对高程
【中文定义】 在局部地区，当无法知道绝对高程时，假定一个水准面作为高程起算面，地面点到该假定水准面的垂直距离称为相对高程，又称为假定高程。

relative stable block

【释义】 相对稳定地块
【例句】 The analysis of the region geology background, the activity of faults close to the dam site, history earthquake in the region, and the earthquake hazard, showed that all of the four power stations stand on the relatively stable block and the region main fractures and activity faults are avoided. 通过对区域地质背景、工程近场区断层活动性、历史地震活动及坝址区地震危险性的分析，认为金沙江下游四座梯级水电站所选定的坝址区域均处于一个地质相对稳定地块，且避开了区域主干断裂和活动断层。
【中文定义】 又称安全岛，是指现今构造活动、岩浆活动、水热活动、地震活动以及区域物理地质作用比较轻微、结构比较完整的地区。

relatively crushed

【释义】 较破碎
【中文定义】 根据《水力发电工程地质勘察规范》(GB 50287—2016) 的规定，当岩体完整性系数 $0.15<K_v\leqslant0.35$ 时，则该岩体的完整程度为较破碎。或根据岩体中结构面发育程度判断，较破碎岩体中结构面发育组数多大于 3 组，结构面间距平均值多小于 0.1m。

relatively developed

【释义】 较发育
【中文定义】 根据《水力发电工程地质勘察规范》(GB 50287—2016) 的规定，当岩体内结构面发育组数 2～3 组，间距介于 10～30cm 时，其岩体结构面发育程度为较发育。

relatively integral

【释义】 较完整
【中文定义】 根据《水力发电工程地质勘察规范》(GB 50287—2016) 的规定，当岩体完整性系数 $0.55<K_v\leqslant0.75$ 时，则该岩体的完整程度为较完整。或根据岩体中结构面发育程度判断，较完整岩体中结构面多发育 1～3 组，结构面间距平均值多介于 0.3～1m。

relax circle testing of surrounding rock mass

【释义】 围岩松弛圈检测

relaxation region

【释义】 松弛区
【例句】 The horizontal displacement of slopes is influenced by excavation, which form excavation relaxation region in excavation process. 边坡中水平方向位移受开挖作用影响较大，边坡在开挖过程中会形成开挖松弛区。
【中文定义】 边坡在开挖过程中所产生的松动区域。

relaxation time

【释义】 松弛时间
【例句】 The effects of shear rate, relaxation time and slip length on extrudate swell ratio were investigated using simulation. 模拟研究了剪切速率、松弛时间和滑移长度对挤压膨胀性的影响。

relaxed rock mass

【释义】 松动岩体
【例句】 According to abundant prospecting data

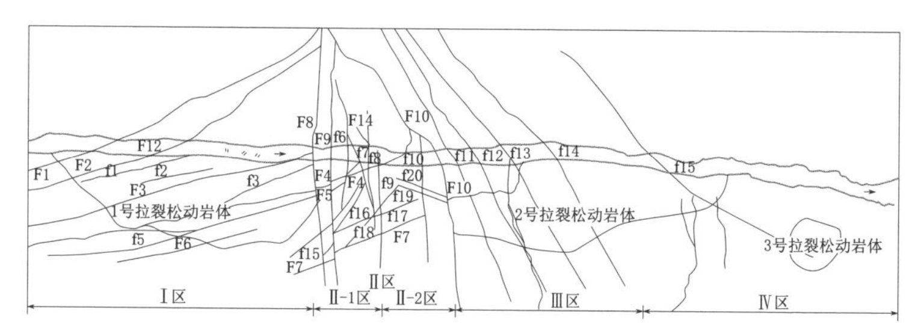

primary partition of structural plane about relaxed rock mass

of irrigation works and hydropower stations and various reports of monographic study it has been testified that there is relaxed rock mass over an extensive area at Daliushu dam site. 根据大量的水利水电勘察资料和各方面的专题研究成果，证明大柳树坝址存在大范围的松动岩体。

relaxed zone of rock mass

【释义】 岩体卸荷带

【例句】 On the base of slope stress, the paper describes carefully the common law of relaxed zone of rock mass, and then gives examples of some projects in China, finally puts forth and summarizes several methods that classifies different relaxed zones of slope rockmass. 以斜坡岩体应力分布为基础，在详细描述了斜坡岩体卸荷带发育的一般规律后，结合中国的几个实际工程，该文提出和总结了划分岩体不同卸荷带的几种方法。

【中文定义】 是由于自然地质作用和人工开挖使岩体应力释放而造成的具有一定宽度的岩石松动破碎带。

reliability analysis

【释义】 可靠度分析

【例句】 Then the design of anchoring length by reliability analysis is proposed. 在此基础上，借助可靠度分析提出锚固长度设计值。

relict texture

【释义】 残余结构

【中文定义】 是原岩在变质作用过程中，由于重结晶、变质结晶不完全，原岩的结构特征被保留下来的部分。

remote sensing

【释义】 遥感

【中文定义】 不接触物体本身，用传感器收集目标物的电磁波信息，经数据处理、分析后，识别目标物、揭示目标物几何形状大小和相互关系及其变化规律的科学技术。

remote sensing (RS) ＋geographical information system (GIS) ＋global positioning system (GPS)

【释义】 3S技术（3S technology）

【例句】 Construction of river pollution monitoring information system based on 3S technology. 基于3S技术构建河流水污染监测信息系统。

【中文定义】 遥感技术（remote sensing，RS）、地理信息系统（geography information systems，GIS）和全球定位系统（global positioning systems，GPS）的统称，是空间技术、传感器技术、卫星定位与导航技术和计算机技术、通信技术相结合，多学科高度集成的对空间信息进行采集、处理、管理、分析、表达、传播和应用的现代信息技术。

remote sensing platform

【释义】 遥感平台

【例句】 With the development of remote sensors, remote sensing platforms, and data communication technologies, remote sensing data is increasing in an incredible speed. 随着传感器、遥感平台、数据通信等相关技术的发展，通过遥感手段获取的数据量急剧膨胀。

remote sensing platform

repetitive load

【释义】 反复载荷

【例句】 Principal stress rotation was shown to be a major factor influencing the behaviour of dry granular material when subjected to repetitive load, the effect of which was to induce permanent deformation while the elastic response remained unchanged. 在反复荷载作用下，主应力旋转是影响干颗粒材料性能的主要因素，表现为弹性响应不变的情况下发生永久变形。

replacement [rɪ'plesmənt]

【释义】 *n.* 置换

【例句】 For stone backfilling area, adopt dynamic replacement method, which simultaneously has dynamic consolidation effect. 对于块石回填区域，采用了强夯置换法加固，同时还具有动力固结效应。

【中文定义】 指用物理力学性质较好的岩土材料置换天然地基中部分或全部软弱土体，以形成双层地基或复合地基，达到提高地基承载力、减小沉降的目的。

replacement method

【释义】 置换法

【例句】 Through hard coarse particle fillers are added into the tamper hole for building drainage channel, dynamic compaction replacement method or dynamic compaction replacement method make the pore water of the foundation soil to divert and to speed up the soil consolidation. Dynamic compaction half replacement method is used to deal with the application of high moisture content of clay foundation in zhaotong dc converter. 强夯置换法或强夯半置换法利用动力固结原理，通过在夯坑中加硬质粗颗粒填料，建立排水通道，使地基土夯坑周围与夯坑底的孔隙水就近转移，加快土层固结。

replacement method

resection method

【释义】 后方交会法

reservoir bank collapse

【释义】 水库塌岸

【常用搭配】 reservoir bank collapse prediction 塌岸预测

【例句】 Reservoir bank collapse is an import geological problem which should have an effect on Three Gorges Reservoir Project. 水库塌岸是影响三峡水库工程成败的重大地质问题。

【中文定义】 水库蓄水后水库周边岸壁发生的坍塌现象。

reservoir bank stability

【释义】 库岸稳定

【例句】 At present, reliability analysis and engineering technic is applied to the reservoir bank stability step by step. 目前可靠性分析及工程技术的研究已逐步在库岸稳定性分析中得以应用。

【中文定义】 水库蓄水过程和运行阶段改变河谷岸坡的自然平衡条件后，库岸岩土体的稳定性。

reservoir leakage

【释义】 水库渗漏

【例句】 This debris might, however, be slowly washed out by percolating water from reservoir leakage. 然而，这些残渣可能被水库渗漏的渗水慢慢地冲刷掉。

【中文定义】 库水向库外低邻谷或向坝下游漏失的现象。

reservoir-induced earthquake

【释义】 水库触发地震

【例句】 In this paper, the mechanism of reservoir-induced earthquake is discussed according to the inner and outer factors. 这篇文章从触发地震的内因和外因两个方面论述水库触发地震形成的机制。

【中文定义】 在特殊的地质背景下，因水库蓄水引起水库及其附近地区内新出现的、与当地天然地震活动性规律明显不同的地震活动。

residual drawdown

【释义】 剩余降深

【例句】 In such a test, a well that has been discharging for some time is shut down, and thereafter the recovery of the aquifer's hydraulic head is measured in the wells; the change in water level during this recovery period is known as the residual drawdown. 在这个测试中，先关闭一个已经抽过一段时间水的井，之后测量水井中含水层的恢复水头；水位在这个

恢复期间的变化被称为剩余压降。

residual soil

【释义】 残积土

【例句】 Noteworthy differences in material properties also became evident when residual soil weathered from different parent materials were considered independently. 当从不同原岩风化的残积土被单独考虑时，材料特性的显著差异也变得明显。

【中文定义】 是岩石风化后残留在原地的碎屑堆积物形成的土。

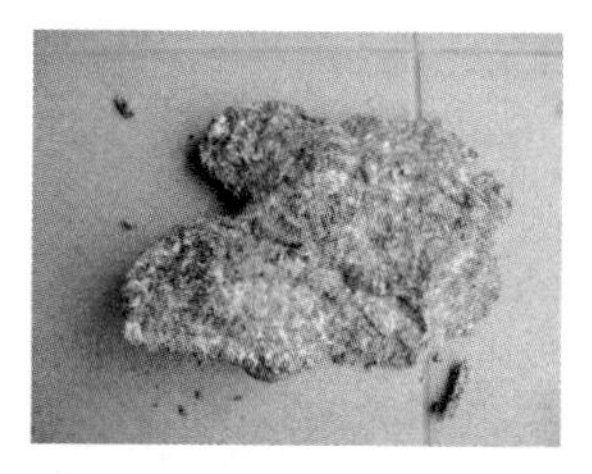

residual soil

residual strength

【释义】 残余强度

【例句】 When shear surfaces are irregular shear bands, residual strength increase after applying cyclic loads. 当土体剪切面为不规则剪切带时，施加循环荷载后出现残余强度上升的现象。

【中文定义】 指岩石在破坏后所残留的抵抗外荷的能力。

residual stress

【释义】 残余应力

【例句】 Cutting condition influences the residual stress of machined surface. 切削条件影响已加工表面的残留应力。

【中文定义】 指消除外力或不均匀的温度场等作用后仍留在物体内的自相平衡的内应力。

resistivity imaging method

【释义】 高密度电阻率法

【例句】 As the resistivity imaging method data of two adjacent arrays have the overlaps, we present a method to combine the arrays and remove the error. 高密度电法集中了电剖面法和电测深法，其原理与普通电阻率法相同。

resistivity method

【释义】 电阻率法

resonant column test

【释义】 共振柱试验

【例句】 To ensure the credibility, the reach points of shear wave velocity are examined through the comparison of resonant column tests. 为保证试验结果的可靠性，同时进行了共振柱试验，比较确定弯曲元接收信号的到达点。

retrogressive landslide

【释义】 牵引式滑坡

【例句】 This complex retrogressive landslide was reactivated in December 1996. 这个复杂的牵引式滑坡于 1996 年 12 月被重新激活。

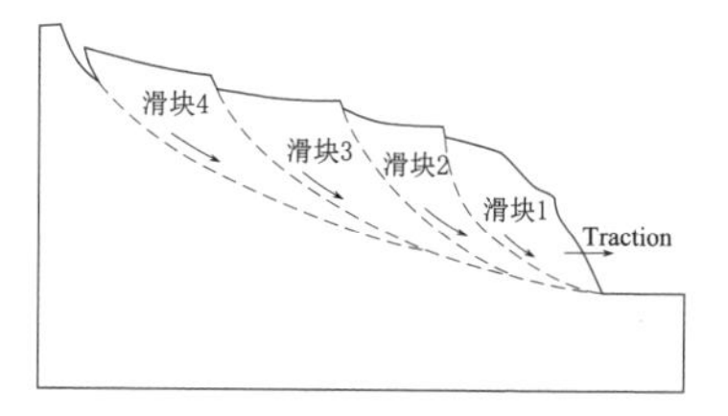

retrogressive landslide

return period

【释义】 重现期，间隔期，逆程周期

【同义词】 recurrence interval

【例句】 The likelihood of earthquake occurrence is often described in terms of a return period. 地震发生的可能性通常被描述为时延重现期。

【中文定义】 指在一定年代的地震记录资料统计期间内，大于或等于某地震强度的地震出现一次的平均间隔时间为该地震发生频率的倒数。

reverse circulation drilling

【释义】 反循环钻进

Reynolds number

【释义】 雷诺数

【例句】 The effect of interfacial shear on critical wave velocity is clearly different under different Reynolds number. 剪切力对临界波速的影响在不同雷诺数下也有所不同。

【中文定义】 一种可用来表征流体流动情况的无量纲数。

rheological behavior

【释义】 流变特性，流变性能

【同义词】 rheological property

【例句】 The fundamental behavior of rock mass can be drastically changed when submerged in water. It is also valid for the rheological behavior of rock mass. 岩石浸没于水中时，其基本特性会发生剧烈变化，其流变性能也不例外。

【中文定义】 流变特性是物体在外力作用下发生的应变与其应力之间的定量关系。

rheological deformation

【释义】 流变

【中文定义】 指在外力作用下，物体的变形和流动。

rheological property of rock

【释义】 岩体流变特性

【例句】 This paper studies and discusses systematically rheological property of rock and its application to large-scale rock engineerings under complex geological and engineering conditions. 本文系统地研究和探讨了岩石的流变性能及其在复杂地质条件和工程条件的大型岩石工程中的应用。

【中文定义】 指岩石的蠕变、应力松弛、与时间有关的扩容，以及强度的时间效应等特性。

rheological property of soil

【释义】 土体流变性能

【例句】 Based on excavation work of foundation ditch in Pudong district, Shanghai, the geological properties of soil was studied and the results were used for monitoring measurement of foundation ditch support. 本文结合上海浦东新区的基坑开挖，研究土体的流变特性及其在基坑支护监控量测中的应用。

【中文定义】 指土体在外力作用下发生的应变与其应力之间的定量关系，由于土体具有流变性，表现出四种主要现象，即蠕变、应力松弛、长期强度和应变率效应。

rhyolite [ˈraɪəˌlaɪt]

【释义】 *n.* 流纹岩

【常用搭配】 porphyritic rhyolite 斑状流纹岩；banded rhyolite 带状流纹岩

【例句】 Topaz occurs in the igneous rock rhyolite. 黄玉通常产于火成流纹岩中。

rhyolite

rhyotaxitic structure

【释义】 流纹构造

【中文定义】 岩浆岩中不同颜色的矿物、玻璃质和气孔等呈条纹状排列的构造。

rhyotaxitic structure

Richter magnitude

【释义】 里氏震级

【例句】 The Wenchuan Earthquake registered 8 on the Richter magnitude scale, making it a huge earthquake. 汶川地震是里氏 8 级地震，是一场大地震。

【中文定义】 里氏震级是两位来自美国加州理工学院的地震学家里克特和古登堡于 1935 年提出的一种震级标度，是目前国际通用的地震震级标准。它是根据离震中一定距离所观测到的地震波幅度和周期，并且考虑从震源到观测点的地震波衰减，经过一定公式，计算出来的震源处地震的大小。

rigid structural plane

【释义】 硬性结构面

【例句】 The rigid structural plane has obvious dilatancy effect and gnawing effect. 硬性结构面具有明显的剪胀效应和啃断效应。

【中文定义】 指摩擦系数较大、延伸较短，且多数无充填的结构面。

rigid structural plane

ripple mark

【释义】 波痕

【例句】 The sedimentary structure mainly includes unidirectional and bidirectional cross bedding, interference and curved ripple mark, and wavy, flaser and lenticular bedding. 沉积构造主要包括双向和单向交错层理、曲线型和干涉波痕以及波状、脉状、透镜状层理。

【中文定义】 沉积岩中的一种层面构造，是由于风、流水或波浪等作用于沉积物表面时，所造成的起伏不平的波纹状痕迹。

ripple mark

risk analysis

【释义】 风险分析

【例句】 The probability distribution type of soil strength affects the results of reliability and risk analysis of geotechnical problems. 土的抗剪强度指标概率分布的类型影响着岩土工程问题的可靠性和风险分析结果。

risk analysis of slope

【释义】 边坡风险分析

【中文定义】 指通过查明系统中的各种不确定性因素，对边坡存在的危险性做出分析与评价。

risk assessment

【释义】 风险评估

【例句】 Flood disaster risk assessment is a new study field. 区域洪水灾害风险评估是一个新的研究领域。

risk assessment for geological hazard

【释义】 地质灾害危险性评估

【例句】 The depth and accuracy of risk assessment for geological hazards varies with the level of construction project evaluation. 地质灾害危险性评估因建设项目评估级别不同，评估深度和精度也不同。

risk control

【释义】 风险控制

【例句】 Causes of decision-making risk analysis, and the risk control method for this kind of more universal. 对决策风险产生原因所进行的分析，以及对这类风险控制方法进行更具有普遍性的探讨。

risk identification

【释义】 危险源辨识

risk of geological hazard

【释义】 地质灾害危险性

【例句】 The appraising range, complex degree of geological environment, classifiable standard in appraising risk of geological hazards and its importance are expounded in this paper. 本文阐述了评估范围、地质环境条件复杂程度、地质灾害危险性在评估工作中的划分标准及其重要性。

risk warning

【释义】 风险预警

risk zone for geological hazard

【释义】 地质灾害危险区

【例句】 The risk level has significantly zoned features. The distribution of risk zone for geological hazard area is consistent with the mined-out area. 地质灾害危险程度有明显的分区性，地质灾害危险区与矿区采空区分布基本一致。

risk zone for geological hazard

river bend

【释义】 河湾，河曲

【例句】 Numerical simulation on supercritical flow at river bend can provide bases for engineering design and safe evaluation. 弯道急流数值计算可以对工程设计及安全评估提供水流模型基础。

【中文定义】 指河流迂曲处。

river bend

river valley

【释义】 河谷

【例句】 Groundwater will be draining towards a river valley and ultimately to the sea. 地下水将朝着河谷排泄，最终流向大海。

【中文定义】 河流所流经的槽状地形称为河谷，它是在流域地质构造的基础上，经河流的长期侵蚀、搬运和堆积作用逐渐形成和发展起来的一种地貌。

river valley

river valley debris flow

【释义】 河谷型泥石流

【中文定义】 以流域为周界，受一定的沟谷制约，泥石流的形成、堆积和流通区较明显，轮廓呈哑铃形。

riverbed ['rɪvərˌbɛd]

【释义】 *n.* 河床

【常用搭配】 old riverbed 古河床；riverbed deformation 河床变形

【例句】 In the slope toe and the bottom, the direction of maximum principal stress parallel to the strike of the bottom or the riverbed after been redistributed. 在坡脚及谷底，重分布后的最大主应力方向与谷底或河床走向近于平行。

【中文定义】 河谷谷底部分河水经常流动的地方称为河床，亦指被河流占据或从前被河流占据的沟槽。

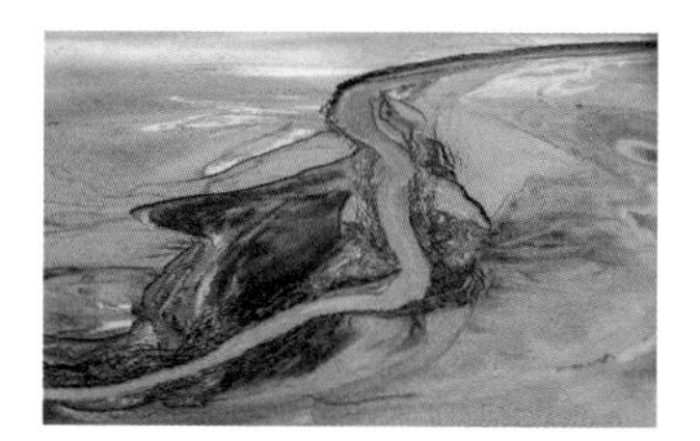

dried up riverbed

rock avalanche

【释义】 山崩

【例句】 Earthquakes can trigger rock avalanche that cause great damage and loss of life. 地震可能触发导致巨大损伤和丧生的山崩。

【中文定义】 大规模的山体崩塌。

rock block density

【释义】 岩石块体密度

【例句】 The rock block density test is necessary. 岩石块体密度测试是必要的。

rock bolt foundation

【释义】 岩石锚杆基础

【例句】 Rock bolt foundation is a type of bolt foundation that fine aggregate concrete and bolt are poured into the rock hole. 岩石锚杆基础是以细石混凝土和锚杆灌注于钻凿成型的岩孔内的锚杆基础。

【中文定义】 由设置于钻孔内、端部伸入稳定基岩中的锚杆所形成的基础。

rock bolt foundation

rock bridge

【释义】 岩桥

【例句】 The proposed approach considers the distribution characteristics of joints and the effects of rock bridge. 改进后的方法考虑了岩体结构面的分布特征和岩桥的作用。

specimen of rock bridge

rock burst
【释义】 岩爆
【中文定义】 在地应力高度集中的岩层中开挖时，围岩应力因突然释放而引起的岩块爆裂向外抛射的现象。

rock core
【释义】 岩芯
【例句】 The geologist must be present while taking out the rock core. 岩芯出筒时，必须有地质人员在场。

rock core

rock dilatancy
【释义】 岩石扩容

rock drillability
【释义】 岩石可钻性
【例句】 It is a simple and feasible approach to calculate rock drillability from log data. 利用测井资料计算岩石的可钻性是一种简便可行的途径。

rock free swelling ratio
【释义】 岩石自由膨胀率

rock lateral constraint swelling ratio
【释义】 岩石侧向约束膨胀率

rock lateral constraint swelling ratio

rock mass
【释义】 岩体
【中文定义】 赋存于一定地质环境，含不连续结构面，且具有一定工程地质特征的岩石综合体。

rock mass

rock mass chimerism
【释义】 岩体嵌合度
【中文定义】 镶嵌结构岩体内，岩块间咬合、镶嵌的紧密程度。

rock mass classifications of slope
【释义】 边坡岩体分类（分级）

rock mass creep
【释义】 岩体蠕变
【例句】 In this paper, structural effects of rock mass creep are studied by using numerical test. 本文采用数值模拟试验方法对岩体结构的蠕变力学效应进行了研究。

rock mass integrity
【释义】 岩体完整程度
【中文定义】 根据岩体内结构面发育程度、结构面结合程度以及结构面类型对岩体的划分，岩体完整程度一般分为完整、较完整、完整性差、较破碎、破碎。

Rock Mass Rating
【释义】 RMR 分类
【例句】 The Rock Mass Rating (RMR) System is a geomechanical classification system for rocks, developed by Z. T. Bieniawski between 1972 and 1973. RMR 分类是一种岩石地质分类系统，由 Z. T. Bieniawski 于 1972—1973 年开发。
【中文定义】 RMR 地质力学分类法是国际通用的岩体质量分类方法之一，由 Bieniasiki 于 1973 年最早提出。

rock mass relaxing
【释义】 岩体松动，岩体松弛
【例句】 The rock mass relaxing due to the earthquake brings about hidden dangers to the surrounding environment. 地震作用造成的边坡岩体松动，给周围环境造成了安全隐患.
【中文定义】 指组成岩体的结构面张开或者连接

强度降低，岩体变形破坏的可能性增大。

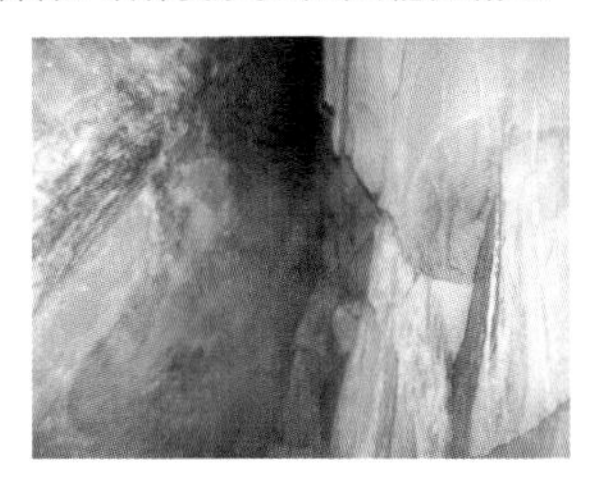

rock mass relaxing

rock mass strength

【释义】 岩体强度

【例句】 Rock fissure as one kind of geological discontinuities is a main factor controlling the rock mass strength and flow characteristics. 岩体中的裂隙是地质作用形成的不连续面，它控制着岩体的强度和渗流特征。

【中文定义】 是指岩体抵抗因受力而破坏的能力。

rock mass stress redistribution

【释义】 岩体应力重分布

【中文定义】 本来一个岩体构件在受一定力的时候变形是一定的，内部应力分布也是稳定的，如果增加外力，那么岩体结构由于几何变形，导致内部应力发生变化，这就是岩体应力重分布。

rock mass weathering

【释义】 岩体风化

【常用搭配】 weathering zoning of rock mass 岩体风化分带

【例句】 Based on the present state of weathering agency research at home and abroad, principal problems are pointed out in quantitative assessment of rock mass weathering. 在分析国内外岩体风化问题研究现状的基础上，指出了目前岩体风化量化评价存在的主要问题。

rock mass weathering

【中文定义】 在水、大气、温度和生物等应力的作用下，地壳上部岩体的物质成分和结构发生变化，从而改变岩体力学性质的过程和现象。

rock pressure

【释义】 岩石压力，山岩压力

【例句】 In mountain city construction, rock pressures between the rock slope and retaining structure can't be calculated accurately up to now. 在山城建筑中，岩石边坡支挡结构上的岩石压力至今尚无法精确地计算。

【中文定义】 又称围压，指的是地壳中岩石所受具有均向性的压力。

Rock Quality Designation (RQD)

【释义】 岩石质量指标（RQD）

【例句】 RQD value, one of the rock quality indices, may become a new and practical way to accurately estimate the strength of broken rock mass. RQD值作为岩体质量指标之一，可为破碎岩体的强度估计提供新的实用途径。

rock strength

【释义】 岩石强度

【例句】 Rock strength is a basic parameter for rock engineering design. 岩石强度是岩体工程设计的基本依据。

【中文定义】 指岩石抵抗因受力而破坏的能力。

rock swelling pressure

【释义】 岩石膨胀压力

rock water absorptivity

【释义】 岩石吸水率

【例句】 Rock water absorptivity is one of the physical properties of rock. 岩石吸水率是岩石的物理性质之一。

rock water enrichment

【释义】 岩石富水性

【例句】 Rock water enrichment refers to water content and drain capacity of rock. 岩石富水性是指岩石的含水和出水能力。

【中文定义】 岩石的含水和出水能力。

rock-mineral identification report

【释义】 岩矿鉴定报告

rock-soil physical properties

【释义】 岩土物理特性

【中文定义】 表示土中三向比例关系的一些物理量。

rockfall [ˈrɒkfɔːl]

【释义】 *n.* 岩崩，落石

【常用搭配】 subaqueous rockfall 水下落石；landslide rockfall 滑坡崩塌

【例句】 The rockfall is a kind of harmful physico-geological phenomenon occurred commonly in mountainous areas. 岩崩是山区常见的一种不良物理地质现象。

【中文定义】 崩塌的一种，主要由岩石形成。

rockfall

roof fall

【释义】 冒顶，塌顶

【常用搭配】 face roof-fall 端面冒顶；fault roof-fall area 断层冒顶区；thrust roof fall 推垮型冒顶

【例句】 Roof fall is the most ordinary accident in coal mine. So analysing the causes, signs and controlling of roof fall is the key of mining in coal mine. 冒顶是煤矿生产中常见的事故，因此认识冒顶的原因及预兆，控制好冒顶是煤矿安全生产的重要保证。

【中文定义】 指洞室或地下采空区顶部围岩发生坠落或塌落的现象

roof fall

root zone

【释义】 根系层

【中文定义】 指土壤剖面中以植物活根系为主的层，是湿地发育的活跃层，物质和能量的迁移转化在此层最为活跃。

rose diagram of joint

【释义】 节理玫瑰图

【常用搭配】 rose stereographic diagram of joint 节理玫瑰赤平图

【中文定义】 是将野外所测裂隙产状要素资料分别以不同组、系予以整理，绘制形似玫瑰花的一种图式，是一种用来表示节理空间方位及其发育程度的图解。

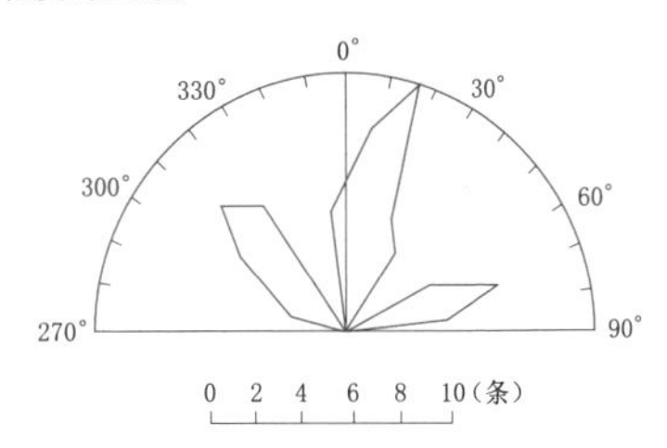

rose diagram of joint

rotary drilling

【释义】 回转钻进

【例句】 At the present time, rotary drilling is the leading position and will be continue to play important role. 目前，回转钻进方法在地质勘探、石油钻井和各种工程勘察施工中仍占统治地位，并将继续发挥重大作用。

rough [rʌf]

【释义】 *adj*. 粗糙的

【例句】 There are still a few rough places on the surface of the table that need to be planed down. 桌子的表面仍有一些粗糙的地方需要刨平。

【中文定义】 是结构面一种粗糙程度的描述。结构面的光滑程度不同，抗剪强度也不同。

roughness [ˈrʌfnɪs]

【释义】 *n*. 粗糙度

【常用搭配】 surface roughness 表面粗糙度；roughness coefficient 粗糙系数

【例句】 The roughness degree of structural plane has an important influence on shear strength. 结构面粗糙度对抗剪强度有重要影响。

【中文定义】 指结构面侧壁的粗糙程度，用起伏度和起伏差来表征。

round gravel

【释义】 圆砾

【例句】 In Nanning basin, the round gravel layer is an ideal bearing stratum of piles, because of it's homogeneous particle composition, compact cohesion and continuous distribution. 南宁盆地圆砾颗粒组成均匀，胶结紧密，分布连续，是较理想的桩端持力层。

【中文定义】 指粒径大于 2mm 的颗粒超过总质

量 50%的砾石。

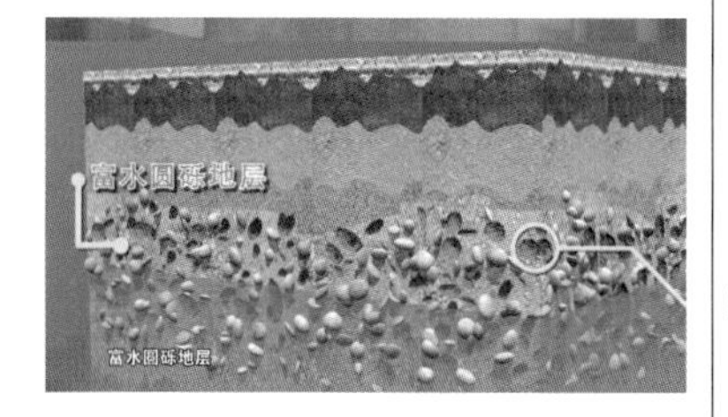

round gravel

rounded ['raʊndɪd]

【释义】 *adj*. 圆状

【中文定义】 指粒径棱角已磨圆，碎屑的原始轮廓已消失。

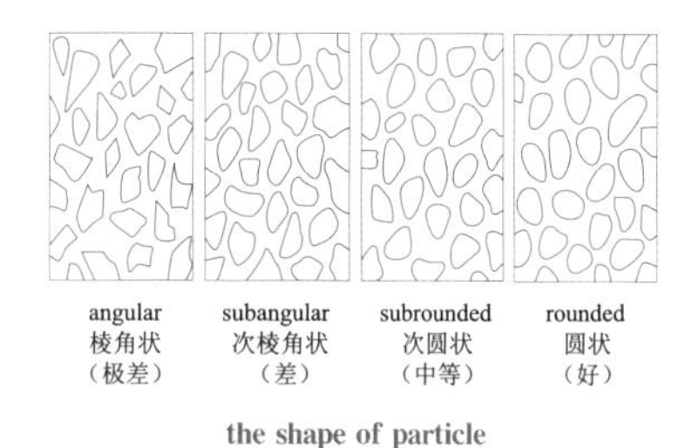

the shape of particle

route survey

【释义】 线路测量

【同义词】 line survey

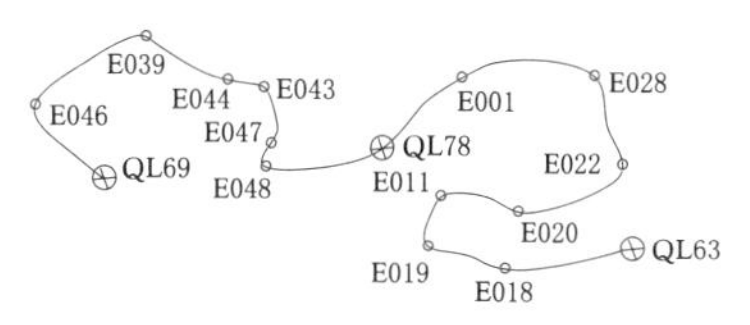

route survey

【例句】 GPS-RTK technology possesses of great technological potentials in transmission route survey, and has wide application and development prospects. GPS-RTK 技术在输电线路测量中蕴含着巨大的技术潜力，其应用及开发的前景十分广阔。

runoff area

【释义】 径流区

【中文定义】 地下水径流区至排泄区的流经范围。地下水的水量与盐量由补给区经该区传输到排泄区。

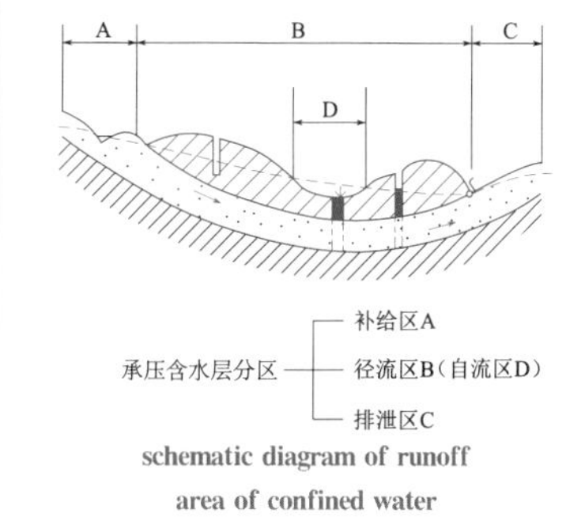

schematic diagram of runoff area of confined water

S

saber tree

【释义】 马刀树（醉林），多见于滑坡体表面。

sabre tree

saddleback ['sædlˌbæk]

【释义】 *n*. 垭口，鞍形山

saddleback of Balang mountain

safety evaluation earthquake (SEE)

【释义】 安全评估地震

【中文定义】 安全评估地震相当于校核标准，是指能产生对设计或评估的大坝而言的最高级别地面运动的地震。对于重要的水坝，可将SEE视为MCE或MDE地面运动。如果不可能对MCE做出真实评估，那么SEE至少应等同于MDE。SEE是用于水坝及相关安全组件的安全评估和抗震设计的控制性地震地面运动，在经受SEE后，水坝及相关安全组件应能正常发挥作用。

safety island

【释义】 安全岛

【例句】 Three evaluation methods of regional crust stability—"safety island"are discussed in detail: ① classification of main indexes, ② zoning, ③expert system and risk analysis. 文章详细探讨了区域地壳稳定——"安全岛"的三种评价方法：①主要指标分级评价法；②分区评价法；③专家系统及风险评价法。

【中文定义】 即相对稳定地块，是指构造活动区内或活动性构造带之间存在的相对稳定地块，并将其作为选择工程建设基地的主要对象。

sag [sæg]

【释义】 *n*. 凹陷，下陷

【常用搭配】 Dongying Sag 东营凹陷

【中文定义】 含油气盆地划分的亚一级构造单元，指一级构造单元内相对下降的地区。其特征是基底埋藏较深，沉积盖层发育完全，主要含油气层系发育完整，是油气生成的地区。凹陷、坳陷是不同的概念，分别表示不同级别的盆地构造单元，坳陷是盆地的次一级构造单元，如渤海湾盆地济阳坳陷，凹陷是盆地的再次一级构造单元，如济阳坳陷东营凹陷，还有更次一级的构造单元如洼陷。隆起和凸起也是不同级别的概念。

sag

saline groundwater

【释义】 地下咸水

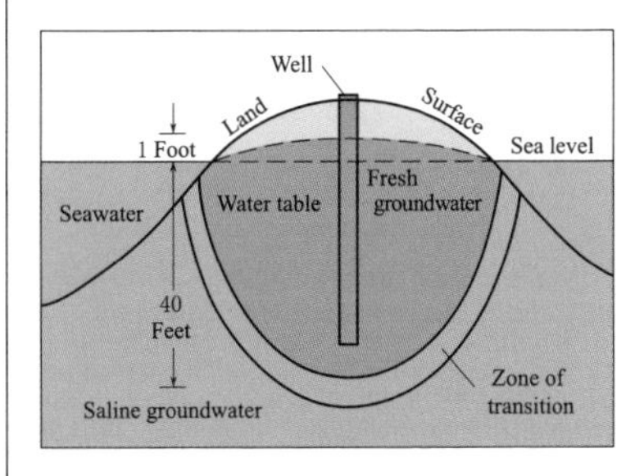

saline groundwater

【例句】 Shallow saline groundwater is a mixture of fresh water and palaeo-seawater. 浅层地下咸水是淡水和古海水的混合物。

【中文定义】 指矿化度>2g/L的地下水。

saline soil

【释义】 盐渍土

【例句】 Compressive deformation of the saline soil means the soil with natural water content will be compressed under the action of load. 盐渍土的压缩变形是指地基土在天然含水量条件下受外荷载作用产生的变形。

【中文定义】 是盐土和碱土以及各种盐化、碱化土壤的总称。盐土是指土壤中可溶性盐含量达到对作物生长有显著危害的土类，盐分含量指标因不同盐分组成而异。碱土是指土壤中含有危害植物生长和改变土壤性质的多量交换性钠。

saline soil

salinization [ˌselɪnaɪ'zeʃən]

【释义】 *n.* 盐化作用，盐渍化，盐碱化

【常用搭配】 salinization indicators 盐碱化指标；salinization effect 盐渍化效应

【中文定义】 潜水位壅高后，毛管水通过蒸发向地表输送的盐分不断积聚，土体演变成盐渍土/盐碱土的过程。

salt efflorescence

【释义】 盐华，盐霜

【中文定义】 是由泉水或地表毛细水经蒸发沉淀而成的盐类沉积物。盐华的主要矿物成分为各种易溶的盐类矿物，如各种芒硝、天然碱、石盐等，有时也可见一些不易溶的盐类矿物，它们常以白色粉末、霜华、被膜沉积于泉口附近及盐碱土的表面，一般规模和厚度不大。

salt efflorescence landscape

sample ['sæmpl]

【释义】 *n.* 试样，样本，例子

【常用搭配】 sample preparation 样品制备；core sample 岩芯样本

sample

sampling plan

【释义】 采样平面图

【例句】 The sampling plan is a diagram showing the sampling position, sample number and analysis result of various samples in the exploration area. 采样平面图是表示勘探矿区各类样品的采样位置、样品编号及分析结果的图件。

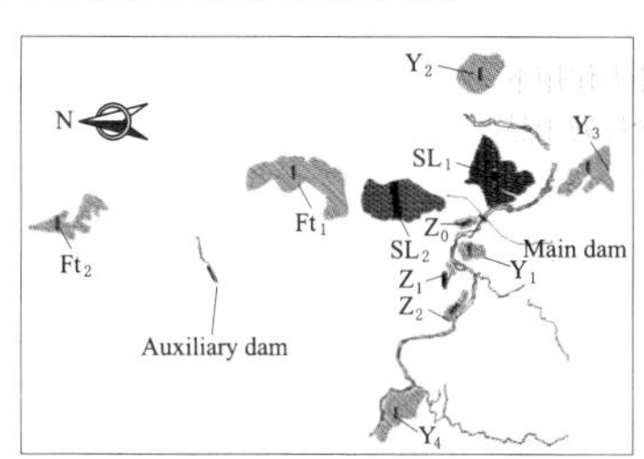

sampling plan

sand boiling

【释义】 砂沸

【中文定义】 指渗透压力引起的砂土液化。当砂土下部孔隙水压力达到或超过上覆砂层和水的重量时，砂土就会因丧失颗粒之间的摩擦阻力而上浮，承载能力也全部丧失。

sanidine ['sænɪdin]

【释义】 *n.* 透长石

【常用搭配】 high sanidine 高温型透长石；natron sanidine 钠透长石；sanidine phenocryst 透长石斑晶

【例句】 The compositions of "Crater group" are dacite and trachy-dacite, and their host crystals are sanidine. "火口组"熔体包裹体成分为英安岩和粗面英安岩，寄主晶多为透长石。

sanidine

sand liquefaction

【释义】 砂土液化

【常用搭配】 evaluation of sand liquefaction 砂土液化评价；index of sand liquefaction 液化指数

【例句】 The propagation of violent earthquake wave and sand liquefaction are two reasons to make underground pipeline damage. 强烈地震波的传播及砂土液化是造成地下管线发生破坏的两类原因。

【中文定义】 饱水砂土在地震、动力荷载或其他外力作用下，受到强烈振动而失去抗剪强度，使砂粒处于悬浮状态，致使地基失效的作用或现象

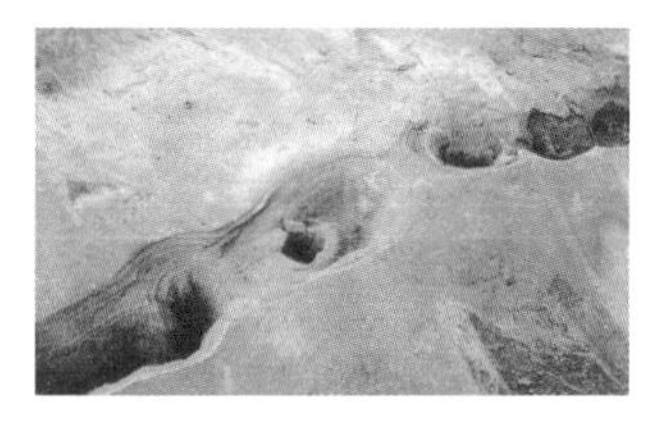

sand liquefaction

sand liquefaction preliminary decision

【释义】 砂土液化初判

【中文定义】 主要应用已有的勘察资料或较简单的测试手段对土层进行初步鉴别，以排除不会发生液化的土层。可采用年代法、粒径法、地下水水位法、剪切波速法进行初判。

sand liquefaction repetitional decision

【释义】 砂土液化复判

【中文定义】 对于初判为可能液化的土层，应进一步进行复判。常用的方法即：标准贯入锤击数法、相对密度法、相对含水量法、液性指数法、剪应力对比法、动剪应变幅法、静力触探贯入阻力法等。

sand texture

【释义】 砂状结构

【中文定义】 碎屑颗粒大小在 2～0.05mm 之间的碎屑结构。

sand-box model

【释义】 砂槽模型（渗流槽）

【常用搭配】 sand-box model test 砂槽模型试验

【例句】 A sand-box confined aquifer model is used to simulate gas transportation process in aquifer after abnormal source gas has formed by selecting argon dissolved gas as study gas. 借助于砂槽承压含水层模型，以氩溶解气为研究对象，模拟气体聚集或稀释的异常源产生后，其在含水层中的运移过程。

【中文定义】 实验槽内装有土壤或砂或玻璃球等、与原型中物理过程完全相同的一种水力学实体模型。

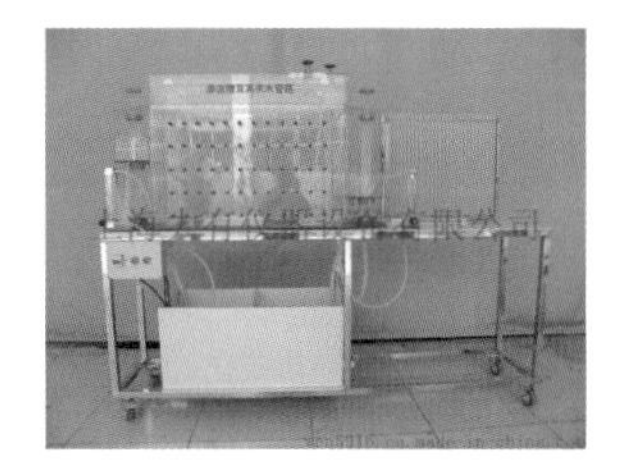

sand-box model sand tank model

sandstone ['sæn(d)stəʊn]

【释义】 *n*. 砂岩

【常用搭配】 kern stone 粗粒砂岩；micaceous sandstone 云母砂岩；sandstone soil 砂岩土

【例句】 Fractured sandstone also has long-term strength, which is determined according to the stress-strain relationship obtained by theoretical analysis. 破裂砂岩也存在长期强度，其值可根据岩石加载过程中的应力-应变关系通过理论分析的方法得出。

【中文定义】 是一种沉积岩，主要由砂粒胶结而成的，其中砂粒含量大于 50%。绝大部分砂岩是由石英或长石组成的，石英和长石是组成地壳最常见的成分。

sandstone

sandy clay

【释义】 砂质黏土

【例句】 As bearing layer of foundation, with special construction properties, the value of bearing capacity and deformation index of granite residual sandy clay is difficult to decide. 花岗岩残积砂质黏性土具有特殊的工程特性，作为地基持力层，其承载力和变形指标的确定存在不小难度。

【中文定义】 砂质黏土这类土壤泛指与砂土性状相近的一类土壤，其物理黏粒含量15%，主要分布于我国西北地区如新疆、甘肃、宁夏、内蒙古、青海的山前平原以及各地河流两岸、滨海平原一带。

sandy clay

sandy gravel

【释义】 砂砾石

sandy gravel

【例句】 Sandy gravel consists of very small stones. It is often used to make paths. 砂砾石中含有粒径很小的石子，通常被用来修路。

【中文定义】 砂砾石是一种颗粒状、无黏性材料。

sandy loam

【释义】 砂壤土

【中文定义】 壤土指土壤颗粒组成中黏粒、粉粒、砂粒含量适中的土壤，砂土是指含砂量占80%，黏土占20%左右的土壤；砂壤土就是介于壤土与砂土之间的土壤。

sandy loam

sandy soil

【释义】 砂土

【例句】 Underground structure in saturated sandy soil might be subject to severe damage due to seismic liquefaction. 饱和砂土地层中的地下结构在地震作用下可能因地基液化而发生严重破坏。

【中文定义】 粒径大于2mm的颗粒质量不超过总质量的50%，粒径大于0.075mm的颗粒质量超过总质量50%的土，定名为砂土。

sandy soil

Sarma's method

【释义】 萨尔玛法

satellite photo map

【释义】 卫星相片图

【中文定义】 用经处理的卫星相片，按一定的几何精度要求，镶嵌成大片地区的影像镶嵌图。

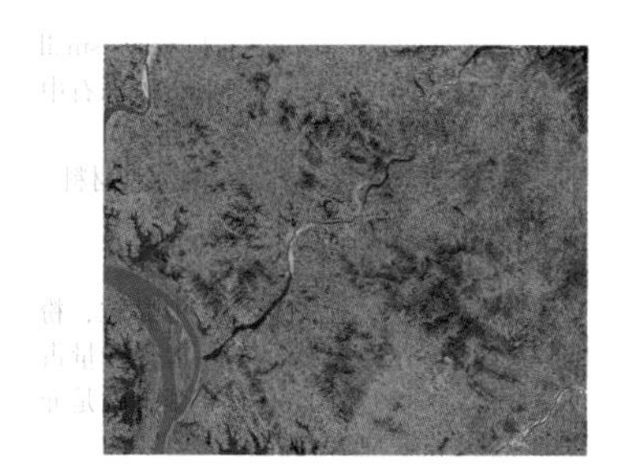

satellite photo map

satellite positioning surveying

【释义】 卫星定位测量法

satellite positioning surveying

saturated compressive strength

【释义】饱和抗压强度

【例句】 Finally, analyzed and compared with various kinds of curing technique on red mud aerated concrete cube shown different maintenance technique directly effected the saturated compressive strength, later development and dry density on performance of products. 最后，通过对采取不同养护工艺的赤泥加气混凝土试块进行分析比较，发现不同的养护工艺对混凝土制品的饱和抗压强度、后期增长强度、干密度等都有影响。

【中文定义】 指材料在饱和状态下的抗压强度。

saturated density

【释义】 饱和密度

【例句】 In the F-H experiment, the temperature affects mercury's saturated density, so it's natural to affect the energy exchange between mercury and electron. 在夫兰克·赫兹实验中，温度影响着汞原子的饱和密度，当然就影响了汞原子与电子之间的能量交换。

【中文定义】 土的孔隙完全被水充满时的密度称为饱和密度。

saturated moisture content

【释义】 容水度（饱和含水率）

【例句】 Soil pressures increase from 0.006MPa to 0.073MPa, while water is accumulated on the top of roadbed and the saturated moisture content of soil increases from optimum value to saturated. 膨胀土路堤堤顶积水时，当路堤中的含水率从最佳含水率增大到饱和含水率左右时，路堤中的土压力增大 0.006～0.073MPa 不等。

【中文定义】 是土孔隙中完全充满水时水的质量与固体颗粒质量之比，以百分率表示。

saturated uniaxial compressive strength

【释义】 单轴饱和抗压强度

【例句】 Experimental results show that the bearing capacity of medium weathered soft slate foundation from saturated uniaxial compression strength of rock is much smaller than that from in-situ test. 室内岩石试验结果表明，按岩石饱和单轴抗压强度乘以折减系数确定的中等风化软质岩石地基的承载力远低于岩石地基的实际承载力。

saturated water absorptivity

【释义】 饱和吸水率

【例句】 The experimental results show that the replacing of the hydroxy groups with acetoxy groups in the resin causes less saturated water absorptivity, but increases the diffusibility of water in the resin. 实验结果表明，该树脂中的羟基改为乙酸基后可降低树脂的饱和吸水率，并增大水在树脂中的扩散能力。

saturated soil

【释义】 饱和土

【例句】 For complexity and diversity of unsaturated soil behaviors, there lies a great difference in permeability characteristics between unsaturated soil and saturated soil. 由于非饱和土的复杂性和多变性，其渗透特性明显不同于饱和土，并且试验难度较大。

saturation swelling stress

【释义】 饱和湿胀应力

scale effect

【释义】 尺度效应，放缩效应

schist [ʃɪst]

【释义】 *n.* 片岩

【常用搭配】 hornblende schist 角闪片岩；mica-schist 云母片岩

【例句】 The shortage of bearing capacity is often encountered in pile foundation construction in complex schist area. 复杂片岩地区的桩基工程常出现桩基承载力不足的问题。

【中文定义】 具明显的片状构造的变质岩。与千枚岩相比，片岩变质程度比千枚岩高，原岩矿物已全部重结晶，矿物颗粒粗大，肉眼均可分辨。粒状矿物主要为石英和长石；石英含量一般大于长石，长石含量常少于25%。以变斑晶出现的特征变质矿物有十字石、蓝晶石、夕线石、堇青石、石榴石等。片岩一般根据主要的片状或柱状矿物进行分类，如云母片岩、角闪片岩、绿泥片岩、滑石片岩等。

schist

schistose structure

【释义】 片状构造

【中文定义】 岩石中的矿物颗粒肉眼可以分辨、连续平行排列成薄片状的一种构造。

schistose structure

scope of investigation

【释义】 勘测范围

【例句】 There's no geological disaster within the scope of investigation. 勘测范围内没有地质灾害。

screen [skrin]

【释义】 *v.* 筛选，筛分；*n.* 筛

【常用搭配】 vibrating screen 振动筛；rotating cleaning screen 旋转清洗筛；rinsing screen 淋水筛；finish screen 二次筛分

【例句】 A vibrating or rotating cleaning screen will be provided and used continuously between the supply mixer and the pump. 在供应浆液的拌浆机及泵之间要有一台振动筛或旋转清洗筛进行连续筛分。

【中文定义】 用带孔的筛面把粒度大小不同的混合物料分成各种粒度级别的作业称为筛分。

screen

screen analysis

【释义】 筛分分析

【中文定义】 指固体颗粒的筛分和分级。

screen size gradation

【释义】 筛分级配

screen test

【释义】 筛分试验

【例句】 The results of the compaction test and the screen test show that the compaction work has a certain effect on the mixture grade. 压实试验和筛分试验结果证实，压实功对混合料级配有一定的影响。

【中文定义】 测定粗集料（碎石、砾石、矿渣等）的颗粒级配的试验。

screening curve

【释义】 筛分曲线

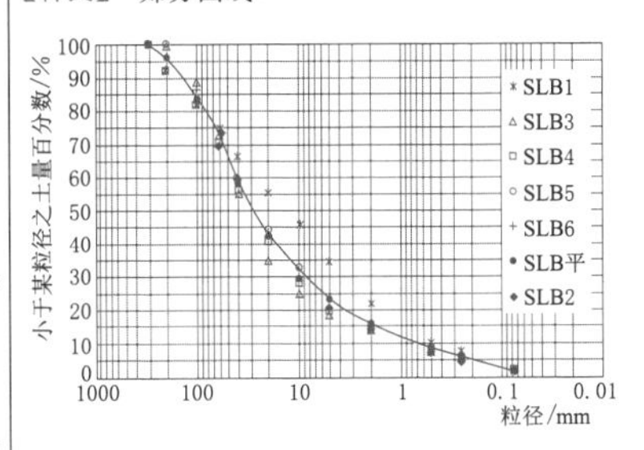

screening curve

screw plate load test

【释义】 螺旋板载荷试验

【例句】 The screw plate load test is one of the in-situ tests. 螺旋板载荷试验是原位试验的一种。

scrotiform weathering

【释义】 囊状风化，囊形风化

【例句】 In the Xiangjiaba dam area, there is the phenomenon of sandstone scrotiform weathering, its formation mechanism is related to its own material composition, geological structure, hydrogeological conditions and hydrochemical effects. 在向家坝坝区，就存在砂岩囊状风化现象，其形成机理与其本身的物质组成，地质构造，水文地质条件及水化学作用有关。

【中文定义】 当岩体中存在规模较大、延展较深的断层破碎带或断裂交汇带、不稳定矿物富集带、岩脉与断裂交叉带及地下水循环交替较强的局部裂隙发育带时，形成的囊状风化结构。

scrotiform weathering

sea-floor spreading hypothesis

【释义】 海底扩张学说

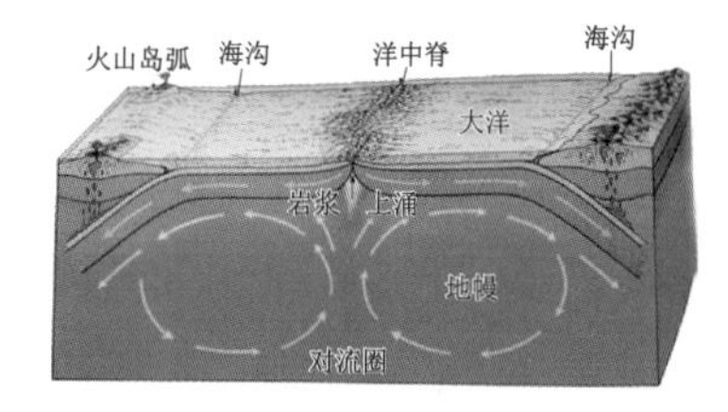

sea floor spreading

【常用搭配】 pulse of sea-floor spreading 海底扩张的脉动

【例句】 This new theory, called sea-floor spreading hypothesis raised by Dietz, explained the distribution of earthquakes and volcanoes on earth quite nicely. 由 Dietz 提出的新学说——海底扩张学说，较好地解释了地球上地震和火山的分布。

【中文定义】 是海底地壳生长和运动扩张的一种学说。洋壳沿大洋中部穿透岩石圈的裂缝或裂谷向两侧扩展并导致新生洋壳的学说。其认为地幔物质在这种裂缝带下因软流圈内的物质上涌、侵入和喷出而形成新的洋壳，随着这个作用的不断进行，新上涌侵入的地幔物质把原已形成的洋壳向裂谷两侧推移扩张，致使洋底不断新生和更新。

seasonal frozen soil

【释义】 季节冻土

【中文定义】 指地表层冬季冻结、夏季全部融化的土（岩），包括季节冻结层和季节融化层（也称活动层）。

seasonal frozen soil

secondary consolidated soil

【释义】 次固结土

secondary consolidation

【释义】 次固结

【例句】 The 1D secondary consolidation tests of soft soils of undisturbed and disturbed through step loading has been carried out. 对原状和重塑软土做了大量的分级加载室内一维次固结试验研究。

secondary consolidation settlement

【释义】 次固结沉降

【例句】 The magnitudes of primary and secondary consolidation settlement could be adequately estimated by the model. 这个模型能估算出主固结和次固结沉降的大小。

【中文定义】 由土体骨架蠕变压缩所产生的地基沉降。

secondary ground fissure

【释义】 次生地裂缝

【中文定义】 位于主地裂缝两侧，错断勘探标志层且有显著垂直位移的地裂缝。

secondary joint

【释义】 次生节理

【例句】 The weathering joint is a kind of secondary joint. 风化节理是一种次生节理.

【中文定义】 是在石成岩以后形成的节理，包括构造节理和非构造节理。

secondary joint

secondary mineral

【释义】 次生矿物

【例句】 Thus primary minerals are wholly or partly changed to secondary minerals including serpentine and montmorillonite(from olivine),calcite, chlorite (from biotite), kaolin (from feldspars),zeolites and other clays. 因此，原生矿物全部或部分转变为次生矿物，包括蛇纹石、蒙脱石（转变自橄榄石)、方解石、绿泥石（转变自黑云母)、高岭土（转变自长石)、沸石和其他黏土矿物。

【中文定义】 在岩石或矿石形成之后，其中的矿物遭受化学变化而改造成的新生矿物，其化学组成和结构都经过改变而不同于原生矿物。

secondary mineral serpentine

secondary salinization

【释义】 次生盐渍化

【中文定义】 在干旱和半干旱地区，人为因素使盐分聚积于地表形成盐渍土的过程。

secondary support

【释义】 二次支护

【中文定义】 根据围岩稳定情况，或初期支护后由监测结果决定的再次支护。

secondary wave (S-wave)

【释义】 横波

【例句】 This paper provides an example of the application of seismic S-wave reflection method and S-wave velocity to engineering investigation. 本文提供了地震横波反射法及横波速度在工程勘察中的应用实例。

【中文定义】 是振动方向和波的传播方向垂直的波。在地壳中横波传播的速度较慢。到达地面时人感觉摇晃，物体为摆动；横波地震是地壳震动在地下几千米的地方碰撞或摩擦，震动后沿着同一平面的底层在地壳传播。震感为太极摇晃，传播距离广泛，对地面破坏很小，只有地震中心会受到影响。

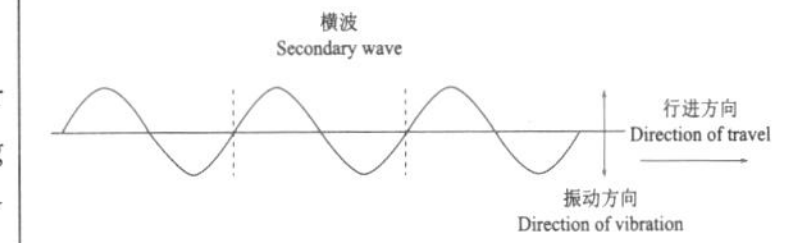

the vibration of S-wave

sediment ['sedɪmənt]

【释义】 *n.* 沉积物，沉淀物

【常用搭配】 abyssal sediments 深海沉积物；calcareous sediments 钙质沉积物；cave sediments 洞穴沉积；chemical sediments 化学沉积物；clastic sediments 碎屑沉积

【例句】 The weight of a deltaic pile of sediment may be great as to depress the earth's crust, thus forming a moat around the depocentre. 三角洲沉积物堆积的重量可以大到使地壳下沉的程度，因此在沉积中心周边形成一条海沟。

【中文定义】 又名堆积物，是陆地或水盆地中沉积的松散矿物质颗粒或有机物质的总称。如砾石、沙、黏土、灰华、生物残骸等。主要是母岩风化的产物，其次是火山喷发物、有机物和宇宙物质等。粗粒的（如沙）具流动性，细粒的（如黏土）具柔性、黏性和可塑性。沉积物经固结成岩作用后，形成沉积岩。

sedimentary rock

【释义】 沉积岩

【例句】 Bedding is the most important structural feature of sedimentary rock. 层理是沉积岩最重要的构造特征。

【中文定义】 三大岩类的一种，又称为水成岩，是三种组成地球岩石圈的主要岩石之一。是在地壳发展演化过程中，在地表或接近地表的常温常压条件下，任何先成岩遭受风化剥蚀作用的破坏产物，以及生物作用与火山作用的产物在原地或经过外力的搬运所形成的沉积层，又经成岩作用而成的岩石。沉积岩主要包括石灰岩、砂岩、页岩等。

sedimentary rock

sedimentation [ˌsedɪmen'teʃən]

【释义】 沉淀作用，沉积作用

【常用搭配】 sedimentation tank 沉淀池，沉积槽

【例句】 Effluent from the sedimentation tank is dosed with disinfectant to kill any harmful organisms. 沉降池中的废水中加入了消毒剂，以消灭所有有害微生物。

【中文定义】 指被运动介质搬运的物质到达适宜的场所后，由于条件发生改变而发生沉淀、堆积的过程的作用。

seepage ['siːpɪdʒ]

【释义】 *n.* 渗流，渗漏，渗水

【常用搭配】 water seepage 渗水；seepage loss 渗漏损失；seepage analysis 渗流分析；seepage control 防渗

seepage

【中文定义】 流体在孔隙介质中的流动称为渗流。通常在地表以下的土层或岩层中的渗流称为地下水。水利水电工程中，汛期高水位历时较长时，在渗压作用下，堤前的水向堤身内渗透，称为渗水。

seepage control material

【释义】 防渗料

【例句】 The dam of Qiaoqi Hydropower Project is 123m high, in which it is the first time for the spreading gradation crushed stone-earth to be adopted as the seepage control material for so high an earth-rock dam in China. 硗碛水电站大坝高123m，采用宽级配碎石土作为高土石坝心墙防渗料。

seepage control material

seepage deformation

【释义】 渗透变形

【例句】 The levee classification provides the method and basis for analysis of levee foundation seepage deformation, levee slope stability and segmented evaluation on levee engineering geology. 堤基结构分类为堤基渗透变形、堤坡稳定性分析和堤基工程地质分段评价提供了方法和依据。

【中文定义】 岩土体在地下水渗透力（动水压力）的作用下，部分颗粒或整体发生移动，引起岩土体的变形和破坏的作用和现象。

seepage deformation test

【释义】 渗透变形试验

【例句】 An indoor seepage deformation test was conducted on the loose deposit soil which was taken from the mountains-gully rivers of the Tibet Plateau. 对选自西藏山南地区高山宽谷河流区松散坝基土体进行了室内渗透变形试验研究。

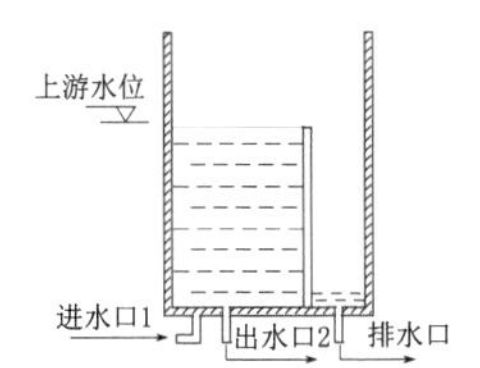

升降式供水箱

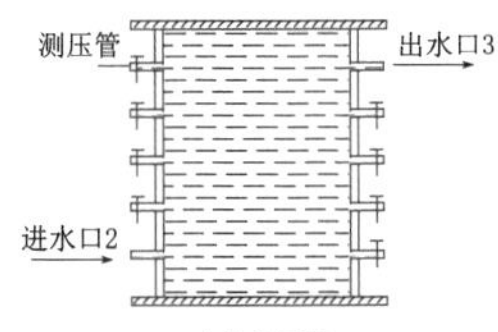

水位分压器

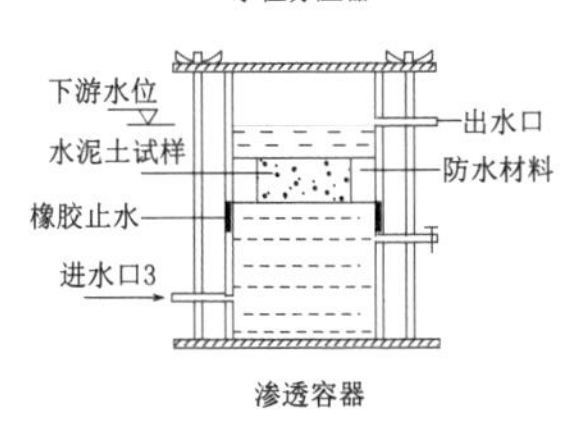

渗透容器

instrument diagram of seepage deformation test of cement soil

seepage failure
【释义】 渗透破坏
【中文定义】 指土工建筑物及地基由于渗流而出现的破坏或变形。

seepage field
【释义】 渗流场
【例句】 The rainfall infiltration will influence the seepage field and stability of landslide . 降雨入渗对滑坡的渗流场和稳定性有一定的影响。
【中文定义】 研究渗流问题时，将压力相等的点连成线即等压线，与等压线垂直的线为流线，等压线和流线所组成的图称为渗流的水动力学场图，也可以称为渗流场。

seepage flow
【释义】 渗透水流（渗流）
【例句】 This paper is dealing with the probability method of stability analysis for soil slope under the conditions of seepage flow and seismic loading. 本文是关于土坡在有渗透水流出并经受地震荷载作用下概率方法校核稳定性的研究。
【中文定义】 假想的充满整个多孔介质的空隙和岩石骨架的全部体积的水流。

seepage force
【释义】 渗透力，渗流压力
【例句】 The settlement is mainly induced by the changes of earth mass weight, and the radial deformation is mainly induced by the changes of radial seepage force. 沉降主要是由土自重的变化引起的，径向变形主要是由渗透力的径向分量的变化引起的。
【中文定义】 指水在土中流动的过程中将受到土阻力的作用，使水头逐渐损失。同时，水的渗透将对土骨架产生拖曳力，导致土体中的应力与变形发生变化。这种渗透水流作用对土骨架产生的拖曳力称为渗透力。

seepage of dam foundation
【释义】 坝基渗漏
【中文定义】 水库蓄水后由于上下游形成的水位差，使库水沿河床坝基和两岸坝肩岩土体之孔隙、裂隙、溶洞、断层破碎带等向下游产生渗漏的现象。

seepage profile
【释义】 渗透剖面图
【常用搭配】 seepage profile of dam foundation (seepage control line) 坝基（防渗线）渗透剖面图
【例句】 In feasibility study stage and construction details design stage of hydropower engineering, the seepage profile of dam foundation shall be submitted in report for engineering geological investigation. . 水力发电工程可行性研究阶段和施工详图设计阶段工程地质勘察报告中应附坝基（防渗线）渗透剖面图。
【中文定义】 指重点反映岩（土）体渗透特性的工程地质剖面图。

seismic acceleration
【释义】 地震加速度
【例句】 The program is applied to calculate the dynamic response of a free horizontal field of saturated sands under seismic acceleration. 该程序用来计算对水平自由场地的饱和沙层在地震加速度作用下的动力响应。
【中文定义】 指地震时地面运动的加速度，可以作为确定烈度的依据。

seismic activities
【释义】 地震活动
【例句】 The scientists positioned the seismic ac-

tivity as being along the San Andreas fault. 科学家们找到了沿着圣安德烈亚斯断层的地震活动位置。

seismic belt

【释义】 地震带

【例句】 The earthquake's location was in what geologists in the area call the East Seismic Belt, where earthquakes are caused by the collision between the Philippine plate and the Eurasian plate. 地震所处位置发生在当地地震学家称作的东方地震带上。由菲律宾板块和亚欧大陆板块相互挤压而产生。

【中文定义】 指地震集中分布的地带，地球上主要有三处地震带。地震带基本上在板块交界处。在地震带内震中密集，在带外地震的分布零散。地震带常与一定的地震构造相联系。

seismic exploration

【释义】 地震勘探

【例句】 Seismic exploration is an important means to survey oil and natural gas resources, and it has been widely used in engineering geological prospecting and regional geological research. 地震勘探是勘测石油与天然气资源的重要手段，在工程地质勘察和区域地质研究等方面也得到广泛应用。

seismic ground motion parameter zonation map

【释义】 地震动参数区划图

【同义词】 seismic zoning map

【例句】 The results of the evaluation of the division are consistent with the results of the seismic intensity zoning and the seismic ground motion parameter zonation map compiled by the China Seismological Bureau. 评价分区结果与中国地震局所编著的地震烈度区划图、地震动参数区划图的结果是一致的。

【中文定义】 以地震参数（以加速度表示地震作用强弱程度）为指标，将全国划分为不同抗震设防要求区域的图件。

seismic hazard analysis

【释义】 地震危险性分析

【例句】 The attitudes of potential rupture surfaces have great influence on the results of probabilistic seismic hazard analysis and seismic zoning. 潜在地震破裂面源的大小、产状，对近震源场点的地震危险性分析和地震区划结果有明显的控制影响。

【中文定义】 对特定区域确定其在未来一定设计基准期内地震参数（烈度、加速度、速度、反应谱等）超过某一给定值的概率方法。

seismic intensity

【释义】 地震烈度

【同义词】 earthquake intensity

【例句】 Seismic Intensity 7 was observed for only the second time since JMA(Japan Meteorological Agency) introduced instrument-based observation for intensity measurements in 1996. 日本气象厅 1996 年刚第二次采用仪器式地震烈度测量观测时就被观测到 7 度的地震烈度。

【中文定义】 用以表示地震对地表及工程建筑物影响的强弱程度。

seismic network

【释义】 地震台网

【例句】 California's official seismic network, a collection of about 450 machines situated throughout the state, relies on the sophistication and placement of the pricey equipment. 加州的官方地震监测网络，是由大约 450 台遍布全州的传感器设备组成的，依靠这些昂贵设备的混合以及连接方式进行工作。

【中文定义】 由各级地震台、站所构成的地震观测网络。

seismic peak ground acceleration zonation map

【释义】 地震动峰值加速度区划图

【例句】 Regional seismic peak ground acceleration can be obtained from seismic peak ground acceleration zonation map. 地区的地震动峰值加速度可以从地震动峰值加速度区划图获得。

【中文定义】 表示地震动峰值加速度区划的地图称为地震区划图。

seismic post-disaster effect

【释义】 地震灾后效应

【中文定义】 指地震过后灾区一定范围地质灾害在一定时间内显著增强的现象。

seismic reflection wave method

【释义】 地震反射波法

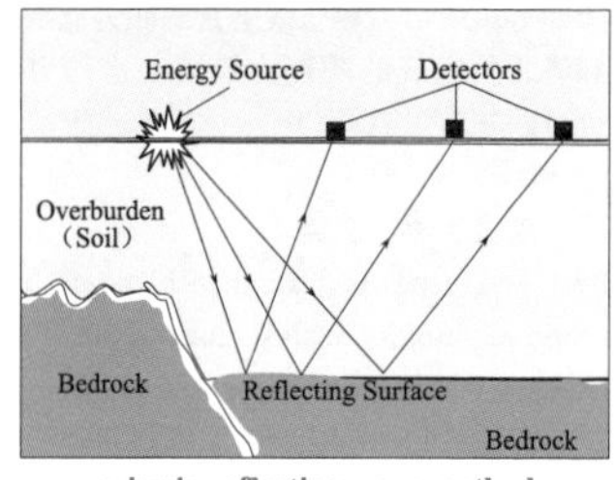

seismic reflection wave method

seismic refraction wave method

【释义】 地震折射波法

【例句】 In the areas with large interference by surface wave, where the seismic reflection wave method is difficult to be applied, the seismic refraction wave method can be adopted for detection. 在面波干扰大、地震反射波法难以开展工作的区域，可尝试利用地震折射波法进行探测。

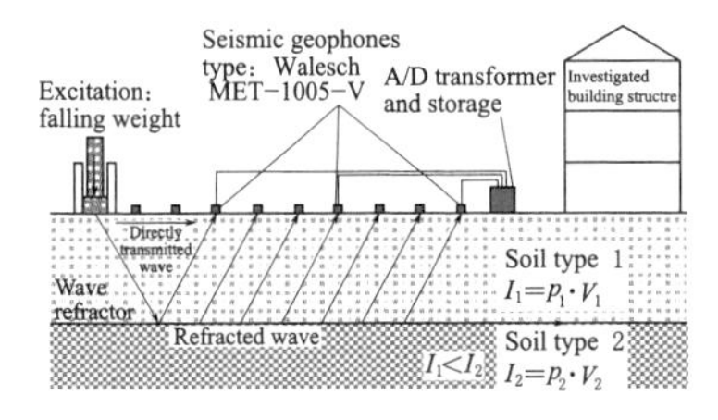

seismic refraction wave method

seismic region

【释义】 地震区

【例句】 The shallow crust structure characteristics of Manas anticline in Manas seismic region are studied using high precision shallow seismic prospecting method and advanced data processing technique. 采用高精度的浅层地震勘探方法和先进的数据处理技术，查明了玛纳斯地震区玛纳斯背斜的浅部地壳结构特征。

【中文定义】 地震活动频繁而强烈的区域称为地震区。我国现有 6 个地震区，分别是天山地震区、青藏地震区、华北地震区、东北地震区、华南地震区及台湾地震区。

seismic response spectrum

【释义】 地震反应谱

【同义词】 earthquake response spectrum

【例句】 The research works are the analysis on the seismic response of high rise structures and tall buildings to vertical ground motion and the vertical seismic response spectrum. 这些研究工作主要是关于高耸结构、高层建筑的竖向地震反应和竖向地震反应谱以及关于竖向地震作用的研究。

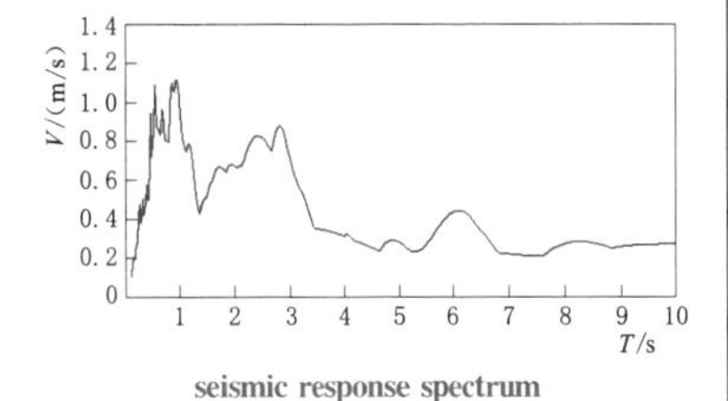

seismic response spectrum

【中文定义】 指单自由度弹性系统对于某个实际地震加速度的最大反应（可以是加速度、速度和位移）和体系的自振特征（自振周期或频率和阻尼比）之间的函数关系。

seismic safety assessment report

【释义】 地震安全性评价报告

seismic safety evaluation

【释义】 地震安全性评价

【例句】 The accumulated basic data and method of this study can provide reference for the seismic safety evaluation of the related marine construction projects close to the study area. 本项工作积累的有关基础资料和研究方法可为研究区域附近有关海洋建设工程的地震安全性评价工作提供借鉴和参考。

【中文定义】 指在对具体建设工程场址及其周围地区的地震地质条件、地球物理场环境、地震活动规律、现代地形变及应力场等方面深入研究的基础上，采用先进的地震危险性概率分析方法和确定性方法，按照工程所需要采用的风险水平，科学地给出相应的工程规划或设计所需要的一定概率水准下的地震动参数（加速度、设计反应谱，地震动时程等）和相应的资料。

seismic source

【释义】 震源

【同义词】 seismic focus

【例句】 The seismic source uses the rotation of centrifugal quality to produce seismic energy. The produced sweep signals are recorded by clockwise and counter clockwise rotation at each source point. 这种地震震源利用离心质量的旋转产生地

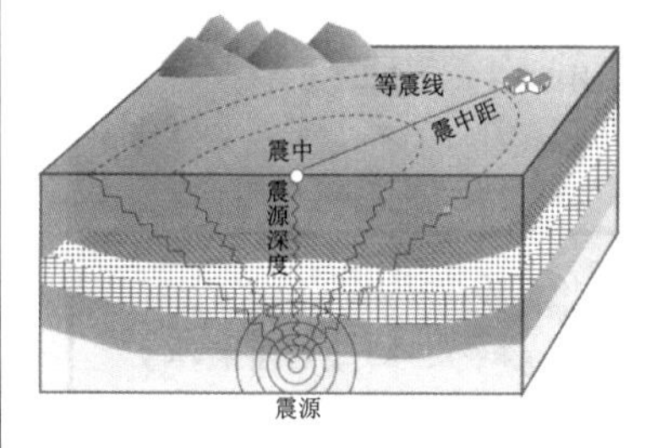

schematic diagram of seismic source

震能量，在每个震源点通过顺时针转动和逆时针转动记录下所产生的扫描信号。

【中文定义】 地球内部岩层破裂引起振动的地方称为震源。它是有一定大小的区域，又称震源区或震源体，是地震能量积聚和释放的地方。

seismic structure

【释义】 地震构造

【例句】 Seismotectonic method was used to research seismic structure, the potential maximum earthquake in the structure and estimate the next limit earthquake risk of engineering site. 地震构造法是通过研究工程场地周围每一发震构造及最大潜在地震来确定场地未来的极限地震危险性。

【中文定义】 指与地震孕育和发生有关的地质构造。通常按照研究内容和角度的不同，将地震构造分为全球地震构造、区域地震构造、震源构造和工程地震构造。

seismic structure zone

【释义】 地震构造区

【中文定义】 具有同样地质构造和地震活动性的地理区域。通常按照研究内容和角度的不同，将地震构造分为全球地震构造、区域地震构造、震源构造及工程地震构造。

seismic symbol

【释义】 地震符号

【例句】 Seismic symbol is used to indicate contents related to earthquakes in the geological map. 地质图中地震符号用以表示地震相关内容。

【中文定义】 地质图中绘制地震相关内容所使用的各种图形符号。

名称	符号	名称	符号
震中及强度 震级 发震时间	7.5 1976	地震烈度(度)	Ⅶ
古地震及震级	7	地震动加速度区划线	0.03g 0.015g

seismic symbol

seismic velocity

【释义】 地震波速

【中文定义】 地震波在岩土体中传播的速度。

seismic velocity logging

【释义】 地震速度测井

seismic wave

【释义】 地震波

【例句】 The propagation characteristics of the amplitude of the blasting seismic wave under the conditions of various topographies are approached by means of experiments. 通过靶场实验，探讨了不同地形条件下爆破地震波幅值的传播特性。

【中文定义】 是由地震震源向四处传播的振动，指从震源产生向四周辐射的弹性波。按波的传播方式可分为纵波（P 波）、横波（S 波）（纵波和横波均属于体波）和面波（L 波）三种类型。地震发生时，震源区的介质发生急速的破裂和运动，这种扰动构成了一个波源。由于地球介质的连续性，这种波动就向地球内部及表层各处传播，形成了连续介质中的弹性波。

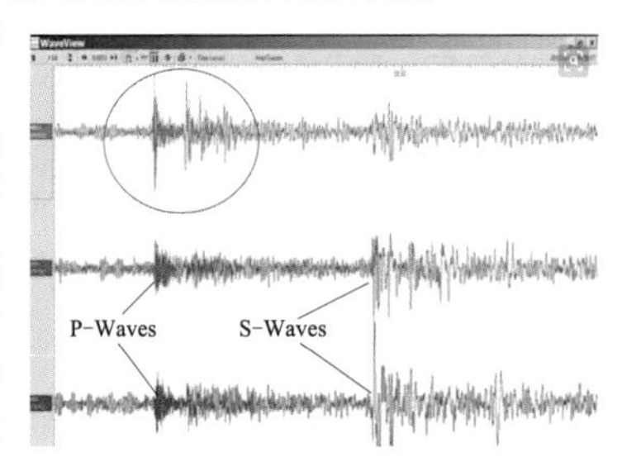

seismic wave

seismogenic fault

【释义】 发震断层

【例句】 The fault movement is the strongest in the tectonic belt, where the seismogenic fault is situated. 断层运动在构造带发震断层位置最为强烈。

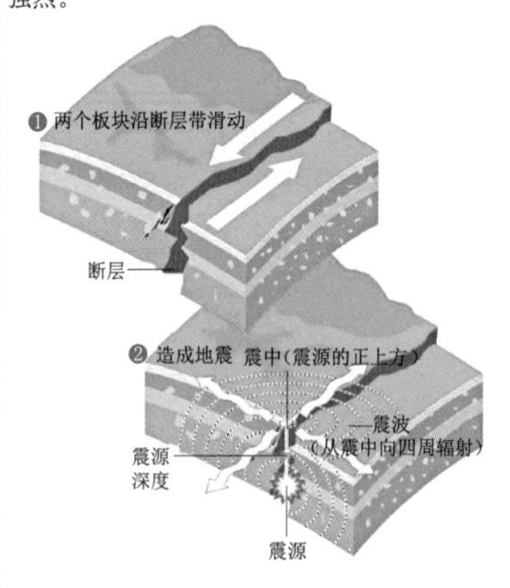

seismogenic fault

【中文定义】 指发生地震的断层。发震断层对某次特定地震的孕育（能量积累）和发生（能量释放）起控制作用，其震源错动或同震地表错动是沿某一先存活动断层的部分段落或全部段落发生的。

seismogenic mechanism

【释义】 发震机制

【例句】 The research via detecting the surface displacement caused by earthquake to study the seismogenic mechanism will provide scientific knowledge to earthquake prediction. 通过探测地震所引起的地表形变，研究地震的发震机制，进而为地震预报提供科学依据的有利途径。

seismogenic structure

【释义】 发震构造

【例句】 To study the seismogenic structure of the Zhangbei-Shangyi Earthquake, in this paper the main shock and its after-shocks with $M_L \geqslant 3.0$ of the Zhangbei-Shangyi Earthquake sequence were relocated using the master event relative relocation algorithm. 本文为研究张北尚义地震的发震构造，将 M_L 大于 3.0 的主地震波与余震的地震波利用主事件相对迁移算法重新定位计算。

【中文定义】 在现代构造条件下，曾发生或可能发生地震的地质构造。

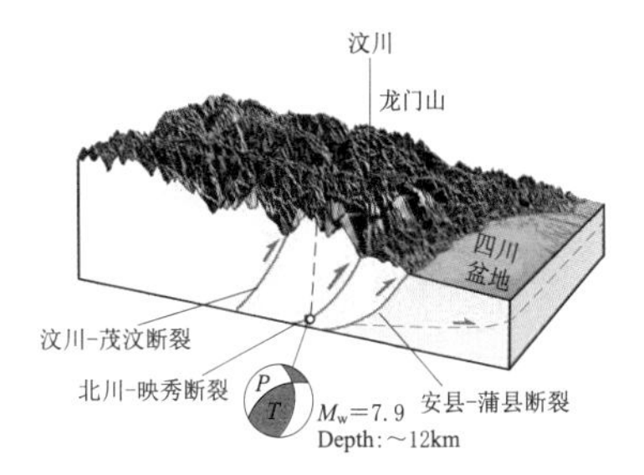

seismogenic structure of Wenchuan Earthquake

seismograph ['saɪzməgræf]

【释义】 *n.* 地震仪

【常用搭配】 seismograph station 地震台站；borehole seismograph 井下地震仪；strain seismograph 应变地震仪

【同义词】 seismometer

【例句】 People always make some misunderstanding for the sampling rate and dynamic range of shallow seismograph. 人们对浅层地震仪的采样频率及动态范围这两个指标往往产生某些误解。

【中文定义】 一种监视地震的发生、记录地震相关参数的仪器。

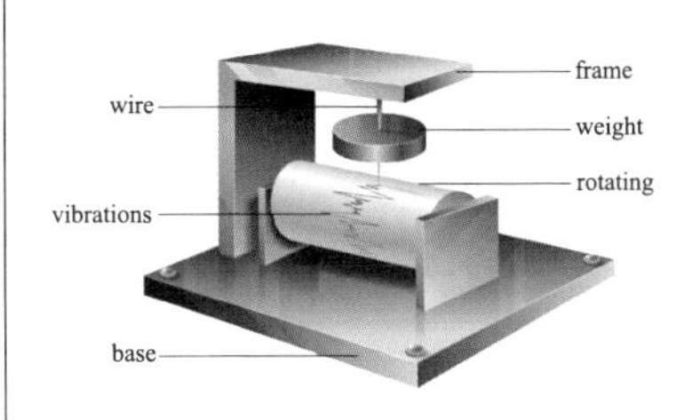

seismograph

seismological observation

【释义】 地震观测

【例句】 An artificial random disturbance is carried out for some seismological observation data in the allowable range of seismological observational error with random disturbance method. 采用随机干扰方法，在地震观测误差容许值范围内，对地震观测资料进行人为随机干扰。

【中文定义】 用地震仪器记录天然地震或人工爆炸所产生的地震波形，并由此确定地震或爆炸事件的基本参数（发震时刻、震中经纬度、震源深度及震级等）。

seismostation [saɪzməs'teɪʃn]

【释义】 *n.* 地震台

【例句】 The basic data of seismostation is the first-hand information for the management of station and earthquake prediction. 地震台站的基础信息是台站管理和地震预测工作的第一手资料。

【中文定义】 指利用各种地震仪器进行地震观测的观测点，开展地震观测和地震科学研究的基层机构。

seismostation

seismo-substation

【释义】 地震子台

self-boring pressuremeter
【释义】 自钻式旁压仪

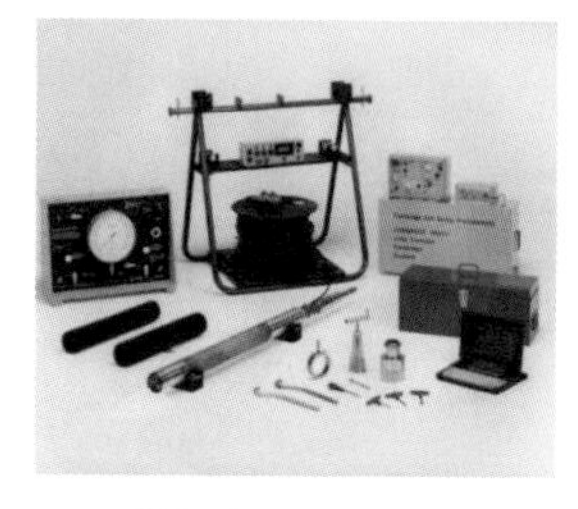

self-boring pressuremeter

self-potential method
【释义】 自然电位法

self-weight collapsible loess
【释义】 自重湿陷性黄土
【例句】 By this method, the influence of property of the self-weight collapsible loess on pile beating performance is analyzed. 运用此方法分析了自重湿陷性黄土湿陷特性对桩基承载性状的影响规律。
【中文定义】 指受水浸湿后在饱和自重压力下发生湿陷的湿陷性黄土。划分自重湿陷性和非自重湿陷性黄土，可按室内或现场浸水压缩试验，在土的饱和自重压力下测定的自重湿陷系数 δ_{zs} 判定。$\delta_{zs}<0.015$ 时，应定为非自重湿陷性黄土；$\delta_{zs}\geqslant0.015$ 时，应定为自重湿陷性黄土。

self-weight collapsible loess

semisolid state
【释义】 半固态

sensitivity [ˈsɛnsəˈtɪvəti]
【释义】 *n.* 灵敏度
【常用搭配】 soil sensitivity 土体灵敏度
【例句】 It is found that the soil yield stress of high sensitivity soil changes greatly when disturbance applied. 研究结果表明，扰动对高灵敏度土体屈服应力衰减影响非常大。

Series [ˈsɪriz]
【释义】 *n.* 统
【常用搭配】 Holocene Series 全新统；Eocene Series 始新统；Upper Jurassic Series 侏罗系上统
【中文定义】 第四级年代地层单位，与地质年代单位“世”对应，往往是某类生物进化显示出阶段性。

serpentinite [səpənˈtiːnaɪt]
【释义】 *n.* 蛇纹岩
【例句】 Serpentinite is a metamorphic rock that is mostly composed of serpentine group minerals. 蛇纹岩是一种变质岩，主要由蛇纹石族矿物组成。
【中文定义】 蛇纹岩主要是由超基性岩受低-中温热液交代作用，使原岩中的橄榄石和辉石发生蛇纹石化所形成。

serpentinite

settlement [ˈsetlmənt]
【释义】 *n.* 沉降
【常用搭配】 settlement observation 沉降观测
【例句】 Secant modulus method (SMM) is a wellspread new method to calculate foundation settlement in recent years. 割线模量法是近几年获得青睐的计算地基沉降的新方法。
【中文定义】 地面高程降低的现象。

settlement curve
【释义】 沉降曲线
【中文定义】 地基沉降量与荷载间的关系曲线。

settlement difference
【释义】 沉降差
【例句】 The control of settlement difference is a key technology of up-side-down construction adopted for soft soil foundation. 沉降差控制是软土地区逆作法施工中的关键技术。

【中文定义】 建筑物两相邻基础间地基沉降量的差值。

settlement observation

【释义】 沉陷观测，沉降观测

【同义词】 subsidence observation

【常用搭配】 settlement displacement observation 沉降位移观测；field settlement observation 现场沉降监测；stratified settlement observation 分层沉降观测

【例句】 The settlement observation of flood protection dike is an essential and important applied technology in the water conservancy project. 防洪堤的沉降的观测，是水利工程中不可缺少的一个重要的应用技术。

settlement observation

settlement observation point

【释义】 沉降观测点

【例句】 Positioning of settlement observation points is very important for correctly analyzing and solving settlement influence on embankment resulted from weak subgrade of high-grade highway. 为正确分析和解决高等级公路软土地基对路堤的沉降影响，布设沉降观测点十分重要。

settlement observation point

settlement of foundation

【释义】 地基沉降，软基沉降，地基沉降量

【例句】 This paper is an analysis and summary about the causes of distortion cracks, including those caused by contraction of concrete and variation of temperature, and uneven settlement of foundation. 本文主要对变形裂缝的成因及控制措施作了分析和总结，其中包括混凝土收缩和温度变化引起的裂缝，以及地基不均匀沉降引起的裂缝等。

【中文定义】 因地层压密或变形而引起的地面标高降低。

shale [ʃeil]

【释义】 *n.* 页岩

【常用搭配】 oil shale 油页岩；shale oil 页岩油（由油页岩中干馏而得的石油）

【中文定义】 是一种沉积岩，成分复杂，但都具有薄页状或薄片层状的节理，主要是由黏土沉积经压力和温度形成的岩石，但其中混杂有石英、长石的碎屑以及其他化学物质。

shale

shallow bored well

【释义】 浅井

【例句】 The cooperating application of P-T packer and MFE tester is easy to make packer loose at closing operation when conducting test in shallow bored well. 浅井测试时，P-T 封隔器与MFE 配合使用容易在开关井操作时使封隔器松动。

shallow landslide

【释义】 浅层滑坡

shallow-focus earthquake

【释义】 浅源地震

【例句】 Because there is a corresponding time relationship between the shallow-focus earthquake and the deep-focus earthquake along the Japan

Islands-Arc System, we can foreshow occurrence times of deep-focus earthquakes in northeastern China using the signs of the shallow-focus earthquakes occurring in the Japan Islands Arc. 鉴于沿日本岛弧系发生的深源和浅源地震之间存在着一种时间对应关系，所以我们可以利用日本岛弧发生的浅震来预测我国东北的深源地震.

【中文定义】 指震源深度在70km以内的地震。它发震的频率最多，对人类的影响也最大，若以释放的能量多少来比较，85%的是浅源地震。在大陆上，浅源地震占95%以上，因此地震灾害主要由浅源地震造成。

shear concrete plug

【释义】 抗剪洞，抗剪置换洞

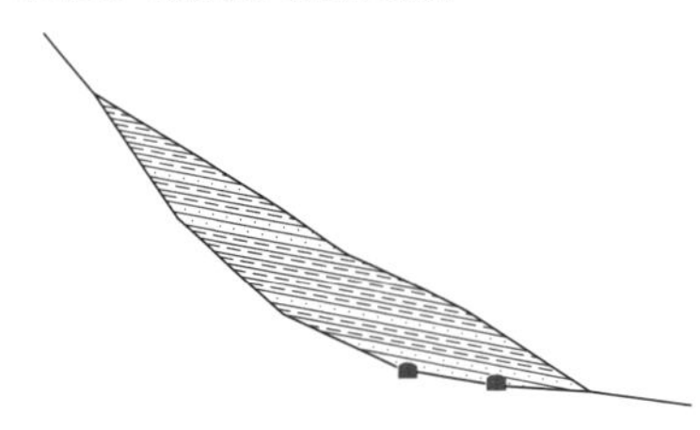

shear concrete plug

shear crack of landslide

【释义】 滑坡剪出口

shear deformation

【释义】 剪切变形

【中文定义】 当杆件在两相邻的横截面处有一对垂直于杆轴，但方向相反的横向力作用时，其发生的变形为该两截面沿横向力方向发生相对的错动，此变形称为剪切变形。

shear dilatation

【释义】 剪胀

【例句】 the assumption of constant shear dilatation angle cannot describe that the shear dilatation is dependent on confining pressure and plastic deformation. 恒定剪胀角的假设不能描述剪胀与围压和塑性变形的关系。

shear fault

【释义】 扭性断层，剪切断层

【例句】 The prospective areas for exploration in future are regions outward the long axis direction of known oil-gas-bearing anticlines, deeper Mushan Formation and below (include the Cretaceous), the upthrow side of shear fault and the high area of the basement, as well as the unconformity and pinchout region around. 下一步勘探区的选择应为已知含油气背斜长轴方向的外延部位，深部木山组及其以下层位（包括白垩系），扭性断层的上升盘，基底高及其周围的地层不整合、地层尖灭区。

shear fault

【中文定义】 又称横移断层、走滑断层，或扭转断层，平移断层作用的应力是来自两旁的剪切力作用，其两盘顺断层面走向相对移动，而无上下垂直移动。

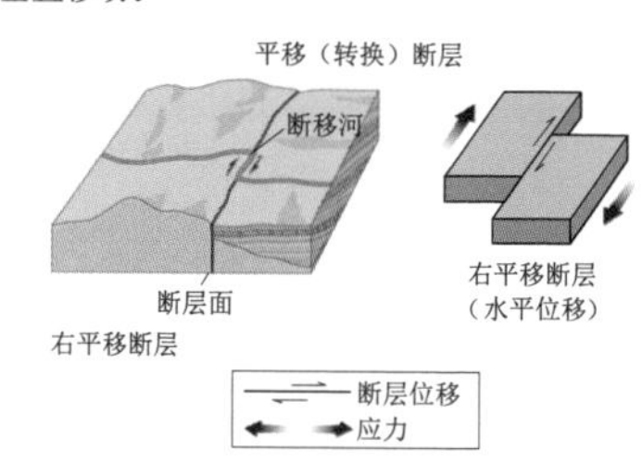

Translation shear fault

shear joint

【释义】 剪节理，剪裂隙

【中文定义】 指岩石受剪应力作用产生的破裂面。

shear strength

【释义】 抗剪强度

【例句】 The research results indicate that the temperature has complicated influence on the

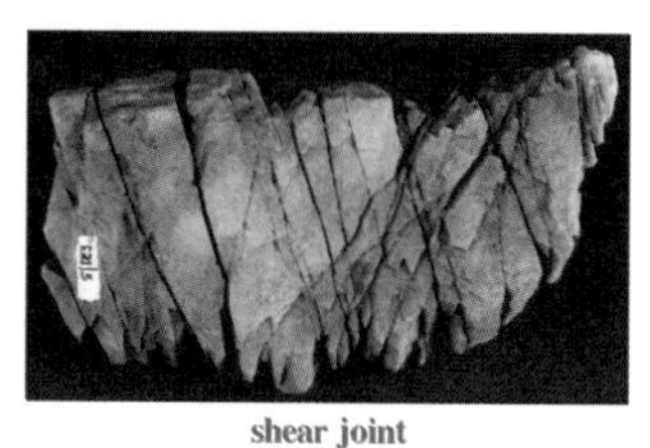

shear joint

shear strength of clayey soil. 试验结果表明：温度对黏性土的抗剪强度有比较复杂的影响。

【中文定义】 指外力与材料轴线垂直，并对材料呈剪切作用时的强度极限。

shear test

【释义】 剪切试验

【常用搭配】 triaxial shear test 三轴剪切试验

【例句】 This paper mainly studies the strength softening characteristic of loess-like soil by triaxial shear test on the basis of primary research productions of loess. 本文在黄土已有研究成果的基础上，通过三轴剪切试验对类黄土（黄土状土）的强度软化特性进行了重点研究。

direct shear apparatus

shear zone

【释义】 剪切带

【例句】 An analysis on the material component variation and volume change of the mylonite in the Yongxiu-Nanchang shear zone has revealed that under the constraint conditions of constant Al_2O_3, the shear zone has Lost 9% mass and volume. The mylonite has an obvious gain in Fe_2O_3 and a loss in SiO_2, K_2O, MgO, P_2O_5, FeO and Na_2O. 永修-南昌剪切带中糜棱岩物质成分变异及体积变化分析结果表明，以 Al_2O_3 守恒为限制条件，该剪切带损失了9%的质量和体积，糜棱岩类 Fe_2O_3 明显带入，SiO_2、K_2O、MgO、P_2O_5、FeO、Na_2O 等组分明显带出。

【中文定义】 发育在岩石圈中具剪切应变的强烈变形带。这一变形带可以是应变不连续的面状构造（断层），或者在露头尺度上见不到几何不连续性而呈连续应变的韧性剪切带。

sheet line system

【释义】 地图分幅

【中文定义】 按一定规格将广大地区的地图划分成一定尺寸的若干单幅地图。图幅编号每幅地图的代号。

shield [ʃild]

【释义】 *n.* 地盾

【常用搭配】 Canadian Shield 加拿大地盾；Guiana Shield 圭亚那地盾；Baltic Shield 波罗的海地盾

【例句】 The Canadian Shield is alikelyg the oldest on earth, with regions dating from 2.5 to 4.2 billion years. 加拿大地盾算是地球上最古老的区域，追溯到25亿～42亿年间。

【中文定义】 地质学名词，构造地貌的术语，是大陆地壳上相对稳定的区域。和造山带相反，在地盾中造山活动、断层及其他地质活动都很少。

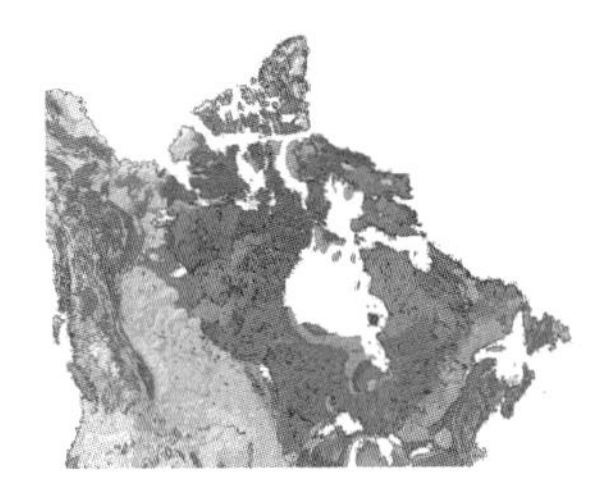

Canadian Shield

short-term stability

【释义】 短期稳定性

【中文定义】 洞室、边坡岩土体，在开挖瞬时及开挖后短期内的稳定性。

shotcrete ['ʃɒtkriːt]

【释义】 *n.* 喷射混凝土

【中文定义】 运用机械设备向围岩或开挖岩坡表面喷射混凝土层以加固围岩的技术。

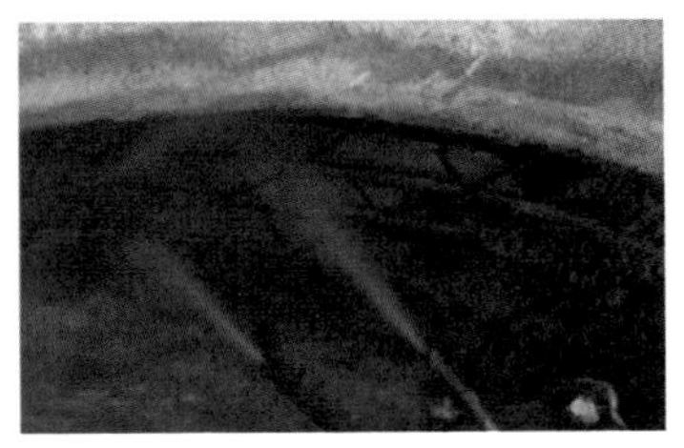

shotcrete

shrinkage limit

【释义】 缩限

shrinkage test

【释义】 收缩试验

【例句】 It is of great practical significance to study the shrinkage performance of semi-rigid base material by choosing suitable shrinkage test method. 选择合适的收缩试验方法对半刚性基层材料的收缩性能进行研究非常具有现实意义。

shrinkage test

sieving analysis

【释义】 筛析法

【例句】 This paper analyzes sandstone coring clastics distribution in different confining pressure using sieving analysis and finds that it is fractal distribution in sandstone coring breakage clastics. 文中用筛析法分析不同围压下砂岩岩心碎屑的分布，发现砂岩岩心破碎碎屑是分形分布。

silicolite [ˈsɪliːkəlaɪt]

【释义】 *n.* 硅质岩

silicolite

【例句】 Discovery of Late Palaeozoic radiolarian silicolite in many places in northeastern Jiangxi province ophiolitic melange belt. 赣东北蛇绿混杂岩带中多处发现含晚古生代放射虫硅质岩。

【中文定义】 指由化学作用、生物和生物化学作用以及某些火山作用所形成的富含 SiO_2（一般大于 70%）的沉积岩，包括盆地内经机械破碎再沉积的硅质岩。

sill [sɪl]

【释义】 *n.* 岩床

【常用搭配】 composite sill 复合集岩床

【例句】 Sills are tabular plutons formed when magma is injected along sedimentary bedding surfaces. 岩床是由岩浆沿层面流动侵入形成的板状岩体。

【中文定义】 又称岩席，是由岩浆沿层面流动铺开，形成与地层相整合的板状岩体。

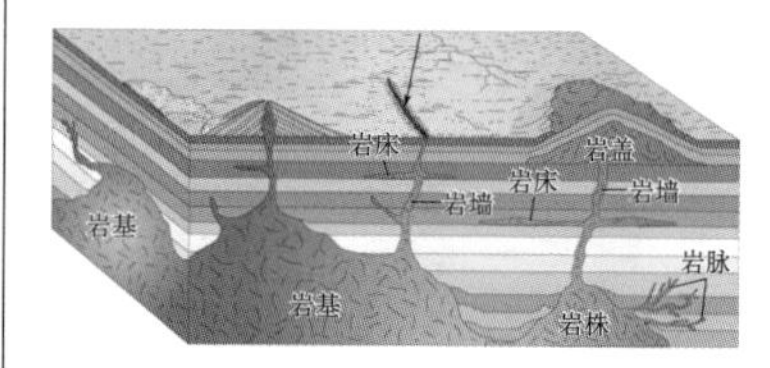

sill

silt [sɪlt]

【释义】 粉土

【常用搭配】 silt loam 粉砂壤土；sandy silt 砂质粉土；silt seam 淤泥层

【例句】 The channel is almost choked with silt. 水渠里淤了很多泥沙。

【中文定义】 指粒径大于 0.075mm 的颗粒质量不超过总质量的 50%，且塑性指数 I_p 小于或等于 10 的土。粉土的工程性质为密实的粉土为良好地基；饱和稍密的粉土，地震时易产生液化，为不良地基。

silt

silt content

【释义】 含泥量

【例句】 Besides cement dose, silt content and construction control, aggregate grade is one of the important factors which affect the shrinkage properties of cement stabilized material bases. 除了水泥剂量、含泥量和施工控制等因素外，集料级配是影响水泥稳定粒料基层收缩特性的重要因素。

【中文定义】 指天然沙中粒径小于 75μm 的颗粒含量。

silt soil

【释义】 淤泥土

【例句】 The Leaning Tower began to tilt almost as soon as it was built on soft silt soil in 1173. 比萨斜塔自从 1173 年建立在软淤泥土上以来就开始倾斜。

【中文定义】 在静水或缓慢流水中堆积而形成的土。

silt soil

silt texture

【释义】 粉砂状结构

【中文定义】 碎屑颗粒大小在 0.05～0.005mm 之间的碎屑结构。

siltstone ['sɪltˌstoun]

【释义】 *n.* 粉砂岩

【常用搭配】 fine siltstone：细粉砂岩；siltstone reservoir 粉砂岩储集层，粉砂岩储层

【例句】 The channel is overlain by siltstone and lenticular beds of limestone. 河道上面覆盖着粉砂岩和石灰岩透晶体。

【中文定义】 主要由 0.1～0.01mm 粒级（含量大于 50%）的碎屑颗粒组成的细粒碎屑沉积岩。

siltstone

silty sand

【释义】 淤泥质粉砂

【中文定义】 粉砂是一种粒径在 0.0625～0.0039mm 的矿物或岩石碎粒，粒度介于砂和黏土颗粒之间。由暴露在地表的各种岩石经风化破碎而成。粉砂的成分与砂相似，但岩屑含量减少。粉砂分布于海滨、河流及湖泊沉积物中，其黏性差，具有透水性。

silty sand

Silurian (S) [sai'ljuəriən]

【释义】 *n.* 志留纪（系）

【中文定义】 古生代的第三个纪，约开始于 4.4 亿年前，结束于 4.1 亿年前。一般说来，早志留世到处形成海侵，中志留世海侵达到顶峰，晚志留世各地有不同程度的海退和陆地上升，表现了一个巨大的海侵旋回。

simple shear test

【释义】 单剪试验

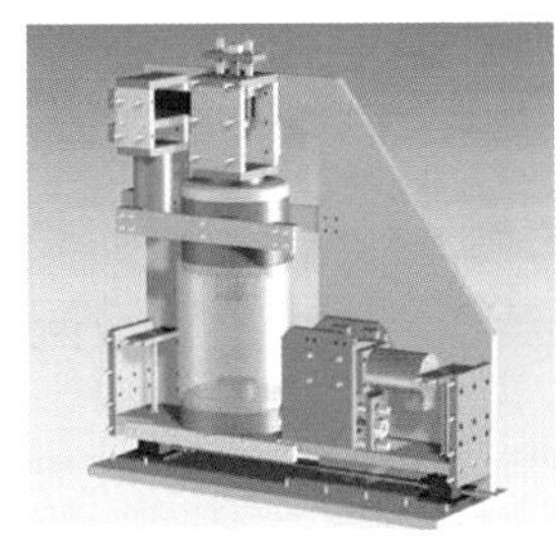

simple shear test apparatus

【例句】 It presents an experimental study on the bond behavior of NSM CFRP strips using a simple shear test set-up. 通过 CFRP 板条嵌入混凝土试块的界面单剪试验，研究黏结长度对界面黏结性能的影响。

simplified water quality analysis

【释义】 水质简分析

【例句】 Simplified water quality analysis is suitable for the initial understanding of the main chemical composition of water. 水质简分析适用于初步了解水的主要化学成分。

single borehole acoustic detection

【释义】 单孔声波检测

【例句】 Single borehole acoustic detection is an important in-situ geophysical prospecting work in hydropower survey, which is vital to detect unfavorable geologic bodies, rock mass integrity and rock mass weathering. 单孔声波测试是水电勘察中重要的原位物探测试项目，在探测不良地质体、岩体完整性、岩体风化中尤为重要。

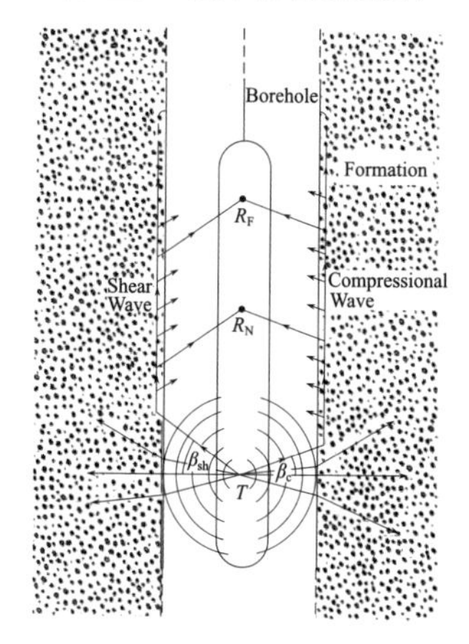

single borehole acoustic detection

single-hole pumping test

【释义】 单孔抽水试验

【例句】 Single-hole or multi-hole pumping test shall be carried out according to the stratified characteristics and hydrogeological structures of overburden, and pumping test for a main pervious layer in dam foundation shall be done in at least 3 sections. 根据覆盖层的成层特性和水文地质结构进行单孔或多孔抽水试验，坝基主要透水层的抽水试验不应少于 3 段。

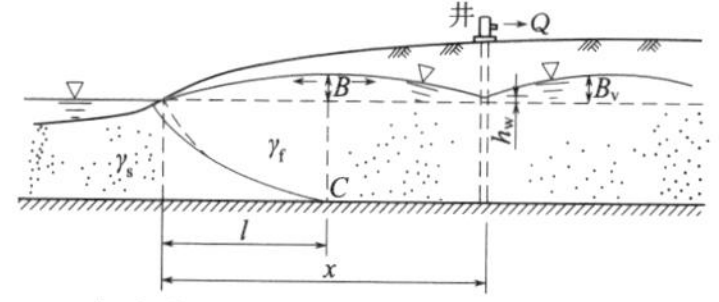

single hole pumping test in coastal zone

Sinian (Z) ['siniən]

【释义】 *n.* 震旦纪（系）

【常用搭配】 Sinian period 震旦纪；Sinian system 震旦系

【同义词】 Ediacaran period 埃迪卡拉纪

【例句】 After the Sinian Great Ice Age, the Phanerozoic started. 震旦纪大冰期后开始了显生宙。

【中文定义】 又称埃迪卡拉纪，为元古宙晚期的一个纪，是在中国命名并向国际推荐的一个地质年代单位。开始于约 8 亿年前，结束于约 6 亿年前，属于新元古代的晚期。

sinistral rotation fault

【释义】 左旋断层

【同义词】 left handed fault

【例句】 Qiulitake structural belt is divided into two segments in E－W direction under the effect of sinistral rotation fault, the western segment is composed of two anticlines, the southern one and the northern one. The fault controlling the southern anticline has better continuity, but the fault controlling the northern one has worse continuity and can be divided to three segments which are sinistrally arranged. 秋立塔克构造带受左旋断层的作用被分为东西两段，控制西段南部背斜生长的断层连续性较好，而控制北部背斜生长的断层连续性较差，大致可分为三段，呈左阶排列。

【中文定义】 如果一个观察者站在断层的一侧，面向断层，另一边的岩块向他左方滑动，那它就叫左滑断层。之所以如此称呼，因为要追索被移动了的地表特征时，该人需沿断层线转向左边，才能在那一边找到与这边相对应的特征。

sinkhole ['sɪŋkhəʊl]

【释义】 *n.* 落水洞

【常用搭配】 sinkhole lake 溶坑湖

【例句】 An area of irregular limestone in which erosion has produced fissures, sinkholes, underground streams, and caverns. 喀斯特是由于腐蚀而产生裂隙、落水洞、潜流和溶洞的不规则的石灰岩地区。

【中文定义】 地表水流入地下的进口，表面形态与漏斗相似，是地表及地下岩溶地貌的过渡类型。它形成于地下水垂直循环极为流畅的地区，即在潜水面以上，落水洞的形成，在开始阶段以沿垂直裂隙溶蚀为主。

sinter ['sɪntə]

【释义】 泉华

【常用搭配】 algal sinter 硅藻泉华

【例句】 Refering to ESR data from other scientists, this paper thinks that the Targjia thermal sinter is formed in five stages: 403～202kaBP, about 99kaBP, 39～25kaBP, 17～4kaBP and the present. 参考前人的 ESR 年龄测定结果，可将搭格架热泉华的形成分为 5 个阶段：403～202kaBP，99kaBP 左右，39～25kaBP，17～4kaBP 与现代。

【中文定义】 溶解有矿物质和矿物盐的地热水和蒸气在岩石裂隙和地表面上的化学沉淀物。

sinter landscape

siphon cylinder method

【释义】 虹吸筒法

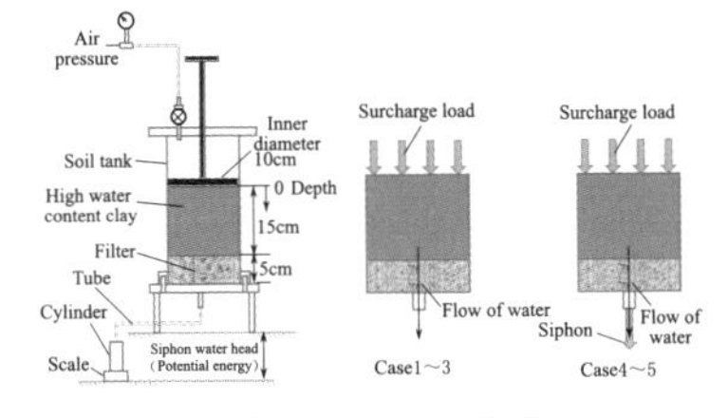

siphon cylinder method

site survey

【释义】 现场查勘

site survey

size effect

【释义】 尺寸效应

【例句】 It is expressed that the high strength concrete cube strength size effect differs from the normal concrete. 表明高强混凝土立方强度尺寸效应不同于普通混凝土。

skarn [skɑːn]

【释义】 *n*. 矽卡岩，硅卡岩

【常用搭配】 skarn mineral 夕卡岩矿物

【例句】 The stratiform skarn of hydrothermal fluid sedimentary origin is well developed in Kendekoke gold deposit, Qinhai province. 热水沉积成因层矽卡岩在青海省肯德可克金矿区非常发育。

【中文定义】 一种变质岩，主要由富钙或富镁的硅酸盐矿物组成的变质岩，一般经接触交代作用形成。矿物成分主要为石榴子石类、辉石类和其他硅酸盐矿物。细粒至中、粗粒不等粒结构，条带状、斑杂状和块状构造。颜色取决于矿物成分和粒度，常为暗绿色、暗棕色和浅灰色，比重较大。

skarn

sketch map

【释义】 示意图，略图，草图

【同义词】 schematic diagram

【中文定义】 指内容概括简略的图件，大体上描述或表示物体的形状、相对大小、物体与物体之间的联系（关系）。

slag material

【释义】 洞渣料

【例句】 The slag material could be used as aggregate in concrete. 洞渣料可以用作混凝土骨料。

slaking test

【释义】 湿化试验

【例句】 Through slaking test, the influence of the soil characteristics and the degree of compaction on the scouring resistance of the soil were analyzed. 通过室湿化试验，分析了土的特性、压实度对试样抗冲刷能力的影响。

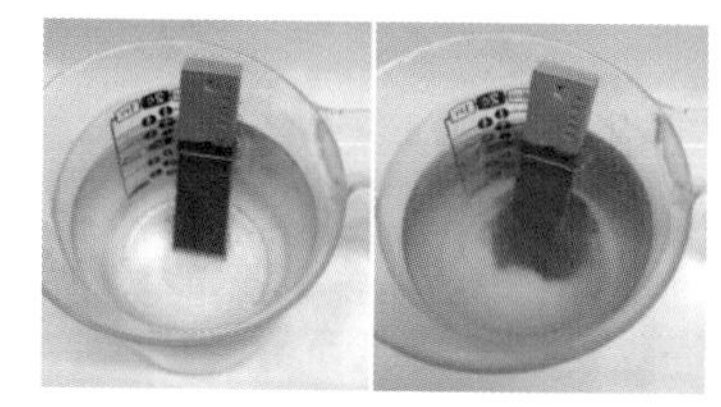

slaking test

slate [sleit]

【释义】 *n.* 板岩

【常用搭配】 clay slate 黏板岩；spotted slate 斑点板岩

【例句】 A concrete bearing pad is being set in the poor quality slate between el. 1725m and 1885m on the left abutment. 左坝肩 1725～1885m 高程间的软弱砂板岩上设置了一个混凝土垫座。

slate

【中文定义】 是具有板状构造，基本没有重结晶的岩石，是一种变质岩。原岩为泥质、粉质及中酸性凝灰岩经轻微区域变质而成。沿板理方向可剥成薄板状，基本上无新生矿物，大部分为隐晶质的黏土矿物及碳质、铁质粉末，常保留残余的泥质、粉砂质岩石结构或变余构造。板岩的颜色随含杂质不同而异，含铁时呈红、黄色，含碳时呈黑、灰黑色，含钙时滴盐酸起泡。板岩进一步命名常以颜色或杂质成分而定，如灰绿色板岩、钙质板岩等。

slaty cleavage

【释义】 板劈理

【例句】 At the late Palaeozoic the study area entered Jinlin stage at which consolidated Tarim platform basement broke apart partially, resulting in a scrips of east-west fault basins The Xingdi Fault accepted the second-phase deformation, characterized by plastic deformation mainly of slaty cleavage and phyllitic cleavage. 晚元古代本区进入晋宁期发展阶段，固结的塔里木地台基底局部裂开，形成一系列近东西向的断陷盆地，发生了第二期变形，形成以塑性变形为主的板劈理和千枚理特征。

【中文定义】 丹尼斯（L. G. Dennis，1964）定义为板岩所特有的连续劈理。它发育在细粒的低级变质岩中，肉眼极难区别出劈理域或微劈石；在显微尺度上，劈理域由平行面状或交织状排列的云母或绿泥石等层状硅酸盐矿物富集成薄膜或薄层，宽约 0.005mm；微劈石由石英、长石等浅色矿物的集合组成，呈薄板状或透镜状，宽约 1～0.01mm 或以下。

slaty cleavage

slice method

【释义】 条块法，条分法

【常用搭配】 Swedish slice method 瑞典条分法；Bishop slice method 毕肖普条分法；Janbu generalized procedure of slices 简布普遍条分法

【例句】 Slice method and finite element method are principal methods of soil slope stability analysis. 条分法和有限元法是土质边坡稳定分析的主要方法。

slickensided

【释义】 有擦痕的

【例句】 The roughness of the structure can be divided into four types, including slickensided, mirror, smooth and rough. 结构面的粗糙程度可以

slickensided

分为四种，包括有擦痕的、镜面的、光滑的和粗糙的。
【中文定义】 结构面的光滑程度不同，抗剪强度也不同。是结构面一种粗糙程度的描述。
slide axis
【释义】 滑坡主轴（主滑线）
slide boundary
【释义】 滑坡周界
sliding stability along weak interlayer
【释义】 软弱夹层抗滑稳定
sliding stability in deep layer
【释义】 深层抗滑稳定
sliding stability in shallow layer
【释义】 浅层抗滑稳定
slightly dense
【释义】 稍密
【中文定义】 介于中密和疏松之间的密实度指标。
slightly developed
【释义】 轻度发育
【中文定义】 根据《水力发电工程地质勘察规范》(GB 50287—2016) 的规定，当岩体内结构面发育组数1～2组，间距介于50～100cm时，其岩体结构面发育程度为轻度发育。
slightly weathered
【释义】 微风化
【例句】 Generally, the foundation of high arch dam is normally required to be placed on fresh or slightly weathered rock. 通常高拱坝建基面要求开挖至新鲜～微风化基岩。
【中文定义】 指岩石组织结构无变化，保持原始完整结构，仅岩石表面或裂隙面有轻微褪色的一种风化现象。

slightly weathered

slip bed
【释义】 滑床，滑坡床
【例句】 So the difference in the engineering properties between the slip masses and the slip beds is evident. The interface is the main slip face of the landslide. 滑体和滑床工程地质性质差异明显，二者界面为滑体下滑的主滑面。
slip cleavage
【释义】 滑劈理
【常用搭配】 strain-slip cleavage 应变滑劈理
【例句】 They are interpreted as shear domains and are geometrically similar to incipient strain-slip cleavage in foliated rocks. 它们被解释为剪切域，在几何上类似于在叶状岩石中初始的应变-滑移劈理。
【中文定义】 又称剪劈理、折劈理或小褶皱，是一类被叠加在板劈理或片理之上的劈理，其劈面平直，平行排列，间距宽窄不一。

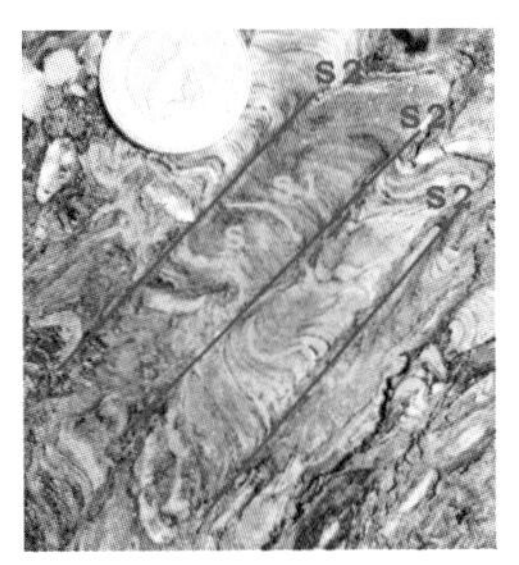

slip cleavage

slip cliff
【释义】 滑坡壁

slip cliff

slip foundation
【释义】 滑坡基座
slip mass
【释义】 滑坡体
【同义词】 landslide mass

【例句】 Some, but not all, of the potential slip masses are labeled. 部分（而非全部）潜在滑坡体已经标记出来。

slip mass detection

【释义】 滑坡体探测

slip surface

【释义】 滑动面

【同义词】 slip plane

【例句】 Further, potential slip surfaces are determined by slope internal displacement and sliding rate. 通过深部位移资料得出边坡的滑动面位置。

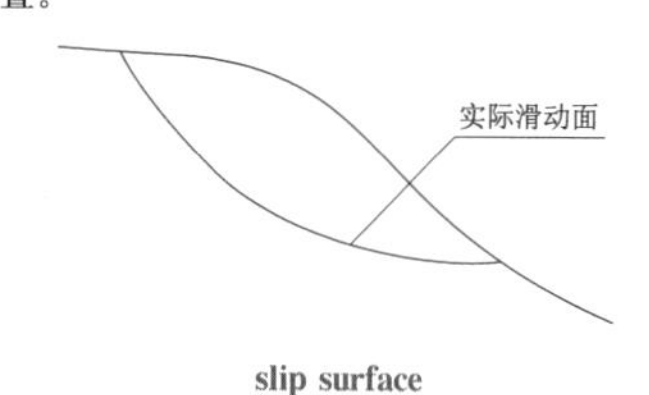

slip surface

slip zone

【释义】 滑动带

slit method test

【释义】 狭缝法试验

slope [sloʊp]

【释义】 *n.* 边坡

【常用搭配】 natural slope 自然边坡; engineering slope 工程边坡

【例句】 The stability of soft soil slope is an complicated engineering problem. 软土边坡稳定是一个复杂的工程课题。

【中文定义】 地壳表面具有侧向临空面的地质体，由坡顶、坡面、坡脚及其下部一定深度内的坡体组成。

slope cracking

【释义】 边坡开裂

slope debris flow

【释义】 山坡型泥石流（坡面泥石流）

【例句】 Slope debris flow is a natural hazard constantly observed in the mountain area. 坡面泥石流是山区常见的一种自然灾害。

【中文定义】 指泥石流无恒定地域与明显沟槽，只有活动周界，轮廓呈保龄球形。

slope deformation

【释义】 边坡变形

slope failure

【释义】 边坡失稳

【中文定义】 指边坡出现裂缝、松动、滑移等现象，严重影响边坡的稳定性。

slope ratio method

【释义】 坡率法

slope retaining

【释义】 边坡支挡

【常用搭配】 slope retaining wall 护坡墙; slope retaining pile 护坡桩

slope retaining

slope structure

【释义】 坡体结构

【中文定义】 指坡面以下一定范围内的岩体，地质结构面和岩石结构体以不同形式的相互组合。

slope structure model

【释义】 边坡结构模型

slow shear test

【释义】 慢剪试验

slug test

【释义】 微水试验

【例句】 Slug tests are applicable to a wide range of geologic settings as well as small-diameter piezometers or observation wells, and in areas of low permeability where it would be difficult to conduct a pumping test. 微水试验广泛适用于小直径钻孔、观测孔以及由于渗透性较低而无法进行抽水试验的地质环境。

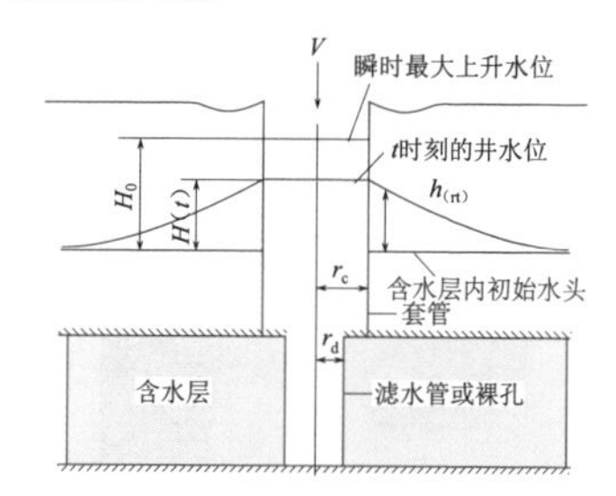

schematic diagram of slug test

slug test in borehole

【释义】 钻孔振荡式渗透试验

【例句】 Slug test in borehole is a site hydrogeological test that is low cost, simple, short duration, and without extracting or injecting large amounts of water. 钻孔振荡式渗透试验是一项成本低、简单、试验历时短、无须抽出或注入大量水的现场水文地质试验。

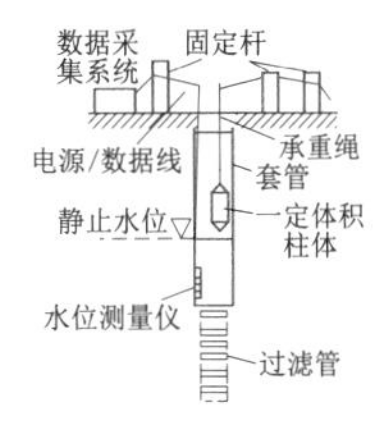

振荡式渗透试验开始前

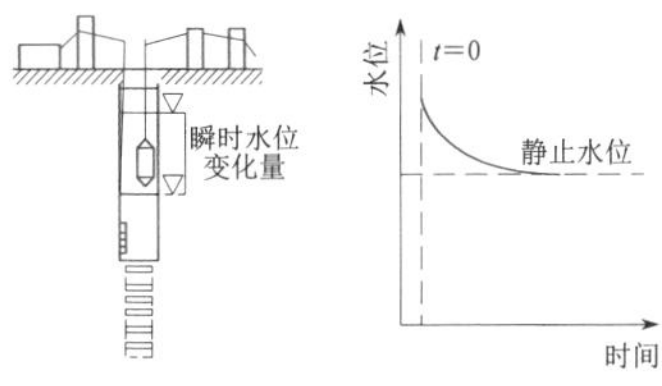

振荡式渗透试验开始后　　水位变化曲线

theory of slug test in borehole and its response curve of water level with time

slump [slʌmp]

【释义】 *n*. 滑移，坍落

【常用搭配】 slump structure 塌滑构造；slump test 坍落度试验，重陷试验

【中文定义】 指岩（土）体沿一定的软弱面，在重力作用下发生滑动的一种破坏模式。

slump test

【释义】 稠度试验

【例句】 Compatibility of cement and flyash with plasticizer was studied through the slump test of concrete from the aspects of cement, flyash and plasticizer. 通过混凝土坍落度试验，从水泥、粉煤灰及塑化剂三方面研究了水泥及粉煤灰与塑化剂的适应性问题。

small-scale landslide

【释义】 小型滑坡

【例句】 The primary types are medium-scale landslide and small-scale landslide in landslide volume classification. 滑坡体体积分类则以中型和小型滑坡为主。

smooth [smuð]

【释义】 *adj*. 光滑的

smooth

soft clay

【释义】 软土

【例句】 The deformation is the dominant factor of the quality of the road constructed on the soft clay foundation. 在软土地基上修建道路，变形是控制道路质量和正常使用的主要因素。

soft clay

【中文定义】 一般是指天然含水量大、压缩性高、承载力低和抗剪强度很低的呈软塑～流塑状态的黏性土。软土是一类土的总称，并非指某一种特定的土，工程上常将软土细分为软黏性土、淤泥质土、淤泥、泥炭质上和泥炭等。

soft grain

【释义】 软弱颗粒

【常用搭配】 soft grain in aggregate 软弱颗粒含量

soft rock

【释义】 软质岩

【例句】 In general, mudstone, shale and shaly sandstone are typical of soft rock。通常，泥岩、页岩、泥质砂岩是典型的软岩

【中文定义】 指单轴饱和抗压强度小于 30MPa 的岩石

soft soil subsidence

【释义】 软土震陷

【例句】 The result of the experiments can supply some basis of the analysis of soft soil subsidence mechanism. 室内试验的结果为软土震陷机理的分析提供依据。

【中文定义】 是软土在地震快速而频繁的加荷作用下，土体的结构受到扰动，导致软土层塑性区的扩大或强度的降低，从而使建筑物产生附加沉降。

softening coefficient

【释义】 软化系数

【例句】 The potential bearing capacity of this rock is relatively higher, having softing coefficient less than 0.75 with low pore rate, water absorptivity, and inferior crack connection. 该类岩石以中-微风化为主，软化系数小于 0.75，孔隙率、吸水率低，裂隙贯通性差，具有较大的地基潜力。

soil [sɔɪl]

【释义】 *n.* 泥土；土地，国土

【常用搭配】 peat soil 泥炭土；red soil 红壤；saline soil 盐土；sandstone soil 砂岩土；sandy soil 砂土

【例句】 The soil washed from the hills is silting up the hydroelectric dams. 从山上冲刷下来的泥土就要让水电大坝淤塞了。

【中文定义】 地球表层由细小泥沙颗粒组成的固体混合物，是岩石风化及成土作用的产物，也称泥土。它是构成土壤、土地、国土、领土的主要部分。

soil

soil borrow area

【释义】 土料场

【中文定义】 指在修建工程附近的适宜合格的土料产地。

soil borrow area

soil consolidation

【释义】 土的固结，土壤固结

soil flow

【释义】 流土

【中文定义】 指在渗流作用下，会出现局部土体隆起，某一范围内的颗粒或颗粒群同时发生移动而流失，这种沙沸现象称为流土。

soil flow on contact surface

【释义】 接触流土

【中文定义】 指渗流垂直于渗透系数相差较大的两相邻土层的接触面流动时，将渗透系数较小的土层中的细颗粒带入渗透系数较大的另一土层的现象。

soil landslide

【释义】 土质滑坡

【例句】 The practice in engineering has proved the important influence of the contaminated underground water on the soil landslide. 已有的工程实践表明，受污染地下水对土质滑坡的变形破坏有重要影响。

soil landslide

soil mass

【释义】 土体

【例句】 Water plays an important role in expansive rock and soil mass, and its distribution varies with space and time. 水在膨胀性岩土中发挥着至关重要的作用，而膨胀岩土体中的水分分布随空间和时间而改变。

【中文定义】 指固体颗粒间无联结或有微弱联结，保持天然结构，通常含有天然结构面的地质体。主要生成于第四纪，并分布在地壳表层，覆盖着陆地和海底的大部分。土体这个概念是20世纪70年代末至80年代初，随着工程地质工作的深入开展而逐步建立起来的。

soil parent rock

【释义】 成土母岩

【例句】 The relation of plantation growth to soil parent material-rock in the dry-hot valleys of the Jinsha river is very tight. 金沙江干热河谷区人工林生长与土壤母质-母岩的关系紧密。

【中文定义】 或称土壤母质，是地表岩石经风化作用使岩石破碎形成的松散碎屑，物理性质改变，形成疏松的风化物，是形成土壤的基本的原始物质，是土壤形成的物质基础和植物矿物养分元素（除氮外）的最初来源。

soil permeability

【释义】 土体渗透性

【例句】 Soil permeability is an important index of evaluating water source conservation function. 土体渗透性是评价土壤水源涵养功能的重要指标。

soil samples

【释义】 土试样

【例句】 The soil samples were collected from forest land and grassland in central Sichuan. 土壤样品采自四川中部的林地与草地。

soil settlement

【释义】 土体的沉降

solid model

【释义】 实体模型

【例句】 A solid model is the fundamental for carrying out finite element analysis and optimizing design. 实体模型是进行有限元分析和优化设计的基础。

【中文定义】 指具有一定体积并赋予特定的模型样式和模型属性的封闭表面代表实际物体位置、大小、形态的模型。实体模型具有各向同性的、密度等特性，可以检查两个几何实体的碰撞和干涉等。

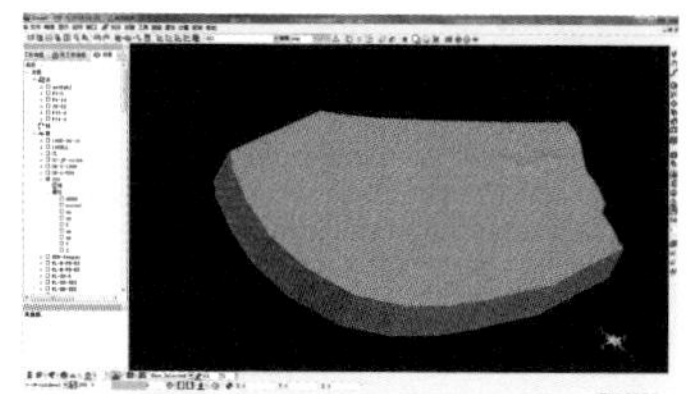

solid model

solid pouring method

【释义】 灌入固化物法

【例句】 There are different treatments in terms of their different structures, including replacement method, draining consolidation method, solid pouring method, pressing and reinforcement method. 针对软弱地基的不同构成有很多不同的处理方法，包括：置换法、排水固结法、灌入固化物法、振密（挤密）法、加筋法等五类。

solid pouring method

solid state

【释义】 固态

【例句】 The process of changing from a gaseous to a liquid or solid state. 由气态向液态或固态的转换过程。

solidify [səˈlɪdɪfaɪ]

【释义】 *v.* 固化，固结

【例句】 The thicker lava would have taken two weeks to solidify. 更厚的熔岩需要两周才会凝固。

【中文定义】 化学上是指物质从低分子转变为高分子的过程。

solubility [sɒljʊˈbɪlətɪ]

【释义】 *n.* 溶解度，溶度，溶解性

【例句】 The solubility of Natamycin is very low in water and in most organic solvents. 纳他霉素在水中和多数有机溶剂中的溶解度很低。

【中文定义】 在一定温度和压强下，某种物质在100g水或其他溶剂中所溶解的最大克数叫作这种物质在这种溶剂里的溶解度。

soluble silicon dioxide

【释义】 可溶性二氧化硅

【例句】 By analyzed the experiments results, we found the optimum technology condition to increase elimination rate of nitrides, starch, phenols and soluble silicon dioxide, to raise apparent purity (A. P), and to decrease color value. 根据极差分析选择最优水平组合，优化工艺条件以提高总

氮物、淀粉、酚类物和可溶性二氧化硅的除去率，提高简纯度，降低色值。

【中文定义】 可以迅速分散的二氧化硅。

solution cap

【释义】 溶帽山

【中文定义】 山顶或接近山顶溶蚀残留的碳酸盐岩体，其下为非可溶性岩层，形成似帽顶的山体。这类地形在山东由中寒武统灰岩及下伏页岩组成，当地称为“崮”，如著名的孟良崮。

solution cap landscape

solution crack

【释义】 溶隙

【例句】 At the exit section of a freeway tunnel in building, karst geological process is strong, solution crack and pore-hole are developed, tectonic deformation and later period eluviation-solution is also strong and rock mass mechanical strength and lithologic character are complicated. 正在建设中的某高速公路隧道出口段岩溶地质作用强烈，溶隙、孔洞发育，且受构造变形及后期淋滤溶蚀作用强烈，力学强度低，岩性复杂。

solution groove

【释义】 溶沟

【例句】 Combined with subsea tunnel engineering at eastern route way in Xiamen, the minimum rock covering thickness of subsea tunnel, water pressure design values, section optimization of lining structure and waterproof and drainage project, construction measures throughout bad subsea geology section, such as fault, solution groove, necessity of service tunnel setting were analyzed. 结合厦门东通道海底隧道工程，对海底隧道最小岩石覆盖层厚度、水压力设计值的确定，衬砌结构断面优化与防排水方案，穿越海底不良地质段（断层、溶槽）的施工措施及服务隧道设置的必要性问题进行了分析。

【中文定义】 指石灰岩表面上的一些沟槽状凹地。它是由地表水流，主要是片流和暂时性沟状水流顺着坡地，沿节理溶蚀和冲蚀的结果。

solution groove

source of release

【释义】 释放源

【中文定义】 指可释放能形成爆炸性气体混合物或有毒气体的位置或地点。

source region of debris flow

【释义】 泥石流物源区

【中文定义】 位于流域上游，又称为形成区，是泥石流主要水源、土源或砂石供给和起始源地。

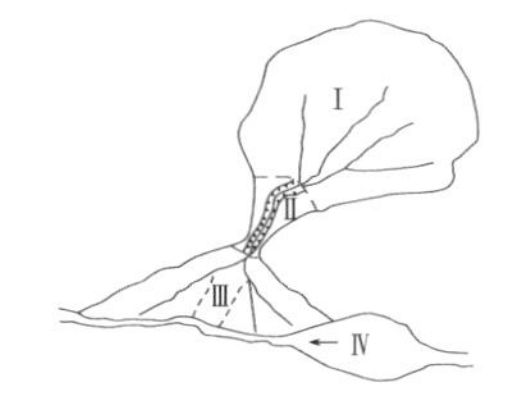

Ⅰ—物源区；Ⅱ—流通区；Ⅲ—沟口堆积区；Ⅳ—主河道

source region of debris flow（see Ⅰ）

special geological model

【释义】 专项地质模型

【中文定义】 为分析和解决专项工程地质问题，以工程区一般三维地质模型为基础，根据有关更详细的地质资料，对专项部分进行更加明细的分析、补充、处理得到的模型，包括深厚覆盖层模型、滑坡模型、地质块体模型、天然建筑材料模型等。

special use map

【释义】 专用地图

【中文定义】 为专门目的制作的地图。

specific gravity

【释义】 比重

【例句】 Specific gravity is the ratio of the density of a substance to the density of water. 比重是某物质的密度与水的密度之比。

specific gravity of soil particle

【释义】 土粒比重

specific gravity test

【释义】 比重试验

【例句】 Laboratory fill test of the project are including: fill water content, grain analysis and specific gravity test were tested. 本工程的填土室内试验包括：填土含水量测定、颗粒分析和比重测定。

specific gravity test apparatus

specific penetration resistance

【释义】 比贯入阻力

spectral induced polarization method

【释义】 频谱激电法

【例句】 Spectral induced polarization method is a kind of frequency domain excitation method in geophysical exploration. 频谱激电法是地球物理勘探中的一种频率域的激电方法。

specific retention

【释义】 持水度，持水率

【常用搭配】 specific capillary retention 毛细水容度

【中文定义】 是指饱水岩石在重力释水后仍能保持的水体积与岩石体积之比。

specific surface area

【释义】 比表面积

【例句】 The physical properties of catalysts such as specific surface area, pore volume, stacking density and crushing strength were tested. 测定了催化剂的比表面积、比孔体积、堆积密度和压碎强度等物理性质参数。

【中文定义】 单位质量物料所具有的总面积。

specific yield

【释义】 给水度

【例句】 Through theoretical analysis, the differences in the basic concept between specific yield and saturation deficiency are specified. 通过对给水度与饱和差的理论分析，明确了给水度与饱和差在基本概念上的差别。

【中文定义】 指饱和介质在重力排水作用下可以给出的水体积与多孔介质体积之比。

spheroidal weathering

【释义】 球状风化

spheroidal weathering

【例句】 The phenomenon of granite spherical weathering is frequently encountered in railway engineering which traverses granite areas. 花岗岩的球状风化现象是途经花岗岩地段的铁路工程中经常碰到的一个问题。

【中文定义】 由于岩体被不同方向裂隙切割成块体后，风化营力使岩块四周风化而残留球状硬块的一种现象。

split test

【释义】 劈裂试验（巴西试验）

【例句】 The present, freeze-thaw split test used to evaluate the resistance to water damage has its limitations. 目前用冻融劈裂试验来评价沥青混合料抗水损害能力有一定局限性。

splitting strength

【释义】 劈裂强度

【例句】 Under the prerequisite of ensuring the splitting strength of concrete, the concrete proportioning was optimized, and the tests on crack resistance of concrete were performed. 在保证混凝土劈裂强度的基础上，优选混凝土配合比进行混凝土抗裂性能试验。

【中文定义】 指在规定的试验条件下，胶接试样产生分离时，单位胶接宽度所需的拉伸载荷。

spoil area

【释义】 弃渣场

【例句】 The constractor shall dispose excavated material from cuts for various tunnel and shaft-portals, as well as preparation of quarries, spoil areas and permanent roads. 承包商应负责各种隧洞，竖井井口明挖部分的弃渣处理，并进行采料场、弃渣场和永久公路所需的准备。

spoil area

spring [sprɪŋ]
【释义】 *n.* 泉
【常用搭配】 thermal/hot spring 温泉；depression spring 低地泉；mineral spring 矿泉
【中文定义】 指含水层或含水通道与地面相交处产生地下水涌出地表的现象。

spring

spring decay coefficient method
【释义】 泉水消耗系数法
spring dynamic analysis
【释义】 泉水动态分析法
SPT blow count
【释义】 标准贯入试验击数
【例句】 Based on the related literatures and data, combining with case histories, this paper discusses the correction method of the SPT blow counts N of the standard penetration test. Concrete suggestions of correction are presented. 本文根据有关文献资料，结合工程实例，对标准贯入试验击数 N 值的修正方法进行了讨论，并提出了具体修正建议。

squeezing action
【释义】 挤压作用
【例句】 From Triassic to Palaeogene, the plate squeezing action weakened and Junggar Basin gradually entered a depression-subsidence phase from the intense compressive phase. 三叠纪至古近纪，板块挤压作用减弱，准噶尔盆地由强烈压陷期逐渐进入坳陷沉降阶段。
stability analysis
【释义】 稳定性分析
stability analysis of slope
【释义】 边坡稳定性分析
【例句】 The limit equilibrium method is widely applied in stability analysis of slope according to the characteristics with the distinct concept and the simple formulation. 极限平衡方法具有概念清晰、数学表达简洁等特点，被广泛用于边坡稳定性分析。
【中文定义】 指采用定性与定量计算等方式对边坡稳定性进行分析和评价。
stability of surrounding rock mass
【释义】 围岩稳定
【例句】 In situ stress is one of the basic parameters for the stability of surrounding rock mass and the design of support structure. 初始地应力是地下工程围岩稳定与支护结构设计的一个基本参数。
【中文定义】 开挖隧道和地下洞室后，围岩不致发生破坏变形的能力。常见的围岩破坏变形有脆性破裂，块体活动与塌落，层状岩体的弯曲折断，碎裂岩体的松动解脱，塑性变形和膨胀等。它主要与岩石性质、地质构造、原岩应力、地下水以及洞室的形状、尺寸、开挖方法、支护类型等因素有关。
stable ['steɪbl]
【释义】 *adj.* 稳定的
【中文定义】 根据《水力发电工程地质勘察规范》（GB 50287—2016），Ⅰ类围岩稳定性多为稳定，表现为围岩可长期稳定，一般无不稳定块体。
stable isotope
【释义】 稳定同位素
【例句】 Stable isotope ratio analysis is an effective method for the identification of fruit juice adulteration. 稳定同位素分析法就是根据这种差异对掺假果汁进行准确鉴别的有效方法。
【中文定义】 指在元素周期表中，原子序数相同，原子质量不同，化学性质基本相同，半衰期大于 10～15 年的元素的同位素。

stable slope

【释义】 稳定边坡

stable slope

Stage [stedʒ]

【释义】 *n.* 阶

【常用搭配】 Substage 亚阶；Maastrichtian Stage 马斯特里赫特阶；Oxfordian Stage 牛津阶

【中文定义】 第五级年代地层单位，与地质年代单位“期”对应，是年代地层学的基本工作单位，以科、属级的生物演化特征划分。

stage loading

【释义】 分级加荷

stalactite [stə'læktaɪt]

【释义】 *n.* 钟乳石

【例句】 Moreover, the saturation degrees of $CaCO_3$ during the formation of the stalactite samples in the two localities were also calculated that has revealed the cause of more developed karst geomorphology in the southwest of China than in the southeast. 另外，计算出两地钟乳石样品形成过程中的 $CaCO_3$ 饱和度，揭示了喀斯特地貌西南地区比东南地区更为发育的原因。

【中文定义】 是碳酸盐岩地区洞穴内在漫长地质历史中和特定地质条件下形成的石钟乳、石笋、石柱等不同形态碳酸钙沉淀物的总称。

stalactite

stalagmite ['stæləgmaɪt]

【释义】 *n.* 石笋

stalagmite

【例句】 Through 31 TIMSU series dating and 543 samples of carbon oxygen stable isotope analysis from 4 stalagmites in Qixing cave, 11000a to 85000a climatic change isotope records are obtained continuously. The study indicates that stalagmite's oxygen isotope records are same with the sea isotope records, which also may be divided correspondingly to MIS2, MIS3, MIS4 and MIS5a climates stages. 通过对贵州都匀七星洞 4 根石笋系统的 31 件 TIMSU 系及 543 件稳定同位素分析，揭示出其气候变化记录时限范围为 1.1 万～8.5 万年，相当于海洋同位素阶段的 MIS2，MIS3，MIS4 及 MIS5a。

【中文定义】 为碳酸钙石灰岩，指在溶洞洞底的尖锥体，是喀斯特地形的一种自然现象。

standard logging

【释义】 标准测井

standard penetration test (SPT)

【释义】 标准贯入试验

【例句】 Pile shaft capacity and main factors are tested by standard penetration test and sit core test. 用标准贯入试验和现场取芯试验可以有效地检测粉喷桩桩身强度和主要影响因素。

state coordinate system

【释义】 国家坐标系统

【例句】 Geometric determination of airborne three line scanner sensors integrated with POS will involve the local tangent plane reference transformation, block adjustment, spatial forword intersection and coordinate conversion between

WGS84 coordinate system and state coordinate system, and so on. 集成 POS 的机载三线阵传感器几何定位将涉及局部切平面坐标系归化、区域网平差、空间前方交会以及 WGS84 坐标转国家坐标系等方面。

static pressure

【释义】 静压力

【例句】 The hydraulics model can predict the properties of journal bearing under dynamic pressure, static pressure and combination pressure. 水力学模型可以对液膜轴承动压、静压和动静压混合状态下的径向滑动性能进行完整预测。

【中文定义】 静止液体内任一点承受的压力，称为该点处液体的静压力。

static water level (head)

【释义】 静水位（头）

【例句】 The synthetic sounding curve will be analyzed to see if it can be used to predict the static water level. 对合成测深曲线进行分析，看是否可以用来预测静水位。

steady creep

【释义】 稳态蠕变

【例句】 The experimental data revealed that the creeping process was divided into three stages: the transient creep, the steady creep and the accelerated creep stages. 试验结果表明，岩石的蠕变破坏过程可以划分为三个阶段：初始蠕变阶段（瞬态蠕变）、稳定蠕变阶段和加速蠕变阶段。

【中文定义】 指应变随时间延续而匀速增加，一般时间较长。

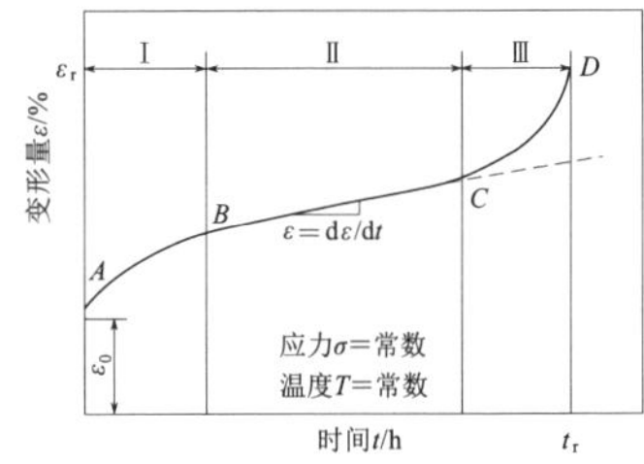

Ⅰ—初始蠕变阶段；Ⅱ—稳定蠕变阶段；Ⅲ—加速蠕变阶段

schematic diagram of steady creep stage

steady water level

【释义】 稳定水位

steady flow pumping test

【释义】 稳定流抽水试验

【例句】 Applying single-hole steady flow pumping test to evaluate hydraulic properties of basement rock aquifers around Liuzhuang western weathering shafts. 利用单孔稳定流抽水试验评价刘庄西区进回风井基岩段地下水径流条件。

steep slope

【释义】 陡坡

【中文定义】 水电工程实践中，习惯将地形坡度大于 55°的岸坡称为陡坡

steep slope

stepped [stept]

【释义】 *adj*. 有台坎的

【例句】 The shear strength is different because of the smoothness of the structural plane. There are mainly three kinds of undulating structures, including planar, undulating and stepped. 结构面的平整程度不同，抗剪强度也不同。结构面的形态有平直的、起伏的和有台坎的三种。

【中文定义】 是一种结构面起伏形态特征的描述。

stepped structural plane

【释义】 台阶结构面

【中文定义】 结构面的形态是台阶状的，反映了结构面的平整程度。

stepped structural plane

stereographic projection

【释义】 赤平极射投影

【例句】 An essential tool in the fields of structural geology and geotechnics, stereographic projection allows three-dimensional orientation data to be represented and manipulated. 作为构造地质学和大地构造学领域必不可少的工具，赤平投影使三维定位数据得以展示和操作。

【中文定义】 简称赤平投影，主要用来表示线、面的方位，相互间的角距关系及其运动轨迹把物

体三维空间的几何要素（线、面）反映在投影平面上进行研究处理。它是一种简便、直观的计算方法，又是一种形象、综合的定量图解，广泛应用于地质科学中。

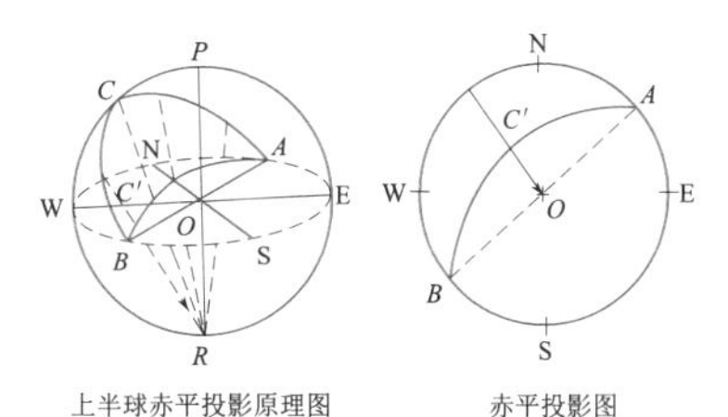

stereographic projection

stereonet

【释义】 投影网

【例句】 The most commonly used stereonets in stereographic projection include Wulff net and Schmidt net. 极射赤平投影常用的投影网包括吴尔福网和施密特网。

【中文定义】 是构造地质学用于极射赤平投影中的投影网格。常用的有吴尔福网（Wulffnet，简称吴氏网，也称等角距网）和施密特网（Schmidtnet，等面积网）。

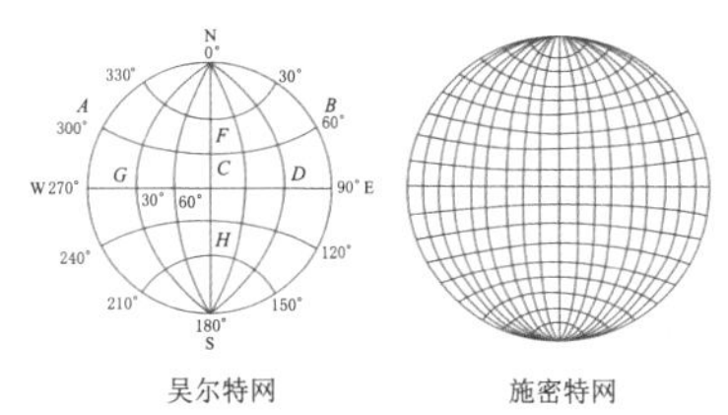

stereonet

still water level (SWL)

【释义】 静止水位

stock [stɑk]

【释义】 *n.* 岩株

【例句】 Damaogou granite stock has a very close relationship with Guangshigou uranium deposit. 大毛沟岩株与光石沟铀矿床形成密切相关。

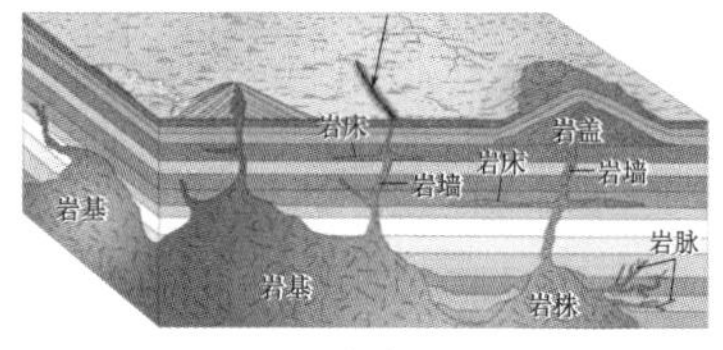

stock

【中文定义】 较岩基为小，平面近于圆形，向下呈树干状延伸的岩浆岩侵入体。

stockpile ['stɑkpaɪl]

【释义】 *n.* 库存，存料场

【例句】 Spoil from the excavations shall be placed in spoil areas or stockpiles. 开挖料应置于弃渣场或存料场。

stone block

【释义】 块石

【例句】 Stone block is heterogeneous and its acoustics property parameters have great dispersion. 岩块具有明显的非均质性，其声学特性参数离散性很大。

【中文定义】 指符合工程要求的岩石，经开采并加工而成的形状大致方正的石块。块石分有多种类型，主要有花岗石块石、砂石块石等。

stone block

stone column

【释义】 石柱

【例句】 The temple, which lies on the banks of the Yellow River in Jiaxian, Shaanxi Province, is supported by a stone column that is more than 20 meters high. It is indeed a marvel. 这所位于陕西省郏县黄河岸上的庙宇，由一条超过 20m 高的石柱支撑，不能不说是一个奇迹。

【中文定义】 溶洞中钟乳石向下伸长、与对应的石笋相连接所形成的碳酸钙柱体。

stone column landscape

stone forest

【释义】 石林

【例句】 Yunnan's prime natural wonder is Shilin, the stone forest, an exposed bed of limestone spires weathered and split into intriguing clusters. 石林是云南的一大自然景观，因暴露的石灰岩经长久的风吹雨打而形成千姿百态的形状。

【中文定义】 由密集林立的锥柱状、锥状、塔状石灰岩体组合成的景观。其间多为溶蚀裂隙，隙坡直立，坡壁上部有平行的溶沟。

stone forest

storage coefficient

【释义】 释水系数，贮水系数

【常用搭配】 specific storativity, specific storage coefficient 单位释水（贮水）系数

【例句】 Under load, there is marked linear correlation between land subsidence and ground waterlevel, and the coefficient of linear correlated function is aquifer's elastic skeleton storage coefficient. 荷载作用下，地面沉降与地下水水位之间具有线性相关关系，且线性相关系数为含水层的骨架弹性释水系数。

【中文定义】 水头（水位）下降（或上升）一个单位时，从底面积为一个单位高度等于含水层厚度的柱体中所释放（或贮存）的水量。

straight channel

【释义】 平直河道

straight channel

【例句】 Because the flow structure in bend is very different from that in straight channel so that their falling distances of rock riprap are also different. 由于弯道水流结构与直段水流结构有着很大的差别，所以弯道抛石落距也有别于直段的抛石落距。

【中文定义】 在平面上，两岸形态总体顺直的河道

strain [streɪn]

【释义】 *n.* 应变

【常用搭配】 plastic strain 塑性变形；strain energy 应变能；strain gauge 应变计

【例句】 Strains measure how much a given deformation differs locally from a rigid-body deformation. 应变测量是将一个给定的变形和刚体变形从不同的局部进行比较。

【中文定义】 指在外力和非均匀温度场等因素作用下物体局部的相对变形。

strain compensation

【释义】 应变补偿

【例句】 The results show that by control of the composition in the materials, the strain can be controlled and good strain compensation can be reached, this is coincident with theoretical calculation. 结果表明，通过对材料中的 As 组分调节，可以对材料中的应变进行控制，达到良好的应变补偿效果，实验结果和理论计算相当吻合。

strain gauge

【释义】 应变计，应变片

【常用搭配】 resistance strain gauges 电阻应变片；vibrating wire strain gauges 振弦式应变计

【例句】 The dynamic test in signal detection of mechanical vibration are realized by using resistance strain gauges. 利用电阻应变计开发的动应变测量系统，成功地实现了机械振动信号的动态检测。

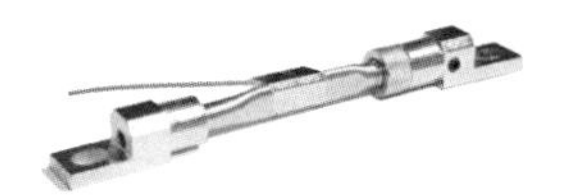

resistance strain gauges

strain hardening

【释义】 应变硬化

【例句】 At the same proportion, the strain hardening of the hydrophobic MMT compound system was more obvious than the hydrophilic MMT compound system. 在配比相同时，疏水性蒙脱土复合体系比亲水性蒙脱土复合体系应变硬化现象明显。

【中文定义】 经过屈服滑移之后，材料重新呈现

抵抗继续变形的能力，称为应变硬化。

strain softening

【释义】 应变软化

【例句】 Property of the strain softening and failure mechanism on different rocks transforms with the increase of confining pressure. 不同岩性岩石的应变软化性态和破坏机制会随侧压的增大而发生转化。

【中文定义】 指材料试件经 1 次或多次加载和卸载后，进一步变形所需的应力比原来的要小，即应变后材料变软的现象。

strain-slip cleavage

【释义】 错动劈理（应变滑劈理）

【例句】 They are interpreted as shear domains and are geometrically similar to incipient strain-slip cleavage in foliated rocks. 它们被解释为剪切域，在几何上类似于在叶状岩石中初始的应变-滑移劈理。

【中文定义】 切过先存面理的差异性劈理面。

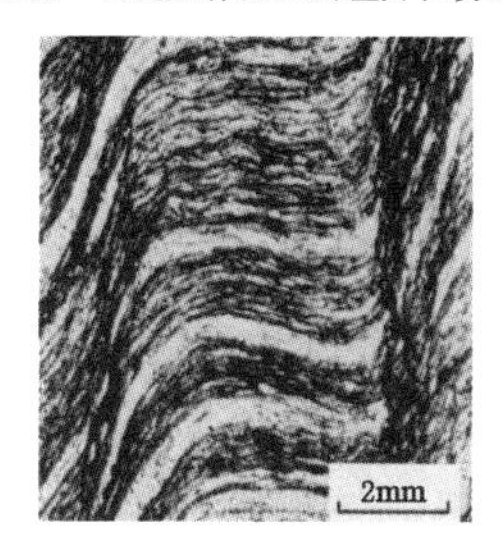

strain slip cleavage

stratified structure

【释义】 层状结构

【中文定义】 又称板状结构，是薄层沉积岩、副变质岩、火山岩岩体被比较发育的层理、片理和节理切割成板状、层形结构体所组成的岩体结构类型。

stratified structure

stratigraphic boundary

【释义】 地层分界线

【例句】 Stratigraphic boundary is used to indicate geological boundary of stratigraphic interface in geological map. 地质图中用地层分界线表示地层分界面。

【中文定义】 地质图中用来表示地层分界面的地质界线。

stratigraphic break

【释义】 岩层间断，地层间断

【中文定义】 指在一套沉积岩的层序中的一个限定层位突然出现岩性的明显变化。

stratigraphic column

【释义】 地层柱状图

【中文定义】 按一定比例尺和图例综合反映测区内地层层序、厚度、岩性特征和区域地质发展史的柱状剖面图。

stratigraphic unit

【释义】 地层单位

【常用搭配】 chronostratigraphic unit 年代地层单位；biostratigraphic unit 生物地层单位；lithostratigraphic unit 岩石地层单位

【例句】 Sequence is the most commonly and most important sequence stratigraphic unit. It is characterized by the features of both lithostratigraphic unit bounded by objective unconformity and chronostratigraphic unit marked by strict property of time. 层序是层序地层学常用的、且最重要的地层单位，它既是以客观不整合面为界面的岩石地层单位，又是有严格时间属性的年代地层单位。

【中文定义】 就是根据岩性、岩相划分的岩石地层单位，根据化石划分的生物地层单位，根据地质时代划分的年代地层单位等的统称。

stratigraphy [strəˈtɪgrəfi]

【释义】 *n.* 地层学，地层情况

【常用搭配】 sequence stratigraphy 层序地层学；descriptive stratigraphy 描述性地层学；interpretative stratigraphy 解释地层学

【例句】 Geological units are established by lithology and stratigraphy. 地质单元是通过岩性和地层建立的。

【中文定义】 是研究层状岩石形成的先后顺序、地质年代、时空分布规律（狭义）和形成环境条件及其物理、化学性质的地质学分支学科。

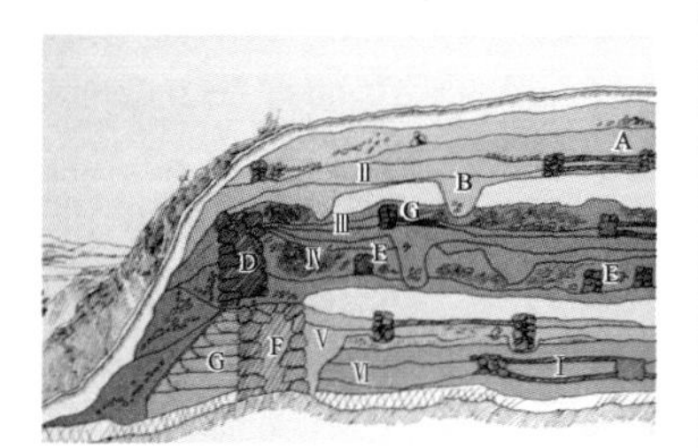

stratigraphy relative dating

stratum ['stretəm]

【释义】 *n*. 地层，岩层

【常用搭配】 attitude of stratum 地层产状；impermeable stratum 不透水层；confining stratum 承压层

【例句】 In addition to the general folds characteristics investigation in the dam site, the stratum reversal phenomenon caused by the strongly fold or fault shall be taken seriously. 坝址区褶皱勘察内容，除一般特征的勘察外，还应注意强烈褶皱或断层造成的地层倒转现象。

【中文定义】 指在地壳发展过程中形成的各种成层岩石的总称。不同于岩层，地层有老有新，具有时间含义。

rainbow sediment stratum

stratum boundary

【释义】 岩层分界线

【例句】 Stratum boundary is used to indicate geological boundary of stratum interface in geological map. 地质图中用岩层分界线表示岩层分界面。

【中文定义】 地质图中用来表示岩层分界面的地质界线。

stratum boundary

strength envelope

【释义】 强度包线

【例句】 Based on Mohr strength envelope characteristics, the dynamic strength parameter of gangue is analyzed. All these provide reference for engineering. 根据试验所得莫尔动强度包络线的特征，分析了煤矸石动强度参数的变化规律，为煤矸石在实际工程中的应用提供参考。

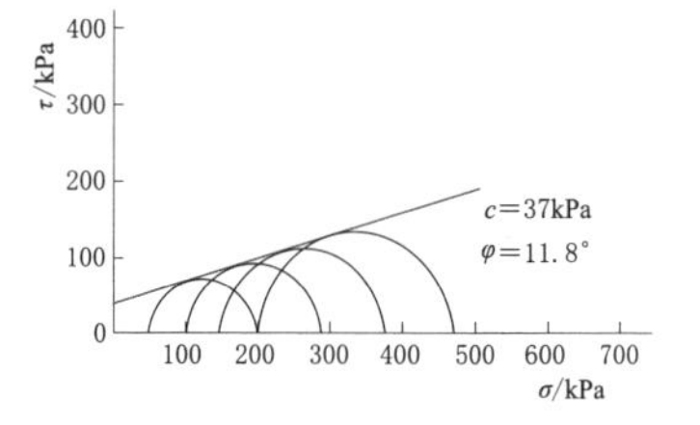

strength envelope

strength reduction method (SRM)

【释义】 强度折减法

【常用搭配】 FEM Strength Reduction Method 有限元强度折减法；Finite Difference Strength Reduction Method 有限差分强度折减法

【例句】 The whole reliability index of a slope is derived through using the strength reduction method in the finite (element) analysis. 在边坡的有限元分析中引入强度折减法求边坡的整体安全系数。

strength-stress ratio

【释义】 强度应力比

【例句】 The far-field stresses are moderate and have few effects on the stability of underground structure according to the modified rock mass strength-stress ratio method. 按照修正后的岩体强度应力比方法评价工程区的应力状态可知，地

应力量值中等，地应力场对地下工程稳定性影响不大。

【中文定义】 是反映岩体初始应力大小与岩体强度相对关系的定量指标。如无岩体初始应力（或称地应力）资料，一般可采用上覆岩体的荷重进行概略估算。

stress [strɛs]

【释义】 *n*. 强调，重音，压力，重力

【常用搭配】 stress concentration 应力集中；stress corrosion 应力腐蚀；bending stress 弯曲应力；triaxial stress 三轴应力

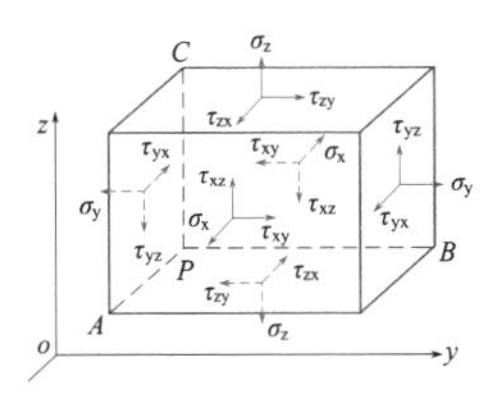

stress

【例句】 Three such simple stress situations, that are often encountered in engineering design, are the uniaxial normal stress, the simple shear stress, and the isotropic normal stress. 工程设计中经常涉及的三种基本应力情况分别是单轴法向应力、剪切应力和各向同性正应力。

【中文定义】 物体由于外因（受力、湿度、温度场变化等）而变形时，在物体内各部分之间产生相互作用的内力，单位面积上的内力称为应力。应力是矢量，沿截面法向的分量称为正应力，沿切向的分量称为切应力。

stress concentration

【释义】 应力集中

【常用搭配】 local stress concentration 局部应力集中

【例句】 Among many factors that cause fatigue fracture of high strength bolts, stress concentration is a dominant reason. 在众多影响高强螺栓疲劳性能的因素中，应力集中是其发生疲劳断裂的主要原因之一。

【中文定义】 指物体中应力局部增高的现象，一般出现在物体形状急剧变化的地方，如缺口、孔洞、沟槽以及有刚性约束处。

stress envelope

【释义】 应力包络线

【例句】 By comparing the stress-strain curves at different temperatures, the lower and upper stress envelope are identified respectively. 通过比较不同温度下的应力-应变曲线，来确定（材料的）上、下应力包络线。

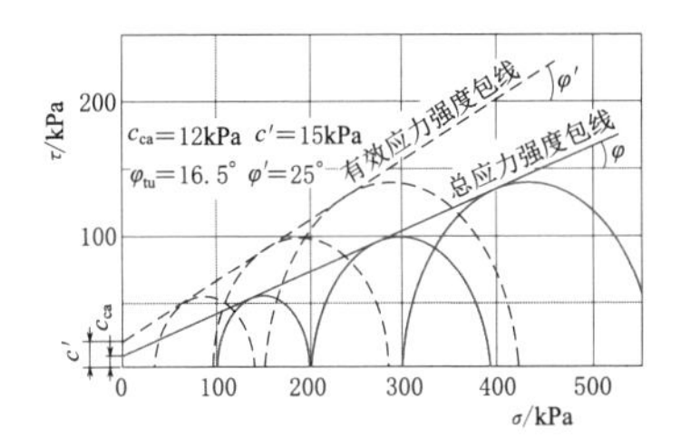

stress envelope

stress history

【释义】 应力历史

【例句】 The effect of stress history under three dimensional stress conditions is one of the important refering factors which have to be evaluated for the accurate prediction of ground deformation. 三向应力历史条件的影响是准确预测地基变形的重要参考因素之一。

【中文定义】 指岩土体的某一点的应力，从开始形成时起至研究它时止，其变化的全部历史过程。

stress level

【释义】 应力水平

【中文定义】 指实际所受应力与破坏强度的比值。

stress path

【释义】 应力路径

【常用搭配】 effective stress path 有效应力路径；unloading stress path 卸荷应力路径

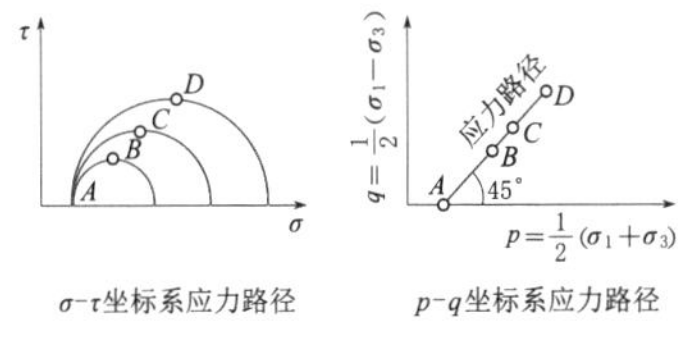

stress path

stress path test

【释义】 应力路径试验

【例句】 A series of stress path tests are carried out for undisturbed mucky clay samples from

Hankou. 对汉口某地的淤泥质黏土进行了一系列应力路径试验。

stress relaxation

【释义】 应力松弛

【例句】 To discover the properties of stress relaxation and creep of patellar ligament(PL) on knee contraction in rabbits. 探讨兔挛缩膝关节髌韧带的应力松弛和蠕变特性。

stress release

【释义】 应力释放，应力消除

【常用搭配】 stress release method 应力释放法；stress release ditch 应力释放沟

【同义词】 stress relief

【例句】 Stress release around the collapse encourages the opening of new fractures with trends that differ from regional fracture patterns. 与区域裂隙模式不同，新的裂隙随着崩塌周围应力释放而扩张。

【中文定义】 指物体内某一点的应力由于释放能量而降低的现象，确切地说是能量释放。

stress restoration method

【释义】 应力恢复法

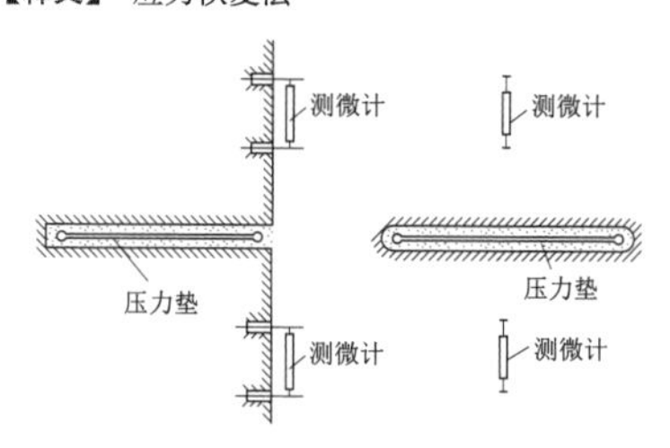

stress restoration method

stress-relief method

【释义】 应力解除法

【例句】 At present, stress-relief method by overcoring is one of the in-situ stress measuring methods with high applicability and reliability. 目前，套孔应力解除法是测定绝对应力适用性最强，可靠性最高的地应力测量方法之一。

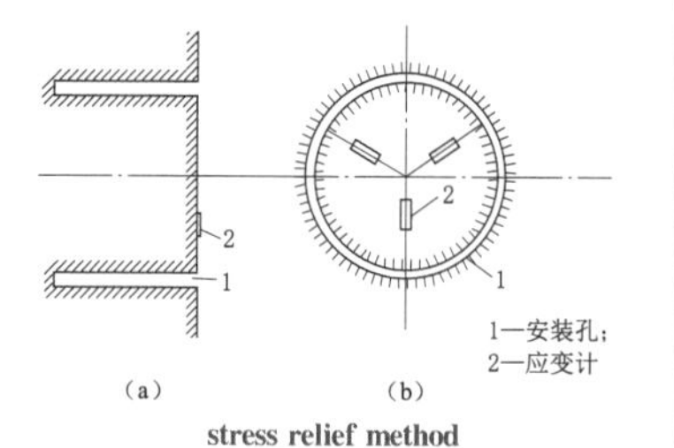

stress relief method

strike [straɪk]

【释义】 *n.* 走向

【常用搭配】 fault strike 断层走向

【例句】 A strike fault is one which strikes essentially parallel to the trend of adjacent rocks. 走向断层就是指那些断层走向与岩体走向一致的断层。

【中文定义】 指岩层层面的延伸方向

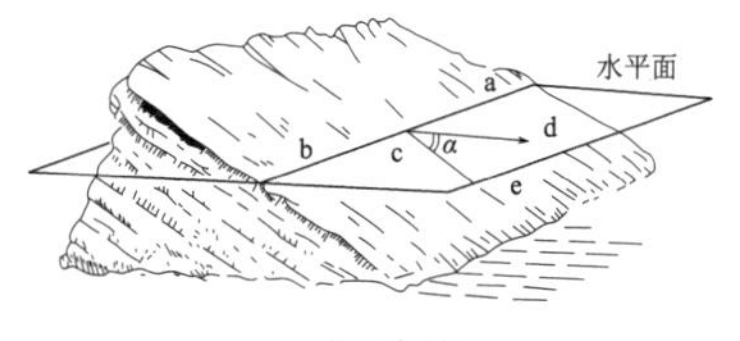

strike (ab)

strike fault

【释义】 走向断层

【例句】 In the structural geology, we have always been to use the method of C. M. Nevin to determine the stratigraphic duplicate and incompletement caused by the strike fault, the method is complicated and trivial. 构造地质学对走向断层造成的地层重复，缺失一直沿用 C. M. Nevin 的方法判断，此法比较烦琐。

【中文定义】 指断层走向与岩层走向基本平行的断层，又称纵断层（longitudinal fault）。

strike-slip fault

【释义】 走向滑动断层，走滑断层

【同义词】 transcurrent fault；flaw

【例句】 The mainshock occurred on a nearly vertical right-lateral strike-slip fault, striking N30°E, with an asymmetric bilateral fracture which propagated 70km northeastward and 45km southwestward with an average velocity of 2.7km/s. 主震是发生在一个近似直立的右旋走滑断层上，走向 N30°E，破裂方式为不对称的双侧破裂，以 2.7km/s 的平均速度向北东传播 70km，向南西传播 45km。

【中文定义】 即规模巨大的平移断层，又称横移断层、走滑断层，亦称为扭转断层。平移断层作用的应力是来自两旁的剪切力作用，其两盘顺断层面走向相对移动，而无上下垂直移动。由于断层面是沿水平方向移动的，所以在野外的观察上经常没有明显的断崖，只会在地面上看到一条断层直线。

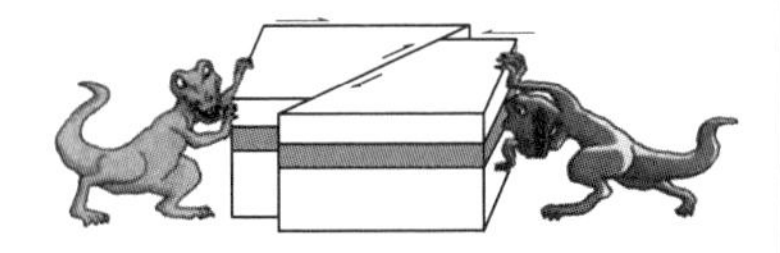

strike slip fault

string spillover

【释义】 串浆

【中文定义】 指在注浆过程中，水泥浆液出现相邻两孔或隔孔冒浆的现象。

strip map

【释义】 带状地形图

【同义词】 belt topographic map

stripped volume

【释义】 剥离量

【中文定义】 指剥离层的数量多少。

stripping layer

【释义】 剥离层

【例句】 The stripping layer is unavailable layer that covers the available layer and needs to be removed. 剥离层是指覆盖于有用层表面需要清除的无用层。

【中文定义】 是覆盖于有用层表面需要清除的无用层。

stripping ratio

【释义】 剥采比

【例句】 There are many complications that influent open-pit limit and the economic stripping ratio is the most important one. 影响露天矿开采境界的因素很多，其中一个最重要的因素就是经济合理剥采比。

【中文定义】 是天然建筑材料料场的无用层剥离量与有用层开采量的比值。

strong capillary water height

【释义】 毛细水强烈上升高度

【中文定义】 指受地下水直接补给的毛细水上升的最大高度。

strong earthquake

【释义】 强震

【例句】 A strong earthquake has jolted the south coast of Indonesia's Java Island. 一次强震震动了印度尼西亚爪哇岛南部海岸。

【中文定义】 指震级等于或大于 6 级的地震。其中震级大于或等于 8 级的又称为巨大地震。

strong motion observation

【释义】 强震监测

【例句】 Strong motion observation is an important part of seismic observation. 强震观测是地震观测的重要组成部分。

strong unloading

【释义】 强卸荷

【例句】 The study shows that the rock mass of the slope can be divided into three zones, strong unloading zone, middle unloading zone and slight unloading zone. 这项研究表明，边坡岩体可划分为强、弱、微三个卸荷带。

【中文定义】 指卸荷裂隙发育较密集，普遍张开，一般开度为几厘米至几十厘米，多充填次生泥及岩屑、岩块，有架空现象，部分可看到明显的松动或变位错落，卸荷裂隙多沿原有结构面张开。岩体多呈整体松弛。

strong vibration accelerometer

【释义】 强震动加速度仪

【例句】 Strong vibration accelerometer is a precision instrument. 强震动加速度仪是个精密仪器

strong vibration accelerometer

strong-motion seismograph

【释义】 强震仪

【中文定义】 是记录强烈地震近地面运动的自动触发式地震仪。一般由拾震系统、记录系统、触发—起动系统、时标系统和电源系统五部分构成。

strong-motion seismograph

structural compressed zone

【释义】 构造挤压带

structural compressed zone

【中文定义】 简称挤压带，是指主要由许多压性构造形迹组成的窄长带状构造，带内岩石部分或全部呈现被搅乱的状态。

structure ['strʌktʃər]

【释义】 *n*. 构造

【常用搭配】 massive structure 块状构造；banded structure 带状构造

【中文定义】 是指岩石中不同矿物集合体之间或矿物集合体与其他组成部分之间的排列、填充方式等。

structure model

【释义】 结构模型

【例句】 Uncertainty in 3D geological structure models has become a bottleneck that restricts the development and application of 3D geological modeling. 三维地质结构模型的不确定性已成为制约三维地质建模发展和应用的瓶颈。

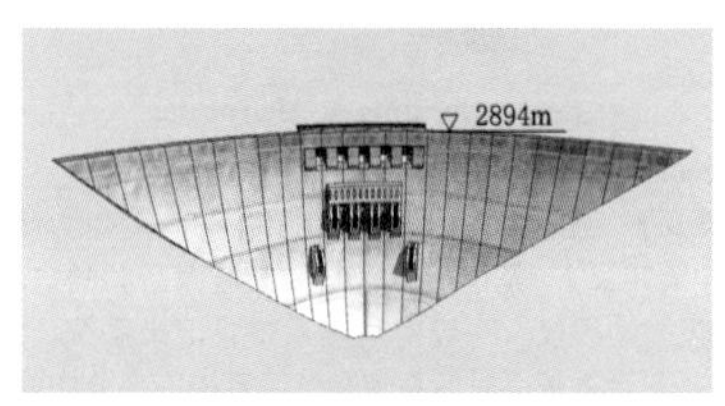

arch dam structure model

stylolite ['staɪləˌlaɪt]

【释义】 *n*. 缝合线

【例句】 There are primary stylolite and secondary stylolite in sedimentary rocks. 沉积岩中的缝合线构造有原生和次生两种。

【中文定义】 沉积岩中的一种构造现象，常见于石灰岩中，火山岩、石英岩中也可见到。它在剖面上呈锯齿状曲线，状如动物头盖骨中的结合缝，平面上是一个起伏不平的面。

stylolite

subangular ['sʌb'æŋgjulə]

【释义】 *adj*. 次棱角状

【中文定义】 碎屑颗粒的棱角稍有磨蚀，尖角并不十分突出。

subaqueous landslide

【释义】 水下滑坡

【例句】 Large earthquake-induced rock avalanches, soil avalanches, and subaqueous landslides can be very destructive. 大地震造成的岩石、土壤崩塌，以及水下滑坡破坏性极大。

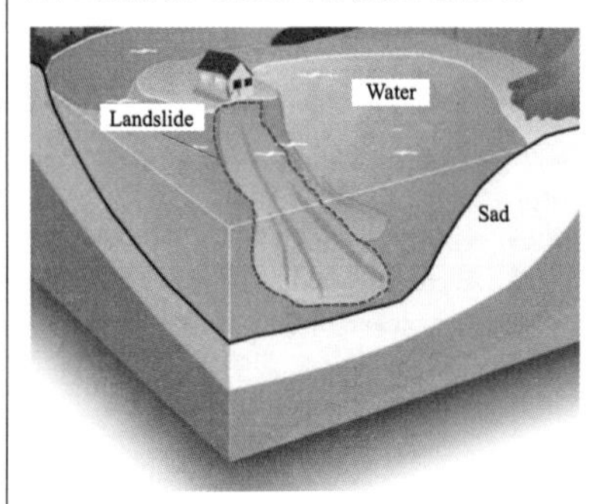

subaqueous landslide

subduction [səb'dʌkʃən]

【释义】 *n*. 俯冲，除去，减法

【常用搭配】 subduction zone 俯冲带，隐没带，消减带；intracontinental subduction 陆内俯冲作用；plate subduction 板块俯冲作用；intercontinental subduction 陆间俯冲

【例句】 Subduction probably occurs to a depth of at least 670 kilometers (400 miles), at which point the plate probably becomes plastic. 俯冲作用可能发生在至少 670km (400mile) 的地下，

此处板块可能已经成为塑性状态。

submarine slumping

【释义】 海底塌陷

【中文定义】 地表岩土在自然或人为因素作用下向下陷落并在海底形成塌陷坑的现象。

submarine weathering

【释义】 海底风化作用，海解作用

【例句】 Glauconite, iron, manganese, calcium nodules mineral zeolite are representative of new minerals formed in submarine weathering. 海底风化作用中形成的代表性新生矿物是海绿石、结核型铁锰、钙质矿物、沸石等。

submarine weathering

【中文定义】 又称海解作用，在沉积作用微弱后完全没有沉积作用的海底上发生的海水与沉积物之间的地球化学反应过程。

sub-massive structure

【释义】 次块状结构

submerged density

【释义】 浮密度

【中文定义】 土单位体积中土粒质量与同体积水的质量之差．也被称为土的有效密度。

subrounded [sʌb'raʊndɪd]

【释义】 *adj*. 次圆状

【中文定义】 碎屑棱角已显著磨损，碎屑的原始轮廓还可看出。说明碎屑经过了较长距离的搬运。

subsidence [səb'saɪdns]

【释义】 *n*. 下沉，沉陷，陷没

【常用搭配】 land subsidence 地面沉陷；subsidence zone 沉陷带

【中文定义】 由于湿陷、地下开采等引起的以竖向为主的地表下沉。

subsidence monitoring

【释义】 地面沉降监测

【例句】 It is strict with stabilization of surveying datums mark for land subsi-dence monitoring. 沉降监测对测量基准点的稳定性有严格要求。

【中文定义】 一般指在发生、发现地面沉降的地区内布设统一的区域性的地面沉降水准网、GPS网和地下水监测网。通过定期的重复观测，可为研究和控制地面沉降提供准确、可靠的资料。

subsidence of pier-abutment foundation

【释义】 墩台地基沉陷

【中文定义】 墩台地基在上部荷载作用下发生的沉降变形。

subsidence of subgrade

【释义】 路基沉陷

【例句】 The measures for preventing the subsidence of subgrade are very useful. 预防路基沉陷的措施很有效。

【中文定义】 软土路基在上部荷载作用下发生的压缩沉降变形。

subsurface runoff

【释义】 地下径流

【中文定义】 由地下水的补给区向排泄区流动的地下水流。

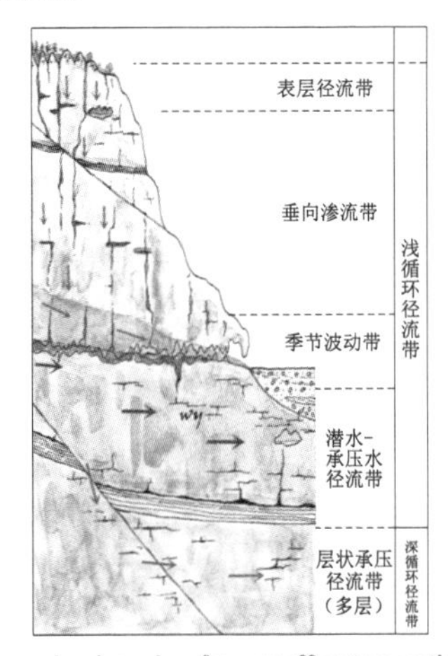

map showing subsurface runoff system vertical zones of karst plateau

sulfur flowers

【释义】 硫华

【中文定义】 现代火山区和高温水热活动区的喷气孔内壁和口垣上的针状或粒状硫磺晶体聚积。

sulfur flowers

sulphate rock

【释义】 硫酸盐岩

【例句】 Sulphate rocks (gypsum and anhydrite) and carbonate rocks (limestone and dolomite) are usually deposited together. 硫酸盐岩（石膏和硬石膏）和碳酸盐岩（石灰岩和白云岩）常共生沉积。

superficial fault-block

【释义】 盖层断块

【例句】 According to their cutting depths, we are of the opinion that fault-blocks cut by various fracture net-works of different depth into the earth may also be divided into 4 major categories, i. e. lithospheric, crustal, foundational and superficial fault-blocks. 根据断裂切割深度的不同，我们将断块划分为四个大类：岩石圈断块、地壳断块、基底断块和盖层断块。

【中文定义】 指被盖层断裂所切割和围限的断块。它存在于现代大陆及大陆边缘，是沉积盖层发育的构造单位。

supply funnel

【释义】 补给漏斗

【中文定义】 是某地区因地下水变化，导致地下水饱和水面（也叫潜水面）以采水点为中心，四周向中心呈梯度下降的现象。

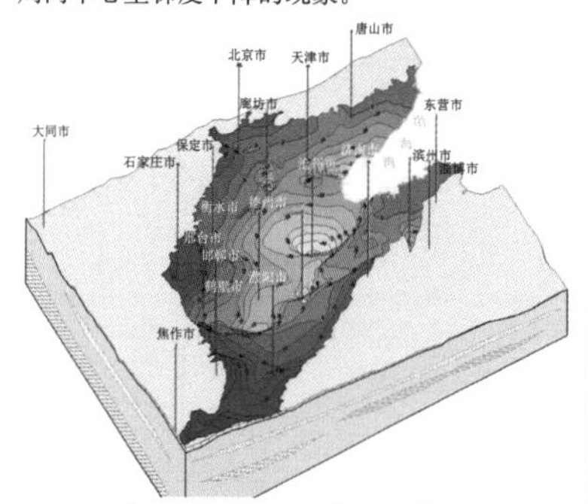

schematic diagram of groundwater supply funnel in North China Plain

support [sə'pɔːt]

【释义】 *n.* 支护

【中文定义】 指采用结构或构件及其材料对围岩进行加固的工程措施。

support

support axis force monitoring

【释义】 支撑轴力监测

【例句】 Support axis force monitoring shall be carried out in deep foundation pit. 深基坑要进行支撑轴力的监测。

support axis force monitoring

supporting and retaining structure

【释义】支挡结构

【例句】As a new supporting and retaining structure, reinforced retaining wall has been applied in various civil engineering due to its prominent predominance. 加筋土挡墙作为一种新型的支挡结构，以其显著的优势被广泛应用于各种土建工程。

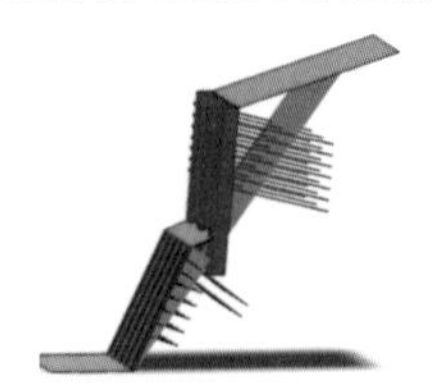

supporting and retaining structure

surface feature

【释义】 地物

【同义词】 ground feature

【中文定义】 地球表面上的各种固定性物体，可分为自然地物和人工地物。

surface measurement of acoustic detection

【释义】 声波平测法

surface model

【释义】 表面模型

【中文定义】 通常用于构造复杂的曲面物体，构形时常常利用线框功能，先构造一线框图，然后用扫描或旋转等手段变成曲面，也可以用系统提供的许多曲面图素来建立各种曲面模型。

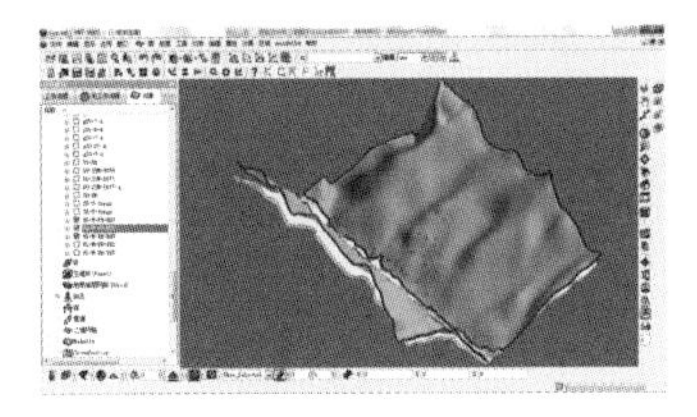

surface model

surface relief

【释义】 地势，地表起伏

【例句】 The adjustment of mantle surface relief ended the rift basin development. 地幔面起伏的调整结束了裂陷盆地的发育。

surface runoff

【释义】 地表径流，地面径流

【同义词】 surface flow

【例句】 The urban greenland is a main component of city ecological system and plays an important role in removing the pollutants in urban surface runoff. 城市绿地是城市生态系统的重要组成部分，对于控制城市地表径流中污染物起到至关重要的作用。

【中文定义】 没有下渗的地表水汇聚流动的过程称地表径流。

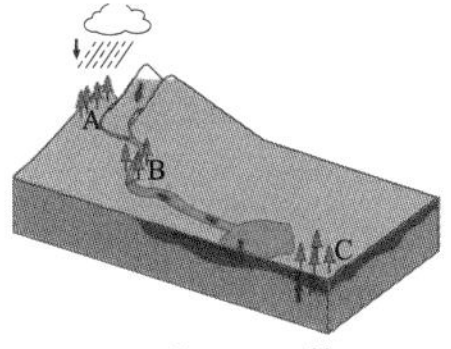

surface runoff

surface wave

【释义】 面波

【例句】 Transient surface wave exploration is a geophysical exploration technique. 瞬态面波勘探就是物探技术中的一种。

【中文定义】 是地震波的一种，主要在地表传播，能量最大，波速约为 3.8km/s，低于体波，往往最后被记录到。如果地震非常强烈，面波可能在震后围绕地球运行数日。面波实际上是体波在地表衍生而成的次生波。面波的传播较为复杂，既可以引起地表上下的起伏，也可以是地表做横向的剪切，其中剪切运动对建筑物的破坏最为强烈。

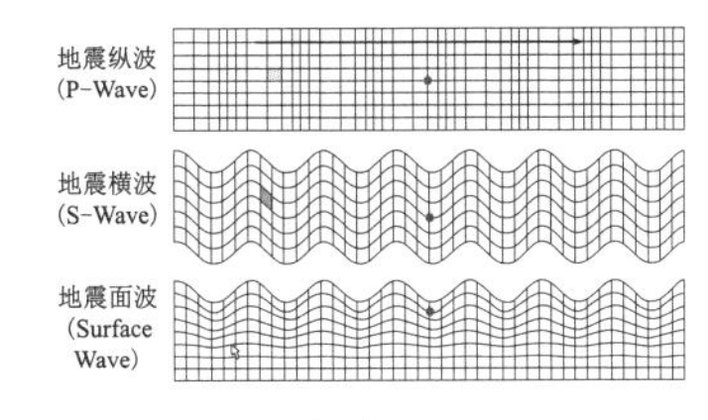

seismic wave

surface wave method

【释义】 面波法

【例句】 As a mature technique, surface wave method could be used as a nondestructive detection method. 作为一种成熟技术，面波法可用作无损探伤。

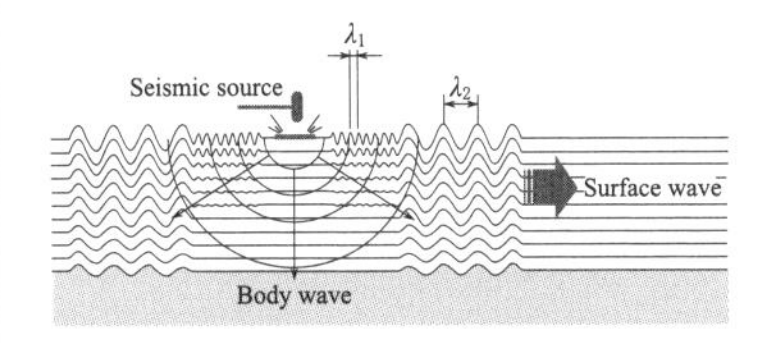

surface wave method

surrounding slope

【释义】 环境边坡

surrounding slope

surrounding rock mass

【释义】 围岩

【例句】 The earth stress is an essential force causing surrounding rock mass deformation and breakage. 地应力是引起围岩变形和破坏的根本作用力。

surrounding rock mass

surrounding rock mass classification

【释义】 围岩分类

【例句】 Quantized study of surrounding rock mass classification is one of the important subjects in the technical fields of underground engineering. 坑道围岩分类的定量化研究是地下工程技术领域中重大课题之一。

surrounding rock mass loose circle

【释义】 围岩松动圈

【中文定义】 指地下洞室开挖引起围岩应力重新分布，岩石强度和岩体内应力变化，在开挖空间周围形成的松弛破碎带。

surrounding rock mass pressure

【释义】 围岩压力

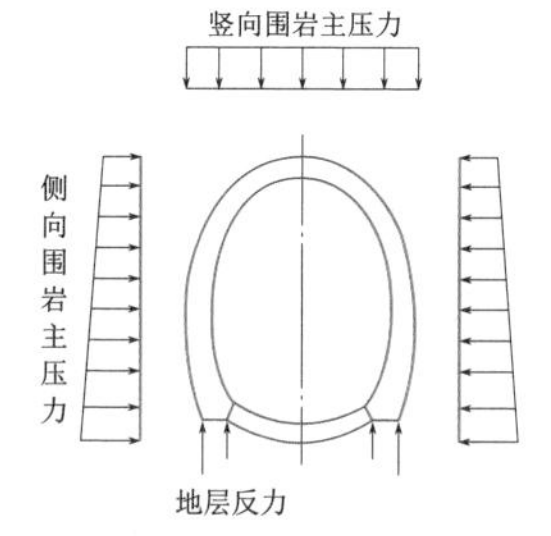

surrounding rock pressure

surrounding rock mass stress

【释义】 围岩应力

【常用搭配】 surrounding rock mass stress field 围岩应力场；redistribution of surrounding rock mass stress 围岩应力重分布

【例句】 The underground terrane keeps balance state before the roadway is excavated, and once the roadway has been excavated, the surrounding rock mass stress will lose its original balance. 地下岩层在巷道开挖前处于平衡状态，巷道一旦开挖，围岩应力将失去平衡。

【中文定义】 围岩在无支护的情况下，经过应力重新调整达到的新的平衡状态即为围岩应力。

surrounding slope hazard

【释义】 环境边坡危险源

survey mark

【释义】 测量标志

【中文定义】 标定地面控制点位置的标石、觇标以及其他标记的通称。

suspended substance

【释义】 悬浮物

【例句】 The methods for analysis of suspended substance in wastewater were studied experimentally. 对污水中悬浮物分析方法进行了实验研究。

【中文定义】 指悬浮在水中的固体物质，包括不溶于水中的无机物、有机物及泥沙、黏土、微生物等。

suspending weigh method

【释义】 浮称法

【例句】 Electron balance can be used for the block density tests by suspending weigh method. 电子秤在浮称法中可用来测块体密度。

suture zone

【释义】 缝合带

【常用搭配】 Yarlung Zangbo suture zone 雅鲁藏布缝合带；Shangzhou Danfeng suture zone 丹凤缝合带；hidden suture zone 隐伏缝合带；continental suture zone 大陆缝合带

【例句】 The Banggong-Nujiang suture zone is underlain by a discontinuous and roughly EW ophiolite belt. 班公湖-怒江缝合带下伏近东西方向、断续延伸的蛇绿岩带。

【中文定义】 即两个碰撞大陆衔接的地方。缝合带通常表现为宽度不大的高应变带，由含有残余洋壳的蛇绿岩混杂堆积和共生的深海相放射虫硅质岩、沉积岩等组成，叠加了蓝片岩相高度变质作用和强烈的构造变形。

swampiness [s'wɒmpɪnəs]

【释义】 *n.* 沼泽化

【常用搭配】 out-of-dam swampiness 坝外沼泽化

【例句】 The effect of Three Gorges Project unon the gleization and swampiness of soil in Dongting Lake area will be studied. 三峡工程对洞庭湖区土壤潜育化和沼泽化的影响将会被研究。

【中文定义】 存在泥炭化的土地长期过湿，在湿生作物作用或厌氧条件下进行的有机质的生物积累和矿质元素还原的过程。

Swedish circle method

【释义】 瑞典滑弧法

【例句】 The Swedish circle method assumes a circular failure interface, and analyzes stress and strength parameters using circular geometry and statics. 瑞典圆弧法假定圆形破坏界面，用圆形几何和静力学分析应力和强度参数。

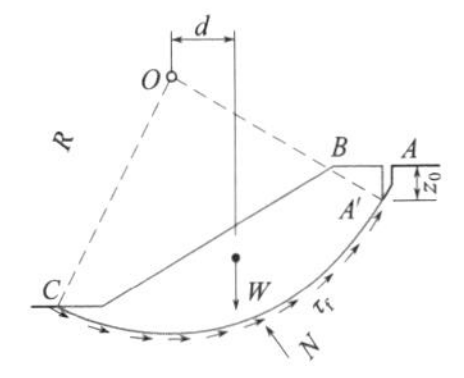

Swedish slip circle method

swelling force

【释义】 膨胀力

【例句】 To compacted samples prepared by same expansive soil material, the smaller the prepared moisture content the higher the swelling force. 对比相同膨胀土材料制备的压实样品，制备的含水量越小，膨胀力越高。

swelling index

【释义】 回弹指数

【例句】 An example using it to analyze the relationship between compression index and swelling index was presented. 举例说明了利用其分析土体压缩指数与回弹指数的关系。

swelling ratio

【释义】 膨胀率

【例句】 Steel and concrete have almost the same rate of contraction and expansion. 钢筋和混凝土有几乎相同的收缩率和膨胀率。

swelling soil

【释义】 膨胀土

【例句】 The methods of modified treatment for swelling soil foundation are summarized and discussed. These methods can be presently divided into physical, chemic and biologic method. 对膨胀土地基改性处理方法进行了总结和评述，目前膨胀土的改性方法可以分为物理改性法、化学改性法和生物改性法三种。

【中文定义】 也称胀缩性土，浸水后体积剧烈膨胀，失水后体积显著收缩的黏性土。由于土中含有较多的蒙脱石、伊利石等黏土矿物，故亲水性很强。这类土对建筑物会造成严重危害，但在天然状态下强度一般较高，压缩性低，易被误认为是较好的地基。对膨胀土地基，应做好地表的防渗与排水措施；也可适当加大基础荷载与基础深度以及提高建筑物的刚度并设沉降缝；或将持力层范围内的膨胀土挖除，用砂或其他非膨胀土回填。

swelling soil

swelling test of rock

【释义】 岩石膨胀性试验

syenite ['saɪənaɪt]

【释义】 *n.* 正长岩

syenite

【常用搭配】 nepheline syenite 霞石正长岩；cancrinite syenite 钙霞正长岩；eleolite syenite 脂光正长岩

【例句】 The experimental rocks were granite, basalt, syenite and gneiss. 实验的岩石有花岗岩、玄武岩、正长岩和片麻岩。

symbiotic debris flow

【释义】 共生型泥石流

symbol ['sɪmbl]

【释义】 *n.* 符号；象征；标志

【常用搭配】 geological symbol 地质符号

【例句】 Two limitations of Manifold V6.0 are manifest in the making of geological maps: the lack of customizable line types, particularly for thrust faults, and the lack of geological symbols. 使用 Manifold V6.0 编制地质图存在两个明显的制约因素：缺乏自定义线型，尤其对于逆冲断层，以及缺乏地质符号。

symbol of tectonic system

【释义】 构造体系符号

【中文定义】 地质图中绘制构造体系对象所使用的各种图形符号。

构造体系			构造形迹					
			压性断裂	压扭性断裂	张性断裂	张扭性断裂	复背斜复向斜	背斜向斜
纬向构造体系								
径向构造体系								
扭动构造体系	近华夏系							
	新华夏系	晚期						
		早期						
	华夏式							
	华夏系							
	河西系							
	西域系							

symbol of tectonic system

symbol of geological structure

【释义】 地质构造符号

【中文定义】 地质图中绘制地质构造对象所使用的各种图形符号。

名称	符号		名称	符号	
	平面	剖面		平面	剖面
实测正断层	50°		推测逆掩断层	20°	
推测正断层	50°		活动断层	60°	
实测逆断层	40°		实测断层线		
推测逆断层	40°		推测断层线		
实测平移断层			掩埋断层		
推测平移断层			断层影响带		
实测逆掩断层	20°		断层破碎带		

symbols of geological structure

symbol of geomechanics

【释义】 地质力学符号

【中文定义】 地质图中绘制地质力学相关内容所使用的各种图形符号。

symbol of plate tectonics

【释义】 板块构造符号

【中文定义】 地质图中绘制板块构造对象所使用的各种图形符号。

名称	符号	名称	符号
蛇绿岩带	OP	活动的扩张脊及转换断层	
混杂堆积	M	不活动的扩张脊	
高压低温变质带		板块的缝合线	
板块俯冲带		裂谷	
板块逆冲带		深断裂	

symbols of plate tectonics

symbol of rock

【释义】 岩石花纹符号

【中文定义】 按岩石类型的基本名称，以结构或特殊构造以及碎屑成分、矿物成分等作为附加名称而设计的不同花纹符号，用于在地质图上表示不同种类和性质的岩石。

岩石名称	花纹	岩石名称	花纹
砾岩		石英砂岩	
角砾岩		硬砂岩	
砂砾岩		铁质砂岩	Fe Fe Fe Fe Fe
砂质砾岩		长石砂岩	N N N N N N N N N

symbols of rock

syncline ['sɪnklaɪn]

【释义】 *n.* 向斜

【常用搭配】 carinate syncline 脊状向斜

【例句】 Geological features of salt deposit in Tianhuan syncline of ordos Basin have been found. 鄂尔多斯盆地天环向斜盐矿床地质特征已被发现。

【中文定义】 属于褶曲的基本形态之一，与背斜相对。为褶曲构造之一部分，两翼指向上方，中央向下屈曲。其在褶弯内之岩层，愈往中央，愈为年轻。

syncline

synclinorium [ˌsɪŋklaɪ'nɔrɪəm]

【释义】 *n.* 复向斜

【例句】 The structure style of Yanshan and the western Liaoning consists of a synclinorium and an antiform. 燕山和辽宁西部的结构类型由复向斜和背斜构成。

【中文定义】 又称复式向斜，是由若干次一级的背斜、向斜组合而成的一个大型向斜构造。

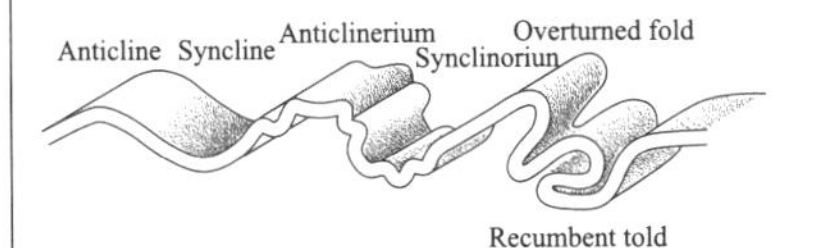

synclinorium

System ['sɪstəm]

【释义】 *n.* 系

【常用搭配】 Cambrian System 寒武系；Jurassic System 侏罗系；Quaternary System 第四系

【中文定义】 第三级年代地层单位，与地质年代单位“纪”对应，根据生物界演化的阶段性划分。系是年代地层单位中最重要的单位，具有全球可对比性。

systematic anchors

【释义】 系统锚杆

【例句】 The differences of whether the powerhouses are supported by systematic anchors are studied with the method of numerical analysis to explore the influence of systematic anchors on the stability of underground plant houses. 以某水电站地下厂房工程为实例，采用数值分析方法，通过对比厂房洞室群是否施加系统锚杆的差别，获得系统锚杆对洞室群稳定性的影响。

systematic anchors

T

tailing edge of landslide

【释义】 滑坡体后缘

talc [tælk]

【释义】 *n.* 滑石，云母；*vt.* 用滑石粉处理，撒滑石粉在……上

【常用搭配】 talc schist 滑石片岩；talc powder 滑石粉

【例句】 Talc is a naturally occurring clay mineral composed of magnesium and silicon. 滑石是一种天然黏土矿物，由镁元素和硅元素组成。

【中文定义】 一种常见的硅酸盐矿物，非常软且具有滑腻的手感。

talc

talus fan

【释义】 坡积裙

【中文定义】 指坡积物沿山麓分布形似裙边的堆积地形。

talus fan

target layer

【释义】 目的层

【例句】 Studying sedimentary facies of target layer is the base of optimum selection of exploration targets. 目的层段沉积相研究是勘探目标优选的基础。

taxitic structure

【释义】 斑杂状构造

【中文定义】 火成岩中，暗色矿物集合体的不均匀分布的一种构造。

taxitic structure

tectogenesis [tektə'dʒenəsɪs]

【释义】 *n.* 构造运动

【常用搭配】 gravitational tectogenesis 重力构造作用；late-tectogenesis 晚期构造作用

【例句】 Tectogenesis was alpinotype, and was accompanied locally by high-grade metamorphism and migmatization. 构造运动是阿尔卑斯式，并伴随着本地的高级变质作用和混合岩化作用。

【中文定义】 地质学专业术语，由地球内动力引起岩石圈地质体变形、变位的机械运动。构造运动产生褶皱、断裂等各种地质构造，引起海、陆轮廓的变化，地壳的隆起和凹陷以及山脉、海沟的形成等。

tectonic basin

【释义】 构造盆地

【例句】 Lanping Basin is a tectonic basin in the Sanjiang Area, which is rich in mineral resources. 兰坪盆地是三江地区的构造沉积盆地，矿产资源丰富。

【中文定义】 指主要由地壳构造运动所形成的盆地，它的形态和分布受构造控制。

tectonic element

【释义】 构造单元，构造要素

【常用搭配】 major tectonic elements 主要构造单元；fracture tectonic element 破裂构造要素

【例句】 Delineation of tectonic elements was the basis of the study on the crustal deformation and development, the distribution of minerals, and the minerogenetic laws. 构造单元的划分是研究地壳变形与发展、矿产分布和成矿规律的基础。

【中文定义】 一个区域尺度的地域，其中的地壳物质组成、构造组合，以及地球物理和地球化学场明显不同于相邻地域，表明它具有自己的地壳演化历史而有别于周缘地区，这样的一个地域就是一个大地构造单元。

tectonic fissure

【释义】 构造节理

【常用搭配】 structural fissure water 构造裂隙水

【例句】 The structural fissure water is dominant in groundwater of Dong'an gold deposit, the inflow of water of the mining pit in the future is not large. 东安金矿区地下水以构造裂隙水为主，未来开采矿坑涌水量不大。

【中文定义】 由构造应力作用形成的裂隙称为构造裂隙或节理

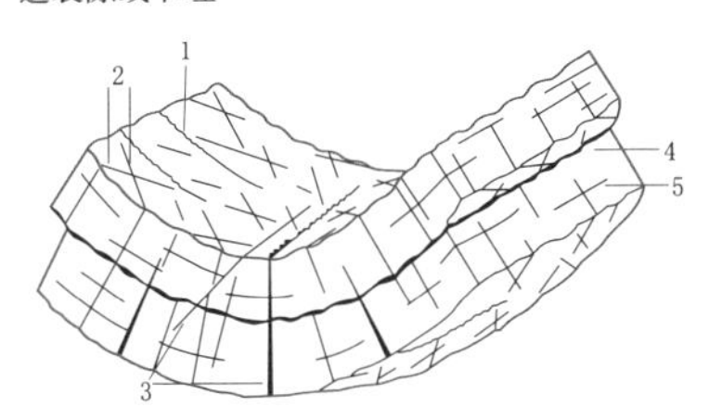

1—横裂隙（与岩层走向垂直）；2—斜裂隙（X节理）；3—纵裂隙（与岩层走向一致）；4—层面裂隙；5—顺层裂隙

tectonic fracture diagram of layered rock mass

tectonic plain

【释义】 构造平原

【中文定义】 指主要由地质构造作用造成的平原，一般指海成平原。

tectonic stable area

【释义】 构造稳定区

【例句】 According to the tectonic upgrading degree, the Middle-Upper Yangze region with Longmaxi shale is divided into the tectonic damage area, close to the denuded area and tectonic stable area. 根据构造改造的程度，将中上扬子地区龙马溪组页岩分为构造改造区、靠近剥蚀区和构造稳定区三种类型。

tectonic plain

tectonic stress

【释义】 构造应力

【例句】 Analysis of the tectonic stress indicates that there is higher tectonic stress at northern wing of the field. 构造应力分析表明，井田北翼有较高的构造应力，它是北翼易于发生冲击地压的重要原因。

【中文定义】 是由于地质构造作用引起的应力。地质构造运动（含地震）归根到底是一个岩层变形与破坏的力学过程，与之对应的应力场叫构造应力场。

tectonic system

【释义】 构造体系

【例句】 Deformation record of the change from Indosinian collision-related tectonic system to Yanshanian subduction-related tectonic system in South China during the Early Mesozoic. 华南早中生代从印支期碰撞构造体系向燕山期俯冲构造体系转换的形变记录。

【中文定义】 指不同形态、不同性质、不同等级和不同序次，但都有成生联系的各项构造要素组成的构造带，以及它们之间所夹岩块或地块组合而成的总体。

tectonic window

【释义】 构造窗

【例句】 The Precambrian Ku'ergan tectonic window has been identified according to the 606 Ma Rb - Sr isochron age of the biotite-tremolite-quartz schist. 根据黑云母透闪石石英片岩等606Ma的Rb - Sr等时线年龄，确认了库尔干前寒武纪构造窗。

【中文定义】 当外来岩块遭受剥蚀，中间剥蚀掉而露出一块新岩层（下伏岩块）来，就称为构造窗。

tectonics [tek'tɒnɪks]

【释义】 *n.* 构造地质学，大地构造学

【常用搭配】 plate tectonics 板块构造；tectonic plate 构造板块；tectonic evolution 构造演化

【例句】 Tectonics also provides a framework for understanding the development of the earthquake and volcanic belts that directly affect much of the global population. 大地构造学为理解影响全球多数人口的火山地震带发育规律提供了一个框架。

【中文定义】 地质学主要二级学科之一，是研究岩石圈内地质体的形成、形态和变形构造作用的成因机制，及其相互影响、时空分布和演化规律的地质学分支学科。

temperature calibration

【释义】 温度标定

【例句】 Finally, the FBG sensing properties have been studied by the experiments of displacement stretch and temperature calibration. 通过微位移平台拉伸实验和温度标定实验研究了光纤光栅的应变和温度传感特性。

temperature logging

【释义】 井温测井

temporary leakage

【释义】 暂时渗漏

【中文定义】 指库水渗入库区未饱和岩土的孔隙和裂隙中，在蓄水初期和运用期每次蓄水时都要发生，由于水没有漏至库外，且库水位降低后还有一部分回归水库，对蓄水不构成严重威胁。

temporary support

temporary support

【释义】 临时支护

【中文定义】 为保证施工安全临时设置的支护。

temporary works

【释义】 临时工程

tensile fault

【释义】 张性断层

【例句】 Rock fractal statistical strength theory using the prediction of tensile fault has been widely used. 张性断层构造预测的岩石分形统计强度理论已广泛使用。

【中文定义】 由断层两盘相对运动引起的派生分支构造，张性分支构造与主干断层所夹锐角指示本盘相对运动方向。

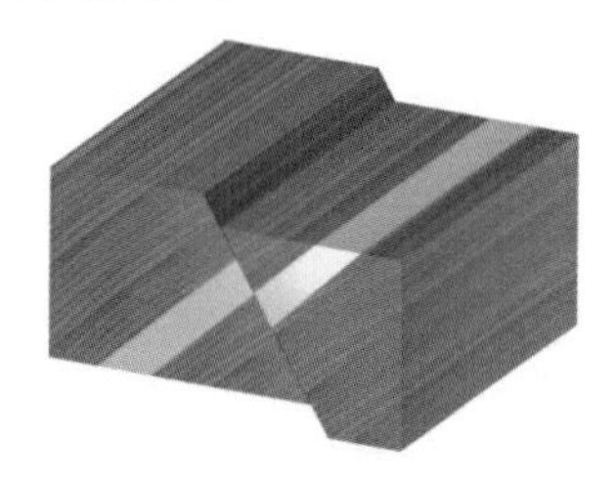

tensile fault

tensile strength

【释义】 抗拉强度

【例句】 So we're going to load it up to the point that it will break, and that allows us to measure the ultimate tensile strength. 我们将继续装物体，直到断裂点，这样我们就能测量到极限抗拉强度的大小。

tensile strength test of rock

【释义】 岩石抗拉强度试验

【例句】 This method provides theoretical and experimental basis for testing tensile strength of rock accurately. 这一结论为岩石抗拉强度的正确测定奠定了理论和试验基础。

tensile structural plane

【释义】 张性结构面

【例句】 Experimental study on parameters of shear strength of tensile structural plane in underground plant. 地下厂房岩体张性结构面抗剪强度参数试验研究。

【中文定义】 简称张裂面，岩块或地块由于引张作用而产生的垂直于主张应力的破裂面，或受挤压而产生的平行于主压应力的破裂面。

tension ['tenʃən]

【释义】 *n.* 张力，拉力；*vt.* 使紧张，使拉紧

【常用搭配】 surface tension 表面张力；interfacial tension 界面张力；axial tension 轴向拉力

【例句】 Geologic tension is also found in the tectonic regions of divergent boundaries. 地质张力也存在于不同边界的构造区域。

【中文定义】 在弹性限度以内，物体受外力的作用而产生的形变与所受的外力成正比。形变随力作用的方向不同而异，使物体延伸的力称拉力或张力。

tension crack

【释义】 张裂缝

【常用搭配】 plump crack 鼓胀裂缝

【例句】 When there are tension cracks and water in the slope, the critical inclination has relations with the position of crack and the level of water. 当边坡中有张裂缝和水时，滑面的临界倾角还与张裂缝的深度和水位高度有关。

【中文定义】 由张应力形成的裂缝。

tension crack

tension fissure

【释义】 张裂隙，张性节理

【常用搭配】 echelon tension fissure 雁行张裂隙

【例句】 In the overlying basalt, there is steep dip and wide tension fissure zone, which is combined with surface of unconformity which is as basal sliding face, to form loosening rock mass.

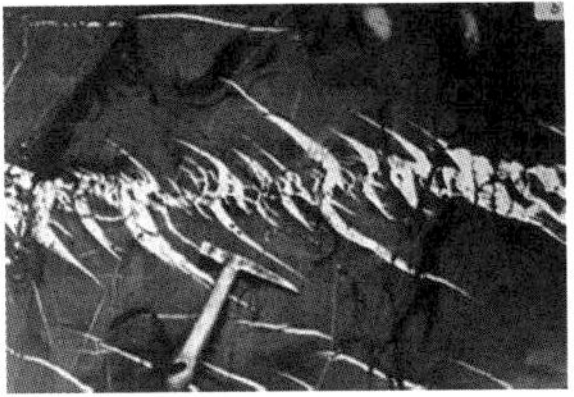

echelon tension fissure

在上覆玄武岩层中见有高倾角的宽张裂隙带，与不整合面一起构成松动岩体，不整合面为底滑面。

【中文定义】 是由张应力产生的破裂面。

tension-cracking

【释义】 拉裂

【常用搭配】 tension-cracking failure 拉裂破坏

【例句】 The plastic-flow and tension-cracking type collapes or soft foundation-type collapse are very common in Southwest mountain region and Three Gorges reservoir area in China. 塑流-拉裂式崩塌或软弱基座型崩塌是我国西南山区和三峡库区常见的一种崩塌类型。

adit reveals cuneiform tensile cracks

tension-cracking plane

【释义】 拉裂面

【例句】 The rock slope may produce tension-cracking plane due to the unloading and rebound. 岩质边坡因卸荷回弹可能产生拉裂面。

【中文定义】 由于拉应力作用形成的破裂面。

tensor-shear fault

【释义】 张扭性断层

【例句】 The Late Cenozoic tensor-shear fault zones around Awati Sag, NW Tarim Basin. 塔里木盆地阿瓦提凹陷周缘的晚新生代张扭性断层带。

tensor-torsion structural plane

【释义】 张扭性结构面

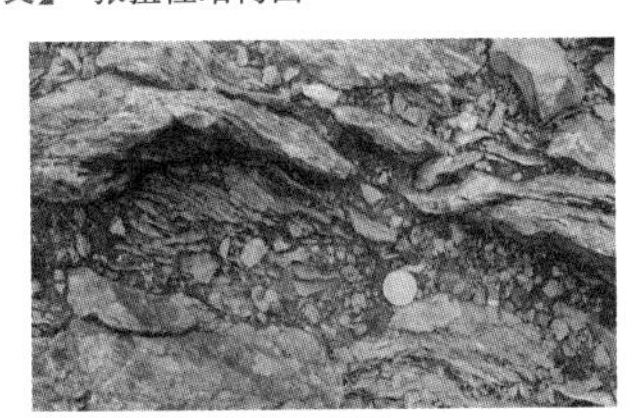

tenso-torsion structural plane

【例句】 In the construction and excavation process of the dam foundation, it is found that multiple tenso-torsion structural planes are developed in the rock mass. 坝基施工开挖过程中发现岩体中发育多组张扭性结构面。
【中文定义】 简称张扭面，指既具张性又具有扭性的结构面。

terrace ['tɛrəs]
【释义】 *n.* 阶地，台地，平台，梯田
【常用搭配】 alluvial terrace 冲积台地；erosion terrace 侵蚀阶地；structural terrace 构造阶地
【例句】 Quaternary deposits may be contained within landforms, such as terraces, drumlins, or barchans that can be recognised by their shape. 第四纪沉积物可能出现在阶地、冰丘和沙丘等容易视觉分辨的地形中。
【中文定义】 指由水流下切侵蚀和堆积作用交替进行而形成的沿河流两岸、湖滨和海滨延伸的阶梯状地貌。

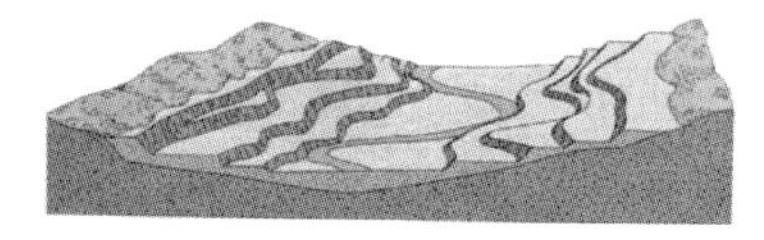

alluvial terraces

terrain model
【释义】 地形模型
【例句】 Digital Terrain Model is the basic and core module of the railway route CAD system. 数字地面模型是铁路线路CAD的基础核心模块。
【中文定义】 用一组有序数值阵列形式表示地面高程的一种地形表面模型，即数字高程模型（digital elevation model，DEM），一般采用散点、等高线、TIN网格、Grid网格采样进行表示。

terrestrial photogrammetry
【释义】 地面摄影测量
【中文定义】 利用地面摄影的像片对所摄目标物进行的摄影测量。

Tertiary (R) ['tərʃɪ'ɛri]
【释义】 *n.* 第三纪（系）
【中文定义】 新生代最老的一个纪（距今6500万～260万年），分为早第三纪和晚第三纪，该时代标志着“现代生物时代”的来临。

test of pore pressure dissipation
【释义】 孔隙压力消散试验

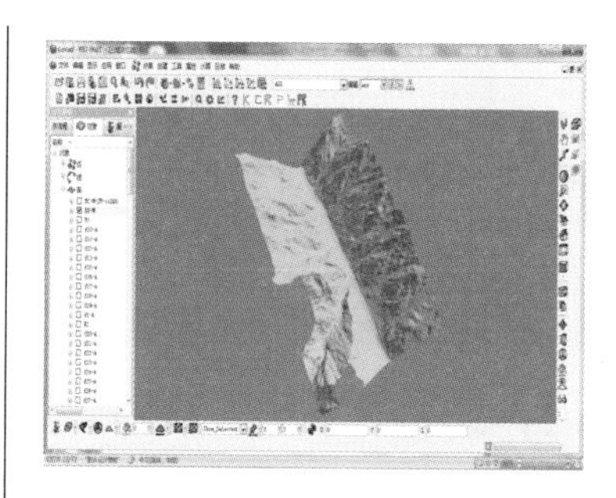

terrain model

test pressure
【释义】 试验压力

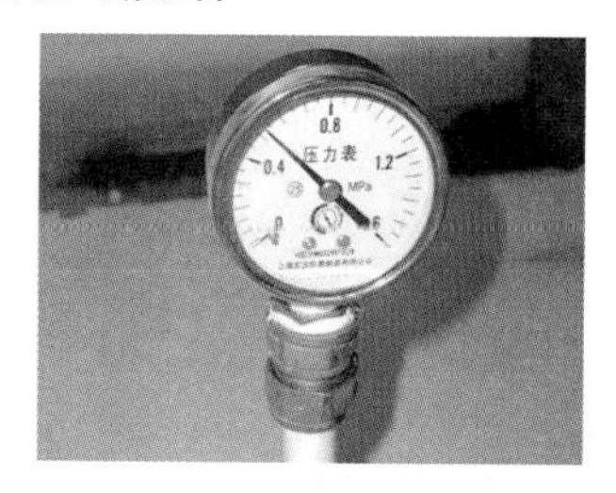

test pressure instrument

testing period
【释义】 检测周期
【例句】 This paper considers a two identical components paralleled repairable system with testing period. 研究了考虑检测周期的两同型部件并联的可修系统。

texture ['tɛkstʃə]
【释义】 *n.* 结构
【常用搭配】 phanerocrystalline texture 显晶质结构；massive texture 块状结构
【中文定义】 指组成岩石的矿物的结晶程度、颗粒大小、晶体的形态、自形程度及矿物之间结合关系等。

thaw collapsibility
【释义】 融陷性

the eigenperiod zoning map of earthquake response spectrum
【释义】 地震动反应谱特征周期区划图

the Q-system of rock mass classification
【释义】 岩体分类Q系统
【例句】 The Q-system of rock mass classifica-

tion expresses the quality of the rock mass in the so-called Q-value，on which are based design and support recommendations for underground excavations. 岩体分类 Q 系统用所谓的 Q 值来反映岩体质量，地下工程在此基础上提出支护建议措施。

【中文定义】 国际通用的岩体质量分类方法之一，由 Barton 等人于 1974 年最早提出。

The Wilson Cycle

【释义】 威尔逊旋回

【例句】 Apart from mineral deposits formed in astroblemes，there are few if any deposit types that form in complete independence of plate tectonics and The Wilson Cycle. 除开陨石坑中的矿藏，几乎没有完全独立于板块构造和威尔逊旋回之外的矿藏。

【中文定义】 大陆岩石圈在水平方向上的彼此分离与拼合运动的一次全过程。即大陆岩石圈由崩裂开始、以裂谷为生长中心的雏形洋区渐次形成洋中脊、扩散出现洋盆进而成为大洋盆，而后大洋岩石圈向两侧的大陆岩石圈下俯冲、消亡，洋壳进入地幔而重熔，从而洋盆缩小；或发生大陆渐次接近、碰撞，出现造山带，拼合成陆的过程。

thematic map

【释义】 专题地图

【中文定义】 着重表示自然现象或社会现象中的某一种或几种要素的地图。

theory of fault-block tectonics

【释义】 断块构造说

【例句】 Development for the theory of fault-block tectonics fully represents Professor Zhang Wenyou's way of academic pursuits-to be open to various ideas，absorb good ones and bring forth new ones. 断块构造说的形成过程充分表现了张文佑先生广蓄并纳继承创新的治学思路。

【中文定义】 是关于地球岩石圈构造及演化的理论。地球表层的第二级断块，也是岩石圈断块内部的次一级断块，比岩石圈断块薄。

thermal spring

【释义】 温泉

【例句】 They planed to go to a thermal spring the next day for a break. 他们计划着第二天去一个温泉休息休息。

【中文定义】 泉水的一种，是从地下自然涌出的自然水，泉口温度显著地高于当地年平均气温而又低于（等于）45℃的地下天然泉水称为温泉。

thermal spring

thermoluminescence method

【释义】 热发（释）光法

【例句】 High precision 230 Th-dating of the tufas deposited on different terraces of Mian River at Niangziguan shows that the ages of the oldest tufas on the terrace Ⅱ are between 407 ka and 466 ka，which are much higher than those dated by the thermoluminescence method earlier. 对山西娘子关绵河不同阶地上沉积的泉钙华进行了高精度的 230Th 定年和碳氧稳定同位素组成测定；结果发现，绵河Ⅱ级阶地沉积的娘子关泉钙华的最老年龄在 407ka～466ka，远老于早前通过钙华中的石英砂热发光法（TL）获得的年龄。

【中文定义】 岩石、矿物在其自身或周围天然放射性物质产生的 α、β、γ 射线的长期照射下，会在其晶体晶格缺陷中累积和储存一定的辐射能。当对岩石、矿石加热时，储存的辐射能以光的形式释放出来，即称为热释光效应，所放出的光叫热释光。利用这种热释光效应进行铀矿普查的方法叫热释光法。

thermoluminescence surveying

【释义】 热释光测量

【例句】 Soil natural thermoluminescence surveying is an accumulation method of measuring radon and it can measure radon in a long time. 土壤天然热释光测量方法是一种累积型测氡方法。

thermomineral spring

【释义】 热矿泉

【例句】 At lava Hot Springs of Idaho，visitors can have a bath in the thermomineral spring. 在爱达荷的溶岩温泉，游客可在热矿泉水中洗澡。

【中文定义】 含矿物盐或气体的地热水称为热矿水，出露于地表的称为热矿泉。溶解的矿物盐或气体使热矿泉具有特殊的味道和较好的疗效。

The thermomineral spring in Shangsi County

thick layer structure

【释义】 厚层状结构

【例句】 During fracturing of thin layer structure with high stress difference of reservoir and restraining barrier, height and width of fracture will decrease compared to thick layer structure. 与厚层状结构相比，在应力差储隔层致裂过程中，薄层状结构缝高、缝宽都应减小。

【中文定义】 层面厚度在 30～100cm 之间的层状结构。

3D geologic system

【释义】 三维地质系统

【例句】 The study of rock weathering and unloading degree in 3D geologic system. 在三维地质系统中研究岩体的风化卸荷程度。

【中文定义】 为提高工程地质勘察的质量和效率，采用地质三维数字化技术作为主要特征，基于计算机辅助设计软件定制开发的计算机辅助勘察设计系统，主要由计算机软硬件、CAD 平台软件、地质应用软件、地质数据、用户等部分组成。

3D geological digitization

【释义】 地质三维数字化

【中文定义】 运用计算机设备和软件实现地质数据模型、图形模型的虚拟创建、修改、固化、分析等一系列过程的数字化操作。地质三维数字化的产品包括三维地质模型和地质数据库。

3D geological model

【释义】 三维地质模型

【中文定义】 在地质专业 CAD 中，根据工程地质勘察设计要求，利用工程区一定范围内的地质勘察资料，按工程对象类别建立的带有图元属性、工程地质属性（含相互约束关系属性）的三维可视化模型，是地质三维数字化工作的主要成果。三维地质模型按工程对象类别划分为地形模型、勘探模型、地质体模型；按 CAD 软件中的

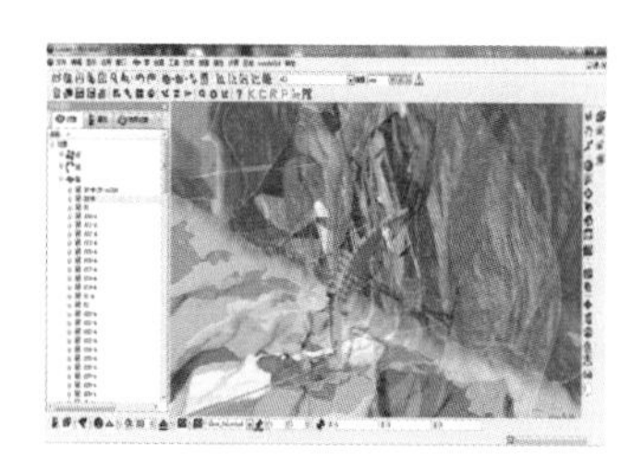

3D geological digitization

图元类别可划分为地质点符模型、地质线框模、地质表面模型、地质实体模型、地质属性模型。

3D geological model accuracy

【释义】 三维地质模型精度

【中文定义】 三维地质模型精度是指依据地质体采集点数据在 CAD 软件中经分析推测、插值拟合后所得到的模拟位置和产状与地质体实际状态的吻合程度，其中位置精度用距离偏差表示，产状精度用角度偏差表示。

3D geological model style

【释义】 三维地质模型样式

【中文定义】 三维地质模型组成要素及其表达规则的集合，包括模型坐标系、单位、方向、图层、显示、颜色、符合、标注、文字、属性等。三维地质模型样式一般在地质专业 CAD 中被定制为可重复使用的标准样式模板，使专门建模工具创建的模型都具有统一的模型样式，可显著提高三维地质模型的建模效率和可利用性。

3D laser scanning

【释义】 三维激光扫描

【例句】 The measurement technology of 3D laser scanning is very useful in aerospace engineering. 三维激光扫描测量技术在航空、航天工程中具有极其重要的应用价值。

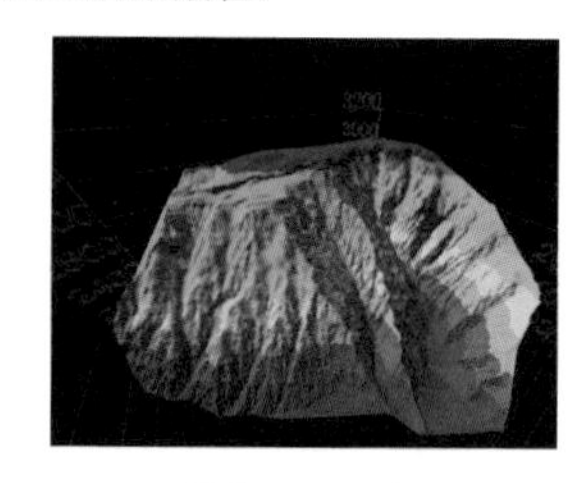

3D laser scanning

3D model

【释义】 三维模型，立体模型

【同义词】 three-dimensional model, stereomodel

【例句】 3D finite element (FE) model of the human lumbar motion segment requires to be qualified both visually and mathematically. 腰椎节段三维有限元模型要求在视觉上和数学求解上均满足要求。

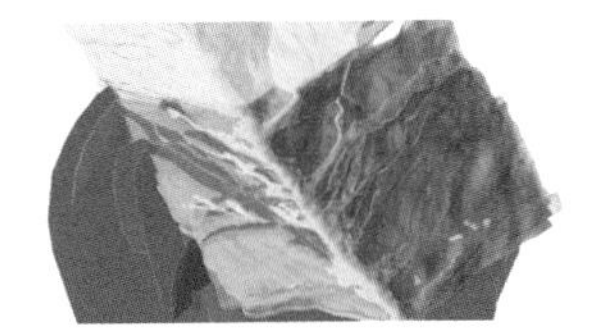

3D model

3D remote sensing interpretation

【释义】 三维遥感解译

3D remote sensing interpretation

3D seismic exploration

【释义】 三维地震勘探

【例句】 3D seismic exploration is a comprehensive application technology integrating physics, mathematics and computer science. Its purpose is to make the image of the underground target more clear and the position prediction more reliable. 三维地震勘探是一项集物理学、数学、计算机学于一体的综合性应用技术，其应用目的是使地下目标的图像更加清晰、位置预测更加可靠。

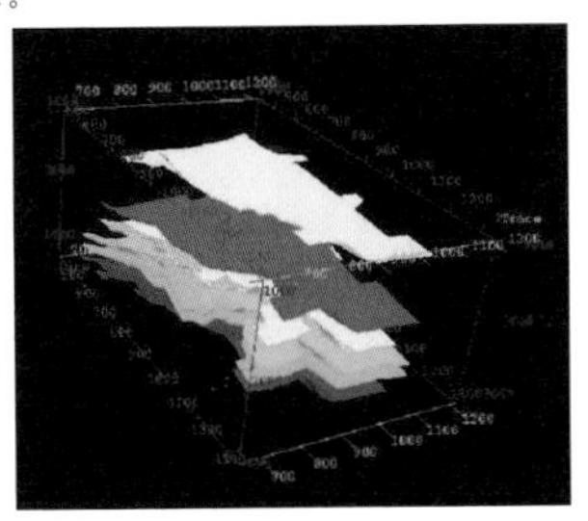

3D seismic exploration

3D true color image system

【释义】 三维真彩色影像系统

three phase diagram

【释义】 三相图

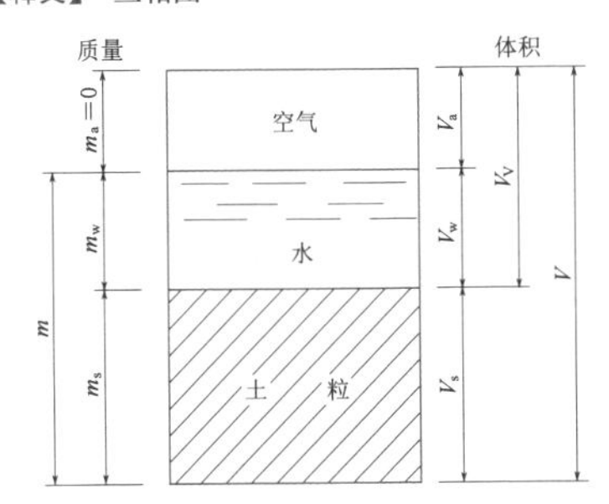

three phase diagram

three-directional layerwise summation method

【释义】 三向分层总和法

【中文定义】 考虑到地基侧向变形对纵向压缩变形的影响，对分层总和法进行修正后得出的计算地基沉降域的方法。

thin layer structure

【释义】 薄层状结构

【中文定义】 层面厚度在1～10cm的层状结构。

thixotropy [θɪk'sɑtrəpi]

【释义】 *n.* 触变性

【例句】 It is normally used where both viscosity and thixotropy are required. 它通常用于要求黏性和触变性的场合。

tilting uplift

【释义】 掀斜隆升

【例句】 Since the late period of Early Pleistocene Tongbai-Dabie Mountains have been tilting uplifting from north to south, which affects the middle Yangtze River environment, especially greatly affects the drainage systems evolution and development. 桐柏-大别山自早更新世晚期以来，不断发生着自北向南的掀斜隆升，从而对长江中游的环境，尤其是水系的演化产生了严重影响。

tilting (toppling) moment

【释义】 倾倒力矩

【例句】 It shall be reckoned in tilting moment resulting from perched water in the stability analysis of tilting failure of bedded rock slope, thus it can only explain many irregular phenomena of toppling mass. 层状岩石边坡倾倒破坏稳定分析

中，应计入上层滞水产生的倾倒力矩，这样才能解释倾倒体的许多不规则现象。

time effect

【释义】 时间效应

time-depth conversion

【释义】 时深转换

【例句】 Time-depth conversion is the process of converting seismic data from the time domain to the depth domain. 时深转换是将地震数据从时间域向深度域转换的过程。

top angle of drilling hole

【释义】 钻孔顶角

top view

【释义】 俯视图，顶视图

【常用搭配】 top view of building 建筑俯视图；top view of dam 大坝俯视图

【同义词】 vertical view, plan view, planform

【中文定义】 也称顶视图，由物体上方向下做正投影得到的视图。

topaz ['topæz]

【释义】 *n.* 黄玉

【例句】 Topaz occurs in the igneous rock rhyolite. 黄玉通常产于火成流纹岩中。

topaz

topazization

【释义】 黄玉化

topographic line

【释义】 地形线

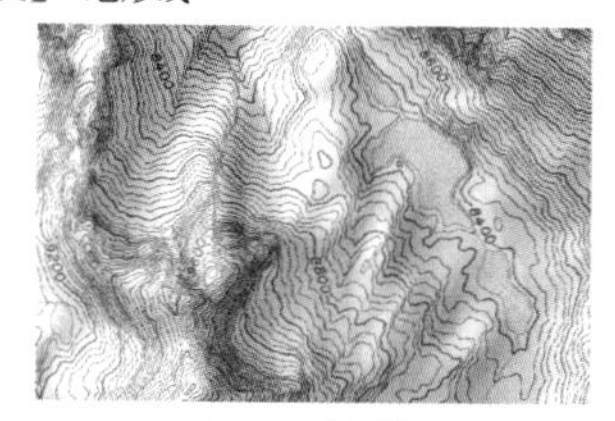

topographic line

【同义词】 form line

【例句】 In the topographic drawing of the mountain, the topographic lines of vertical section and transect have their special graphic information respectively. 山岭地形图中纵断面地形线与横断面地形线具有各自的图形形位信息。

topographic map

【释义】 地形图

【中文定义】 表示地表上的地物、地貌平面位置及基本地理要素，且高程用等高线表示的一种普通地图。

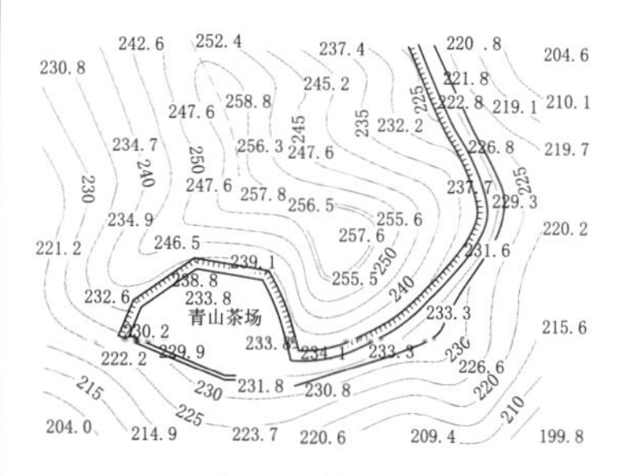

topographic map

topographic map framing

【释义】 地形图分幅

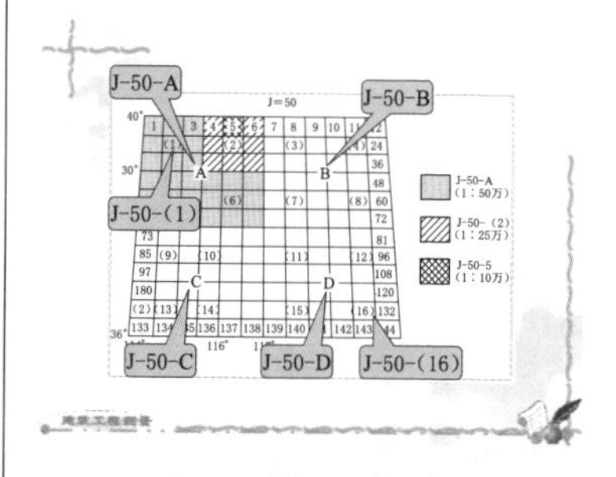

topographic map framing

topographic mapping

【释义】 地形测图，地形图测绘

【例句】 This system of aerial photography in which one vertical and two oblique photographs are simultaneously taken can be used in topographic mapping. 此种航空测绘系统，可同时拍摄一张竖直的和两张倾斜的照片，可用于地形绘图中。

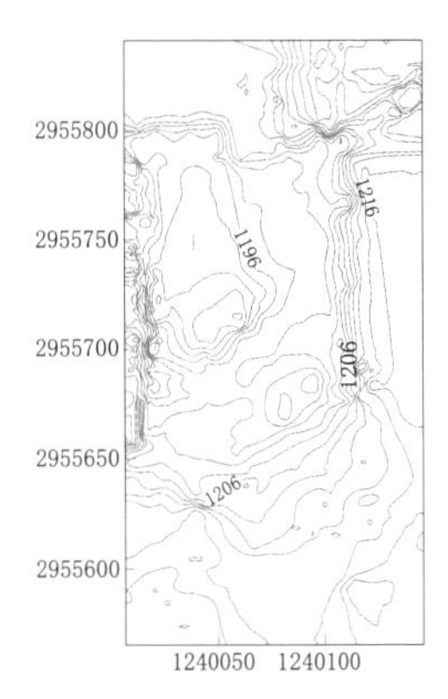

topographic mapping

topographic point
【释义】 地形点

topographic survey
【释义】 地形测量
【中文定义】 根据规范和图式，将地貌、地物及其他地理要素测量并记录在某种载体上的技术。

topography [tə'pɑgrəfi]
【释义】 *n.* 地形，地形学
【例句】 In many complex areas, it is almost impossible to understand the structure only from the study of the topography. 在大多复杂区域，仅根据地形特征来研究地质构造几乎是不可能的。
【中文定义】 地球表面的起伏形态。

the topography of a hydropower station

toppling ['tɔplɪŋ]
【释义】 *n.* 倾倒
【常用搭配】 toppling mass 倾倒体；toppling deformation 倾倒变形；toppling failure 倾倒破坏
【例句】 Therefore, although block toppling may not be kinematically feasible, the slope geometry may still result in a kinematically unstable condition. 因此，虽然块状倾倒未发生运动，但边坡的几何条件仍能提供不稳定运动条件。
【中文定义】 倾倒破坏指层状岩层构成的陡倾逆向坡在重力作用下向临空一侧的弯曲折断。

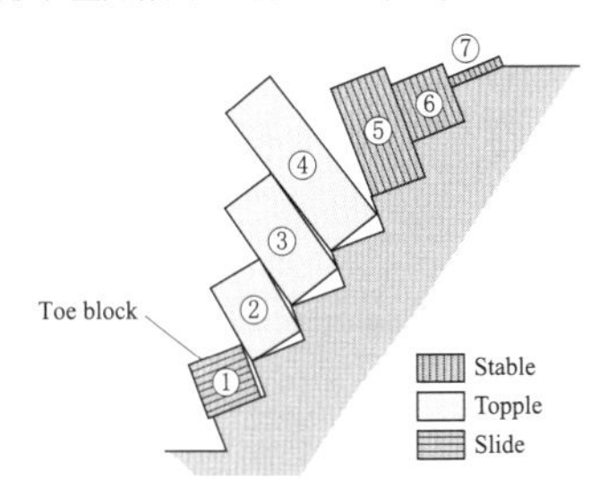

example of a system of toppling blocks on a stepped base

torsional structural plane
【释义】 扭性结构面
【例句】 The torsional structural plane is mainly caused by the stress of twisting or shearing. 扭性结构面主要由扭（剪）应力作用产生。
【中文定义】 简称扭裂面，岩块或地块遭受挤压而产生的一对与主压应力作用面斜交的破裂面。如平移断层面等。

torsional structural plane

torsional shear test
【释义】 扭剪试验
【例句】 Some behaviors of saturated flyash such as pore water pressure, dynamic strength, and dynamic deformation are studied by the dynamic torsional shear test. 通过动扭剪试验研究了饱和粉煤灰的动孔隙水压力特性、动强度特性以及振陷变形特性。

total mineralization of groundwater
【释义】 地下水总矿化度（溶解性固体总量）
【同义词】 total salinity of groundwater
【例句】 Total mineralization of groundwater is the total content of ions, molecules and various compounds in groundwater. It is usually determined by the residue obtained after water drying.

地下水总矿化度指地下水中离子、分子和各种化合物的总含量，通常是以水烘干后所得残渣来确定的。

【中文定义】 是指单位体积地下水中可溶性盐类的质量，常用单位为 g/L 或 mg/L。

total stress method

【释义】 总应力法

【例句】 So the comparison of large strain method and small strain method should be carried out with total stress method. 因此，大小变形法分析结果的比较应采用总应力法。

toxic gas

【释义】 有毒气体

【例句】 Analysis on more than 500 transportation accident of hazard materials show that about 20% of them are concerned with gaseous hazard materials and accidents during transportation of toxic gas often lead to heavy casualty and serious pollution because of their chemical properties. 事故统计分析表明，约 20%左右的道路危险品运输事故涉及气体危险品，其中毒性气体运输过程一旦发生事故，由于其自身特性，往往造成严重人员伤亡和环境破坏。

【中文定义】 指劳动者在职业活动过程中通过机体接触可引起急性或慢性有害健康的气体。常见的有：二氧化氮、硫化氢、苯、氰化氢、氨、氯气、一氧化碳、丙烯腈、氯乙烯、光气（碳酰氯）等。

tracer methed

【释义】 指示剂法

transient electromagnetic method (TEM)

【释义】 瞬变电磁法

transient surface wave method

【释义】 瞬态面波法

【例句】 Transient surface wave method is a new engineering prospecting technology, which makes use of four characteristics of surface wave to deal with problems of rock-earth layer, of bearing capacity of clay layer, antiseismic designing parameters of foundation soil and weak formation, underground caves and Karst collapse, etc. 瞬态面波法是一种新的工程物探技术，利用面波的四种特性，可以解决工程地质中的岩土层分层、厚度、黏土层承载力、地基土抗震设计参数及软弱层、地下洞室、岩溶塌陷等不良地质问题。

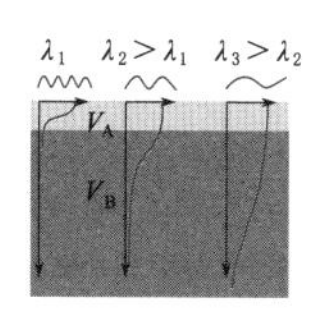

strata media with different wavelengths

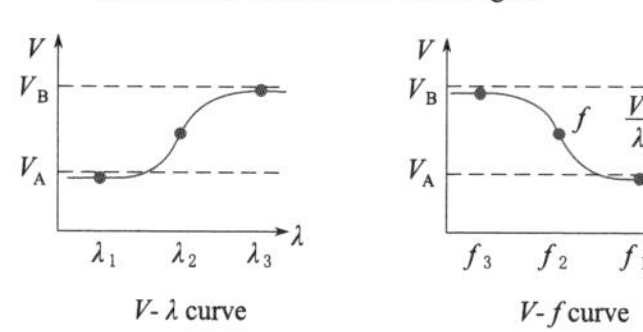

transient surface wave method

translational landslide

【释义】 平移式滑坡

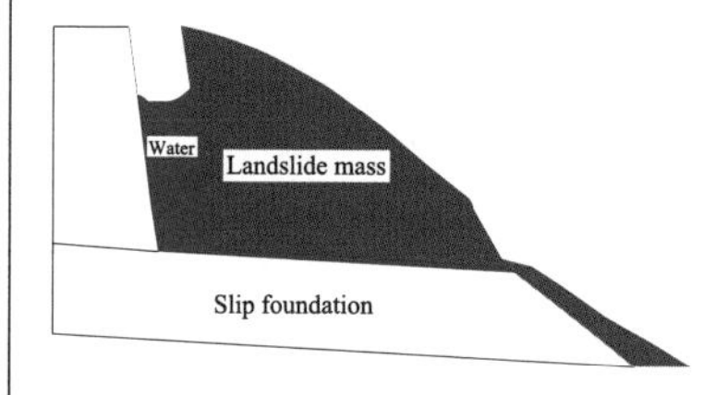

translational landslide

transmissibility coefficient

【释义】 导水系数

【例句】 The main factors affecting the well yield are the transmissibility coefficent and water yield coefficient of the aquifer as well as the permeability coefficient and thickness of the leaky layer. 含水层的导水系数、释水系数和弱透水层的渗透系数、厚度是影响井涌水量的两大重要因素。

【中文定义】 表征含水层全部厚度导水能力的参数，其值等于渗透系数与含水层厚度的乘积。

transversal geological section

【释义】 工程地质横剖面图

【常用搭配】 Transversal geological sections of main structures 主要建筑物工程地质横剖面图

【中文定义】 ①垂直于拟建场地的延长方向或结构物长轴线的工程地质剖面图。②垂直于岩层走向或构造线的工程地质剖面图。

transverse valley

【释义】 横向谷

【中文定义】 延伸方向与岩层走向近正交（60°～90°）的河谷。

traverse survey

【释义】 导线测量法

【例句】 The paper introduced basic principle and a simple calculating method of accuracy evaluation in highway traverse survey. 本文介绍了公路导线测量中精度估算的基本原理和一种简便易行的计算方法。

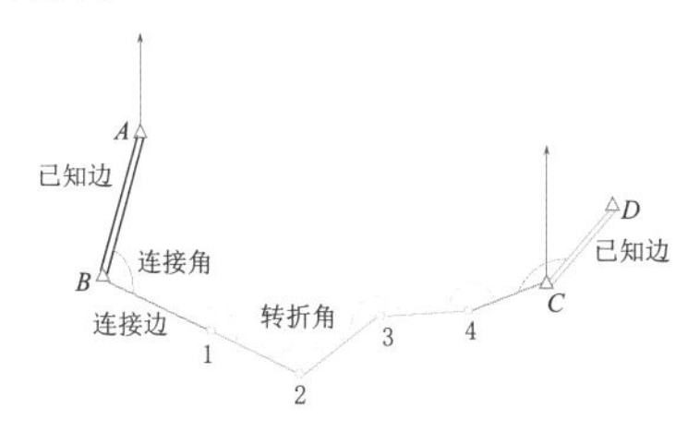

traverse survey

travertine ['trævərˌtin]

【释义】 *n.* 钙华，石灰华

【同义词】 tufa 泉华

【例句】 The Yulongxi travertine landscape in west Gongga Mountains, Kanding, is another sub-nival belt karst landscape besides Jiuzhaigou and Huanglong found. 地处康定贡嘎山西侧的玉龙希钙华景观是继九寨沟、黄龙钙华景观发现之后又一高寒岩溶风景地。

【中文定义】 又称石灰华，是一种碳酸钙的沉淀物，含有碳酸氢钙的地热水在靠近地表时释放大量二氧化碳所形成的。具体成因是岩溶地区的地表水或者地下水在植物等作用的影响下，导致水中的碳酸钙饱和然后沉积下来，有些地区的钙华是因为地热水露出地表后，环境压力和温度的改变导致地热水中的化学物质沉淀出来，这就是泉华。一般碳酸钙积淀形成的钙华比较常见。

travertine

treatment of fault zone

【释义】 断层破碎带处理

【中文定义】 为满足水工建筑物的承载能力、限制变形、抗滑和防渗等要求，对基岩中断层破碎带进行处理的工程措施。

triangular method

【释义】 三角形法

【例句】 In the process of engineering exploration, triangular method is suitable for the borrow area of irregularly arranged exploration points. 在工程勘探过程中，三角形法适用于勘探点布置不规则的料场。

【中文定义】 指勘探点联成三角形网点，各三角形面积乘以其三顶点平均厚度，分别求得三角形部分储量，然后求和各三角形的储量的方法。

triangulated irregular network (TIN)

【释义】 不规则三角网

【例句】 The groundwater flow is simulated with the seepage model of heterogeneous isotrope, and the partitial differential equations are solved with finite difference scheme of triangulated irregular network. 地下水流是以非均质各向同性渗流模型模拟的，并且用不规则三角网有限差分法来求解偏微分方程的。

【中文定义】 也称曲面数据结构，根据区域的有限个点集将区域划分为相等的三角面网络，数字高程由连续的三角面组成，三角面的形状和大小取决于不规则分布的测点的密度和位置。

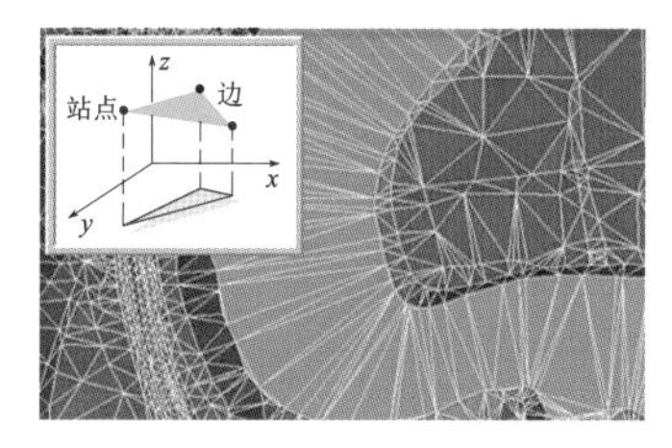

triangulated irregular network (TIN)

triangulation method

【释义】 三角测量法

【例句】 Research has been done on the measurement of surface profile using the laser triangulation method. 文章对激光三角测量法应用于表面形貌的检测进行了研究。

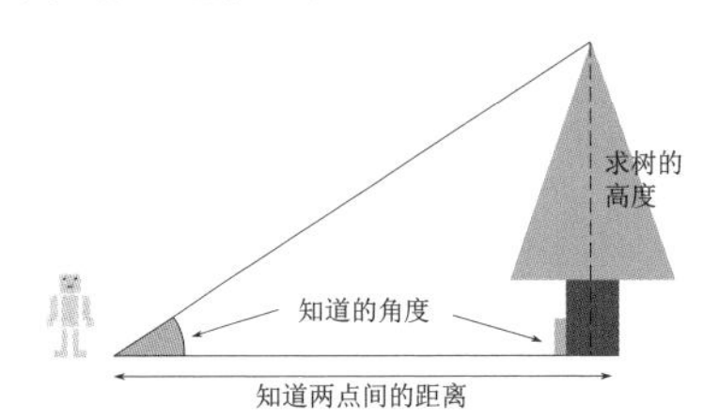

triangulation method

Triassic (T) [traɪ'æsɪk]

【释义】 *n.* 三叠纪(系)

【中文定义】 是公元前2.5亿—公元前2亿年的一个地质时代,是中生代的第一个纪。三叠纪时大多数地球上的大陆形成一块巨大的大陆,称为盘古大陆。

triaxial compression test

【释义】 三轴压缩试验

【常用搭配】 conventional triaxial compression test 常规三轴压缩试验

【例句】 Based on the triaxial compression tests of weathered sand, discussed in this paper was the influence of rubber membrane on strength of soil. 基于风化砂的三轴压缩试验,讨论了橡皮膜约束对土强度的影响。

triaxial compression test system

triaxial compressive strength

【释义】 三轴抗压强度

【例句】 With the deeper drilling operation increased gradually, the triaxial compressive strength model will become one of important parameters to measure the formation strength characteristics. 随着深井钻井作业量的逐步增加,岩石三轴抗压强度的预测模型将成为衡量地层强度特性的重要参数之一。

triaxial extension test

【释义】 三轴伸长试验

triggering factor

【释义】 触发因素

true dip

【释义】 真倾角

【例句】 If the true dip line of one or two joint planes plot within a sliding zone of slope, a sliding failure can be possible. 如果有一两个节理面真倾向线落在滑动区内,滑动破坏才可能发生。

【中文定义】 是某一倾斜构造面的倾斜线与其水平投影线之间的夹角。即在垂直倾斜面走向的横剖面上测定的倾斜面与水平参考面之间的夹角。

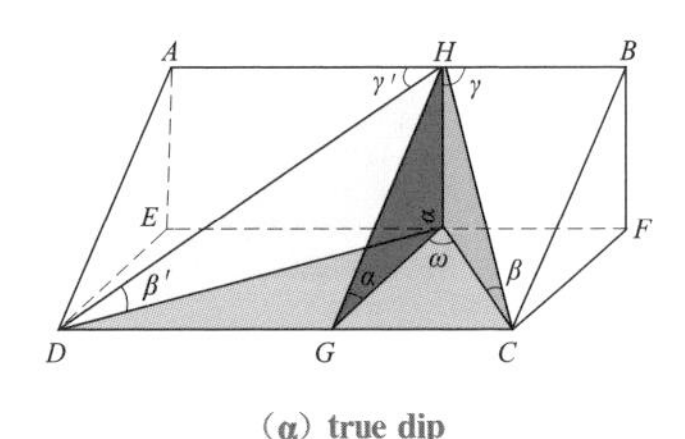

(α) true dip

true triaxial test

【释义】 真三轴试验

【例句】 Finally, a shear yield criteria for coarse-grained soils is proposed based on the results of conventional triaxial and true triaxial tests on these soils. 最后,在粗粒土常规三轴试验及真三轴试验结果的基础上,建立了一个粗粒土剪切屈服准则。

tube logging

【释义】 管测井

tuff [tʌf]

【释义】 *n.* 凝灰岩

【常用搭配】 lithic tuff 石质凝灰岩;岩屑凝灰岩 tuff breccia 凝灰角砾岩;凝灰岩锥 volcanic tuff 火山凝灰岩

【例句】 The contents of clay, tuff, scoria, organic matter shall not exceed certain proportions. 黏土、凝灰岩、渣子、有机质的含量不应超过一定比例。

【中文定义】 是一种火山碎屑岩,其组成的火山碎屑物质有50%以上的颗粒直径小于2mm,成分主要是火山灰,外貌疏松粗糙或致密,有层理

tuff

的称为层凝灰岩；因成分不同导致颜色多样，有紫红色、灰白色、灰绿色等。

tungsten-carbide drilling

【释义】 硬质合金钻进

tunnel advance geological prediction

【释义】 隧洞超前地质预报

【例句】 In recent years, TSP has been used for tunnel advance geological prediction, and it has been well accepted by Chinese technical tunnel personnel. 在近几年的隧道工程施工实践中，利用 TSP 技术进行隧洞超前地质预报，已经得到我国隧道工程技术人员的广泛认同。

tunnel boring machine (TBM)

【释义】 隧道掘进机

【中文定义】 指利用回转刀具开挖，同时破碎洞内围岩及掘进，形成整个隧道断面的一种新型、先进的隧道施工方法。

tunnel boring machine

2D model

【释义】 二维模型，平面模型

【同义词】 two-dimensional model, plane model

【例句】 In 2D model the boundary fitted coordinate transformation were adopted to fit in with irregular area shape. 二维模型采用了边界拟合坐标变换，以适应不规则的区域形状。

2D seismic exploration

【释义】 二维地震勘探

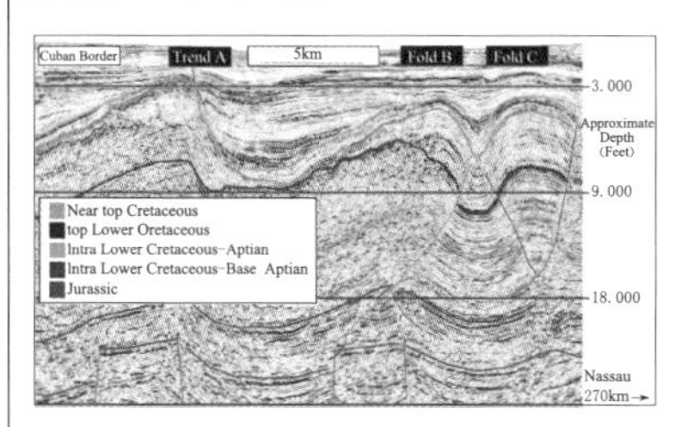

2D seismic exploration

type of rock water content

【释义】 岩石含水类型

typical geological profile

【释义】 典型地质剖面

【常用搭配】 reservoir typical geological profile 水库区典型地质剖面；typical geological profile of landslide 滑坡典型地质剖面

【例句】 In prefeasibility study stage of hydropower engineering, general stratigraphic column and typical geological profile shall be added to reservoir area synthetic geological map in report for engineering geological investigation. 水力发电工程预可行性研究阶段工程地质勘察报告中，水库区综合地质图应附综合地层柱状图和典型地质剖面图。

【中文定义】 能够反映一个区域或地质体典型地质特征的代表性地质剖面。

U

ultimate strength

【释义】 极限强度

【例句】 There are many new problems such as failure mode, ductility, ultimate strength and durability etc in the concrete structures strengthened with FRP. 采用 FRP 材料加固的混凝土结构，存在诸如破坏形式、延性、极限强度、耐久性等一系列的新问题。

【中文定义】 指物体在外力作用下发生破坏时出现的最大应力，也可称为破坏强度或破坏应力。

ultrabasic rock

【释义】 超基性岩

【例句】 Xiaozhangjiakou ultrabasic rocks is the host rock of gold deposit. 小张家口超基性岩体是金矿床的母岩。

ultrasonic image logging

【释义】 超声成像测井

【例句】 Ultrasonic imaging logging gives logging information in the form of acoustic images with high resolution, intuition and easy analysis. 超声成像测井以声学图像形式给出测井资料，它具有分辨率高、直观、便于分析判断的优点。

unavailable layer

【释义】 无用层

【中文定义】 指质量技术指标不能满足水电水利工程天然建筑材料要求的岩土层。

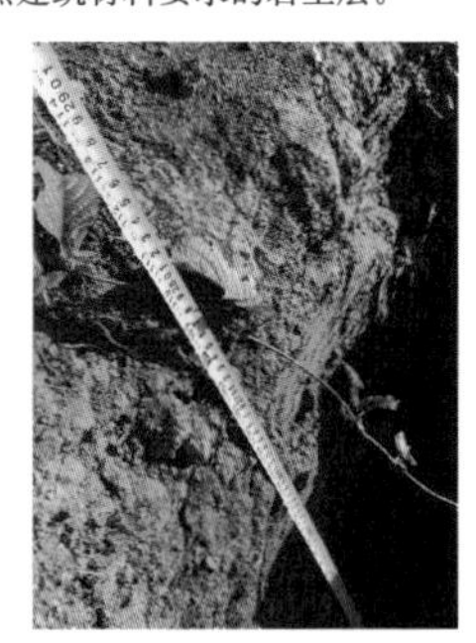

unavailable layer

unconfined compressive strength

【释义】 无侧限抗压强度

【例句】 The unconfined compressive strength of modified red clay increases with the increase of fiber content and age. 改良红黏土的无侧限抗压强度随着纤维掺入量和龄期的增加而增加。

【中文定义】 是指试样在无侧向压力条件下，抵抗轴向压力的极限强度。

unconfined compressive strength test

【释义】 无侧限抗压强度试验

unconfined compressive strength test

unconformity [ˌʌnkən'fɔrməti]

【释义】 *n.* 不整合，不一致

【常用搭配】 topographic unconformity 地形不整合

【中文定义】 是指上下两套不同时代地层之间出现过沉积间断或地层缺失的地层接触关系。不整合表明地层记录的重要间断或缺失。

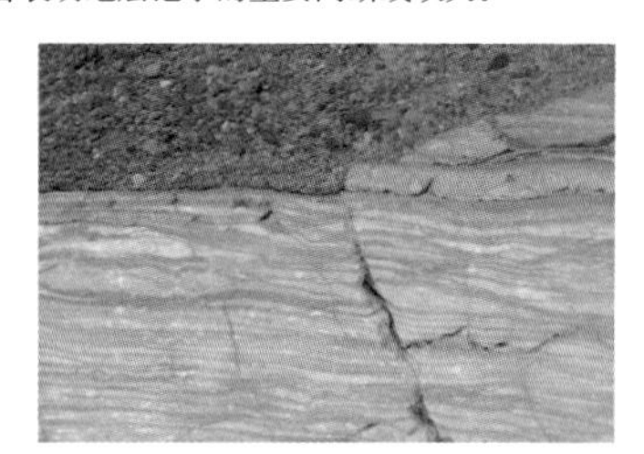

unconformity

unconsolidated-undrained shear test (UU)

【释义】 不固结不排水剪试验

unconsolidated-undrained triaxial test

【释义】 不固结不排水三轴试验

underconsolidated soil

【释义】 欠固结土

underground brine

【释义】 地下卤水

【例句】 The outlook of China's underground brine development and the defects of the pumps now in use was introduced. 介绍了目前我国地下卤水的开发前景、开采用泵存在的一些缺陷。

【中文定义】 矿化很强的地下水。

extraction of underground brine from Lop Nur

underground powerhouse

【释义】 地下厂房

【例句】 Main features of the project include a 305m high double curvature arch dam and a huge underground powerhouse. 本工程主要特点为305m高的双曲拱坝和巨大的地下厂房。

【中文定义】 指建在地面以下洞室中的水电站厂房。

underground river

【释义】 暗河

【例句】 3D nonlinear finite element analysis of blocking body in karst cave of underground river in Kongliang reservoir is conducted. 孔梁水库暗河溶洞堵头三维非线性有限元计算分析得以实施。

【中文定义】 也称伏流，因岩溶作用在大面积石灰岩地区所形成的溶洞和地下通道中，具有河流主要特征的水流。

underground river

underground runoff modulus method

【释义】 地下径流模数法

underground watershed

【释义】 地下分水岭

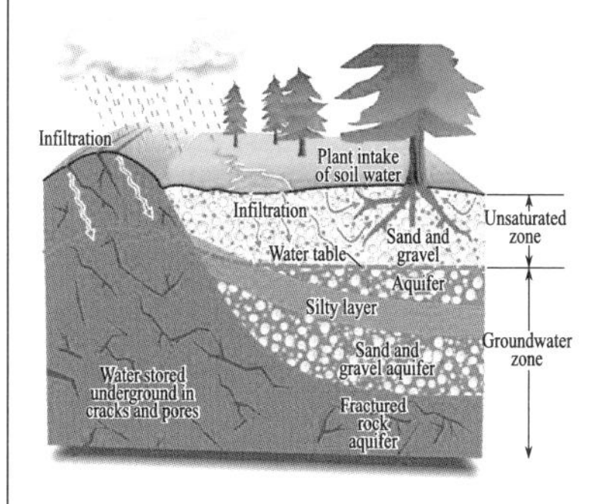

underground watershed

underwater cross-section survey

【释义】 水下横断面测量

underwater longitudinal-section survey

【释义】 水下纵断面测量

undeveloped [ˌʌndi'veləpt]

【释义】 *adj*. 不发育的

【例句】 In this paper further discussion is made on the causes for undeveloped eluvial soil on granite in Hong Kong and its effect on engineering properties. 文章进一步讨论了香港花岗岩残积土不发育的原因和对工程性质的影响。

【中文定义】 根据《水力发电工程地质勘察规范》(GB 50287—2016) 的规定，当岩体内结构面发育组数为1～2组，间距大于100cm时，其岩体结构面发育程度为不发育。

underwater terrain survey

【释义】 水下地形测量

【同义词】 underwater topography survey

underwater terrain survey

【例句】 The underwater terrain survey is playing an important role in the inland river and in the coast engineering construction and the survey precision is always paid key attention to. 水下地形测量在内河和沿海工程建设中发挥着重要作用，测量的精度也历来为人们所重点关注。

undrained consolidation triaxial test

【释义】 固结不排水三轴试验

【例句】 The steady strength of saturated loess is discussed under stress-controlled undrained consolidation triaxial test. 通过对重塑黄土的固结不排水三轴试验，研究了饱和黄土的稳态强度特性。

undulating ['ʌndjəˌletɪŋ]

【释义】 *adj*. 起伏的

【例句】 There are mainly three kinds of undulating structures, including planar, undulating and stepped. 结构面的形态主要有平直的、起伏的和有台坎的三种。

【中文定义】 是结构面一种起伏形态特征的描述。结构面的平整程度不同，抗剪强度也不同。

undulating structural plane

【释义】 波状结构面

【中文定义】 结构面的形态是波状的，反映了结构面的平整程度

undulating structural plane

uniaxial compression deformation test

【释义】 单轴压缩变形试验

【例句】 The digital features of localized deformation of limestone are analyzed by using video images of uniaxial compression deformation test. 根据室内单轴压缩变形试验视频图像对石灰岩局部化变形的数字特征进行分析。

uniaxial dry compressive strength

【释义】 单轴干抗压强度

【例句】 The anisotropy coefficients of the uniaxial dry compressive strength and tensile strength of numerical simulation results are 0.72 and 0.24, respectively. 数值模拟结果得出单轴干抗压强度和抗拉强度的各向异性系数分别为 0.72 和 0.24。

uniform deformation

【释义】 均匀变形

【例句】 The failure process can be divided into three stages, including uniform deformation, strain localization, and post-failure. 破坏过程可分为相对均匀变形阶段、应变局部化阶段和破坏后阶段三个阶段。

【中文定义】 一种变形方式。变形过程中各单薄层之间发生相互滑动，每层相对其邻层之运动都是相同的。

uniform settlement

【释义】 均匀沉降

【例句】 For box foundations on natural ground, it should be taken into account that some differential settlement normally occurs even if the calculation predicts uniform settlement only. 对于天然地基上的箱型基础，即使计算预测为均匀沉降，也应当考虑不均匀沉降的频发。

【中文定义】 基础底面（地基表面）上各部位下沉量均相等的地基沉降。

unit weight

【释义】 容重，单位重量

【例句】 Nondestructive detection and core sample detection show that the unit weight, porosity and permeability of the RCC core are in accord with the design requirement. 经无损检测和钻孔取芯检测，碾压混凝土心墙的容重、孔隙率和渗透系数均满足设计要求。

【中文定义】 单位体积所具有的重量，也称重度。

unloading ['ʌn'loʊdɪŋ]

【释义】 *n*. 卸载，卸荷

variation of thickness induced by unloading

【常用搭配】 unloading effect 卸荷作用；unloading deformation 卸荷变形；unloading crack 卸荷裂隙

【例句】 The unloading cracks were distributed primarily in the east precipice with large aperture and long extension as well as largest destructibility. 卸荷裂隙主要分布在崖体东边，具有张开度大、延伸长的特点，且破坏性最大。

【中文定义】 指把物体承受的力撤掉。

unloading deformation

【释义】 卸荷变形

unloading deformation

unloading effect

【释义】 卸荷作用

【例句】 In recent years, research on hydro-mechanical coupling of fractured rock mass under loading conditions has advanced greatly, but corresponding research on hydro-mechanical coupling considering unloading effect has still been in an early stage. 近年来，加载条件下的裂隙岩体渗流应力耦合研究得到了长足的发展，但卸荷作用下的水岩耦合分析还处于初始阶段。

【中文定义】 又称释重作用、剥离作用，是物理风化的方式之一。地壳深处的岩石被抬升后遭受剥蚀，使其上覆岩石的厚度减薄，以至裸解地表，因失去荷载，体积膨胀，产生垂直岩石表面的张力。

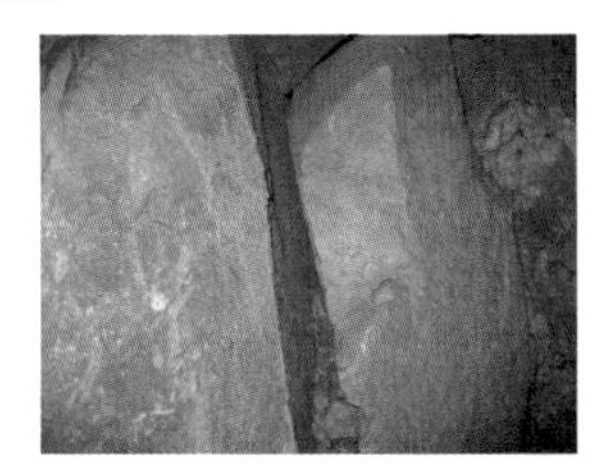

schematic diagram of unloading effect

unloading fracture

【释义】 卸荷裂隙

【同义词】 relief joint; unloading crack; stress-release crack

【例句】 The research on unloading fractures is very significant in rock slope engineering. 对岩体卸荷裂隙的研究对于岩体边坡工程非常重要。

【中文定义】 是由于自然地质作用和人工开挖使岩体应力释放和调整而形成的裂隙。卸荷裂隙往往受重力、风化及岸坡的物理地质作用进一步张开或位移。

unloading rock mass

【释义】 卸荷岩体

【例句】 According to dynamic and static comparative tests of the 35 deformation points, the relation curve between unloading rock mass deformation parameters and wave speed is established; and the rock mass deformation parameters in the unloading zone are evaluated based on results of large-scale acoustic tests. 通过对 35 个变形点的动静对比试验，建立卸荷岩体变形参数与波速之间的关系，并且利用大范围的声波测试结果，评价了卸荷岩体的变形参数。

strong weathering and unloading rock mass on the back wall of landslide

unmanned aerial vehicle (UAV)

【释义】 无人飞行器，无人机

unmanned aerial vehicle (UAV)

unsaturated soil

【释义】 非饱和土

【例句】 Creep of unsaturated soil, a new theory, is put forward based on unsaturated soil mechanics and creep mechanics. 非饱和土蠕变是在非饱和土力学及常规土蠕变力学基础上提出来的一种新理论。

unstable [ʌn'stebl]

【释义】 *adj*. 不稳定的

【中文定义】 根据《水力发电工程地质勘察规范》(GB 50287—2016)，Ⅳ类围岩稳定性多为不稳定，表现为围岩自稳时间很短，规模较大的各种变形和破坏都可能发生。

unsteady groundwater flow

【释义】 地下水非稳定流

【例句】 Groundwater pumping test can be used to determine if the groundwater is steady groundwater flow or unsteady groundwater flow. 地下水抽水试验可以判断地下水的类型是地下水稳定流或地下水不稳定流。

【中文定义】 在所要研究的渗流场区域内任一点，地下水各运动要素（渗流量、渗流速度、压强、水头等）随时间而变化。在非稳定渗流情况下，各要素是时间和空间坐标的函数。

unsteady-flow pumping test

【释义】 非稳定流抽水试验

【例句】 Hydro-geological parameters of phreatic aquifer were solved on the complete penetration of radial well unsteady flow pumping test. 潜水含水层的水文地质参数可以通过辐射井非稳定流抽水试验进行确定。

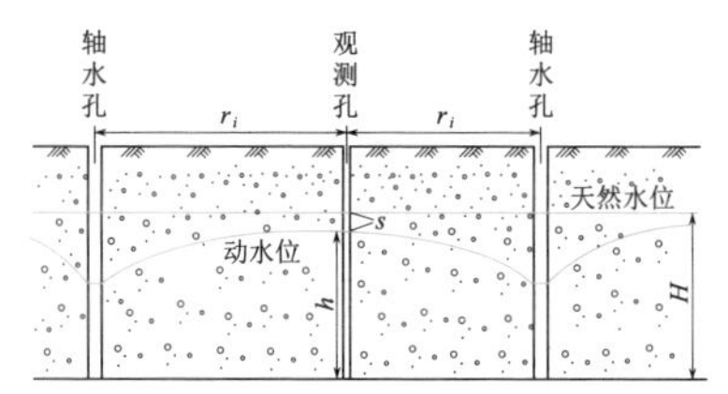

unsteady flow pumping test model of completely interference wells in phreatic aquifer

unsymmetrical pressure

【释义】 围岩偏压

【例句】 Basing on the rock mechanics property of shallow tunnel under unsymmetrical pressure, a load-structure model is used to calculate internal force and displacement of tunnel lining. 根据浅埋偏压隧道围岩的力学特性，采用荷载结构模型计算隧道支护内力及变形。

【中文定义】 指由于客观原因导致围岩压力出现较为明显的不均匀性。

unwatering [ʌn'wɔːtərɪŋ]

【释义】 *n*. 疏干，排水

【常用搭配】 unwatering well 排水井

【例句】 Through cutting off water engineering with high angle, high density, and deep drill in external of mineral mass the unwatering of double-aquifer can be achieved and the problems of water inrush of working face can be solved. 通过实施矿体外围高角度、大密度、大孔深钻孔截水工程，实现了双含水层的疏干，解决了水害威胁及作业面涌水问题。

unwatering yield

【释义】 疏干开采量，枯竭性抽水量

【同义词】 depletion yield

upheaval of road pavement

【释义】 路面隆起

【例句】 Excavation slope and frost heaving could lead to upheaval of road pavement generally. 路堑边坡开挖或冻胀通常会引起路面隆起。

【中文定义】 挖方路面由于卸荷回弹或路堑边坡的压挤而产生的上鼓现象。

uplift pressure monitoring

【释义】 扬压力监测

【例句】 The uplift pressure monitoring refers to the measurement of the uplift pressure carried by the monitoring instruments and equipment to the hydraulic structures. 扬压力监测是指运用监测仪器和设备对水工建筑物所承受的扬压力进行的量测。

Upper

【释义】 *adj*. 上统（晚世）

【常用搭配】 Upper Devonian 泥盆系上统（晚泥盆世）

upper bound solution

【释义】 上限解，上界解

【例句】 The optimum upper bound solution is represented in analytic mode. 该最佳上限解用通用解析式表达了出来。

upper explosion limit (UEL)

【释义】 爆炸上限

【中文定义】 指可燃气体爆炸上限浓度（$V\%$）值。

upper hemisphere projection

【释义】 上半球投影

【例句】 Upper hemisphere projection is one kind of stereographic projection。上半球投影是赤平极射投影的一种。

【中文定义】 投影球上两极的发射点，分上极射点（P）和下极射点（F），由下极射点（F）把上半球的几何要素投影到赤平面上的投影称为上半球投影。

upright fold

【释义】 直立褶皱

upright fold

【例句】 "Xikang-type" folds are developed in a very thick flysch sequence of the Triassic Xikang Group of the Songpan-Garze orogenic belt in southwestern China, they are special syncleavage upright folds. 在松潘-甘孜造山带的三叠系西康群复理石岩系中，普遍发育着一种特殊的同劈理直立褶皱，称为"西康式"褶皱。

【中文定义】 指轴面陡倾至铅直（倾角 80°～90°）的褶皱。

upstream view

【释义】 上游视图

【常用搭配】 upstream view of dam 大坝上游视图

【例句】 The view along the stream direction is called the upstream view. 顺水流方向的视图称为上游立视图。

【中文定义】 顺水流方向的视图称为上游立视图。

U-shaped valley

【释义】 U 形谷

【例句】 Examples of U-valleys are found in mountainous regions like the Alps, Himalaya, the Rocky Mountains, Scottish Highlands, Scandinavia, New Zealand and Canada. U 形谷发育在多山地区，典型的例子有阿尔卑斯山脉、喜马拉雅山脉、落基山脉、苏格兰高地、斯堪的纳维亚半岛 、新西兰及加拿大。

【中文定义】 又称槽谷，是由冰川作用下蚀和展宽形成的典型冰川谷，两侧一般有平坦的谷肩，横剖面近似 U 形。

U-shaped valley at the head of Leh valley

V

valley ['væli]

【释义】 *n.* 山谷，流域，溪谷

【常用搭配】 consequent valley 顺向谷；transverse valley 横向谷；synclinal valley 向斜谷；U-shaped valley U形谷

【例句】 Investigations for structures that may be founded on piles over a buried valley should include defining the shape of the valley. 对桩基可能位于掩埋谷内的建筑物进行调查时，应包含对山谷形状的界定。

【中文定义】 指两山间低凹而狭窄处，其间多有涧溪流过。

deep river valley

vane shear test

【释义】 十字板剪切试验

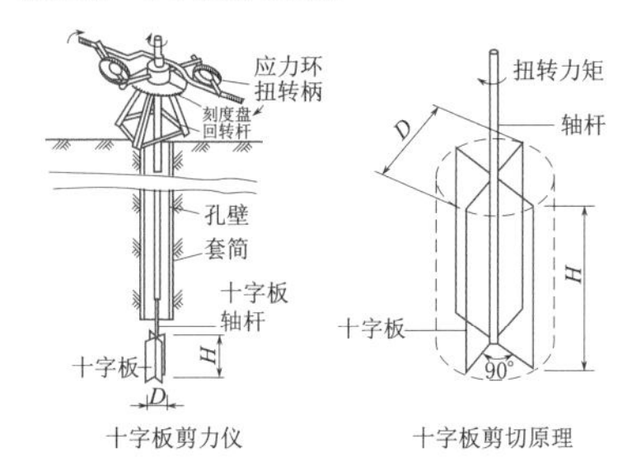

vane shear test

vegetable glue drilling fluid without clay

【释义】 植物胶无固相冲洗液

vein [ven]

【释义】 *n.* 岩脉

【常用搭配】 gash vein 裂缝脉；parallel vein 平行脉

【例句】 An overwhelming majority of the monazites and rutiles are distributed in quartz or hematite veins. 独居石和金红石绝大多数都分布在石英脉或赤铁矿细脉里。

【中文定义】 指充填在岩石裂隙中的脉状侵入岩体。

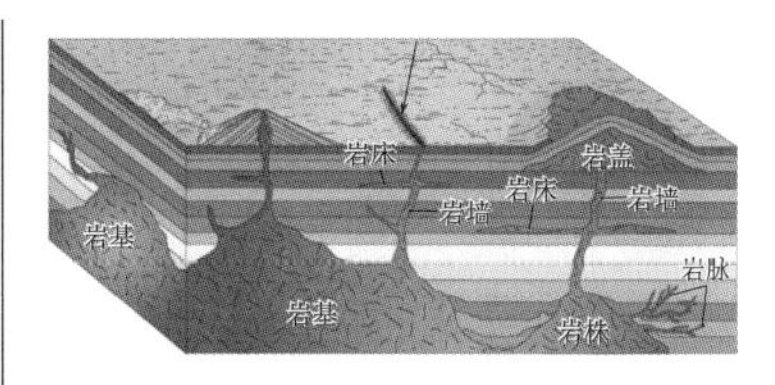

vein

vein rock

【释义】 脉岩

【例句】 Vein rock group in Teli, Western Zhunggar, Xinjiang was formed in middle Variscian, and its Rb-Sr isotopic ages are 277～283Ma. 脉岩群发育于新疆准噶尔西部托里县境内，是海西中期的产物，Rb-Sr 等时线年龄为 277～283Ma。

velocity instrument method

【释义】 流速仪法

【例句】 At present, the most frequently used method for measuring the river discharge is velocity instrument method. 目前测量河流流量最常用的方法是流速仪法。

velocity instrument method

vermiculite [və'mɪkjʊlaɪt]

【释义】 *n.* 蛭石

【常用搭配】 vermiculite aggregate 蛭石骨料；vermiculite concrete 蛭石混凝土

【例句】 Vermiculite is formed by weathering of mica minerals. 蛭石是云母类矿物经风化作用形成的。

【中文定义】 是一种天然、无机、无毒的，在高温作用下会膨胀的黏土矿物，其与蒙脱石相似，为层状结构的硅酸盐。一般由黑云母经热液蚀变或风化形成，因其受热失水膨胀时呈挠曲状，形态酷似水蛭，故称蛭石。

vermiculite

vertical control point

【释义】 高程控制点

vertical control point

vertical displacement measurement

【释义】 垂直位移测量

【例句】 The direction and value of horizontal stray field induced by the toroidal field coils and primary windings are determined by means of vertical displacement measurement. 通过垂直位移测量鉴别了纵向场线圈和加热场绕组引起的水平杂散场的方向和大小。

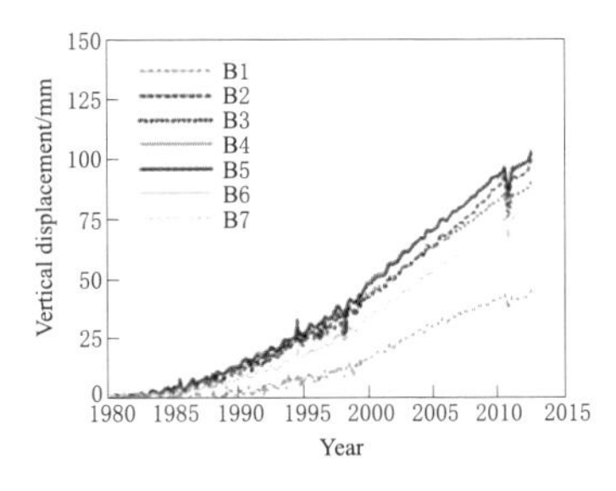

vertical displacement measurement

vertical displacement observation

【释义】 垂直位移观测

【例句】 Based on the general principles of displacement analysis and by studying the structure of the network figure matrix, a practical method for analysing the vertical displacement observation network is put forward. 本文从变形分析的一般原理出发，通过研究沉降监测网的图形矩阵特点，给出了变形分析计算的实用方法。

vertical monitoring control network

【释义】 高程监测网

vertical motion

【释义】 垂直运动

【常用搭配】 vertical motion geophone 垂向运动检波器；vertical motion simulator 垂直运动模拟器；vertical motion seismograph 垂直运动地震仪

【例句】 The vertical motion of the projectile is the motion of a particle during its free fall. 射弹的垂直运动是自由落体运动。

【中文定义】 指大体沿地球半径方向发生的地壳运动。它的大尺度地质表现是地质体发生大规模的垂向升降。

vertical peak ground acceleration

【释义】 地震垂直加速度

【例句】 The empirical attenuation relations of the horizontal and vertical peak ground accelerations for the Chi Chi earthquake are developed by regression method. 用回归分析法对台湾集集地震的加速度峰值数据进行分析，得出了这次地震的水平与垂直向的加速度峰值衰减关系。

【中文定义】 指地震时地面垂直运动的加速度。

vertical recharge

【释义】 垂直补给

【例句】 It is of great importance to do the research on laws of groundwater vertical recharge and discharge, which can lay the ground for correct evaluation of groundwater recharge amount and safe yield, reasonable development of groundwater resources, scientific management and etc. 研究地下水的垂向补排规律，对于正确评价研究区地下水补给资源量和可开采资源量，地下水的合理开发利用和科学管理，有着非常重要的意义。

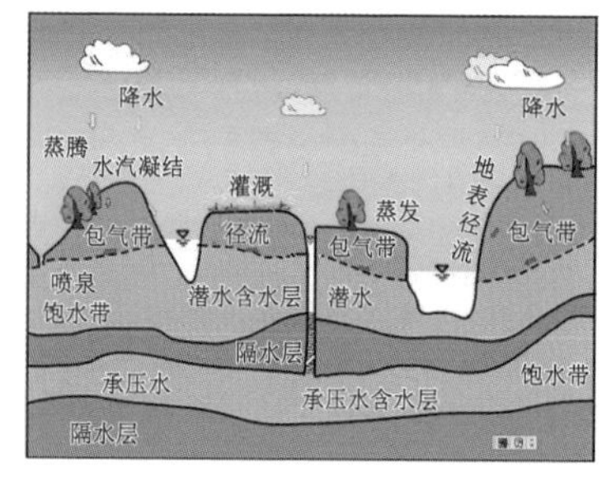

schematic diagram of vertical recharge

vertical seismic profiling (VSP)

【释义】 垂直地震剖面

【例句】 The vertical seismic profiling is a seismic observation method that corresponds to the seismic profile observed on the ground. 垂直地震剖面是一种地震观测方法，它是与地面观测的地震剖面相对应的。

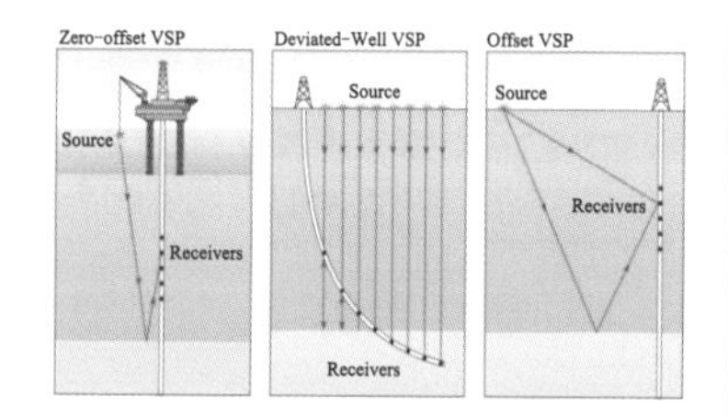

vertical seismic profiling (VSP)

vertical shaft

【释义】 竖井

【例句】 The sampling process of rock will generally be implemented inside of adits, vertical shafts and boreholes, and properties of rock classified in one set shall be basically identical. 岩石试样一般在平洞、竖井、钻孔中采取，同一组岩样的性质应基本相同。

vertical shaft

vertical vadose zone

【释义】 垂直渗流带

【中文定义】 是大气水、地表水与地下水发生联系并进行水分交换的地带。

very poor surrounding rock mass

【释义】 很差围岩

【中文定义】 地下洞室在采用 RMR（rock mass rating system）分类时，当 RMR 评分值小于 21 时，为Ⅴ类围岩，定性评价为很差围岩。

very poor surrounding rock mass

very unstable

【释义】 非常不稳定

【中文定义】 根据《水力发电工程地质勘察规范》(GB 50287—2016)，Ⅴ类围岩稳定性多为非常不稳定，表现为围岩不能自稳，变形破坏严重。

vesicular structure

【释义】 气孔构造

【中文定义】 岩浆岩中具有的近似圆形或椭圆形互不连通的孔洞的构造。

vibrating wire strain gauge

【释义】 振弦式应变计

【常用搭配】 embedded vibrating wire strain gauge 内埋式振弦应变计

【例句】 Vibrating wire strain gauge has excellent property of using and installing. 振弦式应变计具有易使用、易安装等优点。

vibrating wire strain gauge

vibration acceleration monitoring

【释义】 振动加速度监测

【例句】 Vibration acceleration monitoring is an important part of vibration monitoring. 振动加速度监测是振动监测中的重要一项。

vibration amplitude monitoring

【释义】 振动振幅监测

vibration compaction method

【释义】 挤密法

【例句】 From drainage consolidation method, vibration compaction method, reinforcement method and other aspects the design of the bridge abutment in expressway on wet and soft foundation as well as corresponding treatment methods are introduced as well as points for attention in practice to ensure engineering quality. 从排水固结法、振密挤密法、加筋法等，阐述了高速公路湿软地基桥梁桥台设计的方法和湿软地基处理方法，并提出施工中要特别注意的问题，从而确保工程质量。

vibration compaction method

vibration frequency monitoring

【释义】 振动频率监测

vibration velocity monitoring

【释义】 振动速度监测

vibro-densification method

【释义】 振密法，挤密法

【中文定义】 指采用一定的手段，通过振动、挤压使地基土体孔隙比减小，强度提高，达到地基加固目的的一类地基处理方法的总称。

vibro-flotation

【释义】 振冲法

【例句】 In confrontation with a bad geological condition, especially a soft foundation, such a the desert region also with high level groundwater, the compound foundation should be formed with the vibro-flotation method. 在地质条件较差，特别是在软土地基、沙漠地带而地下水水位又较高的情况下，通过振冲法处理形成复合地基。

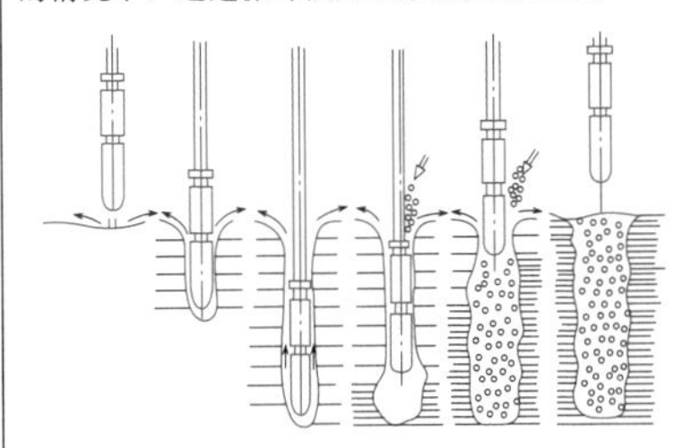

vibro-flotation

vibroseis ['vaibrəsiːs]

【释义】 *n.* 可控震源，可控波源

【例句】 Vibroseis is a non-explosive source that is used in seismic exploration of land and sea. 可控震源是在陆地和海洋的地震勘探工作中都应用的一种非炸药震源。

viscoelastic deformation

【释义】 黏弹性变形

【例句】 Three dimensional finite element simulation can be used to analyze the properties of viscoelastic deformation of materials. 使用三维有限元模拟，可以分析材料的黏弹性变形特性。

【中文定义】 黏弹性材料在加工过程中表现出来的变形形式，具有双重变形特性，既具有固体的

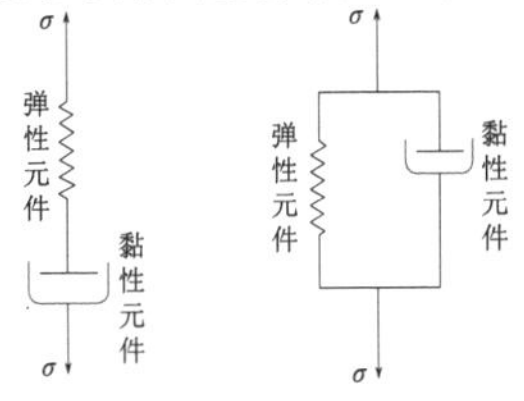

mechanical model of viscoelastic body

弹性特性，也有流体的黏性特性。

viscoelastic modulus

【释义】 黏弹性模量

【例句】 Viscoelastic modulus is an important parameter for evaluating the mechanical properties of viscoelastic media. 黏弹性模量是评价黏弹性介质力学性能的一项重要参数。

visco-plastic deformation

【释义】 黏滞塑性变形

【例句】 Viscous-plastic deformation is a kind of deformation of special material under the condition of force. 黏滞塑性变形是特殊材料在受力状态下发生的一种变形。

viscosity [vɪ'skɑsɪti]

【释义】 *n.* 黏性，黏稠度

【常用搭配】 coefficient of viscosity 黏滞系数；viscosity index 黏度指数；apparent viscosity 表观黏度

【例句】 Viscosity is one of the important characters of soft clay. 软黏土的重要特性之一是其具有黏滞性。

【中文定义】 所有流体在有相对运动时都要产生内摩擦力，这是流体的一种固有物理属性，称为流体的黏滞性或黏性。

viscosity coefficient

【释义】 黏滞系数

【例句】 The formula for calculating liquid viscosity coefficient correctly is deduced in detail. 详细推导了正确计算液体黏滞系数的计算公式。

viscous debris flow

【释义】 黏性泥石流，结构性泥石流

【例句】 The large boulders, the boulder piling up, mud crack, lateral accumulation mound and surge deposits demonstrate the viscous debris flows. 巨大的泥石流漂砾、石背石现象、龟裂现象、侧积堤和龙头堆积证实了这次泥石流为黏性泥石流。

viscous debris flow

【中文定义】 指浆体是由黏性物质组成的泥石流。

visual sketch

【释义】 目测草图

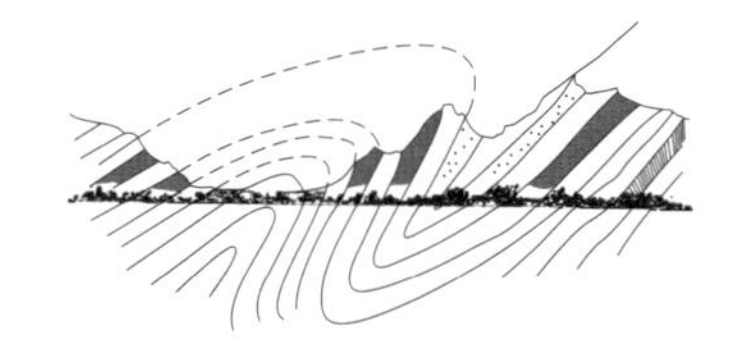

visual sketch

vitreous luster

【释义】 玻璃光泽

【例句】 In mineralogy, vitreous luster is a type of luster, which is seen in glass-like objects such as quartz and most gemstones, caused by reflection of light. 玻璃光泽是矿物学中的一种光泽，光的反射产生的矿物光泽如同玻璃一样。

【中文定义】 矿物表面对光的反射能力弱，如同玻璃表面的光亮。

vitreous luster

vitreous texture

【释义】 玻璃质结构

【例句】 Typical features of subaqueous volcanic rocks include perlite with vitreous texture and pillow structure, lamellar tuff, bentonite. 水下喷发火山岩典型标志为珍珠岩（玻璃质结构、枕状构造）、纹层状凝灰岩和膨润土。

【中文定义】 指全部由玻璃物质所组成的岩石结构。这种结构常见于火山岩中，组分没有结晶，为纯碎玻璃质的结构。

void ratio

【释义】 孔隙比

【例句】 The coarse-grained soil with a small initial void ratio has less volume deformation under the same stress state. 初始孔隙比小的粗粒土在

相同的应力状态下体积变形也较小。

volcanic breccia

【释义】 火山角砾岩

【常用搭配】 sedimentary volcanic breccia 沉积火山角砾岩

【例句】 Pyroclastic rocks are mainly composed of volcanic breccia and tuff. 火山碎屑岩包括火山角砾岩和凝灰岩。

【中文定义】 由直径大于 4mm 的火山岩片所成，所含熔岩碎片以凝灰岩居多，玻璃细片及整石较少。主要由粒径为 2～64mm 的火山角砾组成，也含有其他岩石的角砾及少量的石英、长石等矿物晶屑。

volcanic breccia

volcanic deposit

【释义】 火山堆积

【例句】 In the metallogenic belt there exist copper deposit in volcanic rock and in volcanic deposit, acidic intrusive. 成矿带内铜矿类型有火山岩及火山沉积型铜矿床。

【中文定义】 是由火山喷发物所形成的堆积物，包括熔岩和火山弹、火山砾、火山灰等火山碎屑物质。

volcanic deposit

volcanic rock

【释义】 火山岩

【例句】 LA-ICP-MS U-Pb results of zircon from the volcanic rocks demonstrated that the volcanism started at least since 258 Ma, and continued to 248 Ma. 火山岩锆石 LA-ICP-MS U-Pb 数据表明，岩浆活动至少始于 258Ma，一直持续到 248Ma。

volcanic rock

volume shrinking rate

【释义】 体缩率

volumetric compressibility

【释义】 容积压缩系数，体积压缩系数，体积压缩率

【中文定义】 等温条件下，压力每改变一个大气压时，气体体积的变化率。

volumetric compressibility coefficient

【释义】 容积压缩系数

【例句】 The variation rules of permeability, porosity and volumetric compressibility coefficient with pressure are obtained. 得出了岩石渗透率、孔隙度以及孔隙体积压缩系数随压力变化而变化的规律。

volumetric method

【释义】 容积法

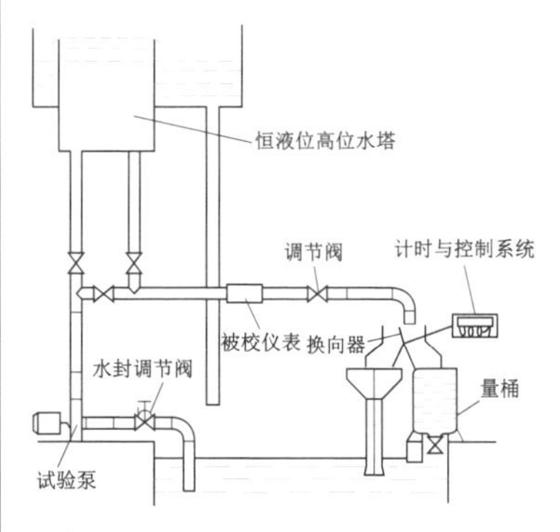

flow measurement with volumetric method

【例句】 The unstructured-finite volumetric method was used to investigate the complex flow and heat transfer state in the products pipeline and hot crude pipeline laying in one ditch by numerical simulation. 采用非结构化有限容积法，对成品油和原油管道内复杂的流动和传热问题进行了数值模拟。

V-shaped valley

【释义】 V形谷

【中文定义】 在河流的上游以及山区河流，由于河床的纵比降和流水速度大，因此河水在垂直方向上的分量也大，就能产生较强的下蚀能力，这样使河谷的加深速度快于拓宽速度，从而形成在横断面上呈V字形的河谷。

V-shaped valley

wall surface

【释义】 壁面

【例句】 The two wall surfaces of the structural plane are separated. There is no connection force between the rock wall surfaces, or only a relatively weak connection force, which is called open structure surface. 结构面的两个岩壁面是分离的，岩壁间无连接力，或仅有相对微弱的连接力，这种结构面被称为张开结构面。

【中文定义】 具有一定宽度的结构面的内侧壁。

wall surface

water balance method

【释义】 水量均衡法

water burst

【释义】 突水，紧急进水

【常用搭配】 water burst mechanism 突水机理；water burst factor 突水因子；water burst passage 突水通道

【中文定义】 又称灾害性涌水，在洞室、巷道等施工过程中，穿过岩溶发育的地段，尤其是遇到地下暗河系统，厚层含水砂砾石层，及与地表水连通的较大断裂带等所发生的突然大量涌水现象。

water burst

water capacity

【释义】 容水量

【例句】 The volume a container can hold is determined by its water capacity and pressure rating. 容器的容量由它的容水量和额定压力来决定。

【中文定义】 指土壤有效水分的上限。

water content

【释义】 含水量，含水率

【中文定义】 试件在 105～110℃下烘至恒量时所失去的水的质量与试件干质量的比值，以百分数表示。

water content of rock

【释义】 岩石含水率

water divide

【释义】 分水岭

【例句】 Water divides can be grouped in three types: continental divide, major drainage divide, minor drainage divide. 分水岭可以分为三种类型：大陆分水岭、大分水岭和小分水岭。

【中文定义】 分隔相邻两个流域的山岭或高地。

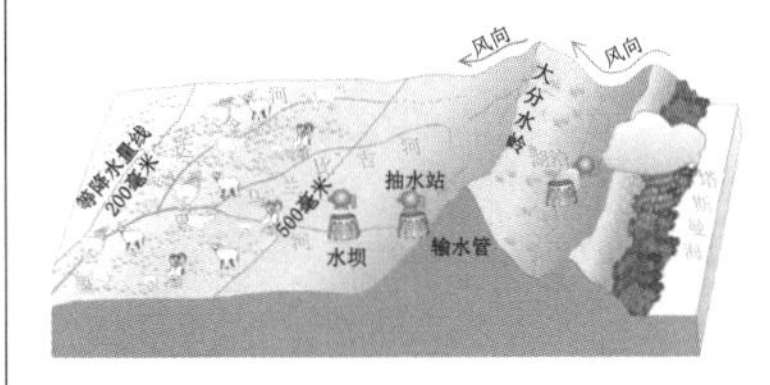

water divide

water gushing

【释义】 股状涌水

【中文定义】 地下水活动状态之一，根据《水力发电工程地质勘察规范》（GB 50287—2016）的规定，当地下洞室内地下水每 10m 洞长水量 q>125L/min 或压力水头 H>100m 时，地下水状态为涌水。

water in aeration zone

【释义】 包气带水

【例句】 The water in aeration zone is an important part of the globe water body. 包气带水是地球水体的重要组成部分。

【中文定义】 指埋藏于包气带中的地下水。

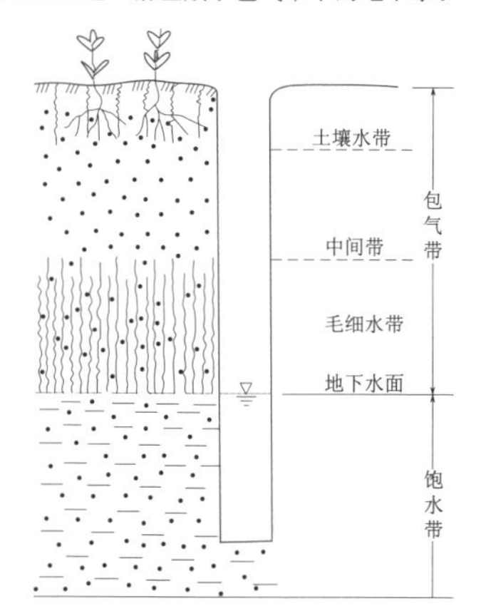

schematic diagram of water in aeration zone

water injecting test

【释义】 注水试验

【例句】 Water injection test is an effective way to check grouting quality. 注水试验是检测注浆质量的有效手段之一。

【中文定义】 往钻孔中连续定量注水，使孔内保持一定水位，通过水位与注水量的函数关系，测定透水层渗透系数的水文地质试验工作。

water injection test in borehole

【释义】 钻孔注水试验

【例句】 The water injection test in borehole is applied to the finite-width cement-soil impermeable wall. 钻孔注水试验应用于有限宽度的水泥土截渗墙。

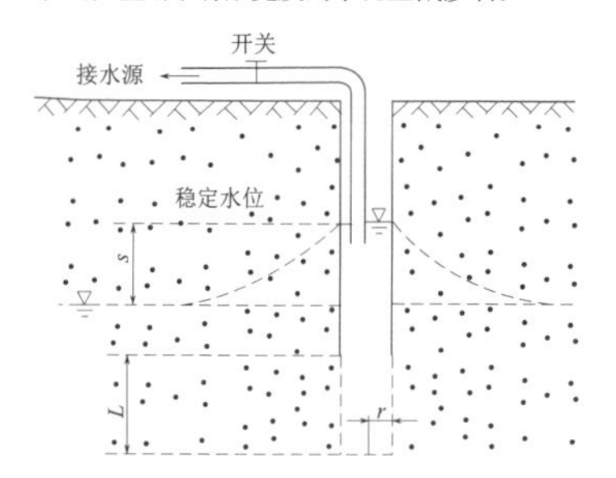

diagram of water injection test in borehole

water injection test in pit

【释义】 探坑注水试验

water inrush

【释义】 涌水

【例句】 The prediction theory and forecast method of tunnel water inrush in karst areas have long been a difficult hydrogeological problem. 岩溶地区隧道地下水涌水预测方法和理论是长期以来难以突破的水文地质难题。

【中文定义】 指在地下洞室施工过程中，穿过含水或透水岩层所发生的地下水向洞内冒出或突然喷出的现象。

water inrush

water inrush in foundation pit

【释义】 基坑涌水

【中文定义】 是建筑时经常遇到的一种现象，多指的是开挖较深时，地下水涌入基坑的现象。

water level recovery method

【释义】 水位恢复法

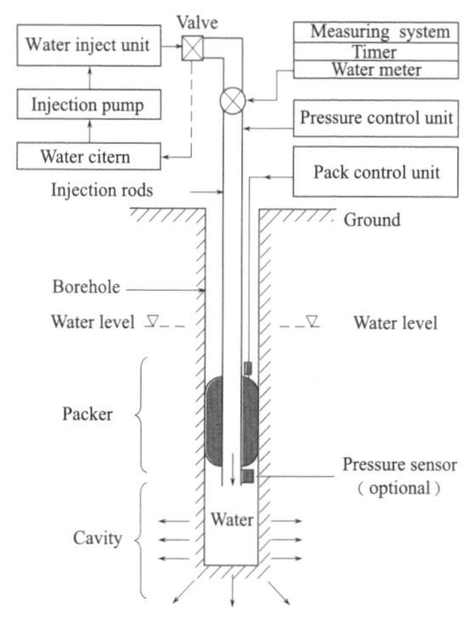

water pressure test

water pressure test

【释义】 压水试验

【例句】 Water pressure test is a special type of groundwater flow test with the features of short duration and test segment. 压水试验历时短、试段短，是一种特殊的地下水流试验类型。

【中文定义】 利用水泵或者水柱自重，将清水压入钻孔试验段，根据一定时间内压入的水量和施加压力大小的关系，计算岩体相对透水性和了解裂隙发育程度的试验。

water pressure test in borehole

【释义】 钻孔压水试验

【例句】 Water pressure test is generally conducted in boreholes in the bed rock, hydrogeological data is collected during the test. 钻孔压水试验一般在基岩中实施，在实施过程中收集水文地质资料。

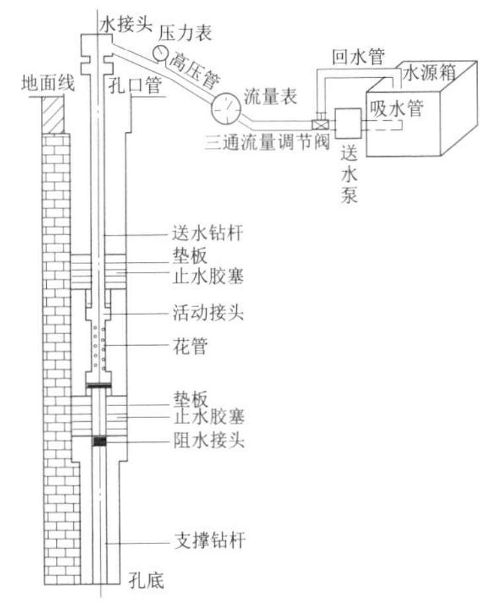

installation diagram of water pressure test in borehole

water quality analysis

【释义】 水质分析，水化学分析

【例句】 The quality control in water quality analysis is the control of the errors of analytical results in substance. 水质分析质量控制实质上是对分析结果误差的控制。

【中文定义】 指用化学和物理方法测定水中各种化学成分的含量。

检测项目	控制指标
pH	8.0～9.0
电导率/(μS·cm^{-1})	≤5000
浊度/(mg·L^{-1})	<10
钙硬度（以 $CaCO_3$ 计）/(mg·L^{-1})	≤800
总碱度（以 $CaCO_3$ 计）/(mg·L^{-1})	150～600
氯离子/(mg·L^{-1})	≤800
总磷/(mg·L^{-1})	5.5～12.0
总铁/(mg·L^{-1})	<1.0
余氯/(mg·L^{-1})	0.1～1.0
化学需氧量（COD）/(mg·L^{-1})	≤30

the quality analysis results of circulating water

water quality analysis report

【释义】 水质分析报告

water resources

【释义】 水资源

【例句】 Several of China's most strategically important regions are predicted to suffer significant water resource shortages as a result of climate change. 据预测，中国最具重要战略地位的几个区域将遭受由于气候变化造成的严重的水资源短缺。

water spraying and sand emitting

【释义】 冒水喷沙

【例句】 The water spraying and sand emitting phenomenon occurred with the earthquake shaking the surface subsidence and the shallow groundwater being squeezed along the cracks to the surface. 地震的晃动使表土下沉，浅层的地下水受挤压会沿地裂缝上升至地表，形成喷沙冒水现象。

【中文定义】 指地震震动诱发地下含水沙土层液化并向上喷发的现象。

water spraying and sand emitting

water storage degree

【释义】 贮水度

waters multi-channel seismic exploration

【释义】 水域多道地震勘探

waters stratum section detection

【释义】 水域地层剖面探测

wave run-up height

【释义】 波浪爬高

【例句】 According to the similarity theory, we have clone physical model experiment in the water channel and find out the regularity of the wave run-up height on the slope. 根据相似理论，我们在波浪水槽中进行物理模型试验，得出了斜坡上波浪爬高的一般规律及其影响因素。

【中文定义】 指波浪在斜坡上发生破碎后水体沿斜坡面上涌、爬升的现象。

wave velocity test

【释义】 波速测试

【例句】 As one of the test items for dynamic characteristics of soil foundation, wave velocity test has been widely used in many geotechnical investigation fields, and has achieved good results. 波速测试作为地基土动力特性测试项目之一，已广泛应用于众多岩土工程勘察领域，并取得了良好的应用效果。

wavy bedding

【释义】 波状层理

【中文定义】 波状层理是层理基本类型之一，由许多呈波状起伏的细层重叠在一起组成，多在振荡的水动力条件下或者风力作用下形成的。层内的层纹呈连续的波状、互相平行或薄的泥纹层和砂纹层成波状互层，细层可连续或断续，总方向平行层理。

wax sealing method

【释义】 蜡封法

【例句】 The wax sealing method is more accurate and reliable and can be used for soil density test. 蜡封法是较为准确和可靠的，可用于测定土体密度试验。

weak earthquake

【释义】 弱震

【中文定义】 震级大于等于1级，小于3级的称为弱震或微震。

weak earthquake observation

【释义】 弱震监测

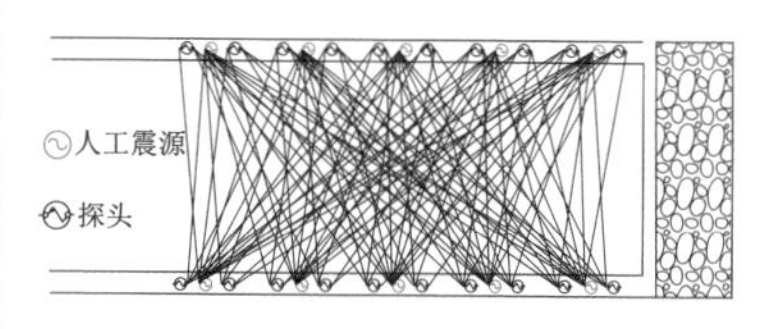

weak earthquake observation

weak intercalation

【释义】 软弱夹层

【例句】 Weak intercalation is the main weak surface which controls the rock stability. 软弱夹层是控制岩体稳定性的主要弱面。

【中文定义】 岩体内存在的层状或带状的软弱薄层。

weak intercalation

weak interlayer detection

【释义】 软弱夹层探测

weak plane

【释义】 软弱结构面

【同义词】 soft structural plane

【例句】 Weak planes such as faults are typical defective geological bodies, which are often encountered and must be solved effectively in tunnel engineering. 断层等软弱结构面是隧道工程中经常遇到而又必须有效处理的不良地质体。

weak plane

【中文定义】 是力学强度明显低于围岩，一般充填有一定厚度软弱物质的结构面，如泥化、软化、破碎薄夹层等。

weathered soil

【释义】 风化土料

【例句】 Because of high water content of all-weathered soil and the influence of rainy and foggy climate in the region, the construction is difficult to a certain extent. 全风化土料含水量高，以及该地区多雨、多雾等因素给施工带来一定难度。

weathered soil

weathering ['weðərɪŋ]

【释义】 *n.* 侵蚀，风化，风化作用

【常用搭配】 physical weathering 物理风化；chemical weathering 化学风化；biological weathering 生物风化

【例句】 Generally speaking, the shallow rock mass of the slope is affected by unloading, weathe-ring and cutting of joints and cracks, and the mechanical properties of the shallow rock mass are unfavorable. 一般而言，边坡浅层岩体受卸荷、风化作用影响及节理裂隙的切割，整体力学性质较差。

【中文定义】 使岩石发生破坏和改变的各种物理、化学和生物作用。

weathering crust

【释义】 风化壳

【例句】 The paleo-weathering crust controls distortion and destruction of spillway rock mass. 古风化壳对溢洪道岩体的变形破坏起控制作用。

【中文定义】 地壳表层岩石风化后部分溶解物质流失，其碎屑残余物质和新生成的化学残余物质大都残留在原来岩石的表层。这个由风化残余物质组成的地表岩石的表层部分，或者说已风化了的地表岩石的表层部分，就称为风化壳或风化带。

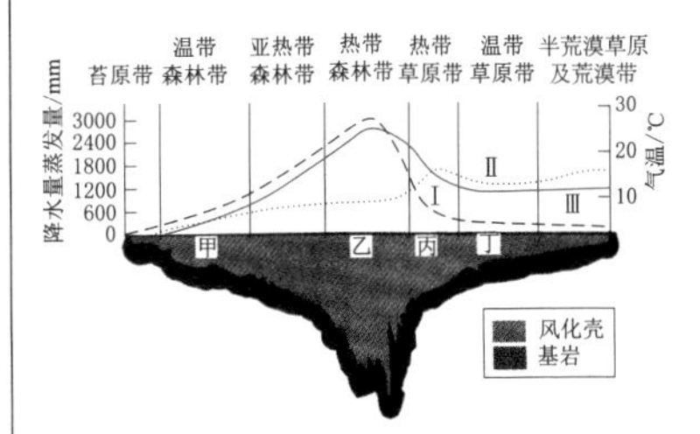

Ⅰ—降雨；Ⅱ—蒸发；Ⅲ—气温

weathering crust

weathering fissure

【释义】 风化裂隙

【中文定义】 是指岩石在风化营力作用下发生破坏而产生的裂隙，主要分布于地表附近。

weathering fissure

weathering zone

【释义】 风化带

【例句】 For arch dam, the division of rock weathe-ring zone influences the choice of foundation surface. 对于拱坝而言，岩体风化带的划分影响建基面的选择。

【中文定义】 指地壳表层岩石按其风化程度，从地壳表层向下分成为全风化、强风化、弱风化（中等风化）、微风化的层带。

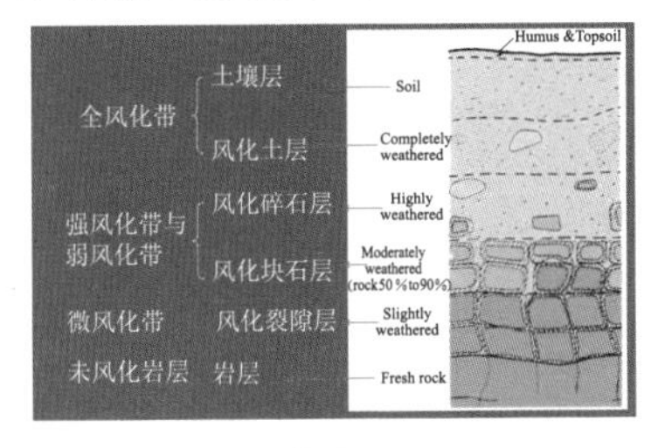

weathering zone

wedge equilibrium method

【释义】 楔形体平衡法

wedge out

【释义】 尖灭

【中文定义】 指沉积层向着沉积盆地边缘，其厚度逐渐变薄直至没有沉积。

wedge slide

【释义】 楔体滑动

wedge slide

weighted average

【释义】 加权平均值

【例句】 The final overall ranking is calculated as a weighted average of all four scores. 最终总体排名计算为所有 4 个分数的加权平均值。

weir method

【释义】 堰测法

【例句】 Weir method can measure water inflow. 堰测法可用来测涌水量。

【中文定义】 是指在井下排水沟中设置测水堰板，逝水流通过一定形状的堰口水流高度，然后计算涌水量。测水堰版通常有三角堰、梯形堰和矩形堰三种。

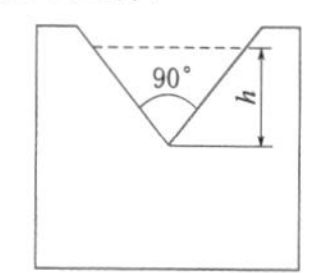

三角堰

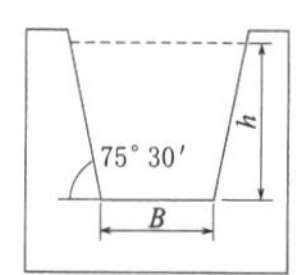

梯形堰

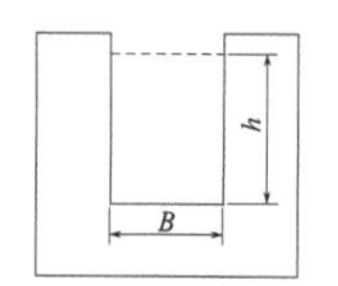

矩形堰

weir method

well function

【释义】 井函数

【例句】 For the well function is a quite complicated function of infinite integral, generally it is difficult to get an accurate solution. 因为井函数是一个复杂的无穷积分函数，一般难以求得其精确解。

well loss

【释义】 井流水头损失

【例句】 These well losses can be reduced by reducing the entrance velocity of the water, which is accomplished by installing the maximum amount of screen and pumping at the lowest acceptable rate. 井损可以通过降低水的流速而减少，一般有两种方法：一种是安装最大数量的过滤器，另一种是采用最低的抽水速度。

well production

【释义】 井产量

【例句】 It also obtains a nice effect in block, arrangement of development controlling wells and prediction of single well production. 应用该技术在区块优选、开发控制井的部署以及预测单井产量方面也取得了较好的效果。

well testing

【释义】 试井

【例句】 The paper introduces the interpretation method, analysis property and influence factor of formation pressure in well testing analysis. 该论文介绍了地层压力在试井分析过程中的解释方法、分析特征及影响因素。

well testing

well screen

【释义】 过滤器

【例句】 A key element of well screen design is the size of the openings, referred to as slot size. 过滤器设计的一个关键要素是开口的大小，即筛孔

大小。

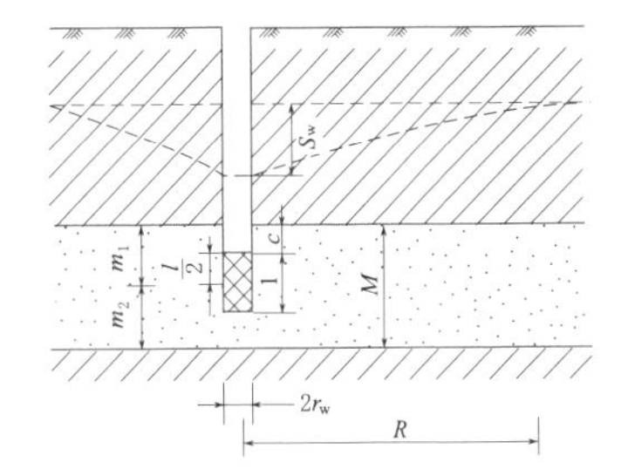

partially penetrating well with well screen in the middle of aquifer

well-graded soil

【释义】 良好级配土

【例句】 The results of the study show that the well-graded soil have low water contents. 研究结果显示，级配良好的土壤含水量较低。

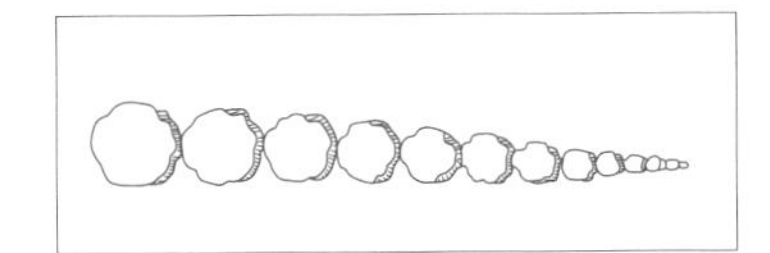

well-graded soil

wireframe model

【释义】 线框模型

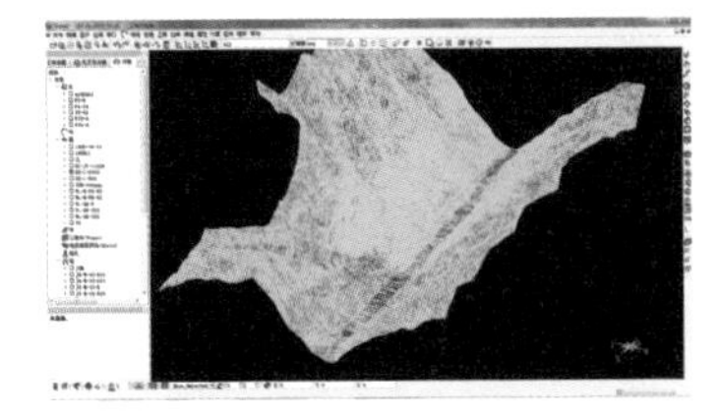

wireframe model

【例句】 The wireframe model is the simplest type of three dimension geometric model and it uses the least amount of computer time and memory, but it is not very useful for complex surfaces. 线框模型是三维几何造型中最简单的一种，使用计算机时间和内存也少，但不能用作复杂表面分析。

【中文定义】 在CAD软件中，由单根或一组线条按照一定的空间方位放置并赋予特定的模型样式和模型属性代表实物外形骨架的模型。线框模型只是用几何体的棱线表示外形，没有表面、体积等几何信息。

wire-line core drilling

【释义】 绳索取芯钻进

【例句】 Wire-line core drilling can lift core from drill rods without picking up coring system. 绳索取芯钻进是一种不提钻而从钻杆内捞取岩芯的钻进。

working benchmark

【释义】 工作基点

【同义词】 working base point

【例句】 This paper presents the consideration of establishing the benchmarks, working benchmarks, marks nets properly for the survey control net of vertical displacement in observation design. 本文介绍了在设计时，为设立完整而合理的垂直位移测控系统，对基准点、工作基点、标志和测控网布设的考虑。

woven geotextile

【释义】 织造型土工织物，有纺型土工织物，织造型土工布，土工织物

【例句】 On the basis of analyzing the character of fabric structure and the demand of geotextile for silt-prevention, the design method of woven geotextile for silt-prevention is studied. 在分析研究织物结构特征及防淤堵用土工织物特性要求的基础上，研究了防淤堵用机织土工布的设计。

woven geotextile

X

xenolith ['zinəlɪθ]

【释义】 *n*. 捕虏体

【例句】 Mafic granulite xenoliths in Cenozoic basanite were newly found in Xikeer, western Tarim Block. 在塔里木西部新生代碧玄岩中新发现了基性麻粒岩捕虏体。

X-ray fluorescence survey

【释义】 X射线荧光测量

Y

Yardang Landform

【释义】 雅丹地貌

【常用搭配】 Yardang trough 风蚀土脊沟

【中文定义】 现泛指干燥地区一种风蚀地貌，河湖相土状沉积物所形成的地面，经风化作用、间歇性流水冲刷和风蚀作用，形成与盛行风向平行、相间排列的风蚀土墩和风蚀凹地地貌组合。

Yardang Landform

yield criterion

【释义】 屈服标准

【例句】 There are three kinds of yield criterion commonly used in construction engineering. 建设工程上常用的屈服标准有三种。

【中文定义】 在一定的变形条件（变形温度，变形速度等）下，只有当各应力分量之间符合一定关系时，质点才开始进入塑性状态，这种关系称为屈服标准。

yield deformation

【释义】 屈服变形

【例句】 The four phases of metallic contacting deformation include the phase of elastic contacting deformation, the phase of yield deformation, plastic contacting deformation and the phase of restoring deformation. 金属材料接触变形过程包括4个阶段：弹性接触变形阶段、屈服变形阶段、塑性接触变形阶段、弹性恢复阶段。

yield limit

【释义】 屈服极限

【例句】 In two and three dimensions the yield limit is reached when a certain function of all the stress components reaches a fixed value. 在二维或三维的情况下，当所有应力分量的某个函数达到某一固定值时即达到屈服极限。

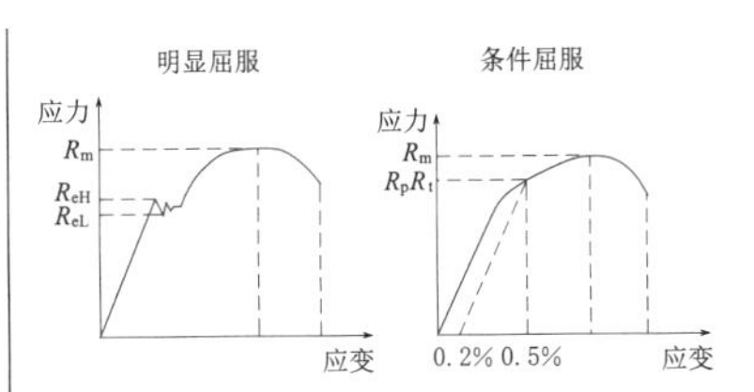

yield deformation

【中文定义】 材料受外力到一定限度时，即使不增加负荷它仍继续发生明显的塑性变形，该限度为屈服极限。

yield strength

【释义】 屈服强度

【例句】 The special technology can obtain high yield strength, elastic limit and high toughness. 特殊的技术可以使材料获得高的屈服强度，弹性极限和较高的韧性。

【中文定义】 是材料开始发生明显塑性变形时的最低应力值。

yield point

【释义】 击穿点，软化点，屈服点

【常用搭配】 lower yield point 下屈服点；upper yield point 上屈服点；shear yield point 剪切屈服点；yielding point 屈服值

【例句】 If a metal is only stressed to the upper yield point, and beyond, Lüders bands can develop. 如果一块金属受到超过上屈服点的压力，吕德斯带将会出现并发展。

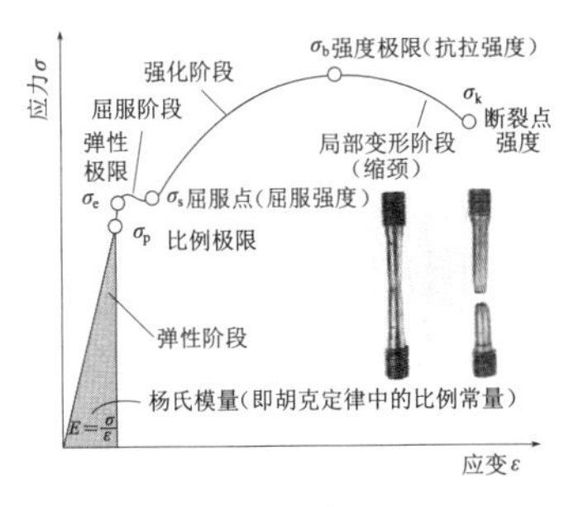

yield point

【中文定义】 屈服值又称塑性值，常指塑性流体开始产生流动所需达到与超过的临界应力值。钢材或试样在拉伸时，当应力超过弹性极限，即使应力不再增加，而钢材或试样仍继续发生明显的塑性变形，称此现象为屈服，而产生屈服现象时的最小应力值即为屈服点。

zig-zag fold

【释义】 锯齿状褶皱

【中文定义】 两翼较平直，转折端急剧转折、甚至成尖顶的褶皱。

zig-zag fold

zone dividing meridian

【释义】 分带子午线

【中文定义】 分带投影中划分投影带的子午线。

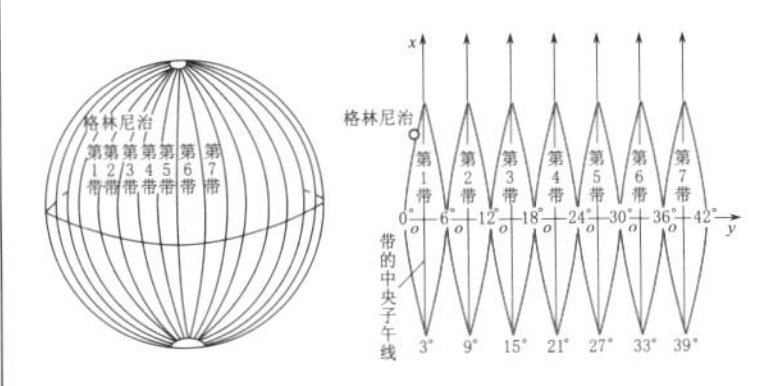

zone dividing meridian

zoning and mapping on geological hazard risk

【释义】 地质灾害危险区划

索　引

A

B

D

E

F

G

H

J

K

L

M

N

O

P

Q

R

S

T

W

X

Y

Z

其他